Semiconductor Optoelectronics

Semiconductor Optoelectronics

Edited by

Marian A. Herman
Institute of Physics,
Polish Academy of Sciences
Warsaw, Poland

A Wiley–Interscience Publication

JOHN WILEY & SONS
Chichester · New York · Brisbane · Toronto

PWN—POLISH SCIENTIFIC PUBLISHERS
Warszawa

Published in co-edition by John Wiley & Sons Ltd. and
PWN—Polish Scientific Publishers, Warsaw

Editorial adjustment of the English: *Jerzy Tomaszczyk*

Copyright © by PWN—Polish Scientific Publishers, Warszawa 1980

All rights reserved.

No part of this publication may be reproduced by any means, nor
transmitted, nor translated into a machine language without the
written permission of the copyright owner.

Library of Congress Cataloging in Publication Data

International School on Semiconductor Optoelectronics,
    2d, Władysławowo, Poland, 1978
    Semiconductor optoelectronics.
    At head of title: Polish Academy of Sciences,
Institute of Physics.
    1. Optoelectronics—Congresses. 2. Semiconductors—Congresses.
I. Herman, Marian A. II. Polska Akademia Nauk. Instytut Fizyki. III. Title.
TA1750.I6 1978   621.38'0414   79-1166
ISBN 0 471 27589 1

Printed in Poland by D.R.P.

Contributing Authors

M. J. ADAMS — Department of Electronics,
The University, Southampton S09 5NH, U.K.

ZH. I. ALFEROV — A.F. Ioffe Physico-Technical Institute,
Academy of Sciences of the USSR,
Leningrad, USSR

F. AUZEL — C.N.E.T.—196, rue de Paris,
92220, Bagneux, France

H. BACHERT — Central Institute of Optics and Spectroscopy,
Berlin, GDR

T. BRYŚKIEWICZ — Department of Optoelectronics,
Institute of Physics, Polish Academy of Sciences,
Warsaw, Poland

M. D. CAMPOS — Instituto de Fisica 'Gleb Wataghin',
Universidade Estadual de Campinas,
Campinas, São Paulo, Brasil

B. CROSIGNANI — Fondazione Ugo Bordoni,
Istituto Superiore Poste e Telecomunicazioni,
Viale Europa, Roma, Italy

P. DI PORTO — Fondazione Ugo Bordoni,
Istituto Superiore Poste e Telecomunicazioni,
Viale Europa, Roma, Italy

J. C. DYMENT — Bell-Northern Research
Ottawa, Canada

P. G. ELISEEV — P. N. Lebedev Physical Institute,
Academy of Sciences of the USSR,
Moscow, USSR

Contributing Authors

D. Z. GARBUZOW — A. F. Ioffe Physico-Technical Institute, Academy of Sciences of the USSR, Leningrad, USSR

H. G. GRIMMEISS — Department of Solid-State Physics, Lund Institute of Technology, Box 725, S-220 07 Lund 7, Sweden

M. A. HERMAN — Department of Optoelectronics, Institute of Physics, Polish Academy of Sciences, Warsaw, Poland

V. I. KOROLKOV — A. F. Ioffe Physico-Technical Institute, Academy of Sciences of the USSR, Leningrad, USSR

T. NISHINAGA — Department of Electrical Engineering and Electronics, Toyohashi University of Technology, 440 Toyohashi, and Department of Electronics, Nagoya University, 464 Nagoya, Japan

J. NISHIZAWA — Research Institute of Electrical Communication, Tohoku University, Sendai 980, Japan

Y. OKUNO — Semiconductor Research Institute, Kawauchi, Sendai 980, Japan

M. H. PILKUHN — Institute of Physics Stuttgart University, Stuttgart, FRG

YU. M. POPOV — P. N. Lebedev Physical Institute, Academy of Sciences of the USSR, Moscow, USSR

E. L. PORTNOY — A. F. Ioffe Physico-Technical Institute, Academy of Sciences of the USSR, Leningrad, USSR

Contributing Authors

M. A. A. PUDENSI — *Instituto de Fisica 'Gleb Wataghin', Universidade Estadual de Campinas, Campinas, São Paulo, Brasil*

J. E. RIPPER — *Instituto de Fisica 'Gleb Wataghin', Universidade Estadual de Campinas, Campinas, São Paulo, Brasil*

L. SOSNOWSKI — *Institute of Physics, Polish Academy of Sciences, Warsaw, Poland*

S. V. SVECHNIKOV — *Institute of Semiconductors, Ukrainian Academy of Sciences, Kiev, USSR*

G. H. B. THOMPSON — *STL, Harlow, Essex, England*

K. UNGER — *Section of Physics, Karl-Marx-University, Leipzig, GDR*

G. R. WOOLHOUSE — *Optical Information Systems, Exxon Enterprises Inc., Elmsford, New York, USA*

PREFACE

Some years ago, in October 1975, the Institute of Physics of the Polish Academy of Sciences started to organize the Cetniewo International Schools on Semiconductor Optoelectronics. The first School was held at the Cetniewo Sports Centre in Władysławowo, Poland, and in May 6–13, 1978, the second School met at the same place. The aim of these Schools is to review the current state of knowledge concerning the physical principles, technological foundations and application perspectives of semiconductor optoelectronic devices and new materials used in optoelectronics. Consequently, only review lectures are presented at these Schools.

The proceeding of the first School were published under the title *Semiconductor Sources of Electromagnetic Radiation*, ed. M. A. Herman, Polish Sc. Publ., Warsaw, 1976, and the present volume does the same for the second School. The chapters of this volume cover a broad range of topics of current significance for semiconductor optoelectronics today. Among the physical problems presented here, the role of deep level impurities in optoelectronic devices and the light-induced dislocation generation processes in semiconductor crystal lattices, seem to be the most important. Some of the technological problems reviewed include the stoichiometric crystallization methods applied to III–V compounds, electroepitaxy and interface morphology in LPE processes. The optoelectronic device problems discussed include the MNOS structures used in optical memories, and the heterostructures used in different electronic and optoelectronic devices, as well as electroluminescent image amplifiers. A number of chapters review the current problems in junction laser physics and optical fibre communication systems. The editor hopes that in spite of the rapid development of semiconductor optoelectronics the present volume will remain of permanent value.

In a text of this nature it is difficult to ensure a uniformity of presentation and notation. Although a certain amount of editing has been done, some differences in notation could not be avoided.

At the end of this volume are attached as an appendix the texts of the opening address given on the occasion of the Second International School on Semiconductor Optoelectronics 'Cetniewo 1978' by Professor Jerzy Kołodziejczak, Director of the Institute of Physics, Polish Academy of Sciences, and the closing address given by Professor Zh. I. Alferov from the Ioffe

Physico-Technical Institute of the Academy of Sciences of the USSR, Leningrad. Both texts seem to be very instructive and may be of interest to the reader.

The editor would like to take this opportunity to thank all the authors for their very comprehensive and instructive contributions.

M. A. HERMAN

CONTENTS

I PHYSICAL PHENOMENA IN OPTOELECTRONIC MATERIALS AND DEVICES

1. IV–VI Semiconductors as Materials for Infrared Optoelectronics, . . . 3
   L. Sosnowski
2. Statistical Considerations of the Influence of Doping and Alloying of III–V Semiconductor Materials on their Optoelectronic Properties 13
   K. Unger
3. Deep-Level Impurities in Semiconductors and their Role in Optoelectronic Devices 33
   H. G. Grimmeiss
4. Generation of Dislocations in Optoelectronic Materials and their Behaviour during Optical Excitation 63
   G. R. Woolhouse

II TECHNOLOGICAL PROBLEMS

5. Stoichiometric Crystallization Method of III–V Compounds for LEDs and Injection Lasers 101
   J. Nishizawa and Y. Okuno
6. Interface Morphology and Inhomogeneity in Impurity Incorporation during Multilayer LPE Processes 131
   T. Nishinaga
7. Quaternary III–V Systems for Semiconductor Sources of Coherent and Non-Coherent Radiation 157
   P. G. Eliseev
8. Electroepitaxy of $A^{III}B^{V}$ Compounds and its Application to Optoelectronic Devices Technology 187
   T. Bryśkiewicz

III OPTOELECTRONIC DEVICES

9. Heterostructures in Semiconductor Electronics 215
   Zh. I. Alferov
10. Processes in Heavily Doped Rare-Earth Materials and their Applications to Optoelectronic Devices 233
    F. Auzel

11 Reversible Optical Memory on MNOS-Structures 277
    Yu. M. Popov
12 Quantum Efficiency and Radiative Lifetimes in GaAs and AlGaAs 305
    D. Z. Garbuzov
13 Electroluminescent Image Transformers and Amplifiers 345
    S. V. Svechnikov
14 Photodetectors Based on Heterostructures 387
    V. I. Korol'kov

IV INJECTION LASERS
15 A Unified Approach to the Problems of Semiconductor Laser
    Theory 419
    M. J. Adams
16 Laser Modes as Eigenfunctions of an Operator Equation . . . 457
    J. E. Ripper, M. D. Campos, M. A. A. Pudensi
17 Investigation of Optical Gain and its Saturation Behaviour in
    Semiconductor Lasers 473
    M. H. Pilkuhn
18 Stripe Geometry Heterojunction Injection Lasers and the Effect
    of Optical and Carrier Confinement 495
    G. H. B. Thompson
19 The Possibilities of Influencing the Spectral Behaviour of Injection
    Lasers 529
    H. Bachert
20 Coherence of the Radiation Emitted by Semiconductor Lasers . . 563
    M. A. Herman

V OPTICAL COMMUNICATION SYSTEMS
21 Properties of Optoelectronic Devices for Optical Communication
    Systems 597
    J. C. Dyment
22 Modal Dispersion in Optical Fibres and the Influence of Mode-
    Coupling 621
    B. Crosignani and P. Di Porto
23 AlGaAs Heterostructures in Integrated Optics 637
    E. L. Portnoy
APPENDIX
    Opening Address 641
    Closing Address 643

Part I

Physical Phenomena in Optoelectronic Materials and Devices

Part 1

Chapter 1

IV-VI Semiconductors as Materials for Infrared Optoelectronics

L. Sosnowski

1.1 INTRODUCTION

IV–VI semiconducting compounds, of which lead sulphide is a typical representative, found applications at a very early date. Natural galena (PbS) had been widely used as a high frequency rectifier long before the beginning of the solid state electronics era. The lead sulphide and lead selenide infrared detectors were developed more than thirty years ago.

Progress in the technology of IV–VI's was however slow, whereas the main effort was directed to silicon and subsequently to III–V compounds.

In recent years all essential techniques developed in semiconductor electronics have been successfully applied to IV–VI compounds and significant progress has been achieved. Vacuum deposition, molecular beam epitaxy, gaseous and liquid phase epitaxy, and ion implantation have all been used to obtain sophisticated multi-layer heterostructures based an IV–VI compounds. Emitting diodes, lasers, detectors and wave guides reached a stage at which integrated optoelectronic circuits are feasible (1).

IV–VI compounds are narrow gap semiconductors. The natural domain of their application is infrared spectroscopy and communication. In this respect they are unique; the only possible competing materials are II–VI compounds based on mercury telluride and mercury-cadmium telluride. However, so far the application of mercury compounds has been restricted to infrared detectors.

The most promising materials for integrated optoelectronics are ternary alloys: $Pb_{1-x}Sn_xSe$ and $Pb_{1-x}Sn_xTe$. To understand the physical basis of their uses in electronics, their prospects and limitations, we have to discuss the basic properties of lead salts, namely PbS, PbSe and PbTe.

1.2 STRUCTURE AND BONDING OF LEAD SALTS

PbS, PbSe and PbTe all crystallize in the rock-salt structure and differ in this respect from the more familiar semiconductors with the diamond, zinc-blende or wurtzite structure. The rock-salt structure exhibits a high degree of ionicity because the coordination number of six for rock-salt cannot be reconciled with the directional valence bonds. According to Phillips' criterion (which applies strictly only to compounds having eight valence electrons to the base) ionicity should exceed 0.785 for rock-salt structure to occur (2).

Table 1.1 Dielectric constants

	PbS	PbSe	PbTe
ε_0	180	231	478
ε_∞	18.4	25.2	36.9
$\left(\dfrac{\omega_{LO}}{\omega_{TO}}\right)^2$	9.9	9.1	13.0

The high degree of ionicity is supported by the evidence provided by the dielectric constant (Table 1.1) (3). Very high value of the ratio of static ε_0 to high frequency ε_∞ dielectric constants, equal to the square of the ratio of frequency of optical longitudinal to optical transverse phonons points to a high degree of ionicity: $\varepsilon_0/\varepsilon_\infty = (\omega_{LO}/\omega_{TO})^2$ equals ten in all these compounds.

Very high values of dielectric constant, higher than in the more familiar semiconductors by several orders of magnitude, influence strongly their properties by delocalizing electronic functions and screening local potentials. High dielectric constant is also one of the reasons of the high mobility of current carriers in all lead compounds.

1.3 ENERGY GAP

The rock-salt structure is based on cubic flat-centred point lattice (Figure 1.1), and hence the Brillouin zone has the same shape as in diamond crystals (Figure 1.2).

It is well established from theoretical calculations and experimental evidence that the top of the valence band and the bottom of the conduction band lie at L point of the Brillouin zone and that both the valence band and the

metry is of D_{3d} point group. Surfaces of constant energy should be rotational ellipsoids with the symmetry axes along the [111] crystallographic direction.

The approximate band structure near the zone edge can be obtained from the $\bar{K} \cdot \bar{P}$ procedure (4).

Let

$$P_\perp = \frac{\hbar}{m_0}|\langle u_c|\hat{p}_\perp|u_v\rangle| \quad \text{and} \quad P_\| = \frac{\hbar}{m_0}|\langle u_c|\hat{p}_\||u_v\rangle|$$

matrix elements of momentum operator between the valence and conduction bands. The bands can be described by a modified Kane formula:

$$E = \pm\sqrt{\left(\frac{E_g}{2}\right)^2 + P_\perp K_\perp^2 + P_\| K_\|^2}$$

Introducing: $m_{\perp 0} = \hbar^2 E_g/2P_\perp^2$ and $m_{\|0} = \hbar^2 E_g/2P_\|^2$, one can obtain expressions for the dependence of masses on energy

$$m_\|(E) = m_{\|0}\left(1+\frac{2E}{E_g}\right), \quad m_\perp(E) = m_{\perp 0}\left(1+\frac{2E}{E_g}\right)$$

$m_{\perp 0}$ and $m_{\|0}$ are, of course, the values of masses at the bottom of the conduction or the top of the valence band. The values of these masses together with the anisotropy factor $K = m_{\|0}/m_{\perp 0}$ are given in Table 1.3 (3).

Table 1.3 Effective masses

		m_t	m_l	K
PbS	c	0.08	0.10	1.3
	v	0.08	0.10	1.3
PbSe	c	0.040	0.07	1.75
	v	0.034	0.068	2.0
PbTe	c	0.024	0.24	10
	v	0.022	0.31	14

From this table one can immediately see the high degree of symmetry between holes and electrons. The masses and mobilities of holes and electrons are almost the same in all three lead salts. This is in striking contrast with III–V and II–VI compounds where the valence band is dominated by heavy holes. The mass of heavy holes in InSb or HgTe exceeds the mass of electrons by one to two orders of magnitude.

IV–VI Semiconductors as Materials for Infrared Optoe

conduction band are non-degenerate. Lead salts are thus dire
ductors. Table 1.2 gives the values of the energy gap for two
near absolute zero and room temperature.

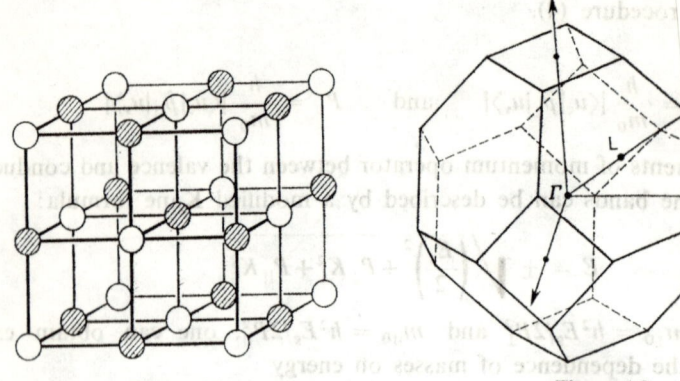

Figure 1.1   Figure 1.2

Table 1.2 Energy gap (in eV)

	PbS	PbSe	PbTe
4 K	0.29	0.16	0.19
300 K	0.41	0.27	0.39

The most striking feature of this table is the positive thermal
of the energy gap. Energy gap increases with increasing tempera
increase is roughly linear for $T > 70$K with $(\partial E_g/\partial T)_p = 4.7 \times 1$
Accordingly, the pressure coefficient is negative: $(\partial E_g/\partial p)_T$
$\times 10^{-6}$ eV/bar.

Thus the energy gap increases with increasing lattice constan
in the case of the more familiar diamond and zinc-blende semico
The temperature and pressure dependence of energy gap is impo
tuning devices, such as lasers and detectors.

1.4 VALENCE AND CONDUCTION BANDS

Both the valence and conduction bands are strongly non-parabolic,
electron and hole masses depend on energy (and concentration of c
The bands are also anisotropic, because at L point of Brillouin zo

IV-VI *Semiconductors as Materials for Infrared Optoelectronics*

Tables 1.2 and 1.3 show striking anomalies in a series of compounds PbS, PbSe, PbTe. First of all we notice that the energy gaps do not decrease regularly from PbS to PbTe; PbSe has the smallest gap. Also the anisotropy which is small for PbS and PbSe increases almost by one order of magnitude in PbTe. Consequently, the longitudinal mass has a minimum value for PbSe. To understand these irregularities and, moreover, to be able to discuss the properties of alloys containing tin, such as $Pb_{1-x}Sn_xSe$ and $Pb_{1-x}Sn_xTe$, we have to discuss more fully the band structure at L point of the Brillouin zone.

1.5 IRREDUCIBLE REPRESENTATIONS AT L POINT

As was mentioned above, the small group in K-space at L is a D_{3d} point group. Its irreducible representations are L_1^+, L_2^+, L_3^+ and L_1^-, L_2^-, L_3^-. L_1 and L_2 are single, and L_3 is doubly degenerate. Representations denoted by $+$ and $-$ sign differ in respect to the symmetry operation of inversion (D_{3d} group contains a centre of symmetry). Taking into account spin leads to the double group $D_{3d} \times D_{1/2}$ whose irreducible representations are L_6^-, L_6^+, L_{45}^- and L_{45}^+. L_6^+ and L_6^- are doubly degenerate, whereas L_{45}^+ and L_{45}^- are four-fold degenerate. The correlations between single and double groups

Table 1.4

D_{3d}	$D_{3d} \times D_{1/2}$	Te(Se)	Pb(Sn)
L_1^+	L_6^+	p	s
L_1^-	L_6^-	s	p
L_2^+	L_6^+	p	s
L_2^-	L_6^-	s	p
L_3^+	$L_6^+ + L_{45}^+$	p	d
L_3^-	$L_6^- + L_{45}^-$	d	p

are shown in Table 1.4. From this table one can see that, for instance, L_6^- representation can be derived from L_1^- and also from L_3^-, which corresponds to quite different orbital functions. We can denote this difference by symbols $L_6^-(L_3)$ and $L_6^-(L_1)$. More generally, a mixing of orbital functions can occur, thus L_6^- might belong to a state described by linear combination $|L_6\rangle = a|L_1\rangle + b|L_3\rangle$.

The last two columns in Table 1.4 show approximate symmetry of atomic functions at anion and cation sites corresponding to irreducible representations of D_{3d}. In this table underlined are representations corresponding to energy levels nearest to the energy gap. It is important to notice that L_1^+ level is of the s-type at the cation site, which means that the wave function has a finite value near the lead (or tin) nucleus.

1.6 BAND STRUCTURE AT L POINT

The band structure calculations by augmented plane waves (APW) (5) or the pseudo-potential method (6) give the ordering of levels at L points as indicated on the left side of Figure 1.3. These calculations do not take into account

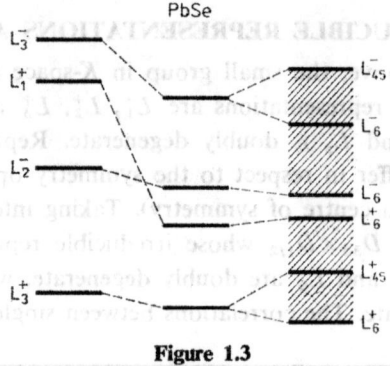

Figure 1.3

the relativistic effects which are of two kinds: the one connected with mass-velocity dependence and the spin-orbit interaction. First of the relativistic corrections suppresses very strongly the L_1^+ level because of its S-characteristic at Pb site. Pb atom with its atomic number $Z = 82$ exhibits a very strong Coulomb field near nucleus and the corresponding relativistic effect is very large. It causes level crossing of L_1^+ and L_1^-, bringing the first one below the last one. After spin-orbit correction we obtain final level ordering as shown on the right side of Figure 1.3.

The energy gap E_g is between L_6^+ and L_6^-. There are altogether six levels close to the gap, and to account properly for the masses and anisotropy of holes and electrons one has to take into account the interaction of all six bands. The two-band model discussed in Section 1.4 gives only a rough approximation.

1.7 TERNARY ALLOYS

We are now ready to understand the role of substitution of some lead atoms by tin atoms in ternary alloys. The properties of all ternary alloys are treated on the basis of the model of virtual crystal. One supposes that at the cation

IV–VI Semiconductors as Materials for Infrared Optoelectronics

site the potential acting on the electron is an intermediate between the atoms forming an alloy. Thus for $Pb_{1-x}Sn_xSe$ or $Pb_{1-x}Sn_xTe$, the potential at the cation site can be written as

$$V_K = (1-x)V_{Pb} + xV_{Sn}$$

Sn atom is much lighter than Pb and its potential does not cause such a drastic depression of L_1^+ level as Pb atom. Thus for sufficiently high Sn content the level crossing will not take place and the order of levels will be the one shown in Figure 1.4.

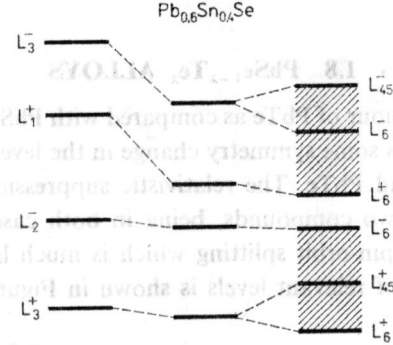

Figure 1.4

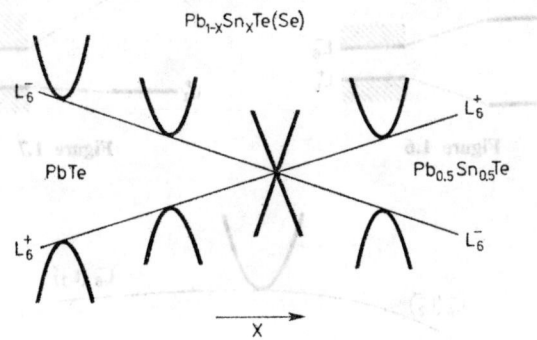

Figure 1.5

The energy gap is now between L_6^- and L_6^+, L_6^+ being at the bottom of the conduction band. It is easy to visualize what happens with the continuous change of x.

Figure 1.5 shows the level crossing with the composition of ternary alloys, together with a schematic diagram of the valence and conduction bands near band edges (7). As the energy gap decreases the non-parabolicity becomes more pronounced. At level crossing $m_{\perp 0}$ and $m_{\parallel 0}$ become zero and the dispersion law becomes a linear function of pseudo-momentum.

The band shapes are not strictly symmetrical on both sides of the crossing point, because higher bands do not cross and influence masses and anisotropy.

For optoelectronics applications the most important is variation of the energy gap with composition, because energy gap determines the spectral response of light detectors and emitters.

1.8 PbSe$_{1-x}$Te$_x$ ALLOYS

The anomalous behaviour of PbTe as compared with PbS and PbSe mentioned in Section 1.4 suggests some symmetry change in the levels framing the energy gap between PbSe and PbTe. The relativistic suppression of L_1^+ level should be the same in the two compounds, being in both cases due to Pb atoms. The difference is in spin-orbit splitting which is much larger for Te than Se atoms. The position of relevant levels is shown in Figures 1.6 and 1.7. Both

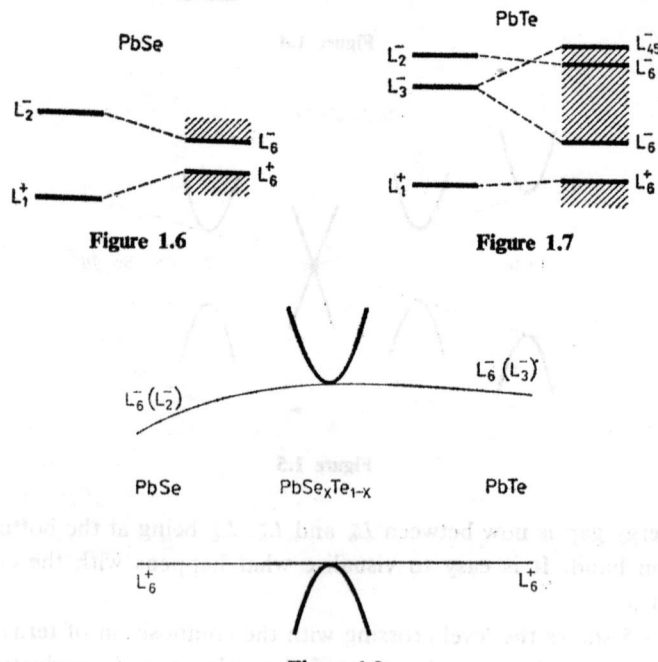

Figure 1.6

Figure 1.7

Figure 1.8

in PbSe and PbTe the conduction band belongs to L_6^- representation. The contribution of L_3^- is, however, dominant in PbTe and negligible in PbSe, where the L_6^- level originates from L_2^-. A recent study of PbSe$_x$Te$_{1-x}$ mixed crystals shows the change of symmetry clearly: the energy gap is a non-monotonous function of the composition x (Figure 1.8) (8).

1.9 CONCLUDING REMARKS

By properly choosing the composition of the Pb$_{1-x}$Sn$_x$Te alloy, photodiodes with cut-off wavelength up to 30 µm were obtained. In the region of 8–14 µm wavelengths background noise limited detectors can be manufactured. This corresponds to detectivity as high as 10^{11} cm Hz$^{1/2}$W^{-1} at 10.6 µm at ambient temperature of 77 K (9).

The frequency response is RC limited, in which respect the high static dielectric constant of band salts is a disadvantage. A frequency cut-off beyond 1 GHz can, however, be obtained.

Laser action up to 32 µm at 4.2 K has been reported. Heterojunction lasers PbTe–Pb$_{0.8}$Sn$_{0.2}$Te have been operated at CW mode up to 120 K in a wavelength range of 8–14 µm (1).

The limiting factor for lasing action at long wavelength side seems to be the polaron coupling between electrons and optical phonons and possibly the Auger recombination. The matter requires still further theoretical consideration.

REFERENCES

1. I. Mengalis, in *Physics of Narrow Gap Semiconductors*, ed. J. Raułuszkiewicz, M. Górska, and E. Kaczmarek, Polish Scientific Publishers, Warszawa 1978, p. 459.
2. J. C. Phillips, *Rev. Mod. Phys.*, **42**, 317 (1970).
3. R. Dalven, *Solid St. Phys.*, **28**, 179 (1973).
4. I. M. Tsidilkovsky, *Electrons and Holes in Semiconductors*, Russian edition, 1972, Polish edition, 1977.
5. J. B. Conklin, Jr., E. Johnson and G. W. Pratt, Jr., *Phys. Rev.*, A, **137**, 1282 (1965).
6. Y. W. Tung and M. L. Cohen, *Phys. Rev. B*, **2**, 1216 (1970).
7. M. Balkanski, *Journal of Luminescence*, **7**, 451 (1973).
8. G. Martinez, D. Guillot, and A. Jędrzejczak, *Inst. Phys. Conf.*, **35**, 51 (1977).
9. I. Mengalis, *Journal of Luminescence*, **7**, 501 (1973).

Chapter 2

Statistical Considerations of the Influence of Doping and Alloying of III-V Semiconductor Materials on their Optoelectronic Properties

K. Unger

2.1 INTRODUCTION

In spite of the fact that reproducible doping of the semiconductor material is a prerequisite for most semiconductor devices, in most *optoelectronic* devices use is made of *high* doping of the semiconductor material, i.e. the impurities are incorporated in such a concentration that wavefunctions and potentials due to impurities show a considerable degree of overlap. High doping is important e.g. to obtain high efficiency LEDs (especially if one is forced to use an indirect semiconductor material like GaP, and impurities participate in the optical transition process) or to obtain laser diodes with low threshold current density at room temperature (this is especially true of homojunction lasers).

Thus, optoelectronics has stimulated the study of the properties of high-doped semiconductor material (1), i.e. the study of the properties of a statistically fluctuating potential perturbing the host crystal.

On the other hand, one has learnt to use semiconductor alloys e.g. for adjusting the mixed material to the required gap width. As a first approximation the properties of this material were considered to be some interpolation of those of the pure end components of the alloy (2) and the term 'band structure engineering' was coined. However, owing to the fact that alloying includes also a fluctuation of the local mole fractions, characteristic properties, such as additional alloy scattering for the current carriers, a decrease of heat conductivity, nonlinearities in the gap, and refraction-index dependences on the mole fraction of the alloys can be observed (3). These additional properties are not always advantageous from the point of view of the device.

These two examples should suffice to argue that semiconductor optoelectronics requires also a thorough knowledge of the influence of impurities (donors, acceptors in high concentrations and alloying atoms), statistically distributed in space, on the properties of the material.

2.2 STATISTICS OF HIGH DOPING

2.2.1 General assumptions

We assume that the fluctuating perturbation potential $V(\mathbf{r})$ is the result of a superposition of contributions $Zv(\mathbf{r})$ due to randomly distributed impurities with valence number Z (relative to the host lattice)

$$V(\mathbf{r}) = \sum_{i,Z} Zv(\mathbf{r}-\mathbf{R}_i); \quad v(\mathbf{r}) = -\frac{e^2}{4\pi\varepsilon\varepsilon_0 r} \cdot e^{-r/L} \quad (2.1)$$

($\mathbf{R}_i$ impurity positions, ε static dielectric constant, L screening length), and that the action of the potential $V(\mathbf{r})$ can be treated within the effective mass approximation. Thus, deep impurities can be included in the analysis only if an adjusted effective mass value is sufficient for their description. The form of the individual potential contribution $v(\mathbf{r})$ corresponds to a Coulomb potential screened according to the semiclassical Thomas–Fermi approximation

$$L^{-2} = \frac{e^2}{\varepsilon\varepsilon_0}\left(\frac{\partial n}{\partial \zeta_n} + \frac{\partial p}{\partial \zeta_p}\right) + \frac{N_\mathrm{I} e^2}{\varepsilon\varepsilon_0 kT^*} \quad (2.2)$$

(n, p concentrations of free electrons, holes; ζ_n, ζ_p corresponding Fermi levels, counted from the band edges into the bands, N_I concentration of impurities, k Boltzmann constant).

$v(\mathbf{r})$ as given by (2.1) is taken as a simple example. In the screening length according to (2.2) also the effect of impurity correlation (4), where the charged impurities show an equilibrium configuration corresponding to some growth or diffusion temperature T^* of the material, can be included. Further refinement and extension of $v(\mathbf{r})$ will follow below.

Now, we imagine an increasing concentration of impurities N_I, the concentration and temperature at the beginning being low enough for carriers to be localized at 'their' impurities. First, in general, the impurity wave functions, having a typical spatial extension $a_\mathrm{B}^* = \hbar^2\varepsilon/m^*e^2$ (effective Bohr radius) will show considerable overlap, allowing carrier transfer from impurity to impurity. At $a_\mathrm{B}^* \geqslant N_\mathrm{I}^{-1/3}$ (mean distance between impurities) the formation

Statistical Considerations of III–V Semiconductor Materials

of an impurity band will take place. Then, further increasing N_I somewhat, the individual potentials $v(\mathbf{r}-\mathbf{R}_i)$ will overlap and screening will be enhanced yielding a situation where no carrier can be bound by an individual $v(\mathbf{r}-\mathbf{R}_i)$, the condition for this being $L \leqslant 0.84 a_B^*$ (5). Since L depends on carrier concentration, the two conditions for the high doping region, i.e. strong overlap of wave functions and of potentials are not independent. In fact, for practically important cases they are fulfilled at not very different impurity concentrations (cf. the examples for n-GaAs ($a_B^* \approx 100$ Å) and GaP ($a_B^* \approx 15$ Å) in Table 2.1).

2.2.2 Thomas–Fermi limit (6, 7, 8)

We now adopt the conceptually simple Thomas–Fermi approximation and consider the eigenvalue spectrum of the host crystal as only shifted according

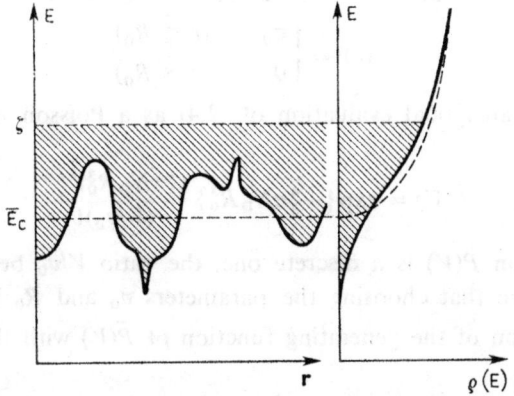

Figure 2.1 Behaviour of the conduction band edge under the influence of a fluctuation potential and the corresponding density of states $\varrho(E)$. The states are populated up to the Fermi level ζ

to the local value of the perturbing potential: The density of states for a spherical parabolic band ϱ_0 has to be formulated simply as follows:

$$\varrho_0(E) = \frac{1}{2\pi^2}\left(\frac{2m^*}{\hbar^2}\right)^{3/2}\sqrt{E} \to \varrho(\mathbf{r},E) = \frac{1}{2\pi^2}\left(\frac{2m^*}{\hbar^2}\right)^{3/2}\sqrt{E-V(\mathbf{r})}$$

$$\varrho(E) = \frac{1}{2\pi^2}\left(\frac{2m^*}{\hbar^2}\right)^{3/2}\int_{-\infty}^{\infty}\sqrt{E-V}P(V)\,dV \qquad (2.3)$$

The spatial averaging procedure to get $\varrho(E)$ is sketched in Figure 2.1.

Thus, the probability $P(V)dV$ of meeting a given value of the total potential within the interval $V \ldots V+dV$ is of principal interest. $P(V)$ can be derived by using generating functions of mathematical statistics (cf. e.g. (9)), which yields in terms of individual potentials $v(\mathbf{r})$

$$P(V) = \frac{1}{2\pi i} \int_{\mu-i\infty}^{\mu+i\infty} \exp\left\{4\pi N_D \int_0^\infty r^2(e^{-vt}-1+vt)dr \right.$$
$$\left. +4\pi N_A \int_0^\infty r^2(e^{vt}-1-vt)\,dr + Vt\right\}dt \quad (2.4)$$

The integration has to be performed to the left of all singularities. Two kinds of impurities, donors with concentration N_D and $Z = 1$, and acceptors with concentration N_A and $Z = -1$ have been assumed. Although numerical evaluations of expression (2.4) have been performed (10), it is more transparent to replace v approximately by a potential well (11)

$$\tilde{v}(r) = \begin{cases} v_0 & (r \leqslant R_0) \\ 0 & (r > R_0) \end{cases} \quad (2.5)$$

permitting the analytical evaluation of (2.4) as a Poisson distribution, e.g. for $N_A = 0$:

$$\tilde{P}(V) = \exp\{-\tfrac{4}{3}\pi N_D R_0^3\} \frac{(\tfrac{4}{3}\pi N_D R_0^3)^{V/v_0}}{(V/v_0)!} \quad (2.6)$$

(The distribution $P(V)$ is a discrete one, the ratio V/v_0 being an integer.) It can be shown that choosing the parameters v_0 and R_0 by equating the Taylor expansion of the generating function of $\tilde{P}(V)$ with that of the exact $P(V)$, i.e.

$$v_0 = -1.213 \frac{e^2}{4\pi\varepsilon\varepsilon_0 L}, \qquad R_0 = 0.698\,L \quad (2.7)$$

(12), yields the best agreement with the exact numerical values. Furthermore, it can be shown that this approach can be systematically refined by using the Laguèrre integration technique for arbitrary $v(\mathbf{r})$ in (2.4).

Characteristic parameters of the V-distribution are its first moments. We find for the first moment, i.e. the averaged potential

$$\overline{V(\mathbf{r})} = -(N_A - N_D) \int_0^\infty 4\pi r^2 v(r)\,dr = [(N_A - N_D)L^3] \cdot \frac{e^2}{\varepsilon\varepsilon_0 L} \quad (2.8)$$

and for the second moment

$$\overline{[V(\mathbf{r})-\overline{V}]^2} = \frac{Q^2}{2} = [(N_A + N_D)L^3] \cdot \left(\frac{e^2}{\sqrt{2}\varepsilon\varepsilon_0 L}\right)^2 \quad (2.9)$$

where the results are arranged in such a manner to show that $(N_A \pm N_D)L^3$, the number of impurities within a cube of volume L^3, occurs and that $e^2/(\varepsilon\varepsilon_0 \cdot L)$ is a typical value of the potential energy (cf. also the fitting conditions (2.7)).

Note that the different charge signs of donors and acceptors tend to cancel each other only in the first moment, whereas their contributions has to be added to get the second moment. It can be seen that the distribution, in general similar to the Poisson distribution, reaches as a limiting case a Gaussian distribution $(\bar{V} = 0)$

$$P(V) = \frac{1}{\sqrt{\pi}Q} \cdot e^{-V^2/Q^2} \qquad (2.10)$$

if $(N_A + N_D)L^3 = N_I L^3 \gg 1$.

We take GaAs and GaP as typical examples of a direct and indirect semiconductor material, respectively, to analyse the fulfilling of this condition (averaged for $T = 0 \ldots 300$ K):

Table 2.1 Conditions for Gaussian statistics

n-GaAs	$(n = 10^{17}$ cm$^{-3})$	$L \approx 100$ Å	$N_I \geqslant 5 \cdot 10^{19}$ cm^{-3}
	$(n = 10^{18}$ cm$^{-3})$	$L \approx 55$ Å	$N_I \geqslant 3 \cdot 10^{20}$ cm^{-3}
GaP, p-GaAs	$(p, n = 10^{17}$ cm$^{-3})$	$L \approx 70$ Å	$N_I \geqslant 1.5 \cdot 10^{20}$ cm^{-3}
	$(p, n = 10^{18}$ cm$^{-3})$	$L \approx 35$ Å	$N_I \geqslant 1.2 \cdot 10^{21}$ cm^{-3}

It can be seen from Table 2.1 that this limiting case cannot be adopted in the cases of practical interest for optoelectronic devices. Halperin and Lax (13), assuming a Gaussian distribution for the potential, calculated the corresponding density of states by a variational procedure, thus avoiding the Thomas–Fermi approximation (cf. also (14)). They found a shape of the tail of $\varrho(E)$ like $\exp\{-a|E|^m\}$ with increasing $m = \frac{1}{2}, \ldots, 2$ for increasing concentrations, $m = 2$ performing the Gaussian limit and $m \approx 1$ (exponential shape of the tail) for concentrations of interest to us. Although in general there is no Gaussian, but a more slowly decaying potential distribution, we can conclude that also the density of states decays essentially more slowly in the tail region than according to a Gaussian shape—the second moments being considered fixed in this discussion.

Indeed, exponential tails are observed in many cases and it is therefore reasonable to postulate the following distribution for V

$$P(V) = \frac{1}{E_0(1+e^{V/E_0})(1+e^{-V/E_0})} \qquad (2.11)$$

with the energy parameter E_0 characterizing an exponential tail (8). E_0 is best determined by equating the second moment of (2.11) with the exact one:

$$E_0^{TF} = \frac{Q}{2.565} = \frac{e^2}{2.565 \times 4\pi\varepsilon\varepsilon_0} [4\pi(N_D+N_A)L]^{1/2} \qquad (2.12)$$

The practical advantage of this approach is that it is possible to evaluate (to a good approximation) all the broadening integrals which occur in semiconductor statistics, usually containing Fermi integrals

$$F_n(\zeta/kT) = \frac{1}{n!} \int_0^\infty y^n (1+e^{y-\zeta/kT})^{-1} dy.$$

These integrals have to be shifted with respect to the Fermi energy and then averaged:

$$G_n(\zeta/kT, E_0) = \int_{-\infty}^\infty F_n\left(\frac{\zeta-V}{kT}\right) P(V) dV \approx \frac{1}{\lambda^{n+1}} F_n\left(\frac{\lambda\zeta}{kT}\right)$$
$$\lambda = \left(1+\left(\frac{E_0}{kT}\right)^2\right)^{-1/2} \qquad (2.13)$$

In effect, kT is to be replaced by $\sqrt{(kT)^2+E_0^2}$.

For example, the density of states according to (2.3) is obtained (using (2.13) in the limit $kT \to 0$) as

$$\varrho(E) = \frac{1}{2\pi^2} \left(\frac{2m^*}{\hbar^2}\right)^{3/2} \cdot \frac{\sqrt{\pi}}{2} E_0^{1/2} F_{-1/2}\left(\frac{E}{E_0}\right) \qquad (2.14)$$

which represents a square root behaviour with an exponential tail. The generalization of other well-known expressions (e.g. the expression for the screening length (2.2)) for the region of high doping is straightforward (8, 15). Also the expression for the density of states in a magnetic field H

$$\varrho_H(E) = \frac{1}{2\pi^2} \left(\frac{2m^*}{\hbar^2}\right)^{3/2} \frac{eH\hbar}{c} \cdot \frac{\sqrt{\pi}}{2} E_0^{-1/2} \sum_n F_{-3/2}\left(\frac{E-E_n}{E_0}\right)$$
$$E_n = \left(n+\frac{1}{2}\right) \hbar\omega_c, \quad F_{-3/2}(x) = \frac{d}{dx} F_{-1/2}(x) \qquad (2.15)$$

(ω_c the cyclotron frequency) obtained in this way, is convenient for treating magneto-optical effects in high-doped material (16) (cf. also (17)).

It should be noted that the distribution law (2.11) is a symmetrical one, and thus especially appropriate for situations with high compensation

Statistical Considerations of III–V Semiconductor Materials

$N_D \approx N_A$, i.e. for the p–n junction of a light-emitting device. Information about the actual degree of compensation is included in the third moment of $P(V)$, which is proportional to $(N_D - N_A)$, and is known as the 'skewness' of the distribution (9): In the case of low compensation ($N_D \gg N_A$ or $N_A \gg N_D$) asymmetric distributions and asymmetrically broadened line shapes occur.

2.2.3 Refinement with respect to quantum corrections (18)

The Thomas–Fermi approximation used up to this point, needs an important refinement for the fast varying part of the fluctuation potentials. Only for wavelengths larger than the quantum mechanical carrier wavelength is the Thomas–Fermi approximation justified. In the opposite case we have to take into account the uncertainty principle. The lowest states for electrons near a sharp and deep potential well lie higher than its bottom; on the other hand, electrons can tunnel through sharp potential peaks. This means that the band edge of, say, the conduction band does not follow fully the short range part of the fluctuating potential (cf. Figure 2.2), the spatial variation

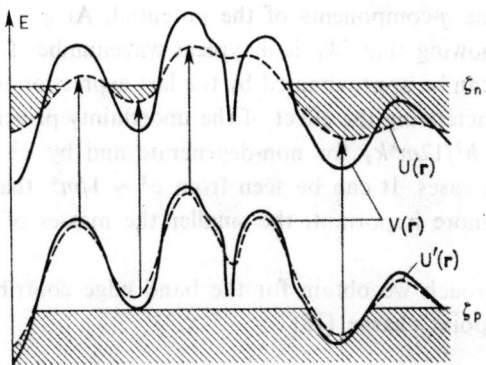

Figure 2.2 Behaviour of the effective band edges $\mathscr{U}(\mathbf{r})$ and $\mathscr{U}'(\mathbf{r})$ (dashed lines) under the influence of a fluctuation potential $V(\mathbf{r})$ (continuous line). Some optical transitions of minimal energy are indicated by vertical arrows (neglecting population effects). The limits of population of states are given by the Fermi levels ζ_n, ζ_p.

of the band edge $\mathscr{U}(\mathbf{r})$ being somewhat smoothed in comparison with $V(\mathbf{r})$. To speak about the Fourier transforms $\mathscr{U}(q)$ and $V(q)$ we have to expect that $\mathscr{U}(q) \approx V(q)$ for the limiting case of low wavenumbers q (Thomas–Fermi limit), but that $\mathscr{U}(q)$ is somewhat 'filtered' in comparison with $V(q)$ for higher wavenumbers. To put this picture on steady ground, we *define* a band edge $\mathcal{U}(\mathbf{r})$ by the following electron density relation:

$$n(\mathbf{r}) = n_0(\zeta - \mathcal{U}(\mathbf{r}), kT) \qquad (2.16)$$

$n_0(\zeta, kT)$ being the usual expression for the unperturbed electron density. Shifting the energy scale in this expression locally according to the potential value $V(\mathbf{r})$ would correspond to the Thomas–Fermi approximation, but shifting it by the smoothed band edge function $\mathcal{U}(\mathbf{r})$ involves now the quantum corrections discussed above.

Treating the fluctuation potential in the *linear* Hartree perturbation theory, one obtains for a highly degenerate electron gas

$$\frac{\mathcal{U}(q)}{V(q)} = \frac{1}{2} + \frac{1}{4}\left(\frac{2k_F}{q} - \frac{q}{2k_F}\right)\ln\left|\frac{q+2k_F}{q-2k_F}\right| = \varphi(q) \approx \frac{1}{1+c^2q^2} \qquad (2.17)$$

(k_F the Fermi wavenumber). This result is closely related to the Lindhard dielectric function (cf. (19)) and can easily be generalized to arbitrary degrees of degeneracy (20). We call $\varphi(q)$ a 'filter function', for it shows, beginning with $\varphi(0) = 1$, how the higher q-components of the band edge are filtered with respect to the q-components of the potential. At $q = 2k_F$ from (2.17) $\varphi = \frac{1}{2}$ follows, showing that $2k_F$ is a typical wavenumber for the electrons involved. This φ can be approximated by the last expression in (2.17), c being some length characterizing the effect of the uncertainty principle. This length is given by $c^2 = \hbar^2/12m^*k_T$ for non-degenerate and by $c^2 = \hbar^2/24m^*\zeta$ for highly degenerate cases. It can be seen from $c^2 \sim 1/m^*$ that quantum corrections are the more important, the smaller the masses of the carriers involved.

Using this approach we obtain for the band edge contribution $\mathcal{U}(\mathbf{r})$ due to one screened point charge (18)

$$\mathcal{U}(\mathbf{r}) = -\frac{e^2}{4\pi\varepsilon\varepsilon_0 r}\exp\left\{-\frac{r}{2c}\sqrt{1+\frac{2c}{L}}\right\}$$

$$\times \frac{2}{\sqrt{1-4c^2/L^2}}\sinh\frac{r}{2c}\sqrt{1-\frac{2c}{L}} \qquad (2.18)$$

which is to be compared with $v(\mathbf{r})$ (cf. (2.1)).

The advantage we have now is that we can use all the former statistical considerations for the effective band edge $\mathcal{U}$, define $P(\mathcal{U})d\mathcal{U}$, and so on.

The second moment for the $\mathcal{U}$-distribution, for example, has an additional factor $(1+2c/L)^{-3/2}$ in comparison with (2.9). Clearly, the quantum corrections tend to decrease the second moment, E_0 becoming less than according to (2.12). Because of the dependence of c and L on the position of the Fermi level (this in turn being influenced by the broadening of the density of states) fairly complex calculation is necessary to get the required self-consistent solutions. Consequently, for practical purposes, one is confronted with the task of constructing numerical relations or graphs for the quantities one is interested in (see below).

Because screening itself is a problem of the spatial variation of carrier concentrations, also $v(\mathbf{r})$ given in (2.1) has to be somewhat modified (18). The reason is that electrons, which do not react to the short range parts of the potential, cannot screen them. On the other hand, it is clear that electrons, bound on some sufficiently deep impurities, participate also in the screening of the fluctuation potential (21), but these electrons are inefficient in the screening of potential contributions of longer range than the extent of their wavefunction, short range parts being screened equally by all electrons, free or bound (22). Fortunately, the screening length L does not vary much in the concentration and temperature ranges of interest (cf. Table 2.1), nor does it have much influence on the quantities of final interest. Thus, some roughly estimated L-values can be used for an iterative calculation, showing fast convergence in most cases.

2.2.4 Broadening of the combined density of states

The spectra of direct optical transitions are determined mainly by the combined density of states, and the matrix elements can be considered to be almost constant for transitions between the band edges of the valence and conduction band. Using the Thomas–Fermi approximation now for both bands, we would arrive at the result, i.e. that a fluctuating potential would shift both bands about equally and no broadening of the optical spectra would occur. Obviously, to discuss such broadenings, quantum corrections are essential. According to the different effective masses for electrons and holes in direct semiconductors, the effective band edges $\mathcal{U}(\mathbf{r})$ and $\mathcal{U}'(\mathbf{r})$ (cf. Figure 2.2) are smoothed to a different extent (note that $c^2 \sim 1/m^*$). The broadening of the optical spectra is then given by the probability distribution $P(\mathcal{U} - \mathcal{U}')$. All the above considerations can therefore be transferred to the treatment of optical transitions (18).

Varying effective band edges means mixing of different wave vectors **k** into one energy state. Thus, if only states near the band edges are involved, this approach corresponds to the case of no-**k**-selection rule which is often used but which involves also a gradually increasing importance of **k**-selection for transitions to or from states far away from the band edge.

The combined broadened density of states, for example, is of the same shape as in (2.14), where m^* is now to be replaced by the reduced effective mass, and E_0 characterizes the exponential tail, usually observed in optical transitions. However, one point should be stressed: there is only one fluctuating potential acting on both bands, but producing two different effective band edges and two different densities of states (characterized by E_{on} and E_{op}). These densities of states govern the precise degree of screening of the bare ion potentials by the carriers (cf. relation (2.2)) and thus influence the screened potentials $v(\mathbf{r})$. On the other hand, the broadening of optical spectra is determined by $P(\mathscr{U}-\mathscr{U}')$ and characterized by an 'optical' E_0, this quantity revealing no simple relation to E_{on} and E_{op}.

The discussion above is quite transparent, but evaluation of concrete data is complicated by the self-consistency loops, to be treated numerically. The following points should be discussed as extensions and refinements of the model.

1. Impurities can also be the cause of remarkable lattice distortions yielding, in accordance with the usual deformation potential concept, two band edges fluctuating against each other (23) and not in phase like those caused by electric potentials. In this way broadening of optical spectra already occurs without taking into account quantum corrections.

2. If $m_n^* = m_p^*$ (occurring eventually in indirect semiconductors), $\mathscr{U}(\mathbf{r})$ and $\mathscr{U}'(\mathbf{r})$ would fluctuate in the same way as in the linear treatment given above. Then an extension to non-linear V-relations (cf. 2.2.5) or the use of other approaches (1, 24) is necessary. It is in order to mention here the treatments of Dow and Redfield (25) and of Esser and Herzog (26), where the probability distribution of electric field strengths is considered and its influence on the optical spectra is calculated along the same lines as for the Franz–Keldysh effect.

3. Actually, the case of equal effective masses is unimportant. Instead, there is the problem of properly accounting for the two degenerate hole bands (and eventually the multivalley structure of the conduction band). Obviously, the density-of-states effective mass is then not the correct one to use in determining the extent of the quantum corrections. These corrections (cf. the uncertainty length c) are reasonably described with an effective mass value, fitted to yield the correct effective Bohr radius and activation energy for shallow impurities, but a more detailed analysis would be welcome.

4. Although exciton lines can scarcely be observed in high-doped materials, final state interactions have a remarkable influence on the optical properties of these materials; since electron-hole pairs, which are coupled by Coulomb interaction, are produced by optical absorption, it is necessary to improve the one-electron treatment by multiplying the absorption coefficient $\alpha(\hbar\omega - E_g)$ by the probability of finding electron and hole at the same position (27). For free excitons of energies higher than the gap (correlated electron-hole pairs) this is

$$|\psi(0)|_{\hbar\omega > E_g} = \frac{2\pi \sqrt{R}}{\sqrt{\hbar\omega - E_g}} \left(1 - \exp\left\{-2\pi \sqrt{\frac{R}{\hbar\omega - E_g}}\right\}\right)^{-1} \quad (2.19)$$

($R = \mu e^4/2\hbar^2\varepsilon^2$ exciton binding energy, μ reduced effective mass.)

For electron-hole pairs in a high-doped material the electric microfields tend to separate electrons and holes (25), thus decreasing the factor $|\psi(0)|^2 > 1$, given above. In this way, exponential absorption tails result, but this is again slightly altered, if bound-to-free transitions are considered (28).

In spite of the complications mentioned, the model working with two effective band edges yields results which can be used to solve a number of practical problems (see below).

Note that also the analysis of pair recombination in semiconductors like GaP requires a lot of statistical considerations (29, 30), but we exclude this subject from our discussion, for no high-doping situations occur, the carriers being localized at the impurities involved.

2.2.5 Non-linear effects

Extending the perturbation series for the density of states of free carriers up to second order in the perturbation potential V, Milchev and Pickenhain (31) obtained the following formula (slightly altered to satisfy also the Thomas–Fermi limit):

$$\varrho(E) = \frac{1}{2\pi^2} \left(\frac{2m^*}{\hbar^2}\right)^{3/2} \sqrt{E - V(0) + \sum_{q \neq 0} \frac{|V(q)|^2}{\hbar^2 q^2/2m^* - 4E}} \quad (2.20)$$

which contains $V(\mathbf{q})$, the Fourier transform of $V(\mathbf{r})$. This formula shows clearly that the parabolic density of states is shifted about the mean value of the potential $V(0)$ (32), and also about a term of the second order in the potential. Whereas a linear perturbation theory is sufficient to describe broadening, only in second order does a shift of the line centres of gravity occur. The numerical evaluation of this shift for n-GaAs and p-GaAs using a Debye potential (2.1) for the impurities, accounting for the quadratic effects on both band edges, is in good agreement with gap shrinkage effects, observed e.g. by Sverev et al. (33) The expression for the band shrinkage in GaAs, given by Stern (34) to represent experimental data,

$$\Delta E_g = -1.6 \times 10^{-8} \text{ eV cm} \cdot (p^{1/3} - n^{1/3}) \qquad (2.21)$$

yields related values. It should be noted, however, that the explanation of the band shrinkage effect by using the quadratic response function (see above) is based on the fluctuation potential of charged impurities, and not on the effect of free carriers. On the other hand, via their screening effects on the crystal potential and according to self-energy corrections free carriers also yield a band shrinkage, discussed recently by Inkson (35), the numerical values obtained being different from those given above. The many-body effects on the band shrinkage are highly relevant also for an electron-hole liquid, generated optically in very clean semiconductor crystals. Clearly, this depends only on the free carrier concentrations, whereas the band shrinkage mechanism given above should be present in each case. The extent of the shrinkage is via their screening of the impurity potentials also indirectly dependent on free carrier concentrations. Further investigation of these points is obviously necessary, for band shrinkage effects have to be considered if one wants to explain e.g. the details of laser diode operation.

2.2.6 Transport properties

The mobility of carriers in high-doped semiconductor material can usually be analysed with the help of the Brooks–Herring formula for the impurity scattering, generalised for accounting for the concrete degree of degeneracy and broadening effects (see above). Quantum effects considered in 2.2.3 alter the screened potential of the impurities and thus also the mobility. An investigation of this (36) has shown that typically only 10%-effects appear. An interesting behaviour occurs for material of high doping, but almost exact compensation. In such a case only few carriers, say electrons, have to travel through the deepest states, produced by the fluctuation potential. (The screening length L and therefore E_0 are in this case especially high.) This gives rise to very low mobility values and a frequency dependence of the conductivity, which is approximately proportional to the frequency (37). This behaviour is quite similar to that of amorphous material. Although states at decreasing energy within the conduction band tail become more and more localized, no abrupt changes of the mobility at a 'mobility edge' were found.

Statistical Considerations of III–V Semiconductor Materials 25

In a light-emitting p–n junction almost exact compensation exists, with injected carriers often recombining through states, also deep in the tail of the density of states. Therefore, if the characteristic time of the optical transition is low enough (like in a direct material, especially during laser action, e.g. < 1 ns), the validity of the concept of two quasi-Fermi levels to describe the population of tail states becomes questionable. Comparable capture times for the carriers into deep tail states will limit their population and the optical transitions through these states; thus the optical spectra can show less broadening than the combined density of states. Indeed, E_0 values, measured by absorption or by emission from the same sample, often show differences which could be explained by such a mechanism.

2.2.7 Some conclusions on practical problems

2.2.7.1 Homogeneous n-GaAs material

Measuring the optical absorption in the tail region (e.g. References 23, 38) one obtains E_0, the optical broadening parameter. According to (2.12), modified to include quantum corrections, E_0 depends on the total number of charged impurities, the self-consistently calculated L values (cf. Table 2.1) varying little in the concentration region of interest ($n = 10^{17}, \ldots, 10^{19}$ cm^{-3}). From numerical values we found the following approximate relation:

$$E_0 = A[(N_D+N_A) \cdot 10^{-18} \text{ cm}^3]^{1/2} \tag{2.22}$$

with $A = 4.9$ meV for 80 K, and $A = 6.3$ meV for 300 K.

Thus, at a given free electron concentration (measured independently, say, by plasma reflexion), E_0 measurements can be used to determine the compensation degree of the sample (39). On the other hand, via the generalized Brooks–Herring formula (see above) the total concentration of charged impurities determines the magnitude of the impurity scattering and that of the measurable mobility μ ($\mu_I \sim (N_D+N_A)^{-1}$). This fact can be used to determine independently the total impurity concentration or the parameter E_{on}. The agreement has been found to be very good (39), making it thus possible to determine N_D and N_A by measuring $n = N_D - N_A$ and $E_0 \sim (N_D+N_A)^{1/2}$ or n and μ. The first possibility permits the use of optical measurements only and no problems with contacts arise.

2.2.7.2 GaAs-homojunctions

Using the theory sketched above and performing all the self-consistent calculations, it is easy to obtain the emission spectra of luminescence diodes, the gain curves and threshold characteristics of homojunction laser diodes

(15). (The extension to heterojunction lasers is, of course, also possible, but requires more concrete data on the junction structure.) As an application of our statistical considerations ending in calculable E_0 values, we will only state that E_0 determines the temperature dependence of the threshold current density and it is therefore possible to calculate the optimum doping concentration leading to minimal threshold current density for a given working temperature of the laser diode (15):

$$N_D = 1.8 \cdot 10^{18} \text{ cm}^{-3} \quad (80 \text{ K}) \quad \text{and} \quad N_D = 6.0 \cdot 10^{19} \text{ cm}^{-3} \quad (300 \text{ K}).$$

The last value clearly shows the necessity to use heterojunctions for room temperature laser diodes.

2.2.7.3 Current-voltage characteristics of GaP–LEDs

Using the well-known theory for the current-voltage characteristics of a p–n junction, where the current is diffusion-limited, and accounting for the broadening effects via parameter λ (or E_0) as shown above, one obtains the relation

$$j = j_s \left(\exp\left\{ \lambda \frac{eV_a}{kT} - 1 \right\} \right) \tag{2.23}$$

(j current density, j_s current density at saturation, V_a applied voltage). Thus from the exponential part $kT/\lambda = \sqrt{(kT)^2 + (E_{0D})^2}$ is available.

We find for nondegenerate GaP

$$L = 39.6 \text{ Å} \left(\frac{10^{18} \text{ cm}^{-3}}{n} \right)^{1/2} \cdot \left(\frac{T}{300 \text{ K}} \right)^{1/2} \tag{2.24}$$

and

$$E_{0D} = 11.4 \text{ meV} \left(\frac{N_D}{10^{18} \text{cm}^{-3}} \right)^{1/4} \cdot \left(\frac{T}{300 \text{ K}} \right)^{1/4} \tag{2.25}$$

The parameter E_{0D} describes the conduction band broadening at the p–n junction. It is assumed that this junction is produced by Zn-diffusion, Zn-incorporation yielding only acceptors. Then at the junction $N_D^+ = N_A^-$ is valid and the broadening E_{0D} of the current-voltage characteristics depends only on the donor concentration of the active GaP-substrate, irrespective of its content of acceptors, these acceptors influencing only the precise position of the p–n junction. Detailed examination of (2.25) and comparison with direct N_D-determination of the substrate (besides that of $n = N_D - N_A$) has shown good agreement (40). Thus, the current-voltage characteristics of the diodes and

Statistical Considerations of III–V Semiconductor Materials 27

the knowledge of n from the substrates used makes it possible to measure the compensation degree of the substrate material (or of the LPE layer if this is the active material with the p–n junction).

2.3 STATISTICAL EFFECTS IN MIXED III–V COMPOUNDS

2.3.1 General considerations

To get the properties of a mixed III–V compound the virtual crystal approximation (VCA) can be used as a starting point; the potentials due to the atoms, substituting each other, are averaged according to the mole fraction x of the alloy. Thus, an exact periodic crystal is regained and band structure calculations can easily be performed.

Deviations from this virtual crystal arise from the fact that in reality there are no averaged atoms and the mole fraction x is a random variable bound to the limits $x = 0$ and $x = 1$. If we have a mixed crystal $A_xB_{1-x}C$ where the atoms A and B give rise to potentials v_A and v_B, it is an easy task to show that the second moment of the fluctuation potential is given by

$$\overline{(V-\overline{V})^2} = N_s \cdot x(1-x) \int_0^\infty (v_A - v_B)^2 4\pi r^2 dr \qquad (2.26)$$

(N_s concentration of lattice sites, available for atoms A and B). Except for the factor $(1-x)$, formula (2.26) corresponds exactly to (2.9): as is to be expected, in the limit $x \ll 1$ the 'impurity' concentration corresponds to $N_s \cdot x$ and the additional factor $(1-x)$ guarantees the correct behaviour for $x \to 1$, where disorder effects should disappear.

The disorder effects on the energetic structure of mixed crystals can be taken into account starting from the virtual crystal by the second order perturbation theory (cf. (41) for GaP_xAs_{1-x}) or using the quadratic response of the virtual crystal (cf. 2.2.5). Note that some problems arise if the end components of the alloy have very different lattice parameters a_0: even if a linear interpolation of a_0 (Vegard's rule) is possible, there are slightly different ways to account for the a_0 change in constructing the VCA potential (cf. (42), for $Ga_xIn_{1-x}As$, non-linear screening problems arise) and by the construction of the disorder potential, some positional disorder of the atoms cannot be excluded (not even in the case of $Al_xGa_{1-x}As$ (43) with very little change of the lattice parameter).

Generally, already in the VCA non-linear dependences of gaps on the mole fraction appear; disorder tends to decrease the gaps (by an amount, proportional to $x(1-x)$, cf. (2.26)); a semi-empirical explanation of this 'bowing' of gaps has also been attempted (44).

This is illustrated in Figure 2.3 which depicts the gap variation of $Al_xGa_{1-x}As$ (43), where the crossover from direct to indirect material is obvious. The experimental E_g data for the direct gap are systematically lower than the values of linear interpolation.

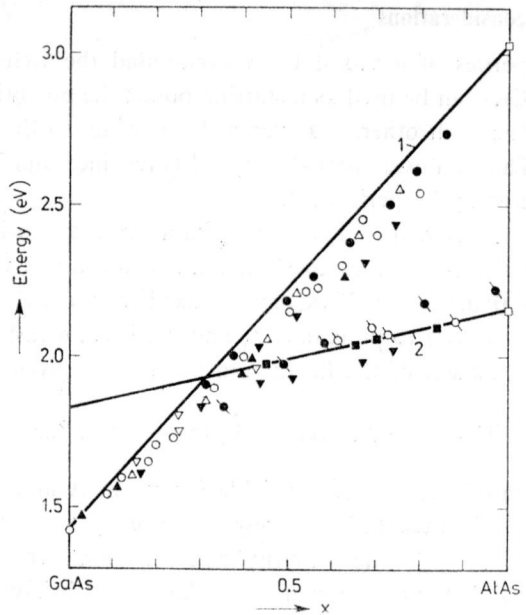

Figure 2.3 Energy gaps (direct: 1; indirect: 2) for $Al_xGa_{1-x}As$: experimental points of different origin and straight lines interpolating the gaps of the end components (43)

This fact can easily be accounted for in terms of a quadratic formula for $E_g(x)$.

Refraction indices n^*, on the other hand, are related via the non-linear relationship

$$n^{*2} = 1 + \frac{4\pi n_v e^2 \hbar^2}{m\bar{E}_g^2} \qquad (2.27)$$

to the band gap ($\bar{E}_g$ is the Penn gap, averaged over the Brillouin zone, and n_v is valence electron concentration), thus a 'blowing' of refraction indices corresponds typically to the 'bowing' of energy gaps (45).

2.3.2 A simple model and its application

Whereas in the case of high-doped material the simple effective-mass approximation can be used quite safely, the perturbing potentials in a mixed crystal, e.g., $v_A - [xv_A + (1-x)v_B]$ if an atom A with v_A replaces the VCA atom with $xv_A + (1-x)v_B$, are short-ranged and the knowledge of full band structure is necessary to account for this perturbation. Of course, the concentration of these (short-ranged) potentials in the crystal is very much higher than in a high-doped material.

To treat disorder effects in a mixed crystal within the effective-mass approximation one can try to use differences of electron affinities V_0 or differences of absolute energies of the corresponding band edges for the pure end components instead of pseudopotentials. If we take R_0 corresponding to the volume of one formula unit AC ($\frac{4}{3}\pi R_0^3 = a_0^3/4$) as the radius of this perturbation, we will have an interpolation of energies of the pure end component. The long-range part of the corresponding fluctuation potential, made up of potential pockets with depth V_0, can certainly be treated within the effective mass approximation. For the fluctuations of smaller range some counterbalance of quantum corrections (filtering these contributions) and corrections to the simple effective mass equation (increasing effectively m^*) can be hoped for.

It is now a simple matter to show that the second moment for the distribution of this fluctuating electron affinity is given by $M_2 = x(1-x)V_0^2$. Taking $E_0 = \sqrt{2M_2}/2.565$ (cf. (2.12)), we arrive at

$$E_0^m = 0.551|V_0|\sqrt{x(1-x)} \quad (2.28)$$

E_0^m, which characterizes the broadening due to alloying, is statistically independent of the broadening due to high doping. Thus, we have as the combined broadening parameter of, say, the conduction band states

$$E_{comb} = \sqrt{(E_{on})^2 + (E_0^m)^2} \quad (2.29)$$

To get the necessary differences of absolute energies of the bands, information about ionization potentials (experimental, semi-empirical values (46, 47) or those calculated with the help of pseudopotentials (48)) is needed.

To test the model discussed we take $GaP_{1-x}As_x$, as an example, LEDs from which have been investigated (40). The current-voltage characteristics was used to determine E_{comb} (cf. 2.2.7.3). By comparing with E_{on}, caused by the doping alone, it is possible to extract E^m for the conduction band (in the indirect region, $x < 0.5$). x was taken from electron beam microanalysis (ESMA) of the material. On the other hand, exploiting the theoretical relations given, and taking $|V_0| = 0.24$ eV as the mean value of different estimates $|V_0| = 0.18, \ldots, 0.29$ eV (46, ..., 48), we arrive at the relation

$$E_0^m(\text{As in GaP}) = 132 \text{ meV} \sqrt{x(1-x)} \tag{2.30}$$

which agrees with the experimental data astonishingly well. This shows that broadening effects due to alloying can be extracted from investigations of LEDs, permitting the determination of the mole fraction. Of course, other alloying systems have yet to be tested.

2.4 CONCLUDING REMARKS

It was the aim of this paper to show that valuable information about important mechanisms of disorder (broadening and shifting by randomly distributed potentials) and about relevant material parameters can be obtained by applying statistical methods, especially in real space.

The models to be used are in most cases fairly straightforward, but the relations between the different quantities are often very complex. Thus, for concrete problems of optoelectronic materials complex numerical work is needed to get experience on possible approximations and to obtain gauge curves directly useful for comparison with experimental data.

REFERENCES

1. V. L. Bonch-Bruevich, *The Electronic Theory of Heavily Doped Semiconductors*, Elsevier, New York, 1966.
2. C. Hilsum, in *Semiconductors and Semimetals*, Willardson and Beer, Ed., **1**, 3 (1966).
3. N. A. Goryunova, F. P. Kesamanly, D. N. Nasledov, in *Semiconductors and Semimetals*, Willardson and Beer, Ed., **4**, 413 (1968).
4. F. Stern, *Phys. Rev.*, **B3**, 3559 (1971).
5. V. L. Bonch-Bruevich, V. B. Glasko, *Opt. i. Spektr.*, **14**, 495 (1963).
6. E. O. Kane, *Phys. Rev.*, **131**, 79 (1963).
7. A. L. Efros, *Usp. Fiz. Nauk*, **111**, 451 (1973).
8. K. Unger, *Wiss. Zeitschr. Karl-Marx-Univ., Math.-Naturw. R.*, **20**, 93 (1971).
9. T. N. Morgan, *Phys. Rev.*, **139**, A 343 (1965).
10. H. Schmid, *Diploma Work*, Section of Physics, Karl-Marx-Univ., Leipzig, 1971.
11. I. M. Zidilkovsky, *Electrons and Holes in Semiconductors* (in Russian), Nauka, Moscow, 1972.
12. O. Breitenstein, K. Unger, *phys. stat. sol.*, to be published.
13. B. I. Halperin, M. Lax, *Phys. Rev.*, **148**, 722 (1966).
14. R. Eymard, G. Duraffourg, *J. Appl. Phys.*, **6**, 66 (1973).
15. K. Unger, *Z. Phys.*, **207**, 332 (1967).
16. W. Thielemann, *phys. stat. sol.*, **34**, 519 (1969).

17. M. I. Dyakonov, A. L. Efros, D. L. Mitchell, *Phys. Rev.*, **180**, 819 (1969).
18. K. Unger, *Ann. Phys. (Lpz.)*, **27**, 161 (1971).
19. K.-R. Schulze, K. Unger, *phys. stat. sol. (b)*, **66**, 491 (1974).
20. W. Eisenberg, K. Unger, *Ann. Phys. (Lpz.)*, **31**, 125 (1974).
21. D. M. Larsen, *Phys. Rev.*, **B11**, 3904 (1975).
22. B. Gobsch, *Thesis*, Department of Physics, Karl-Marx-Univ., Leipzig, 1975.
23. J. I. Pankove, *Phys. Rev.*, **140**, A 2059 (1965).
24. H. van Cong and G. Mesnard, *phys. stat. sol. (b)*, **52**, 553 (1972).
25. J. D. Dow, D. Redfield, *Phys. Rev.*, **B5**, 594 (1972).
26. B. Esser, F.-N. Herzog, *phys. stat. sol. (b)*, **71**, 63 (1975).
27. R. J. Elliott, *Phys. Rev.*, **108**, 1384 (1957).
28. J. D. Dow, D. L. Smith, F. L. Lederman, *Phys. Rev.*, **B8**, 4612 (1973).
29. F. Williams, *phys. stat. sol.*, **25**, 493 (1968).
    A. E. Yunovich, *Fortschr. Physik*, **23**, 317 (1975).
30. D. G. Thomas, J. J. Hopfield, W. M. Augustyniak, *Phys. Rev.*, **140**, A 202 (1965).
31. A. Milchev, R. Pickenhain, *phys. stat. sol. (b)*, **79**, 549 (1977).
32. This is a trivial fact and corresponds to the Thomas–Fermi limit, but this term has no influence on optical spectra.
33. P. L. Sverev, *Fiz. Tekh. Poluprov.*, **11**, 1017 (1977).
34. F. Stern, *J. Appl. Phys.*, **47**, 5382 (1976).
35. J. C. Inkson, *J. Phys.*, **C9**, 1177 (1976).
36. W. Eisenberg, *Ann. Phys. (Lpz.)*, **31**, 131 (1974).
37. D. Redfield, *Adv. Phys.*, **24**, 463 (1975).
38. D. Redfield, M. A. Afromowitz, *Phil. Mag.*, **19**, 831 (1969).
39. W. Hörig, *Thesis*, Department of Physics, Karl-Marx-Univ., Leipzig, 1971.
40. H. Schaefer, K. Unger, *phys. stat. sol.*, to be published.
41. A. Baldareschi, K. Maschke, *Solid State Commun.*, **16**, 99 (1975).
42. K.-R. Schulze, H. Neumann, K. Unger, *phys. stat. sol. (b)*, **75**, 492 (1976).
43. A. Baldareschi, E. Hess, K. Maschke, H. Neumann, K.-R. Schulze, K. Unger, *J. Phys.*, **C10**, 4709 (1977).
44. J. A. van Vechten, T. K. Bergstresser, *Phys. Rev.* **B1**, 3351 (1970).
45. G. Kühn, K. Löschke, K. Unger, L. Hildisch, *Kristall und Technik*, **11**, 1065 (1976).
46. J. A. van Vechten, *Phys. Rev.*, **182**, 891 (1969).
47. K. Hübner, K. Unger, *Nachrichtentechnik-Elektronik*, **23**, 387 (1973).
48. K. Unger, E. Heß, H. Neumann, K.-R. Schulze, paper given at the *Annual Solid-State Physics Conference*, Warwick, 1978, *phys. stat. sol.*, to be published.

Chapter 3

Deep-Level Impurities in Semiconductors and their Role in Optoelectronic Devices

H. G. Grimmeiss

3.1 INTRODUCTION

If an atom of the host lattice is replaced by an atom belonging to one of the adjacent groups in the periodic table, the potential binding the extra electron or hole at the impurity atom can be approximated by a hydrogen-like potential. This gives rise to an energy level in the bandgap for the impurity ground state and, in addition, a series of excited states. In most semiconductors, the energy separation between the ground state and the nearest band edge is typically less than 100 meV. Such impurities are therefore called *shallow impurities* and are widely used in semiconductor technology for modifying the degree and type of electrical conductivity.

Impurities which do not belong to one of the nearest groups in the periodic table will in general create energy levels farther away from the band edge. They are therefore called 'deep'-level impurities. In this chapter some recent results concerning the electronic properties of several 'deep'-level impurities are discussed. The emphasis will be placed on isolated point defects. We will start with defects which are at the present time considered to be single substitutional impurities, even though it might turn out that this picture is too simple. Substitutional defects are easier to handle, both experimentally and theoretically, than are the more complicated defects such as complexes. In spite of this fact, however, the understanding of even simple substitutional deep impurities is still rather modest compared with shallow impurities.

Deep-level impurities seem to be present in all known semiconductor materials. One of their most important properties is the ability to control the carrier lifetime even in small concentrations. According to Shockley–Read–Hall statistics (1, 2) the lifetime τ of excess charge carriers in a semiconductor with a single impurity level is given by

$$\tau = \frac{c_p(p_0+p_1)+c_n(n_0+n_1)}{c_n c_p N_{TT}(n_0+p_0)} \tag{3.1}$$

where c_n and c_p are the average values of the capture constants of electrons and holes over the states in the bands, n_0 and p_0 are the free-carrier concentrations in thermal equilibrium, and n_1 and p_1 are the electron and hole concentrations for the case where the Fermi level E_F falls at E_T, the energy position of the impurity level. N_{TT} is the concentration of the impurity energy levels. Consider for the sake of illustration an n-type semiconductor with an impurity level above the Fermi level. In this case, we have

$$n_1 \gg n_0 \gg p_0 \gg p_1 \qquad (3.2)$$

a condition, for example, which exists in the space-charge region of a junction or a Schottky barrier. The above equation then reduces to

$$\tau = \frac{n_1}{N_{TT} c_p n_0} = \frac{N_c}{N_{TT} c_p n_0} \exp\left(-\frac{E_c - E_T}{kT}\right) \qquad (3.3)$$

showing that the free-carrier lifetime changes exponentially with E_T. The significant influence on the carrier lifetime not only of a large value of $E_c - E_T$ but also of a large capture constant is very important for the fabrication of devices. This lifetime can be affected both by desired dopants, such as in fast switching diodes, and by undesired 'killer centres' in light emitting devices. This technological importance is one of the reasons for the current interest in deep impurities.

The energy position and the capture constant are important parameters for the characterization of deep-level impurities. These parameters are in turn related to the thermal emission rate e^t by the detailed balance relationship

$$e_n^t = c_n N_c e^{\frac{-\Delta G_n}{kT}} \qquad (3.4)$$

$$e_p^t = c_p N_v e^{\frac{-\Delta G_p}{kT}} \qquad (3.5)$$

Hence, the electronic properties of impurities are often characterized by their emission rates, capture constants and energy position. These parameters can easily be converted into more basic physical parameters. The capture cross section, for example, is obtained by dividing the capture constant by the thermal velocity, while the optical cross sections σ^o can be calculated by dividing the respective optical emission rates by the photon flux φ of the incident light:

$$\sigma = \frac{c}{\langle v \rangle} \qquad (3.6)$$

$$\sigma^o = \frac{e^o}{\varphi} \qquad (3.7)$$

Capture constants and thermal and optical emission rates are most commonly measured by junction space charge techniques such as dark-current transients, photocurrent and photocapacitance. The energy position of a deep level may be obtained either from the temperature dependence of the thermal emission rates or from the spectral distribution of the optical cross sections. During the last few years, a number of different techniques have been developed which make it possible to measure the electronic properties of deep level impurities very accurately.

3.2 MEASURING TECHNIQUES

3.2.1 Constant photoconductivity technique (CPCT)

Most of the recently developed measuring techniques are based on excitation processes in the space charge region of junction barriers. One of the reasons why bulk material is used less frequently than more complicated structures such as p–n junctions and Schottky barriers is the fact that recombination processes can in most cases be neglected in junction space charge techniques. The analysis of data obtained from junction techniques is therefore more straightforward than, for example, bulk photoconductivity measurements obtained with constant photon flux. In general, optical emission rates cannot be obtained from bulk photoconductivity measurements. This is easily seen if, for the sake of illustration, an n-type semiconductor is considered with a single impurity level at energy E_T in the upper half of the bandgap and concentration N_{TT}. Illuminating the sample with photons of energy $h\nu < E_T - E_v$ and a photon flux such that $\Delta n \gg n_0$ the concentration of free electrons in the conduction band can be calculated as follows:

$$e_n^0 n_T - c_n n(N_{TT} - n_T) =$$
$$e_n^0 N_{TT} - e_n^0 n - c_n n^2 = 0 \qquad (3.8)$$

From Equation 3.8 one obtains for the free electron concentration n

$$n = -\frac{e_n^0}{2c_n} \pm \sqrt{\left(\frac{e_n^0}{2c_n}\right)^2 + N_{TT}\frac{e_n^0}{c_n}} \qquad (3.9)$$

where $n_T = N_{TT} - p_T$ is the concentration of energy levels occupied by electrons. Because c_n and N_{TT} are in most cases not known, the optical emission rate can be obtained from bulk photoconductivity measurements with constant photon flux only under very special conditions. In addition, the response and decay times of the photoconductivity current in semiconductors containing deep energy levels are often large due to the recharge of these levels (3). In semi-insulating GaAs:O, for example, decay times of several hours have been observed (4, 5). Compared with junction space charge techniques where emission rates and capture rates are measured directly, it is quite clear why bulk measurements are less attractive. However, in spite of the many advantages of junction techniques, there is one inherent problem which makes these techniques inferior to bulk measurements. All excitation processes investigated by junction space charge techniques occur in regions of high electric fields.

High fields may, however, influence the excitation and capture processes investigated. Hence, when using junction techniques, careful attention must be paid to whether or not the electric field has any influence on the data. If the electric field influences the measurements, junction techniques cannot be used to measure emission rates and capture constants. In order to be able to investigate emission rates in this case, we developed the constant photoconductivity technique (CPCT) (6). The important feature of this technique is the fact that the photon flux of the incident monochromatic light is adjusted so as to keep the photoconductivity constant with photon energy. Keeping the photoconductivity constant means that the occupancy of the energy levels involved is not changed, and, hence, that both n and n_T are constant. From Equations 3.7 and 3.8 it then follows that

$$\sigma^0 = B \cdot \frac{1}{\varphi} \qquad (3.10)$$

where B is independent of photon energy. Thus, the spectral distribution of the photoionization cross section σ^0 is obtained by plotting the inverse of the photon flux φ necessary to keep the photocurrent constant, as a function of photon energy.

3.2.2 Thermal emission rates

For all cases where high electric fields do not influence emission rates and capture constants, junction space charge techniques are most commonly used to investigate the electronic properties of deep energy levels. Because the photocapacitance technique has been described quite often, we will report in this chapter briefly on two other techniques: the dark current transient technique and the dual light source photocurrent technique.

It has been shown previously (7) that the dark current of a junction is not proportional to the total width w of the space charge region, but to an effective width $w-w_0$. The quantity w_0 is that part of the outer transition region in which the net excitation rate can be neglected because of competing recombination processes due to free-carrier tails extending from the neutral regions into the space charge region. Denoting the boundaries of the effective gener-

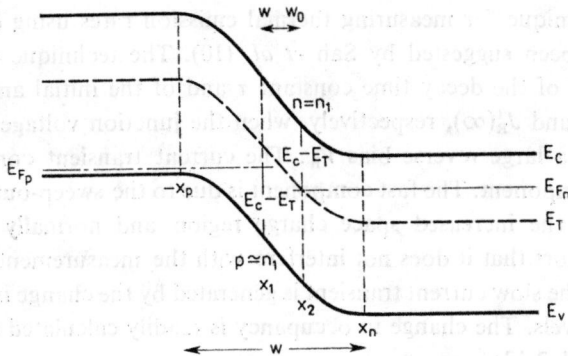

Figure 3.1 Band diagram of a p-n junction with deep-energy levels at E_T

ation region $w-w_0$ by x_1 and x_2, respectively, (see also Figure 3.1) the dark reverse current density J_R is then given by

$$J_R^t = q \int_{x_1}^{x_2} \left[\frac{1}{D} U_n + \left(1 - \frac{1}{D}\right) U_p\right] dx$$

$$= q(w-w_0)\left[\frac{1}{D} U_n + \left(1 - \frac{1}{D}\right) U_p\right] \quad (3.11)$$

where U_n and U_p are the total net rates of electron and hole emission, respectively, and D is the ratio between the excited charge within the effective generation region $w-w_0$ and the measured charge in the external circuit (8, 9). Because capture processes can be neglected in reverse biased p-n junctions, the net rate of electron and hole emission is given by

$$U_n = e_n^t n_T(t) \quad (3.12)$$

and

$$U_p = e_p^t p_T(t) = e_p^t (N_{TT} - n_T(t)) \quad (3.13)$$

From Equations 3.11, 3.12, and 3.13, we obtain for the dark reverse current density

$$J_R^t(t) = q(w-w_0)\left\{\frac{1}{D} e_n^t n_T(t) + \left[1 - \frac{1}{D}\right] e_p^t [N_{TT} - n_T(t)]\right\} \quad (3.14)$$

At steady state, $U_n = U_p$, and hence

$$n_T(\infty) = \frac{e_p^t}{e_n^t + e_p^t} N_{TT} \qquad (3.15)$$

A simple technique for measuring thermal emission rates using dark reverse currents has been suggested by Sah *et al.* (10). The technique involves the measurements of the decay time constant τ and of the initial and final dark current $J_R^t(0)$ and $J_R^t(\infty)$, respectively, when the junction voltage is switched from zero to a large reverse bias V_R. The current transient contains a fast and a slow component. The fast component is due to the sweep-out of majority carriers from the increased space charge region and normally has a time constant so short that it does not interfere with the measurement of the slow component. The slow current transient is generated by the change in occupancy of the deep levels. The change in occupancy is readily calculated using Equations 3.12 and 3.13:

$$\frac{dn_T}{dt} = U_p - U_n = e_p^t N_{TT} (e_n^t + e_p^t) n_T(t)$$

By integration one obtains

$$n_T(t) = \frac{e_p^t}{e_n^t + e_p^t} N_{TT} + \left[n_T(0) - \frac{e_p^t}{e_n^t + e_p^t} N_{TT} \right] e^{-(e_n^t + e_p^t)t} \qquad (3.16)$$

where $n_T(0)$ is the concentration of occupied impurity level at $t = 0$. If $n_T(0) = N_{TT}$ (a situation which, for example, is achieved by zero biasing a p$^+$–n junction prior to the application of a large reverse bias) one obtains for the time dependence of the slow dark current transient by substitution of Equation 3.16 into 3.14

$$J_R^t(t) = q(w - w_0) N_{TT} \left\{ \left(1 - \frac{1}{D}\right) e_p^t + \right.$$

$$\left. + \left[\frac{e_p^t}{e_n^t + e_p^t} + \frac{e_n^t}{e_n^t + e_p^t} \exp\left(-\frac{t}{\tau}\right) \right] \left[\frac{1}{D} e_n^t - \left(1 - \frac{1}{D}\right) e_p^t \right] \right\} \qquad (3.17)$$

where

$$\tau = \frac{1}{e_n^t + e_p^t} \qquad (3.18)$$

Equation 3.17, at $t = 0$, reduces to

$$J_R^t(0) = \frac{1}{D} q(w - w_0) N_{TT} e_n^t \qquad (3.19)$$

while for the steady state it gives

$$J_R^t(\infty) = q(w-w_0)\frac{e_p^t}{e_n^t+e_p^t} N_{TT} e_n^t \qquad (3.20)$$

From a single transient wave form given by Equation 3.17, e_n^t and e_p^t of an impurity level at E_T can be calculated using Equations 3.18, 3.19 and 3.20

$$e_n^t = \frac{1}{\tau}\left(1 - \frac{1}{D}\cdot\frac{J_R^t(\infty)}{J_R^t(0)}\right) \qquad (3.21)$$

$$e_p^t = \frac{1}{\tau}\cdot\frac{1}{D}\cdot\frac{J_R^t(\infty)}{J_R^t(0)} \qquad (3.22)$$

3.2.3 Optical emission rates

Most of the data presented in this chapter have been obtained with the dual light source spectroscopy technique (DLSS). By analogy to the dark current technique, the optical emission rates e_n^0 and e_p^0 of an energy level at E_T can be obtained from a single photocurrent transient wave form. In contrast to the dark current technique, e_n^0 and e_p^0 may, however, also be obtained from steady state photocurrent measurements, using two light sources simultaneously. Steady state measurements are more straightforward than transient techniques and are especially suitable for highly compensated diodes.

For the sake of simplicity, it is assumed that the sample is below the freeze-out temperature and that the energy level is in the upper half of the bandgap. This implies that thermal excitation processes can be neglected ($e_n^t = e_p^t = 0$). If the sample is illuminated with photons of energy $h\nu$ smaller than the bandgap E_g (extrinsic excitation), the resultant photocurrent is then obtained by simply replacing the thermal emission rates in Equation 3.14 by optical emission rates

$$J_R^0(t) = q(w-w_0)\left\{\frac{1}{D}e_n^0 n_T(t) + \left[1-\frac{1}{D}\right]e_p^0[N_{TT}-n_T(t)]\right\} \qquad (3.23)$$

Similarly, the time dependence of the change in occupancy of the deep energy level can then be expressed as

$$n_T(t) = \frac{e_p}{e_n^0+e_p^0}N_{TT} + \left[n_T(0) - \frac{e_p}{e_n^0+e_p^0}N_{TT}\right]\exp[-(e_n^0+e_p^0)t] \qquad (3.24)$$

Choosing again the initial condition such that $n_T(0) = N_{TT}$, the photocurrent density at $t = 0$ is obtained by substitution of Equation 3.24 into 3.23, giving

$$J_R^0(0) = \frac{1}{D}q(w-w_0)N_{TT}e_n^0 \qquad (3.25)$$

At steady state, Equation 3.24 reduces to

$$n_T(\infty) = \frac{e_p^0}{e_n^0 + e_p^0} N_{TT} \qquad (3.26)$$

and the corresponding photocurrent density is therefore given by

$$J_R^0(\infty) = q(w-w_0) \frac{e_p^0}{e_n^0 + e_p^0} N_{TT} e_n^0 \qquad (3.27)$$

Hence, by measuring the initial and final photocurrent density $J_R^0(0)$ and $J_R^0(\infty)$, respectively, together with the time constant $\tau^{-1} = e_n^0 + e_p^0$, the optical emission rates e_n^0 and e_p^0 can be calculated from a single photocurrent transient wave form analogous to the dark current technique. It should be noted, however, that absolute values of e_n^0 and e_p^0 can be obtained only if D is known. D may change with the experimental conditions (8, 9) and, hence, may give rise to substantial errors. It is therefore often more convenient to first measure e_n^0 and e_p^0 in relative units and then to calibrate the spectrum at a particular photon energy by a transient measurement.

The spectrum of optical emission rates can be measured in many ways. It follows from Equation 3.19, for example, that under the experimental conditions assumed so far, the spectral distribution of e_n^0 is obtained by plotting the initial photocurrent $J_R^0(0)$ as a function of photon energy. The absolute value of e_n^0 is then obtained from the decay time τ for an energy for which e_p^0 is vanishing. This means that together with the spectrum of e_n^0 the optical emission rate of electrons is known for all energies.

It is somewhat less straightforward to measure e_p^0 from transient techniques under the above conditions. However, even in the case of e_n^0 steady state measurements provide a very simple method for investigating optical emission rates. Steady state measurements are best performed by illuminating the sample with an additional light source of different photon energy hv_s. Each of the two light sources generate excitation processes which are described by a particular emission rate (11). The change in the number of occupied deep energy levels due to the illumination with two light sources is then given by

$$\frac{dn_T}{dt} = (e_p^0 + e_{ps}^0)N_{TT} - (e_n^0 + e_{ns}^0 + e_p^0 + e_{ps}^0)n_T(t) \qquad (3.28)$$

By integration and substitution into Equation 3.24, one obtains for the steady state photocurrent density

$$J_R^0(\infty) = q(w-w_0)N_{TT} \frac{(e_p^0 + e_{ps}^0)(e_n^0 + e_{ns}^0)}{e_n^0 + e_{ns}^0 + e_p^0 + e_{ps}^0} \qquad (3.29)$$

Taking the photon energy $h\nu_s$ of the second light source such that $E_c - E_T < h\nu_s < E_T - E_v$ (i.e., $e_{ps}^0 = 0$) and choosing the photon flux φ_s such that $e_{ns}^0 \gg e_n^0 + e_p^0$ (i.e., $\varphi_s \gg \varphi$) (11), Equation 3.29 reduces to

$$J_{Rh}^0(\infty) = q(w - w_0)N_{TT}e_p^0 = c_1 e_p^0 \tag{3.30}$$

where c_1 can be considered as constant, because the occupancy and, hence, $w - w_0$ does not change during the measurement. This is readily seen by considering that $e_{ns}^0 \gg e_n^0 + e_p^0$, which gives $N_{TT} \gg n_T$ (cf. Equation 3.26). The low occupancy is due to the high intensity of the second light source and will therefore not change during the measurement.

Equation 3.30 which shows that J_{Rh}^0 is proportional to e_p^0 is valid for all energies. The spectrum of e_p^0 is therefore obtained by plotting J_{Rh}^0 as a function of photon energy $h\nu$.

The spectrum of e_n^0 in the whole extrinsic energy region is obtained by choosing $h\nu_s$ and φ_s such that $E_T - E_v < h\nu_s < E_g$ and both e_{ns}^0 and e_{ps}^0 are much larger than e_n^0 and e_p^0 for variable photon energies (11). The increase of the photocurrent due to the illumination with the second light source $\Delta J_R^0 = J_{R1}(\infty) - J_R^0(\infty)$ is then given by

$$\Delta J_R^0 = q(w - w_0)N_{TT}[e_p^0(1-b)^2 + e_n^0 b^2]$$
$$= (1-b)^2 J_{Rh}^0 + q(w-w_0)N_{TT}b^2 e_n^0$$
$$= (1-b)^2 J_{Rh}^0 + c_1 b^2 e_n^0 \tag{3.31}$$

where

$$b = \frac{e_{ps}^0}{e_{ns}^0 + e_{ps}^0} = \frac{J_{Rh}^0(\infty) - J_R^0(\infty)}{J_{Rh}^0(\infty)} \tag{3.32}$$

is the electron occupancy of the deep energy level when the junction is illuminated with photons of energy $h\nu_s$ alone (see Equation 3.26). Because $h\nu_s$ is kept constant during the measurements, b is a constant. Hence, plotting $\Delta J_R^0 - (1-b)^2 J_{Rh}^0$ versus $h\nu$ gives the spectrum of e_n^0.

For energy levels which are not close to the middle of the forbidden energy gap, e_p^0 will vanish for a large part of the spectrum of e_n^0. In the energy region in which $e_p^0 = 0$, also J_{Rh}^0 will be zero. Equation 3.31 then reduces to the simple relation

$$\Delta J_0^R = q(w - w_0)N_{TT}b^2 e_n^0 = c_2 e_n^0 \tag{3.33}$$

where c_2 is constant.

The spectra of e_n^0 and e_p^0 can be correlated to each other by measuring $J_{Rh}^0(\infty)$ and the steady state photocurrent $J_R^0(\infty)$ at a particular photon energy

$h\nu_s$. It follows from Equation 3.32 that

$$\frac{e_{ps}^0}{e_{ns}^0} = \frac{J_{Rh}^0(\infty)}{J_R^0(\infty)} - 1. \qquad (3.34)$$

Once the spectra of e_n^0 and e_p^0 are correlated to each other, the absolute values of e_n^0 and e_p^0 are obtained for all energies by measuring the absolute value of e_n^0 at a particular energy at which $e_p^0 = 0$ using the transient technique as previously described.

As soon as e_n^0 and e_p^0 are known, the concentration N_{TT} of the energy level is readily obtained using, for example, Equation 3.33. Plotting ΔJ_R^0 for constant photon energy against $w - w_0$ by varying the reverse bias, a straight line is obtained. From the slope of the straight line N_{TT} can be calculated.

Finally, without going into details, it should be mentioned that similar techniques are available to measure capture constants. The basic idea of these techniques (12) is that the reverse bias V_R of a one-sided junction is momentarily reduced for a short time t_{sc}. The depletion region then contracts, thereby making free-charge carriers available for recombination processes. The time available for these processes is equal to t_{sc}, while the spatial location in which recombination can take place is the region through which the depletion layer is moved. The amount of charge and, hence, the capture constant can then be determined by capacitance methods or by any of the above-mentioned techniques.

Difficulties which arise in the analysis of deep-level data are therefore seldom caused by the lack of adequate measuring techniques, but rather by the uncertain chemical nature or structure of the defect under investigation. For example, changes in the electronic properties of a sample may not necessarily be caused by the deliberately introduced impurity. These difficulties are, of course, well known. Nevertheless, a proper identification of the defect giving rise to a particular energy level is usually not attempted since the required analysis is at best very time consuming and may not even be possible. Even if a chemical analysis shows that the energy level is caused by the intentionally introduced impurity, it is extremely difficult to determine whether the underlying defect is a single substitutional impurity or a more complex centre. However, knowledge of the nature of the defect giving rise to the energy level is of utmost importance for successful theoretical models. Such models may make possible a more elaborate and precise analysis of the experimental data, which in turn may provide the key to a detailed understanding of the basic physics of deep levels.

3.3 EXPERIMENTAL RESULTS

3.3.1 Copper-doped ZnSe

In this chapter three examples which question the chemical identity of some previously investigated impurities will be discussed. It will be shown that the apparent disagreement among published data in these cases is most likely caused by incorrect identifications of the investigated defects. The discrepancies are easier to interpret, however, once the nature of the different defects is known.

The first example is a defect centre in ZnSe, which was originally supposed to be due to Mn, but which later was shown to be due to Cu (13, 14). The earliest Mn doped ZnSe samples which we studied had absorption properties which were clearly different from those not deliberately doped with Mn (15)

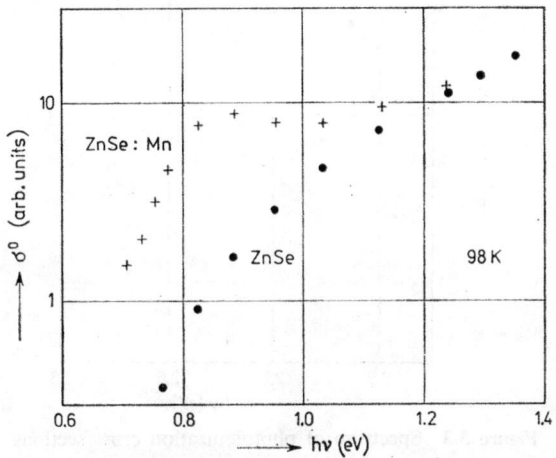

Figure 3.2 Spectra of photoionization cross sections measured in undoped and Mn-doped ZnSe, respectively (15)

(Figure 3.2). The spectra of photoionization cross sections for the excitation of holes from energy levels in the lower half of the bandgap to the valence band indicated a threshold energy of about 0.7 eV in Mn-doped samples whereas the threshold in samples not containing Mn was at about 0.8 eV. The interesting feature of this 0.7 eV energy level (M-centre) was that the spectrum of the optical cross section was different from that commonly observed for deep levels. Another property of the M-centre was that its concentration was not always correlated with the Mn concentration, even though there was a rather large amount of Mn in the samples as determined by EPR measurements (13). In fact, it turned out that the M-centre was sometimes observed even in ZnSe crystals which were not intentionally doped, depending on how carefully the Zn-annealing of the samples was performed. The similarity of the Mn-emission in ZnSe with the 'red'-copper emission very soon suggested that the M-centre might be correlated with copper. This idea was further supported by the observation that the photoluminescence excitation spectrum for the copper-red emission was similar to that of the photoionization cross section for electrons at the M-centre (Figure 3.3) and that the spectrum of

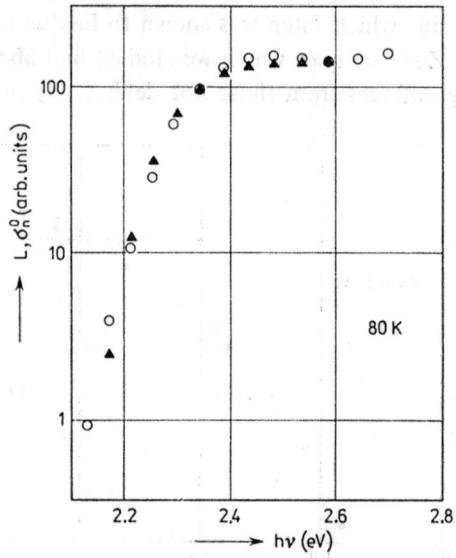

Figure 3.3 Spectrum of photoionization cross sections for electrons (○) and photoluminescence excitation spectrum for the Cu-red emission (▲) in ZnSe:Cu (13)

optical quenching of the Cu-red photoluminescence agreed very well with optical cross sections for holes on the M-centre (Figure 3.4) (13). In addition, the concentration of the M-centre increased by about two orders of magnitude in ZnSe due to doping with copper (13). The spectra and absolute values of the photoionization cross sections in these Cu-doped samples were in good agreement with previous results obtained for the M-centre. We therefore believe that the M-centre is not caused by Mn but by copper and that the M-centres observed in Mn-doped ZnSe are due to residual copper contamination introduced together with the Mn-doping (13).

Deep-Level Impurities in Semiconductors

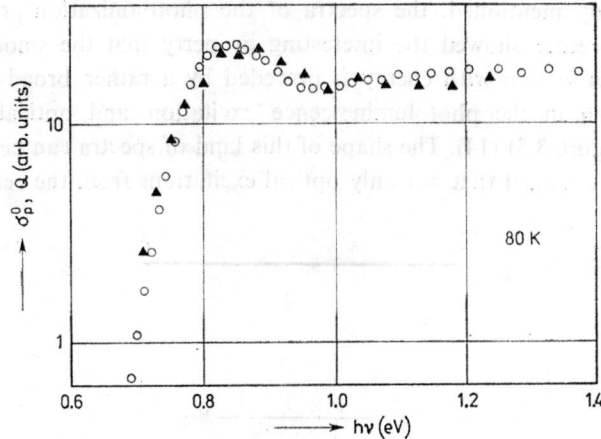

Figure 3.4 Spectrum of photoionization cross sections for holes (○) and spectral distribution of optical quenching of the Cu-red emission (▲) in ZnSe:Cu (13)

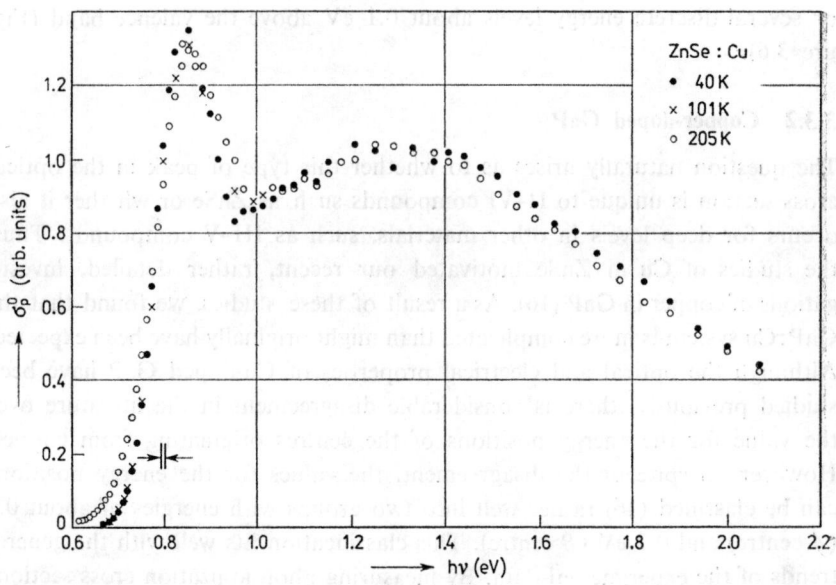

Figure 3.5 Spectra of the optical cross section for holes at different temperatures (14)

As already mentioned, the spectra of the photoionization cross sections for the M-centre showed the interesting property that the smooth increase of the cross section with energy is preceded by a rather broad peak which is even seen in the photoluminescence excitation and optical quenching spectra (Figure 3.5) (14). The shape of this kind of spectra can be understood (14) if it is assumed that not only optical excitations from the centre directly

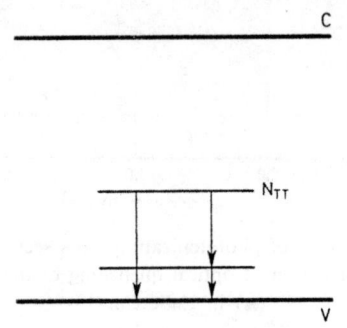

Figure 3.6 Suggested model for hole excitation processes in ZnSe:Cu (14)

into the valence band are observed, but also excitations to the band via one or several discrete energy levels about 0.1 eV above the valence band (Figure 3.6).

3.3.2 Copper-doped GaP

The question naturally arises as to whether this type of peak in the optical cross section is unique to II–VI compounds such as ZnSe or whether it also occurs for deep levels in other materials, such as III–V compounds. Thus, the studies of Cu in ZnSe motivated our recent, rather detailed, investigations of copper in GaP (16). As a result of these studies, we found that the GaP:Cu system is more complicated than might originally have been expected. Although the optical and electrical properties of Cu-doped GaP have been studied previously, there is considerable disagreement in the literature over the value for the energy positions of the centres originating from copper. However, in spite of the disagreement, the values for the energy positions can be classified (16) rather well into two groups with energies of about 0.5 (A-centre) and 0.7 eV (B-centre). This classification fits well with the general trends of the experimental data. By measuring photoionization cross sections in our diodes before diffusion of copper, only one deep level centre was observed. This centre had a threshold energy of about 0.62 eV as deduced from the spectral distribution of the optical cross section and was in agreement with published data on undoped GaP.

After copper diffusion, two other centres were observed. Their concentrations were typically two orders of magnitude larger than that for the 0.62 eV centre. From photocapacitance measurements we could show that both centres were in the lower half of the bandgap. The spectrum of the photoionization cross sections at 85 K as determined by photocurrent transients showed two threshold energies at about 0.51 eV and 0.72 eV, respectively (Figure 3.7).

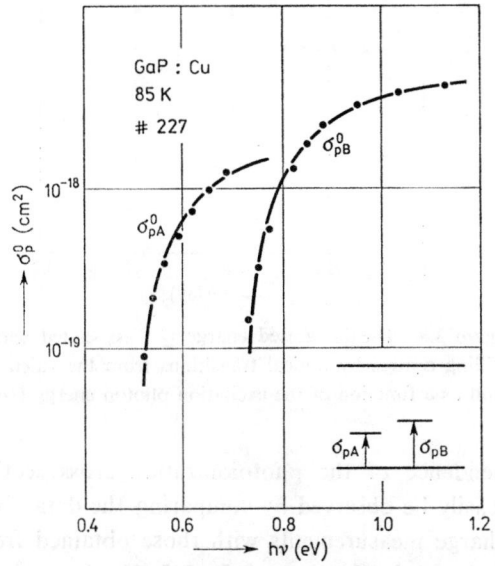

Figure 3.7 Spectra of photoionization cross sections for holes in GaP:Cu (16)

The spectra of the two centres were measured independently of each other and the threshold energies obtained are in good agreement with published data (16). The fact that these two copper centres are not coupled, can be shown by integrating the charge Q_∞ associated with refilling the centres by optical transitions from the valence band as a function of the excitation photon energy. If the initial occupancy of the centres is constant for each such measurement, Q_∞ is proportional to the number of refilled centres and will therefore change for every additional deep centre which is refilled (16). The experimental data show two steps with threshold energies of about 0.5 eV and 0.7 eV, in agreement with the results obtained for the photoionization cross sections (Figure 3.8). Because the ratios of the Q_∞ observed for the two centres were different in different samples, it has to be assumed that the two centres are not coupled.

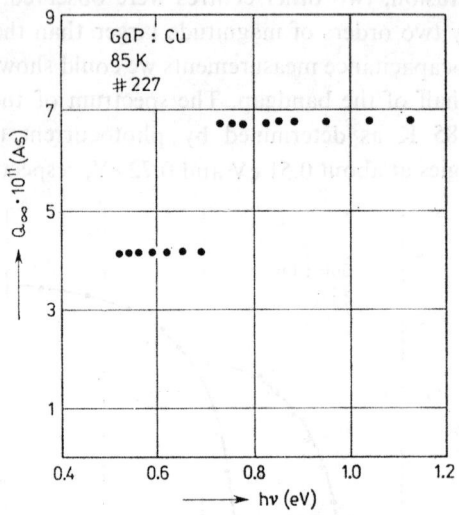

Figure 3.8 The integrated charge Q_∞ associated with refilling centres by optical transitions from the valence band as a function of the excitation photon energy (16)

Any field dependence of the photoionization cross sections in copper-doped GaP can easily be observed by comparing the data obtained from the junction space charge measurements with those obtained from bulk photoconductivity techniques. In the case of the 0.5 eV centre, similar spectra are measured irrespective of whether the photoionization has been determined by optical quenching of the photoconductivity, the photocurrent transient technique, or the optical quenching of photoluminescence (Figure 3.9). In addition, it should be noted that these data on the 0.5 eV centre were obtained in copper-doped GaP samples with different shallow impurities as background doping. No influence whatsoever of shallow impurities on the spectra could be observed.

An interesting feature of this 0.5 eV centre is its large capture cross section for holes. From photoluminescence measurements we have to conclude that the temperature dependence of the energy separation of this level from the valence band is very similar to that of the energy bandgap (17). Hence, by measuring the temperature dependence of the thermal emission rate for

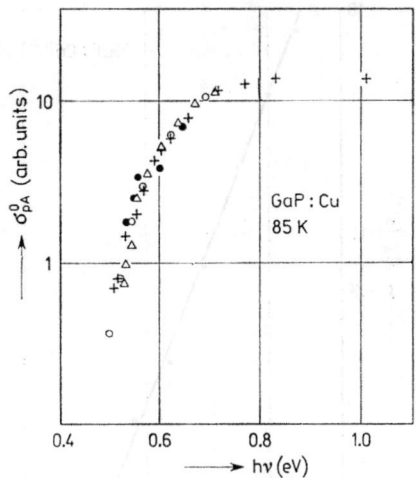

Figure 3.9 Spectrum of the photoionization cross section for holes as determined by optical quenching of the photoconductivity ($\triangle$), photocurrent transient technique ($\circ$) and optical quenching of photoluminescence (+) (16, 17)

holes, a thermal activation energy is expected which is larger than the value obtained for the optical threshold energy of this 0.5 eV centre. Plotting the logarithm of thermal emission rate e_{pA}^t as a function of $1/T$, an activation energy (enthalpy ΔH_{pA}) of 0.547±0.020 eV is obtained (Figure 3.10). Knowing the temperature dependence of the bandgap, these data can be used in conjunction with the equation of detailed balance to estimate the capture cross section for holes. At 200 K a value of about 6×10^{-14} cm^2 is obtained. If instead of the enthalpy the optical threshold energy of 0.50 eV (Gibbs free energy ΔG_{pA}) is used in the equation of detailed balance (cf. Equation 3.5), a value of 15×10^{-14} cm^2 is estimated for the capture cross section of holes σ_{pA}^t. Both values although obtained by rather indirect measurements, nevertheless indicate that the 0.5 eV energy level in copper-doped GaP is quite an effective hole trap (16).

Although the other energy level caused by copper doping in GaP behaves in some respect similarly as the 0.5 eV centre, it nevertheless shows properties which are very different from the A-centre. The spectra of the photoionization cross section obtained in bulk photoconductors and diodes were very similar also for the 0.7 eV centre, hence indicating that no drastic field-dependence of the optical cross section occurs (16).

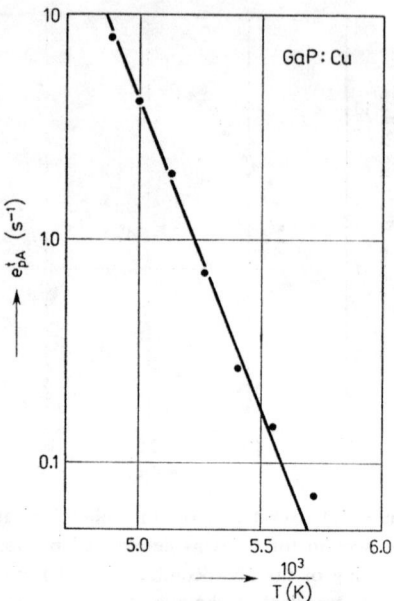

Figure 3.10 Variation with temperature of the rate of thermal excitation of electrons from the valence band into the A-centre, plotted as e^t_{pA} versus T^{-1} (16)

To demonstrate that thermal broadening of the impurity absorption edge is rather small in the temperature region between 85 K and 300 K, the spectra of the photoionization cross sections obtained at 85 K and 200 K were normalized to the 300 K data (Figure 3.11). This was carried out by shifting the curves to lower energies by 35 meV at 200 K and 60 meV at 85 K. Although this shift is very close to the shift of the bandgap with temperature, it should be pointed out that the absolute value of the photoionization cross section σ^0_{pB} increased by almost a factor of 8 when the temperature was raised from 85 K to 300 K.

The small thermal broadening of the impurity absorption edge suggests that the Franck–Condon shift for the 0.7 eV centre might be rather moderate. This is indeed confirmed by low temperature photoluminescence measurements (18) and is in agreement with the observation that the sum of the two energy separations of the centre from the valence band and conduction band, respectively, $\Delta E_{nB} + \Delta E_{pB}$ is rather close to the value of the bandgap.

The change of the photoionization cross section of the B-centre by almost a factor of 8 between 85 K and 300 K is quite unusual. To the best of our knowledge, such a strong temperature dependence of a photoionization cross section has not previously been observed. In earlier work, a number of complex centres composed of lattice defects and other impurities have been observed in GaP (19). Therefore, it is possible that for the case of the 0.7 eV centre copper might also form some kind of complex. The unusual temperature dependence of the photoionization cross section might then be due to such a complex. The experiments described so far were performed

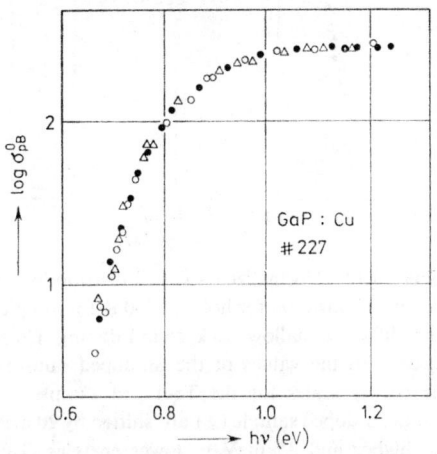

Figure 3.11 Normalized spectral distribution of the photoionization cross sections for holes at different temperatures. Compared with the 300 K (●) values, the energy scales for the 85 K (△) and 200 K (○) values are shifted to higher energies by 60 and 35 meV, respectively

on copper-diffused GaP, where the starting material was Sn-doped GaP. When performing similar measurements with Te-doped and S-doped GaP, it turned out (16) that the spectral shape of the photoionization cross sections did not change but that different threshold energies for the excitation of holes from B-centres to the valence band were obtained (Figure 3.12). Comparing these data with those obtained for Sn-doped samples, it can readily be seen that the threshold energy of the S-doped GaP:Cu sample is about 35 meV larger than for the Sn-doped sample, while for the Te-doped GaP:Cu sample it is about 20 meV smaller. To make sure that the differences in optical threshold energies are not due to experimental errors, the spectral distributions of the photoionization cross section σ_{pB}^0 were carefully measured in two diodes which were both doped with Sn but prepared from

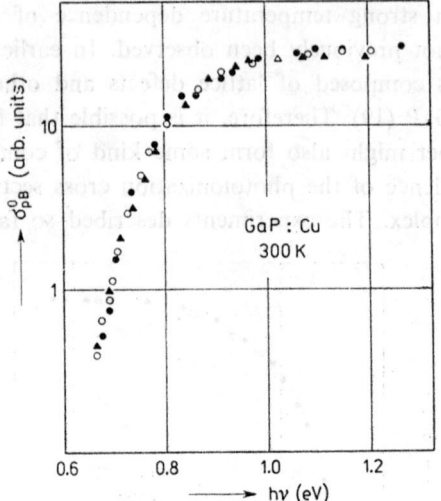

Figure 3.12 Normalized spectral distribution of the photoionization for holes at 300 K for samples with different shallow background doping. Compared with the values of the Sn-doped samples, the energy scales for the Te-doped sample (○) and the S-doped sample (▲) are shifted by 20 meV to higher and 35 meV to lower energies (16)

two different LPE layers. Within the experimental error, no differences in the spectra could be observed. These results suggest that copper forms complexes with other impurities and might be the reason for the disagreement in published values for the energy position of this centre. Indeed, a normalized plot of optical data published by Allen & Cherry and Goldstein & Perlman are in good agreement with our data obtained with Te-doped samples (Figure 3.13). Our results on S-doped samples also agree well with previously published data on S-doped GaP:Cu. This gives further support for the idea that copper forms some kind of complex in GaP and shows at the same time that the discrepancies in the published data are mostly due to the methods of analysis used and to the influence of other impurities rather than to inconsistencies in the experimental data.

It is not fully understood why the shifts of the optical threshold energies for the Cu B-centres are so small. We should note, however, that similar small shifts of about 40 meV are also observed for the well-known self-activated luminescence centres in ZnSe and ZnS, corresponding respectively to a donor atom being either a first or second neighbour to a zinc vacancy.

However, somewhat larger shifts of the optical threshold energies can probably be generated with proper doping. For example, in p-type Mg-doped GaP into which Cu has been diffused, two energy levels are found in the

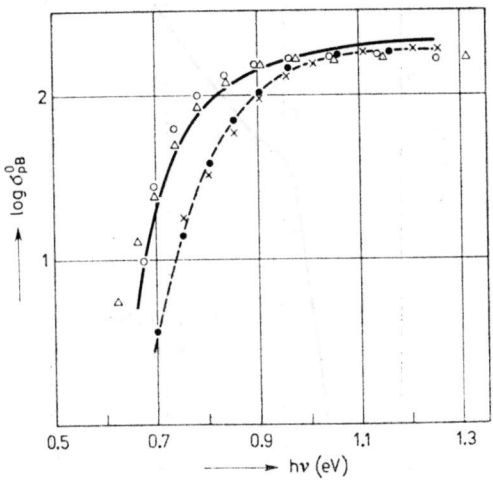

Figure 3.13 Comparison of published data for photoionization cross sections of holes with our own results at 300 K (○): J. W. Allen, R. J. Cherry, *J. Phys. Chem. Solids*, **23**, 509 (1962). (△): B. Goldstein, S. S. Perlman, *Phys. Rev.*, **148**, 175 (1966). (×): H. G. Grimmeiss, H. Scholz, *Philips Res. Rep.*, **21**, 246 (1966). (●): H. G. Grimmeiss, M. O. Ottosson, *Phys. Status Solidi* (a), **5**, 481 (1971). (– –): S-doped GaP. (–): Te-doped Gap (16)

lower half of the bandgap which are not seen in reference samples without copper diffusion (20). One of these energy levels has an energy position about 0.5 eV above the valence band, in agreement with the previously observed A-centre. The other centre, which might be considered analogous to the B-centre, has an optical threshold energy of about 0.84 eV, which is close to the value published by Wessels, but quite different from the energy of the B-centre in Te, Sn or S-doped GaP (Figure 3.14). The most striking feature of the deeper centre, however, is that the form of the spectrum for the optical cross section of holes is quite different from that in Te, Sn and S-doped GaP but is very similar to that observed for copper in ZnSe. If it is again assumed as in the case of Cu-doped ZnSe, that excitation of holes into the valence band is possible not only by direct transitions but also by transitions via

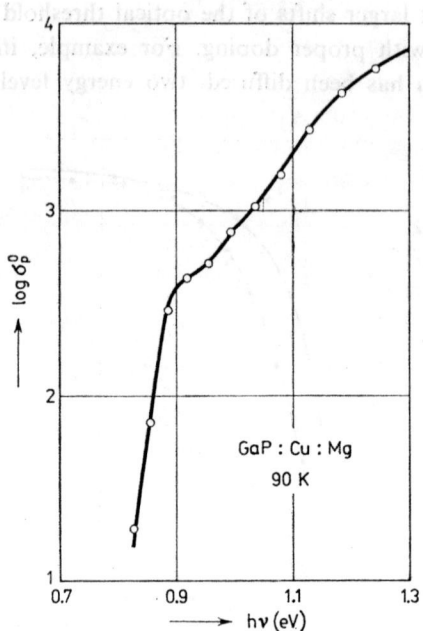

Figure 3.14 Spectrum of the photoionization cross section for holes in Cu-doped GaP:Mg (20)

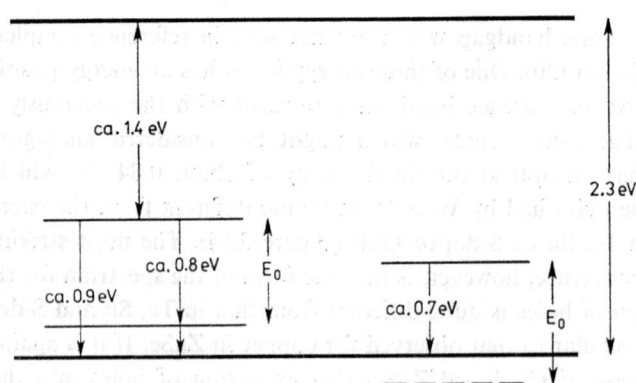

Figure 3.15 Possible energy band model for GaP:Cu

one or several discrete energy levels close to the valence band, one may understand why the spectrum of the deeper copper centre is different in Mg-doped samples compared with that in Te, Sn and S-doped GaP. If we assume that the energy separation between the shallow energy state and the ground state of the centre is independent of the background doping, as might be the case for crystal-field-split intra-d-level transitions, then the shallow energy state will move into the valence band with decreasing energy separation between the ground state of the centre and the valence band, such as in the Te, Sn and S-doped samples (Figure 3.15). From photocurrent excitation spectra in the Mg-doped samples the optical threshold between the ground state and the conduction band is found to be of the order of 1.4 eV. Hence, the ground state of the centre formed with Mg doping should be about 0.9 eV above the valence band and therefore almost 0.2 eV deeper than in Te, Sn and S-doped samples in agreement with the observed 0.84 eV optical threshold for exciting holes to the valence band via an assumed shallow state in the Mg-doped sample.

This example shows that doping a material with a certain impurity—in this case Cu—under apparently similar conditions will not necessarily result in the same energy level if the impurity associates itself with crystal defects or with other impurities. This sort of situation may be quite common in solids and means that one must be extremely careful to measure all properties of the deep level as well as to carefully document the history of the sample.

3.3.3 Gold-doped silicon

A third example of how such small energy shifts, presumably due to the formation of different complexes, may confuse the experimental situation, is the system of gold in silicon. It is commonly assumed that gold creates two energy levels in silicon, one level about 0.55 eV below the conduction band and another level about 0.35 eV above the valence band. However, as in the case of copper-doped GaP published values on the energy position of these centres show quite a remarkable scattering (21). In the case of the deeper gold centre the values vary between 0.49 eV and 0.58 eV. We remeasured these data some years ago (21) and obtained thermal activation energies from the temperature dependence of the thermal emission rates for electrons and holes the sum of which agrees well with the enthalpy of the bandgap which for the temperature involved is about 1.195 eV (Figure 3.16). The thermal data were in good agreement with optical threshold energies obtained from the spectra of the photoionization cross sections (11) if it was assumed that the energy level is pinned to the conduction band. This assumption was supported by the fact that the optical threshold for the holes showed a much larger temperature shift than that for electrons (22). From these results degeneracy factors and thermal capture rates could be estimated. In addition, it was assumed that the capture cross section for electrons is temperature independent. Later Nagasawa and Schulz published data on ion implanted

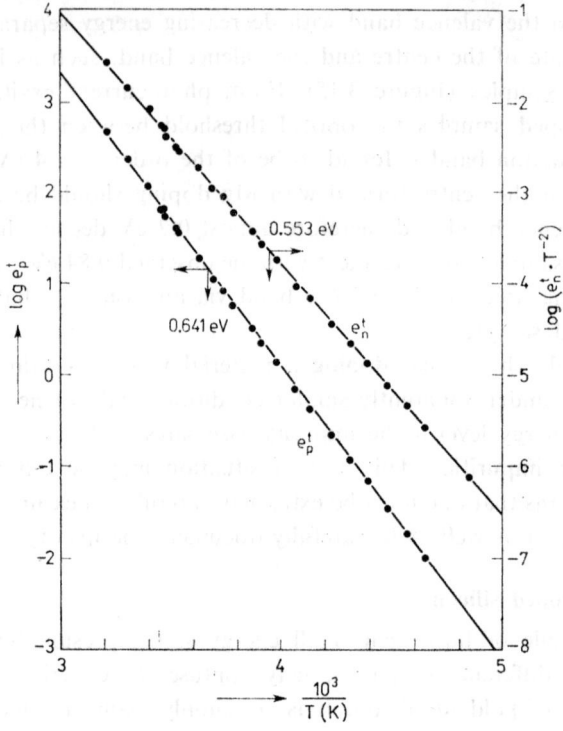

Figure 3.16 Temperature dependence of $\log(e_n^t T^{-2})$ and $\log(e_p^t)$ in Si:Au (21)

gold in silicon (23) and their results agreed fairly well with our data, especially their directly measured value of the capture cross section was rather close to our estimated value (see Table 3.1). In 1978, however, Brotherton and Bicknell published a detailed investigation of gold-doped silicon (24) which agreed with previous results for the thermal emission rate and its thermal activation energy but which revealed a capture cross section for electrons which was a factor of about 25 smaller (see Table 3.1). From these results they concluded that the gold level must be pinned to the valence band. On the other hand, Lang and Jaros found electron capture cross sections in their gold-doped

Table 3.1

	e_n^t (250 K) [s^{-1}]	σ_n^t (250 K) [cm^2]	ΔH_n [eV]	X_n (calc.)
Sample A	10	$6.9 \cdot 10^{-17}$	0.553	48
Sample B	10	$1.8 \cdot 10^{-16}$	0.553	15.7
Brotherton & Bicknell	10	$9 \cdot 10^{-17}$	0.555	50
Nagasawa & Schulz	8.5	$2.2 \cdot 10^{-15}$ (295 K)	0.54	0.87

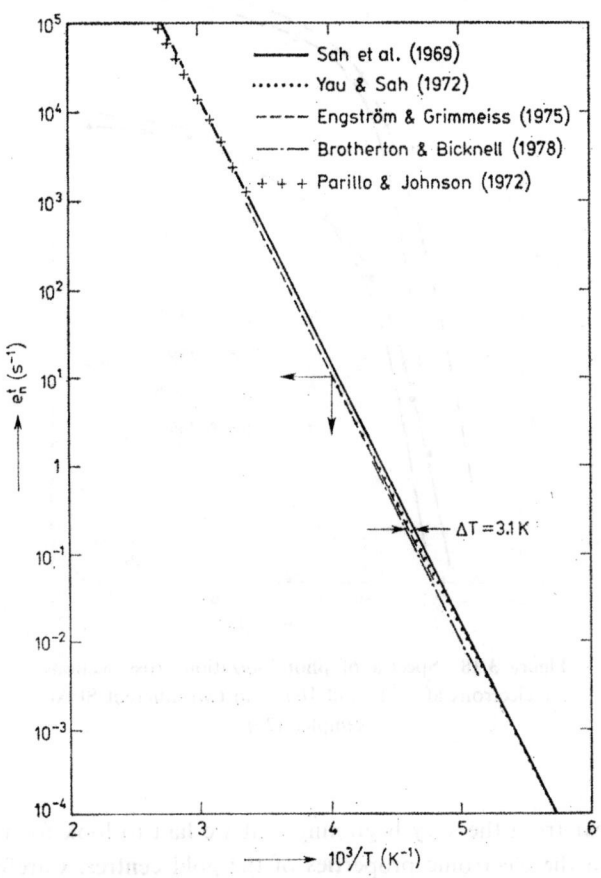

Figure 3.17 Published data of thermal emission rates for electrons in Si:Au

silicon sample which were in reasonable agreement with the results of Brotherton and Bicknell (25). However, there is one feature of gold-doped silicon which is very similar in all the published data. When the temperature dependence of the thermal emission rate for electrons is plotted (Figure 3.17), it is easily seen that they all agree within experimental error.

Because of this rather confusing situation, we started a joint project together with Lang and Jaros (26). After exchanging samples, it turned out that the results obtained were independent of the measurement technique. It was

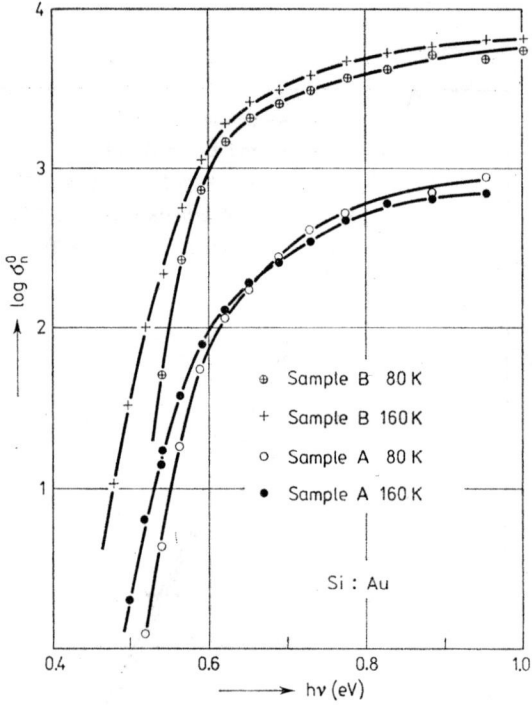

Figure 3.18 Spectra of photoionization cross sections for electrons at 80 K and 160 K in two different Si:Au samples (26)

therefore clear from the very beginning that we had to look for rather small differences in the electronic properties of the gold centres. Careful measurements of the capture cross section for electrons nevertheless showed that the capture cross section differs by a factor of 2.6 in our two samples, labelled A

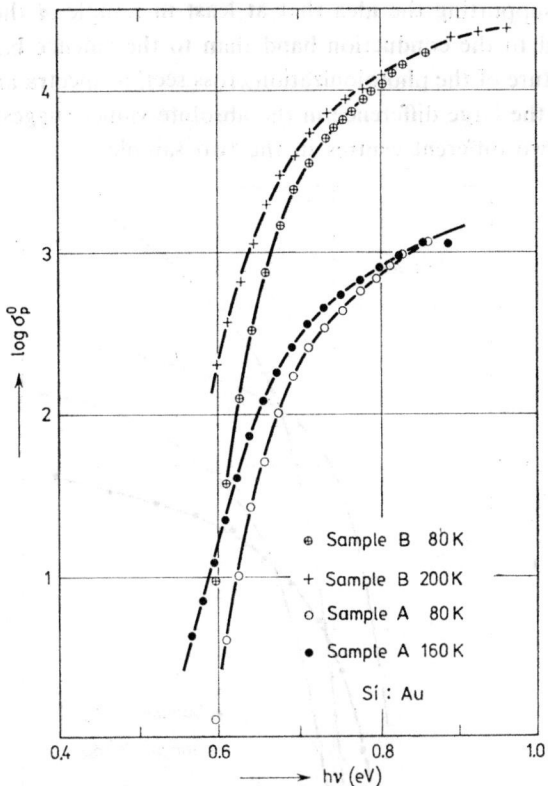

Figure 3.19 Spectra of photoionization cross sections for holes at different temperatures (26)

and B (see Table 3.1). For sample A these data resulted in a value for X_n larger than 100 which is not in agreement with an energy level pinned to the conduction band. We therefore remeasured the temperature dependence of the optical cross section spectra and found that although there was a slight temperature shift in both samples, the shift for the photoionization cross section for electrons (Figure 3.18) was rather small for sample A, at least smaller than in sample B, in agreement with the thermal data. The most surprising result, however, was that the absolute values of the cross sections were different by a factor of 9 in sample A and B. For the case of the photoionization cross section for holes (Figure 3.19) the ratio of the cross sections in the two samples was even larger. Furthermore, it should be noted that the temperature shift of the cross section for holes is much larger than that for electrons supporting the idea that at least in sample A the energy level is more pinned to the conduction band than to the valence band. Although the overall feature of the photoionization cross section spectra are very similar (Figure 3.20), the large difference in the absolute values suggests that we are dealing with two different centres in the two samples.

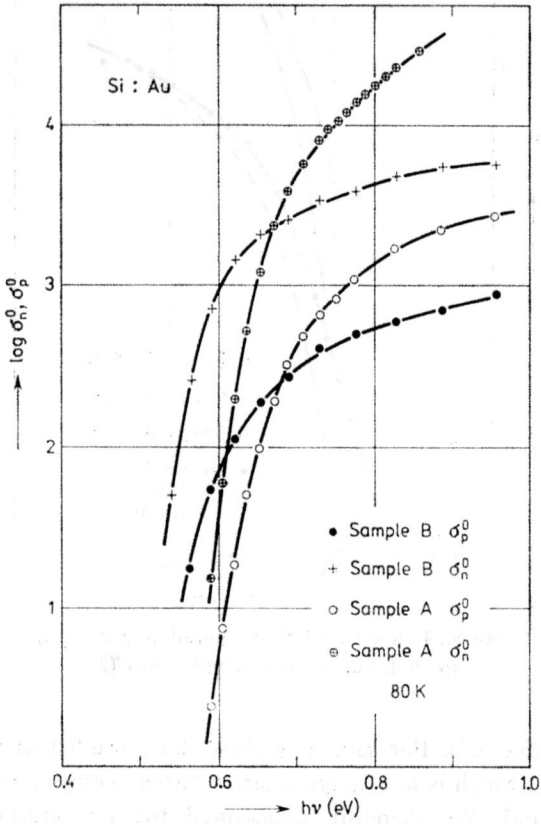

Figure 3.20 Spectra of photoionization cross sections for electrons and holes, respectively, in two different samples of gold-doped Si (26)

3.4 CONCLUSION

With respect to all published data, we are therefore inclined to believe that there is not a single, well-defined gold centre in silicon, as had previously been assumed, but rather a family of gold-related defects with very similar

Deep-Level Impurities in Semiconductors

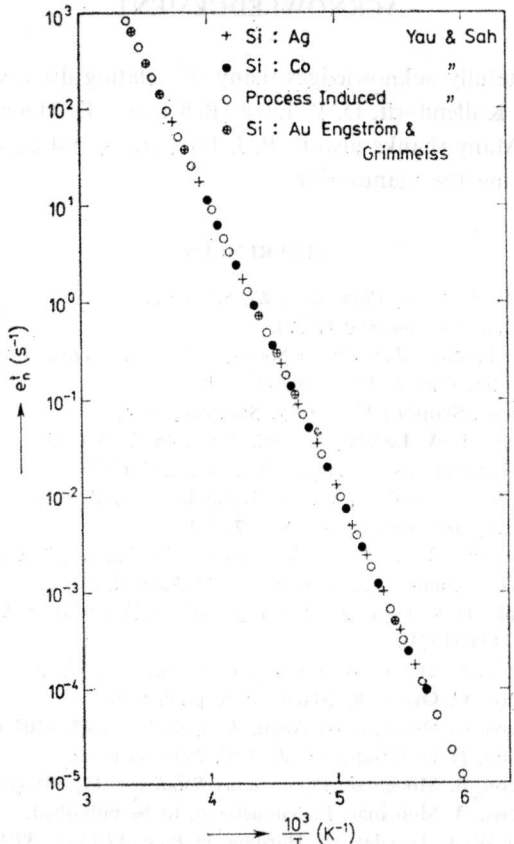

Figure 3.21 Published data of thermal emission rates for electrons in differently doped silicon

properties. Different defects of the same general type may be distinguished by paying careful attention to the small shifts in the optical and thermal properties, just as in the case of gold in silicon and the Cu-centres in GaP. This might not always be easy. For example, plotting published values of the thermal emission rate for electrons as a function of temperature for the Ag, Co, Au and the 'process induced' centre in silicon, it is seen that all values lie on the same straight line (Figure 3.21). Thus, there may well be a family of 0.50–0.55 eV centres in silicon which is even broader than the several gold-related centres reported in this chapter (26).

ACKNOWLEDGEMENT

The author gratefully acknowledges many stimulating discussions with P. O. Fagerstrom, N. Kullendorff, D. V. Lang (Bell Labs), E. Meijer, B. Monemar, and C. Ovren. Many thanks also to B. J. Fitzpatrick and G. F. Neumark for critically reviewing the manuscript.

REFERENCES

1. W. Shockley, W. T. Read, *Phys. Rev.*, **87**, 835 (1952).
2. R. N. Hall, *Phys. Rev.*, **86**, 600 (1952).
3. R. H. Bube, in Photoconductivity of Solids, Wiley, New York, 1960.
4. F. Prat, E. Fortin, *Can. J. Phys.*, **50**, 2551 (1972).
5. A. W. Lin, *Thesis*, Stanford University, Stanford, 1974.
6. H. G. Grimmeiss, L.-Å. Ledebo, *J. Appl. Phys.*, **46**, 2155 (1975).
7. S. Braun, H. G. Grimmeiss, *J. Appl. Phys.*, **44**, 2789 (1973).
8. O. Engström, *Thesis*, Lund Institute of Technology, Lund, 1976.
9. H. G. Grimmeiss, *Ann. Rev. Mater. Sci.*, **7**, 341 (1977).
10. C. T. Sah, L. Forbes, L. L. Rosier, A. F. Tasch, *Solid State Electron.*, **13**, 759 (1970).
11. S. Braun, H. G. Grimmeiss, *J. Appl. Phys.*, **45**, 2658 (1974).
12. See, for example, D. V. Lang, *J. Appl. Phys.*, **45**, 3023 (1974); J. A. Pals, *Solid State Electron.*, **17**, 1139 (1974).
13. H. G. Grimmeiss, C. Ovrén, W. Ludwig, R. Mach, *J. Appl. Phys.*, **48**, 5122 (1977).
14. H. G. Grimmeiss, C. Ovrén, R. Mach, to be published.
15. H. G. Grimmeiss, C. Ovrén, J. W. Allen, *J. Appl. Phys.*, **47**, 1103 (1976).
16. P. O. Fagerström, H. G. Grimmeiss, *J. Appl. Phys*, in press.
17. H. G. Grimmeiss, B. Monemar, *Phys. Status Solidi (a)*, **19**, 505 (1973).
18. H. G. Grimmeiss, B. Monemar, L. Samuelsson, to be published.
19. D. R. Wight, J. W. A. Trussler, W. Harding, in *Proc. XI Conf. SEM.*, Warsaw, 1972.
20. H. G. Grimmeiss, N. Kullendorff, H. Tietze, to be published.
21. O. Engström, H. G. Grimmeiss, *J. Appl. Phys.*, **46**, 831 (1975).
22. O. Engström, H. G. Grimmeiss, *Appl. Phys. Lett.*, **25**, 413 (1974).
23. K. Nagasawa, M. Schulz, *Appl. Phys.*, **8**, 35 (1975).
24. S. D. Brotherton, J. Bicknell, *J. Appl. Phys.*, **49**, 667 (1978).
25. D. V. Lang, M. Jaros, private communication.
26. D. V. Lang, H. G. Grimmeiss, E. Meijer, and M. Jaros, to be published.

Chapter 4
Generation of Dislocations in Optoelectronic Materials and their Behaviour during Optical Excitation

G. R. Woolhouse

4.1 INTRODUCTION

It is by now well established that the presence of dislocations in optoelectronic materials is deleterious to their performance in fabricated devices. Not only do the dislocations *per se* reduce the luminescence efficiency but furthermore the operation of the optoelectronic device usually results in an increase in the line length of individual dislocations and thus in a time dependent reduction in the device efficiency (*degradation*). We begin this chapter by reviewing some elementary dislocation concepts and some evidence that dislocations are deleterious to optoelectronic device operation. We go on to review the manifold methods of observing dislocations in optoelectronic materials and the mechanisms by which they can be generated. We will see that they can originate in the substrate and grow into the device through replication, or they can originate during epitaxial growth due to the presence of incorporated inclusions or through the operation of misfit stress. Alternatively they can be generated after the complete growth of the device through mechanical damage ('fresh dislocations').

Lastly we will go on to review the behaviour of dislocations during optical excitation. We find that in general grown-in dislocations behave distinctly differently from 'fresh' dislocations although in a certain regime of excitation (sufficiently low) the behaviour is similar.

4.2 DISLOCATIONS

Since, as will be discussed, dislocations play such a crucial role in the degradation of optoelectronic devices and their avoidance is so essential to the fabrication of a good device and since furthermore, many scientists and engineers involved in the semiconductor optoelectronics field have not had the basic background necessary for their understanding it seems to be appropriate to review briefly some basic concepts associated with dislocations and then to discuss their effect on the performance of fabricated devices.

4.2.1 Dislocation concepts

Figure 4.1 reminds us of a simple example of a crystal dislocation. We can think of it as an otherwise perfect crystal which has been cut and had an *extra half plane* of atoms (indicated by solid circles) inserted into it. Far

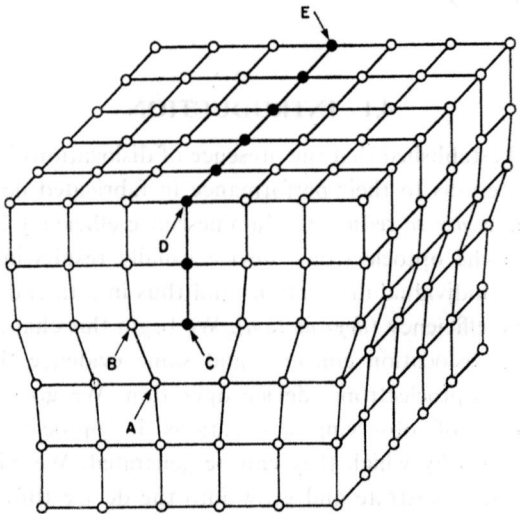

Figure 4.1 An edge dislocation in a simple cubic crystal. The dislocation line runs parallel to *DE* through *C* and the Burgers vector is *CB*. See text for further explanation

from the edge of the extra half plane the crystal remains undistorted whereas close to it the distortions are quite severe. The centre of the distortion is known as the *dislocation line* and because in this case it corresponds to the edge of an extra half plane this dislocation is known as an *edge dislocation*.

We can think of a dislocation as an entity which is free to move large distances in a crystal while moving the actual atoms only small distances. Thus if the plane of atoms (indicated *AB* in Figure 4.1) were de-bonded between *A* and *B* and re-bonded between *A* and *C*, the dislocation would be displaced by the vector *CB* while the individual atoms surrounding the dislocation would have moved distances considerably less than this. This unit vector is a characteristic of the dislocation known as the *Burgers vector* and this motion involving only rearrangements of the atoms is called *slip* or *glide*. The plane on which the motion takes place is called the *slip* or *glide plane*. The edge dislocation can also move perpendicular to the glide plane by emitting or absorbing vacancies. This is known as *climb* motion and because it involves bulk diffusion of atoms it is usually a much slower process than glide. Also note that for the edge dislocation the *Burgers vector* is *normal* to the *dislocation line*. (Note that the dislocation line runs through *C* parallel to *DE*.)

Another kind of dislocation is shown in Figure 4.2. In this case after cutting the perfect crystal the lips of the cut are sheared with respect to one another before re-welding. This is known as a *screw dislocation* and it is evident that

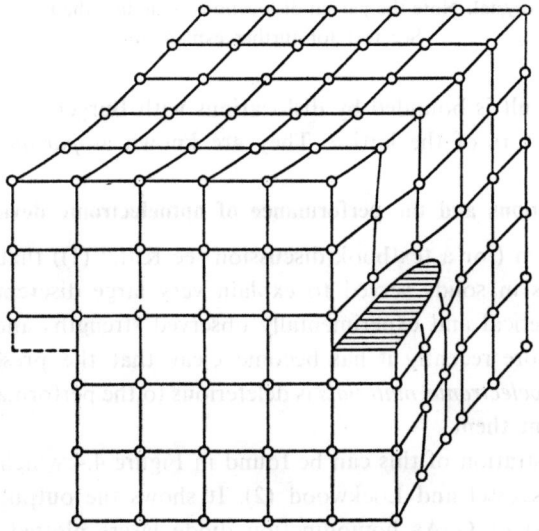

Figure 4.2 A screw dislocation in a simple cubic crystal. In this case the Burgers vector and the dislocation line are parallel as explained in the text

in this case the *Burgers vector* is *parallel* to the *dislocation line*. In general, a dislocation can have its Burgers vector at an arbitrary angle to the dislocation line in which case it is known as a *mixed dislocation*.

Another related kind of crystalline defect of importance in optoelectronic materials is exemplified by Figure 4.3 which shows the (*ABC* ... type) [111] sequence of stacking of close-packed (or quasi-close-packed) planes in a F.C.C. metal or the diamond or sphaelerite structures. The dotted region represents a fault in the stacking sequence and is known as a *stacking fault*. Note that

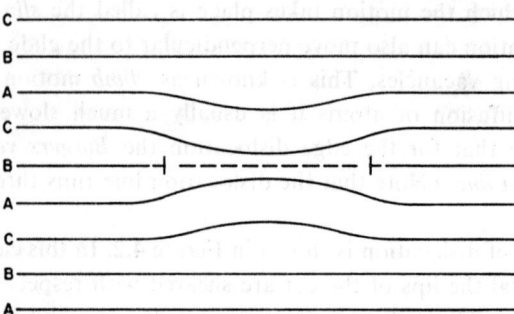

Figure 4.3 A stacking fault in a F.C.C. (or quasi-F.C.C.) crystal. Note the partial dislocations bounding the fault. See text for further explanation

the stacking fault is bounded by dislocations with Burgers vectors which are not repeat vectors of the lattice. They are known as *partial dislocations*.

4.2.2 Dislocations and the performance of optoelectronic devices

It is well known (for a textbook discussion see Kittel (1)) that the discovery of dislocations in solids served to explain very large discrepancies existing between theoretical and experimentally observed strengths and growth rates of crystals. More recently it has become clear that the presence of dislocations in *optoelectronic materials* is deleterious to the performance of devices fabricated from them.

A good illustration of this can be found in Figure 4.4 which is taken from the work of Kressel and Lockwood (2). It shows the output (measured in arbitrary units) of GaAs homojunction diode lasers plotted as a function of device operating time (in hours). Two sets of data are shown. The solid lines are for diodes prepared from dislocation-free material whereas the dotted lines represent diodes prepared from material deliberately plastically deformed to introduce dislocation densities in the 10^4–10^5/cm^2 range. It is apparent that not only are the dislocated diodes less efficient at the outset (by 30% on the average) they became even less efficient with time until after 140 hours of operation they are 75% less efficient on the average than the dislocation-free diodes. This time dependent reduction in efficiency is known as *degradation*.

Generation of Dislocations in Optoelectronic Materials 67

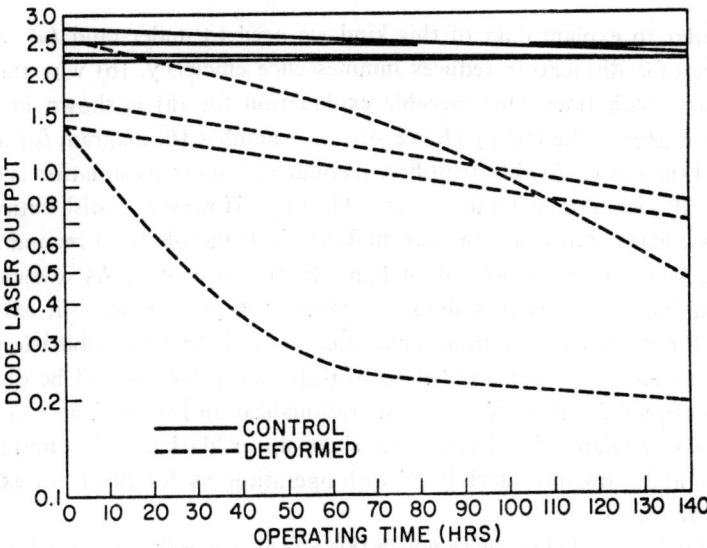

Figure 4.4 The output of dislocated and dislocation-free diode lasers measured as a function of operating time (after Kressel and Lockwood (2))

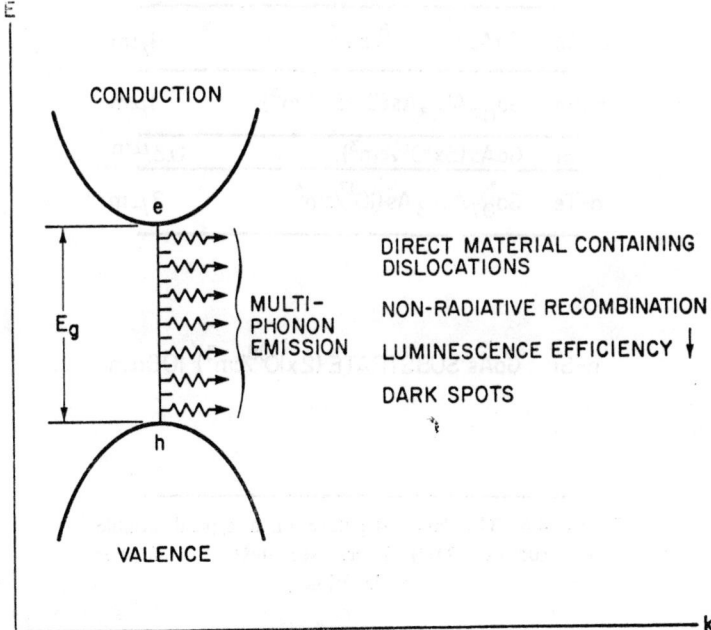

Figure 4.5 A schematic energy versus wave-number diagram for a direct band gap semiconductor (such as GaAs) containing dislocations or other crystalline defects. See text for further explanation

In order to explain data of this kind we need to understand (a) why the presence of a dislocation reduces luminescence efficiency, (b) why the effect gets worse with time. One possible explanation for (a) is shown in Figure 4.5 which depicts the energy (E) versus wave-number (k) diagram for a direct material such as GaAs. Electron-hole recombination in this situation is usually associated with a high luminescence efficiency. However, a dislocation may introduce extra states into the gap making electron-hole (e-h) recombination a much less efficient generator of light. In the case of GaAs it seems that the band gap energy (E_g) is about the same as or greater than the activation energy for dislocation motion. Thus this non-radiative recombination tends to lead to an increase in the length of dislocation lines as will be discussed later (Section 4.5). This effect is some rationalization for (b). These considerations also explain why dislocations show up as 'dark spots' in luminescent material and grow into 'dark lines' with operation, as described, for example, by Ito et al. (3).

One kind of diode laser with which this review is mainly concerned is shown in Figure 4.6. It is the double heterostructure laser. It is usually grown by

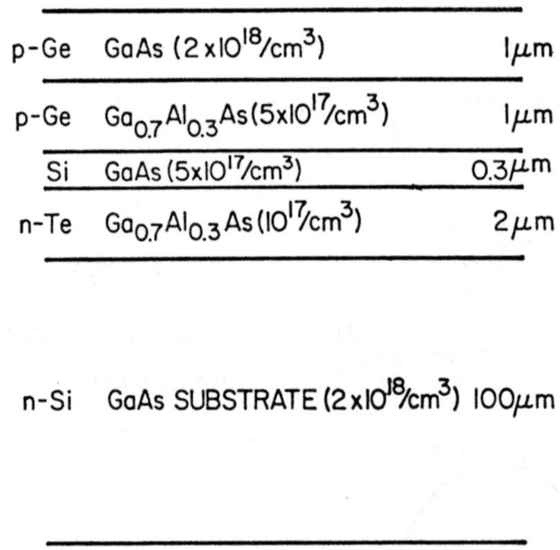

p-Ge	GaAs (2×10^{18}/cm^3)	1 μm
p-Ge	Ga$_{0.7}$Al$_{0.3}$As (5×10^{17}/cm^3)	1 μm
Si	GaAs (5×10^{17}/cm^3)	0.3 μm
n-Te	Ga$_{0.7}$Al$_{0.3}$As (10^{17}/cm^3)	2 μm

n-Si	GaAs SUBSTRATE (2×10^{18}/cm^3)	100 μm

Figure 4.6 The layer structure of a typical double heterostructure (DH) laser. See text for further explanation

liquid phase epitaxy and can be thought of as a sandwich: the thin GaAs active region is sandwiched between two GaAlAs confining layers. A p-type GaAs cap layer is grown for contacting purposes and the entire structure is grown on an n-type substrate the layers being suitably doped to create a p–n junction in the neighbourhood of the active region.

When the device is operated in forward bias electrons and holes recombine in the active region to give light which is confined by refractive index differences to such an extent that a lasing condition develops. If a dislocation threads through the active region, however, it is able to grow and eventually 'choke off' the laser output.

4.3 METHODS OF OBSERVATION

Since it is clear that dislocations play an important role in the operation of optoelectronic devices, we will discuss some of the many methods available for observing them. We will begin with a general synopsis of methods followed by a detailed discussion of some methods used successfully for GaAs Double Heterostructure (DH) material.

4.3.1 Synopsis of methods

4.3.1.1 Methods relying on the scanning electron microscope

4.3.1.1a Electron Beam Induced Current (EBIC) Method

This requires that the sample contain a p–n junction. The electron beam induces carriers in the sample and leads to a current which crosses the junction and can be collected. Crystal defects give rise to local increases in carrier recombination and so to dark contrast in the EBIC signal. Excellent reviews and descriptions of the method have been given by McMullin (4), Leamy et al. (5), and Wells (6).

4.3.1.1b Schottky Barrier EBIC (SBEBIC)

This can be applied to semiconductor samples which are homogenous with respect to carrier type (no p–n junction). The internal electric field is provided by Schottky barrier metallization which, of course, must be thin enough for primary beam penetration and therefore a few hundred Å typically. A current review of SBEBIC can be found in Leamy et. al. (5).

4.3.1.1c Cathodoluminescence

This technique utilizes the light emitted through recombination of carriers generated by the electron beam. Dislocations are imaged by a mechanism similar to EBIC. Cathodoluminescence images of dislocations have been recorded by Esquivel et al. (7) in GaAs and by Rasul and Davidson (8), Titchmarsh et al. (9) and Werkhoven et al. (10) in GaP.

All three SEM methods are limited to a dislocation resolution on the order of the minority carrier diffusion length ($\sim 1\mu$). These techniques while largely non-destructive are thus limited to materials with dislocation densities less than $\sim 10^8/cm^2$ which of course includes most semiconductors which have not been intentionally deformed.

4.3.1.2 X-ray topography

This is a widely used means of studying defects in semiconductors. It is a good technique for large area samples with low dislocation densities ($< 10^4$ cm) and it is non-destructive. However, it has the disadvantage, with conventional X-ray sets at least, of being rather slow (exposure times can be ~ 12 hours), although the advent of synchrotron radiation topography promises to ameliorate that situation as demonstrated by, for example, Petroff and Sauvage (11). Studies of dislocations in III–V compounds using conventional X-ray topography have been carried out by Howard and Dobrott (12), Bartels and Nijman (13), and Bartels et al. (14).

4.3.1.3 Chemical etching

This is probably the oldest and most widely used method of observing dislocations in all solid materials. There are many different etching solutions for each material. The most commonly used ones for GaAs are probably Molton KOH (15) and the A–B etch developed by Abrahams and Buiocchi (16). Etching techniques are at their most powerful when combined with other defect observation techniques (see, for example, Woolhouse et al. (17)). They suffer from the disadvantage of being destructive of the examined material. More will be said about chemical etching in Section 4.3.2.

4.3.1.4 Photoluminescence Topography (PLT)

This appears to have been developed simultaneously and independently by Johnston and Miller (18) in the United States and by Ito et al. (19) in Japan. It is particularly useful when applied to double heterostructure material with which it is possible to select an exciting wavelength which is transparent to the uppermost ternary (passive) layer but which excites strong luminescence from the binary (active) layer. As with scanning electron microscopy and for similar reasons dislocations in the active region act as non-radiative recombination centres and give images (~ 1 µm) in diameter. We will have more to say about PLT in Section 4.5.

4.3.1.5 Transmission Electron Microscopy (TEM)

This is probably the ultimate analytical tool for semiconductors as far as resolution is concerned with the possible exception of Field Ion Microscopy (49). It is destructive of the sample (requiring thinning to ~ 1 µm) and in general terms it is useful only for high dislocation density material ($\sim 10^6/\text{cm}^2$). The very high resolution (~ 10 Å typically) does make the technique useful for analysing dislocation clusters and we will deal with some of these situations in Sections 4.3.2 and 4.5.

4.3.1.6 Arrowheads

'Arrowheads' are surface growth pits seen after liquid phase epitaxy (LPE). Provided the crystal defect density is not too high they correspond one-to-one with both stacking faults and undissociated dislocations as will be discussed futher in Section 4.3.2.

4.3.2 TEM, Arrowheads and A-B etching applied to AlGaAs DH material

When AlGaAs DH material is examined by TEM, three classes of defects are commonly observed: stacking faults, dislocation dipoles and dislocation groups. Figure 4.7 shows a stacking fault together with its bounding partial dislocations *AB*, *CD*. The black and white fringes, running parallel to the intersection of the fault with the sample surface represent the characteristic electron interference effect first observed and accounted for in metals by Whelan and Hirsch (20). Figure 4.8 shows a dislocation dipole consisting of two closely spaced accurately parallel dislocations of opposite sign. Figure 4.9 shows a group of undissociated dislocations (not related to stacking faults). Projecting the dislocations back, as indicated, reveals that they probably had a common source in a region of the material removed by the chemical etching. This observation is strongly relevant to the discussion in Section 4.4.3.

Clearly, TEM is very useful for highly detailed pictures of complex dislocation arrangements and for dislocation densities $\sim 10^6/\text{cm}^2$ or higher. It is not too useful if the dislocation density is in the $10^3/\text{cm}^2$ range, which, taking into account typical device dimensions, is about the upper limit of dislocation density for useful device material. Fortunately, as mentioned

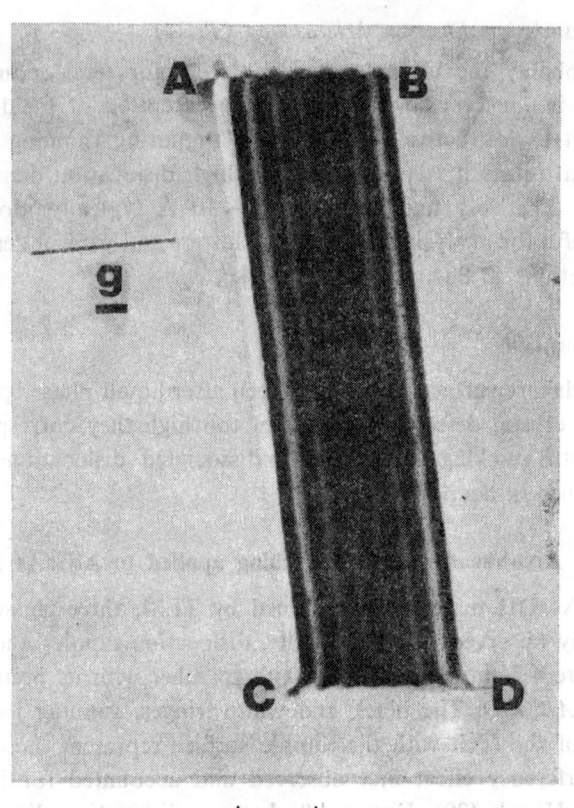

Figure 4.7 Electron micrograph of a stacking fault in AlGaAs DH material. Note the characteristic fringes and the bounding partial dislocations. See text for further explanation. g = 220. Reprinted from Reference 27 by courtesy of the Institute of Physics

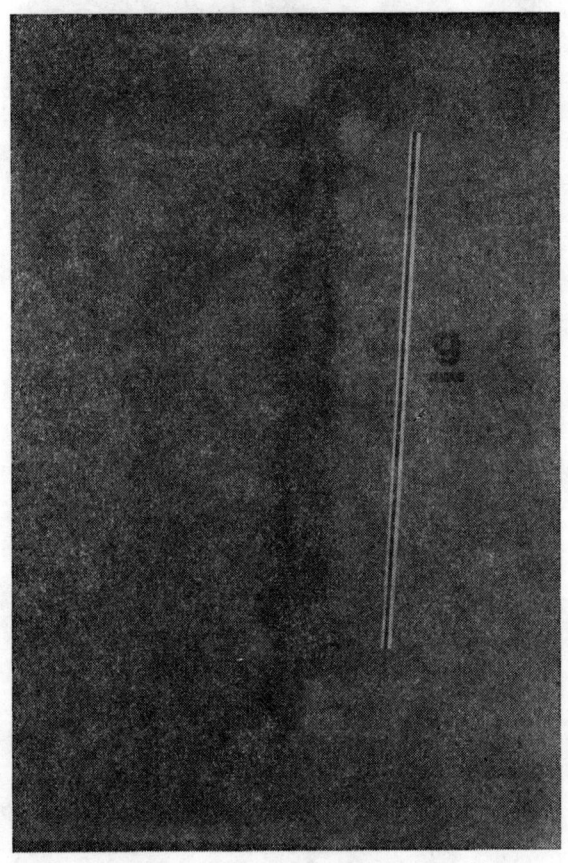

├─ 0.1 μm ─┤

Figure 4.8 Electron micrograph of dislocation dipole in AlGaAs DH material. g = 220. Reprinted from Reference 27 by courtesy of the Institute of Physics

74   G. R. Woolhouse

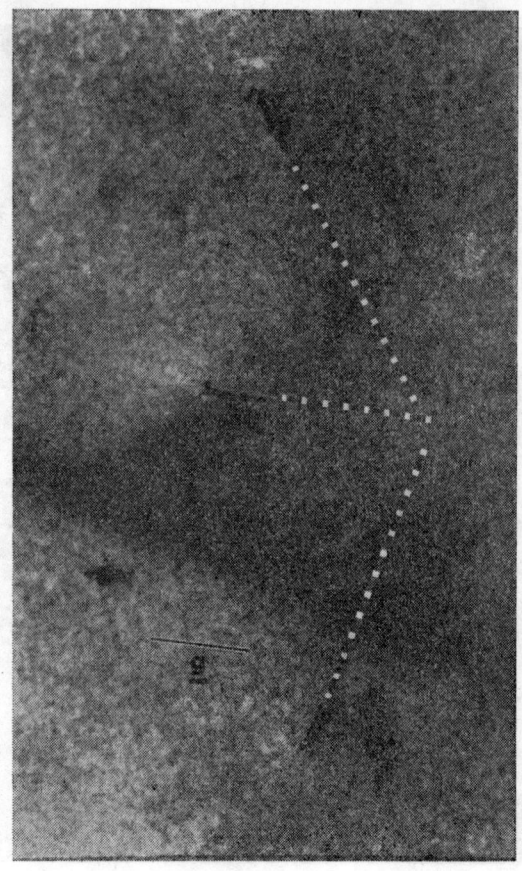

|— 0.5μm —|

Figure 4.9 Electron micrograph of a group of dislocations in AlGaAs DH material. Projection as indicated suggests a common source for these defects. See text for further discussion. g = 220. Reprinted from Reference 27 by courtesy of the Institute of Physics

Generation of Dislocations in Optoelectronic Materials 75

Figure 4.10 Nomarski optical micrograph of arrowheads on a DH LPE wafer surface (after Woolhouse and Anderson (47))

in Section 4.3.1.6, such low dislocation density material can be characterized using 'arrowheads'. These shallow pits, in many cases perhaps only a few tens of Å in depth probably form by a thermal grooving or etching mechanism at the intersections of crystal defects with the LPE surface and can be rendered visible using Nomarski Interference Microscopy (NIM) or its equivalent. Figure 4.10 shows an example of arrowheads on a DH LPE wafer surface (47). The characteristic triangular shape is clearly visible. Figure 4.10 also exhibits a set of lines known as meniscus lines. These have been described in the literature by Shih *et al.* (45) and Small *et al.* (46). The discovery of arrowheads was first reported by Woolhouse *et al.* (17) who also demonstrated, using a combination technique (comprising TEM and NIM applied to the same area of material), that arrowheads (also referred to as 'diffuse' pits)

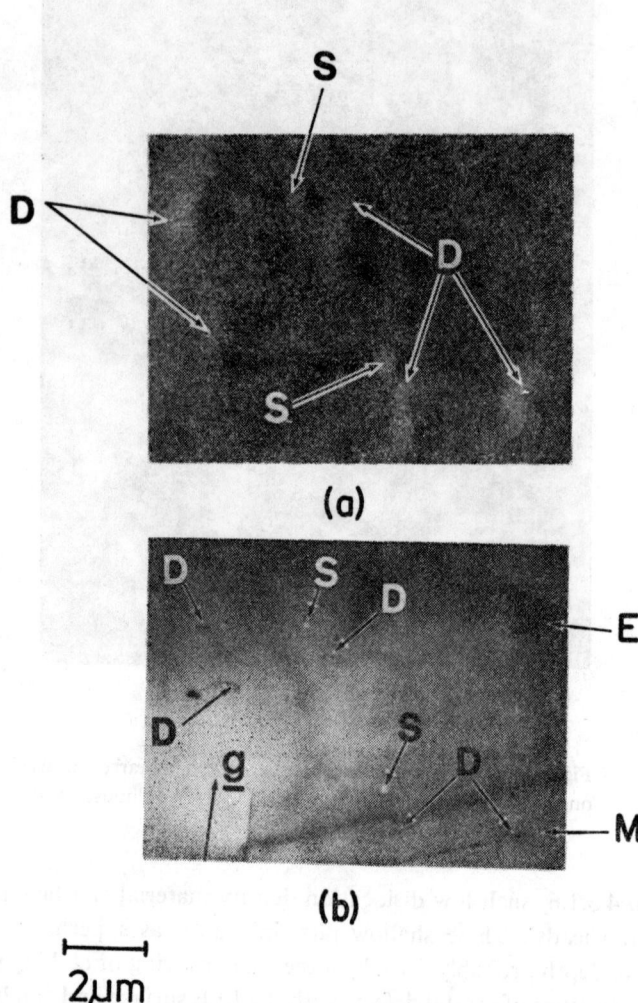

Figure 4.11 Nomarski optical (a) and transmission electron (b) micrographs of the same region of AlGaAs DH material. Sharp images (S) correspond to straight-sided craters in the material surface. Diffuse images (D) (also known as arrowheads) correspond one-to-one with dislocations. **g** = 220. Reprinted from Reference 17 by courtesy of the American Institute of Physics

Generation of Dislocations in Optoelectronic Materials 77

correspond one-to-one both with stacking faults and undissociated dislocations. The dislocation correspondence is illustrated in Figure 4.11.
DH GaAlAs material can also be characterized by A–B etching as mentioned in Section 4.3.1.3, provided a thick capping layer (several microns) is grown to permit etching patterns to form. A typical etching pattern is shown in

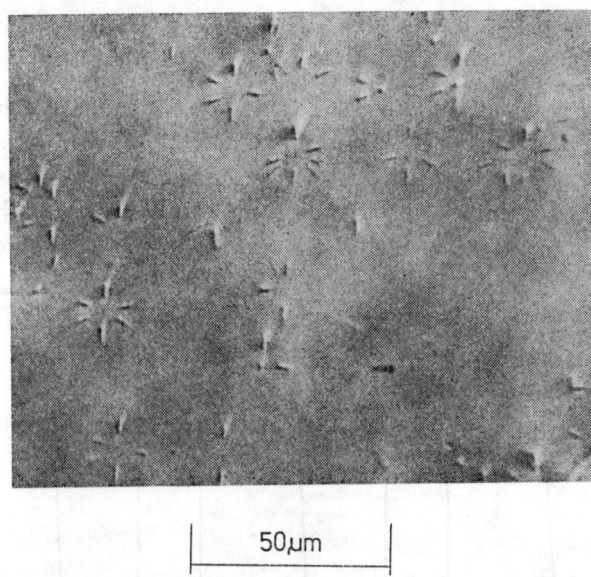

Figure 4.12 A–B etching pattern on AlGaAs DH material with extra thick capping layer. See text for further discussion. Reprinted from Reference 17 by courtesy of the American Institute of Physics

Figure 4.12. The dislocation pits show up as crystallographically oriented lines arranged in 'star' patterns. This is due to the presence of dislocation clusters in the material which will be discussed further in Section 4.4.3.

4.4 MECHANISMS FOR THE GENERATION OF DISLOCATIONS IN OPTOELECTRONIC MATERIALS

As discussed for example in Kittel's text (1), crystal dislocations are not present in thermal equilibrium. Of necessity they are introduced by artifacts and invariably result in an increase in the crystalline free energy. In and of itself this physical fact constitutes a rationale for why the world future for optoelectronic technology is a bright one: there are good reasons for thinking that optoelectronic device material can be prepared in a virtually dislocation free state. However, it is prudent for us to understand the several fold mechanisms by which dislocations can be generated in those materials. These are reviewed in the following section.

4.4.1 Replication from the substrate

This is the most obvious mechanism by which dislocations can be generated in epitaxial material and it is illustrated in Figure 4.13. The mechanism follows directly from one of the most elementary of dislocation properties: a dislocation cannot end in the interior of an otherwise perfect crystal. For a discussion of this property see Friedel (21). Consider the case, illustrated in Figure

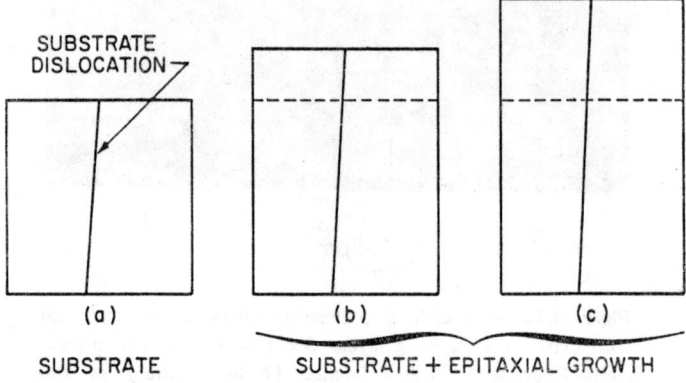

Figure 4.13 Illustrating dislocation replication from a substrate. See text for further explanation

4.13 (a), of a substrate dislocation intersecting an interface with epitaxially growing material. To the extent that the epitaxial layer represents a simple continuation of the substrate crystal the dislocation has but two alternatives: (1) To grow out to the side of the material. (2) To propagate through the epitaxial layer. Case (1) has been considered in some detail in the literature. It represents the formation of a misfit dislocation (for a review see Matthews (22)). It turns out that this mechanism is only operative when the misfit in lattice parameter between the substrate and the epitaxial layer is greater than a critical value. In the absence of such a misfit, case (2) obtains, as illustrated in Figure 4.13 (b) and (c): the dislocation propagates through the

Generation of Dislocations in Optoelectronic Materials

epitaxial layer or layers. This can be thought of as due to the existence of *epitaxial template*: in the absence of sufficient misfit stress to perturb the situation, the epitaxial layer precisely replicates the dislocation content of the substrate. This effect limits the acceptable substrate dislocation density for typical device applications to $< 10^3/\text{cm}^2$. The property of epitaxial template precludes, furthermore, the possibility that the grown layer can contain more dislocations than the substrate except in situations of high misfit stress and in the presence of incorporated inclusions. These situations are discussed in 4.4.2 and 4.4.3.

4.4.2 Operation of a misfit stress

Even if the substrate is dislocation-free, dislocations can be generated in an epitaxial layer due to the action of a misfit stress. This situation has been considered in some detail by Matthews (22) and a schematic of the process can be seen in Figure 4.14. This shows dislocation arcs being nucleated at

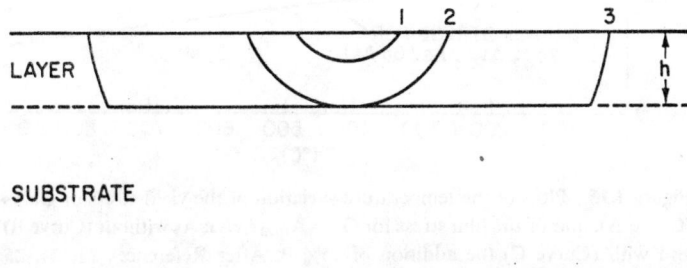

Figure 4.14 Illustrating the generation of misfit dislocations from surface nucleated dislocation arcs through the operation of a misfit stress.
(after Matthews (22))

the surface of an epitaxial layer and assuming successive configurations (1–3) in response to the misfit stress existing in the layer. In the final configuration (3) the dislocation has a misfit portion in the interface between layer and substrate and two threading portions. It can be shown from energy considerations that a critical thickness (h_c) for the epitaxial layer exists and that this can be written:

$$h_c \propto \frac{b}{f}$$

where b is the Burgers vector of the generated dislocation and f is the fractional difference in lattice parameter between layer and substrate.

The existence of a layer with thickness greater than h_c is a necessary but not sufficient condition for dislocation generation. It is also necessary for the strongly temperature dependent yield stress of GaAs compounds to be exceeded. The significance of this can be seen in Figure 4.15. Here curve A

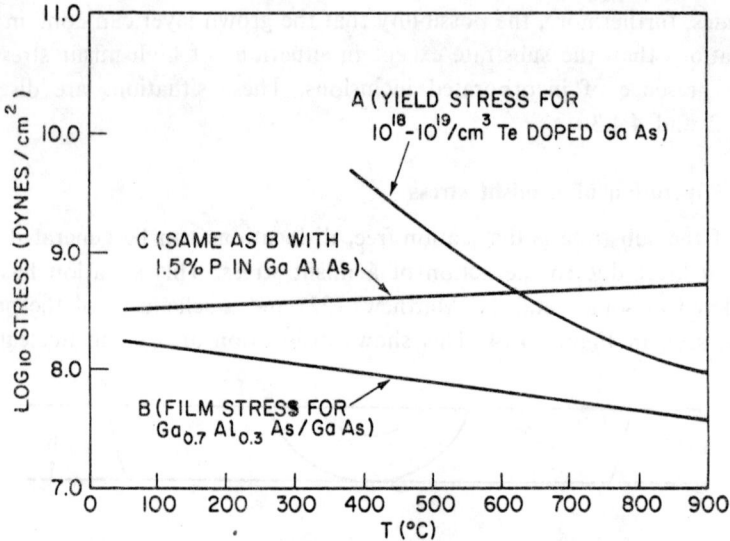

Figure 4.15 Plots of the temperature variation of the yield stress of GaAs (Curve A), and of the film stress for $Ga_{0.7}Al_{0.3}As$/GaAs without (Curve B) and with (Curve C) the addition of 15% P. After References 23, 24, 25 and 26

represents the temperature variation of the yield stress of GaAs (taken from the data of Sazhin et al. (23)) and curve B represents the temperature variation of the misfit stress in a film of $Ga_{0.7}Al_{0.3}As$ grown on GaAs (calculated from the lattice parameter data of Ettenberg and Paff (24) and Pierron et al. (25)). It is evident that even though the yield stress of GaAs drops very rapidly with increasing temperature, the stress in the $Ga_{0.7}Al_{0.3}As$ film also drops with temperature and dislocations are not generated. It has been shown by Rozgonyi et al. (26), however, that if a small amount ($\sim 1.5\%$) of phosphorus is added to the epitaxial film this situation is drastically altered. Whilst compensating the film stress at room temperature the phosphorus gives rise to a large film stress at the growth temperature (typically ~ 800 C) and this can be sufficient for misfit dislocations to be generated at the interface as their X-ray topographs indicate. Their data is shown also in Figure 4.15 (curve C).

Generation of Dislocations in Optoelectronic Materials 81

Contrariwise in situations involving purely GaAlAs and GaAs the writer is not aware of any experimental evidence for the generation of misfit dislocations consistent with the data shown in Figure 4.15.

4.4.3 Incorporation of inclusions

One might well ask, given that the misfit stress in the GaAlAs/GaAs system is apparently insufficient for dislocation generation, and given, furthermore, that substrates with acceptably low dislocation densities ($\sim 10^2/\text{cm}^2$) are commercially available, why dislocation degradation remains a concern for DH laser scientists and engineers? The most likely reason for this relates to the strong possibility of incorporating inclusions in the LPE growth. The sources of these inclusions are probably manifold: graphite particles (Woolhouse (27)) and oxygen precipitation (Ishii et al. (28)) being two referenced candidates.

Figure 4.12 (mentioned briefly in Section 4.3.2) shows one of the first pieces of published data indicating that inclusions could be incorporated in the LPE growth of DH material. It represents an A–B etching pattern on a wafer grown with an extra thick capping layer. The pattern contains characteristic 'star' arrangements of etch lines, each line corresponding to a dislocation or a stacking fault. It was found that as the etching time increased the stars tended to converge to a central point suggesting buried sources of dislocations (inclusions) in the material.

This idea was confirmed by transmission electron micrographs of the type shown in Figures 4.16 and 4.17. These represent inclusions together with a stacking fault and other associated dislocations (Figure 4.16) and an inclusion with two associated dislocation dipoles (Figure 4.17).

Figures 4.16 and 4.17 show examples of complex inclusion related dislocation clusters. However, many much simpler clusters have also been found. These fall into four distinct categories as indicated in Figure 4.18. The categories are distinguished by the nature of the Burgers vector diagram—found by conventional Burgers vector analysis of the cluster followed by placing the vectors end-to-end rather like the polygon of forces in mechanics.

By way of example, Figure 4.19 shows electron micrographs of the type II cluster. The fact that all of the Burgers vector diagrams in Figure 4.18 form closed figures indicates the existence of a Burgers vector sum rule which can be written:

$$\sum_i b_i = 0$$

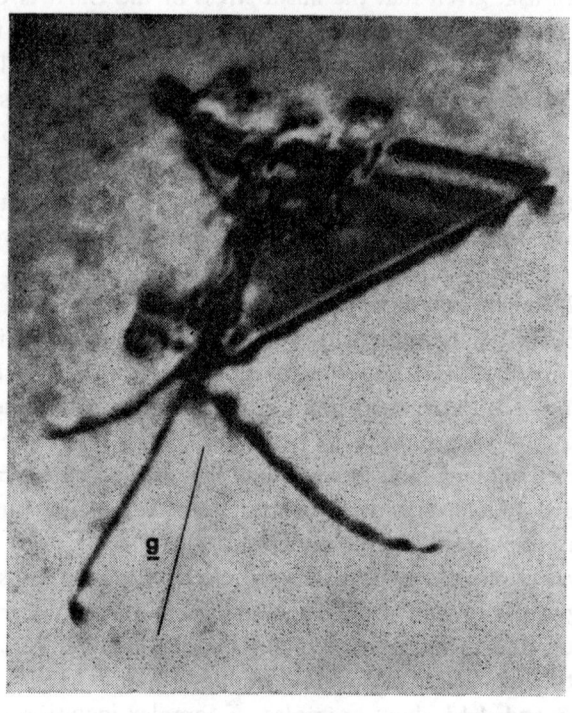

Figure 4.16 Transmission electron micrograph of inclusion-related stacking fault and dislocations in AlGaAs. g = 220. Reprinted from (17) by courtesy of the American Institute of Physics

Generation of Dislocations in Optoelectronic Materials 83

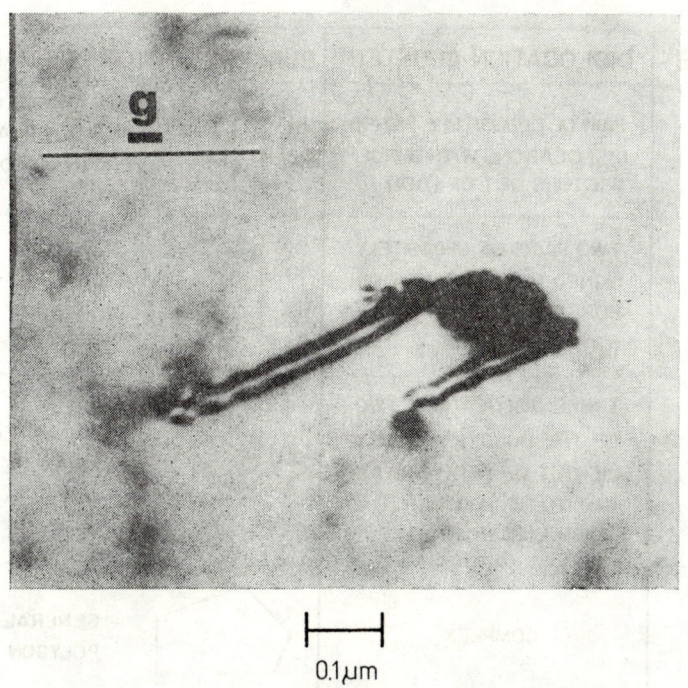

Figure 4.17 Transmission electron micrograph of inclusion-related dislocation dipoles in AlGaAs material. g = 220. Reprinted from (27) by courtesy of the Institute of Physics

TYPE	DISLOCATION CLUSTER	BURGERS VECTOR DIAGRAM	
I	PAIR OF OPPOSITELY SIGNED DISLOCATIONS WITH BURGERS VECTORS OUT OF (001)	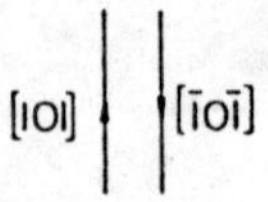 $[10\bar{1}]$ $[\bar{1}0\bar{1}]$	ANTIPARALLEL LINES OF EQUAL LENGTH OUT OF (001)
II	TWO PAIRS OF OPPOSITELY SIGNED DISLOCATIONS WITH BURGERS VECTORS OUT OF (001)	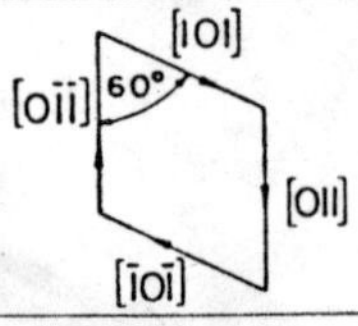 $[0\bar{1}\bar{1}]$ $[101]$ $60°$ $[011]$ $[\bar{1}0\bar{1}]$	RHOMBUS ON $(1\bar{1}\bar{1})$
III	THREE DISLOCATIONS. TWO OF THE BURGERS VECTORS LIE OUT OF (001) AND AT 60° TO EACH OTHER. THE THIRD LIES IN (001)	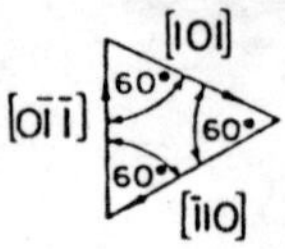 $[0\bar{1}\bar{1}]$ $[101]$ $60°$ $60°$ $60°$ $[\bar{1}10]$	EQUILATERAL TRIANGLE ON (111)
IV	COMPLEX		GENERAL POLYGON

Figure 4.18 The four types of dislocation clusters observed together with their Burgers Vector Diagrams. Reprinted from (29) by courtesy of Taylor and Francis Ltd.

Generation of Dislocations in Optoelectronic Materials 85

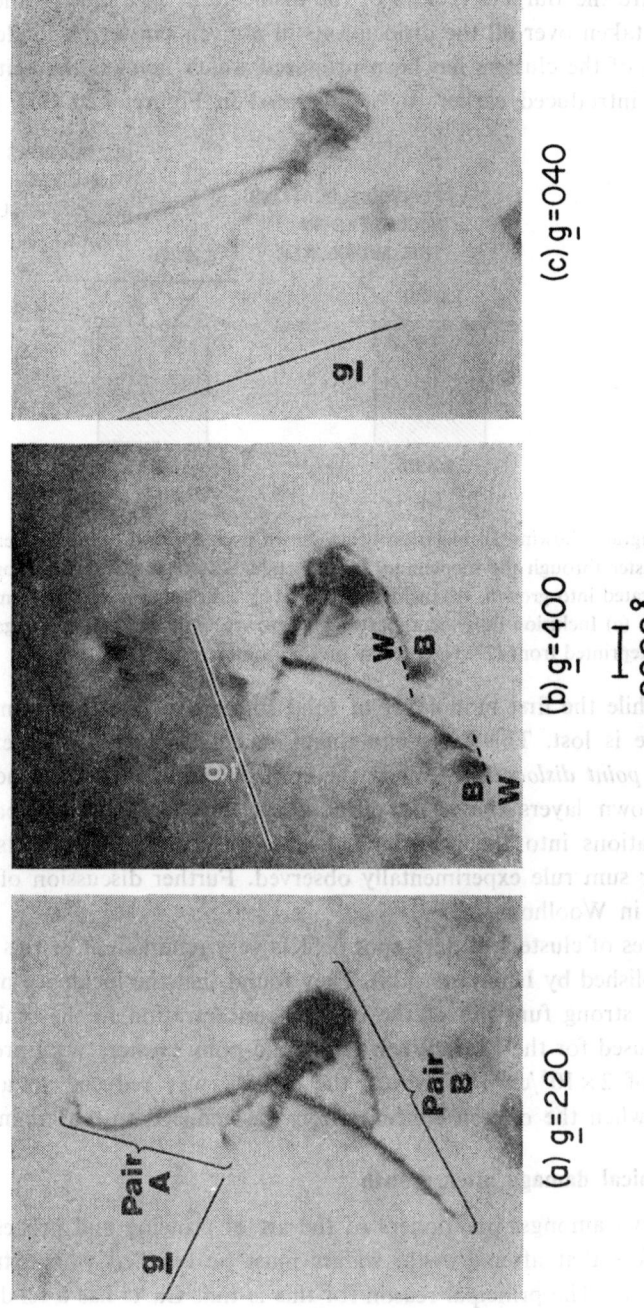

Figure 4.19 Transmission electron micrographs of the Type II dislocation cluster in AlGaAs material. Reprinted from (29) by courtesy of Taylor and Francis Ltd.

where the b_i are the Burgers vectors of the dislocations in a cluster and the summation is taken over all the dislocations in a given cluster. A model for the generation of the clusters has been proposed which invokes the epitaxial template idea introduced earlier. It is illustrated in Figure 4.20. The basic

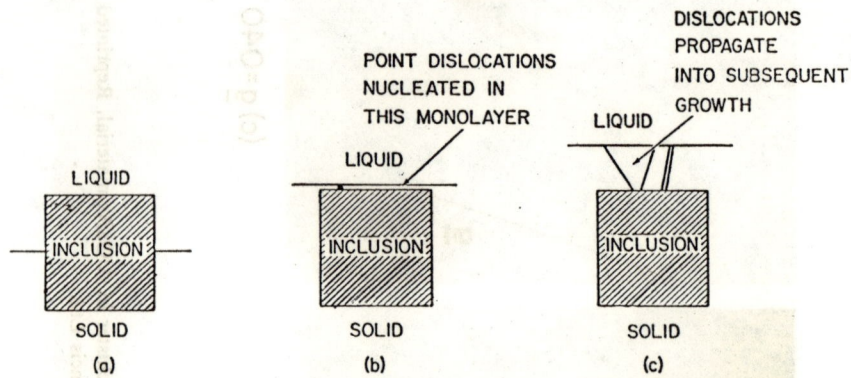

Figure 4.20 Diagram showing an inclusion being grown over during LPE and generating a dislocation cluster through the momentary loss of epitaxial template. (a) Inclusion not yet fully incorporated into growth. (b) Inclusion covered by a monolayer. Epitaxial template momentarily lost. (c) Inclusion fully incorporated into growth. Epitaxial template regained.
Reprinted from (29) by courtesy of Taylor and Francis Ltd.

idea is that while the first monolayer of solid forms over the inclusion epitaxial template is lost. This layer can therefore be defective to the extent of containing *point dislocations*. When the epitaxial template is regained in subsequent grown layers the point dislocations thus nucleated propagate as line dislocations into the growth. This model successfully predicts the Burgers vector sum rule experimentally observed. Further discussion of this can be found in Woolhouse (29).

EBIC pictures of clusters of dark spot defects very reminiscent of this idea have been published by Ishii *et al.* (28). They found that the incidence of the clusters was a strong function of the oxygen concentration in the ambient hydrogen gas used for the LPE: When this was 5 ppm. clusters were present in a density of $2 \times 10^5/cm^2$. However, the density was reduced to a few hundred/cm^2 when the oxygen concentration was reduced to 0.03 ppm.

4.4.4 Mechanical damage after growth

It is well known amongst practioners of the art of growing and processing DH laser wafers that after growth, wafers must be handled with extreme delicacy and care. The principal reason for this is that GaAs has a tendency to abrade very easily, such abrasions usually generating dislocations which can degrade the devices very rapidly during operation. Evidence for this statement is shown in Figure 4.21 which represents a TEM picture of an abrasion-related linear array of dislocation arcs.

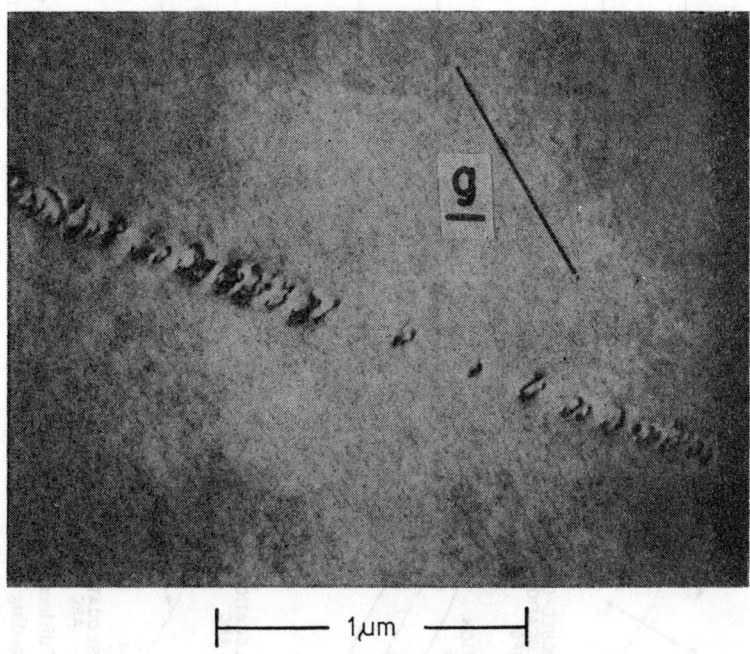

Figure 4.21 Transmission electron micrograph of an abrasion-related linear dislocation array in AlGaAs DH material. g = 220. Reprinted from (33) by courtesy of the Institute of Physics

Arrays of this kind can be thought of as being produced by a modification of the loop punching mechanism first proposed by Seitz (30) and first observed in silver chloride by Jones and Mitchell (31). Seitz's model was appropriate to the case of a stationary indenter which generated dislocation loops along an axis equivalent to the indentation direction. An abrasion clearly represents a moving indenter. This can obviously generate quite complex arrangements of dislocations which can reduce to the simple type of array shown in Figure 4.21 by a sequence of the kind shown in Figure 4.22.

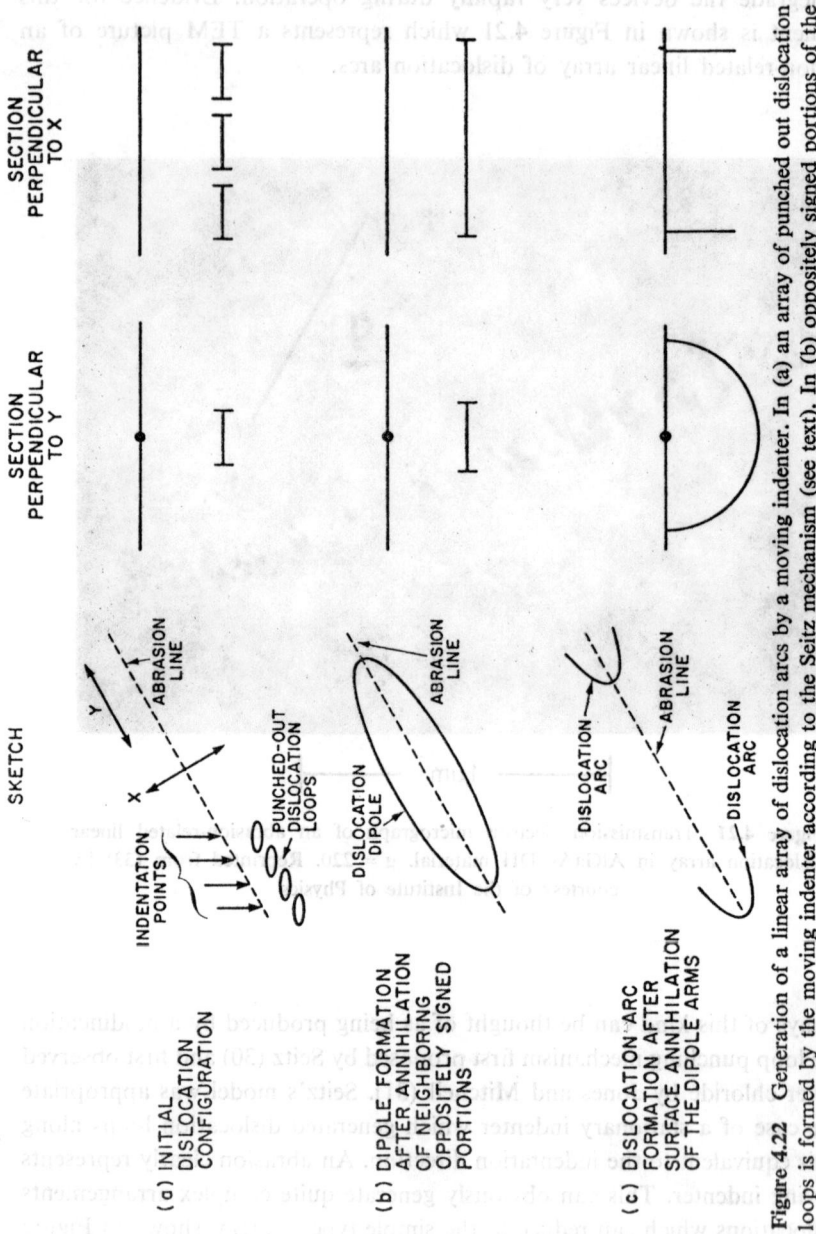

Figure 4.22 Generation of a linear array of dislocation arcs by a moving indenter. In (a) an array of punched out dislocation loops is formed by the moving indenter according to the Seitz mechanism (see text). In (b) oppositely signed portions of the dislocation loops have annihilated to form one dislocation dipole whose axis runs parallel to the abrasion line. In (c) the long arms of the dipole have annihilated at the surface leaving two dislocation arcs with the abrasion line as a common axis

Such abrasion-related dislocation arc arrays are exceedingly deleterious to device performance because of the ease with which such 'fresh' dislocations can glide and grow as will be discussed in the next section.

4.5 BEHAVIOUR OF DISLOCATIONS UNDER OPTICAL EXCITATION

While the behaviour of dislocations under real lasing conditions—i.e., electrical excitation—is of obvious primary concern, it turns out to be very useful to study the behaviour of dislocations under optical excitation from an external laser source. This method, known as *photoluminescence topography* (PLT), was described briefly in Section 4.3.1.4.

In order to perform PLT it is usual to prepare special DH material. This is shown in Figure 4.23 and is characterized by having the cap layer removed,

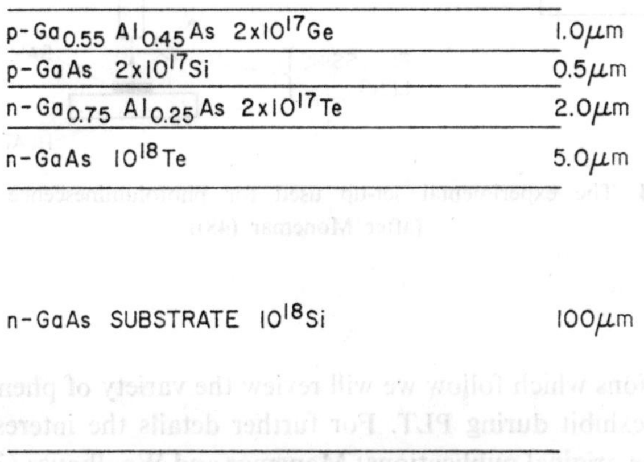

p-$Ga_{0.55}Al_{0.45}As$ 2×10^{17}Ge	1.0 μm
p-GaAs 2×10^{17}Si	0.5 μm
n-$Ga_{0.75}Al_{0.25}As$ 2×10^{17}Te	2.0 μm
n-GaAs 10^{18}Te	5.0 μm

n-GaAs SUBSTRATE 10^{18}Si	100 μm

Figure 4.23 The special AlGaAs DH material usually used in PLT experiments. See text for further explanation

an extra thick active region (up to 0.5 μm) and a larger-than-usual amount of Al in the p-ternary layer. These features enable an incident wavelength to be chosen which is well transmitted by the p-ternary layer but strongly absorbed in the active region, exciting photoluminescence from it. Defects in the active region show up as dark spots (DSDs) and growth of these into dark line defects (DLDs) can be followed. Figure 4.24 shows the very simple equipment used for PLT. The sample can be translated in a plane perpendicular to the diagram. The filters block out any diffusely scattered radiation from the Krypton laser, passing only radiation from the active region of the DH sample which is then detected and displayed on the TV monitor.

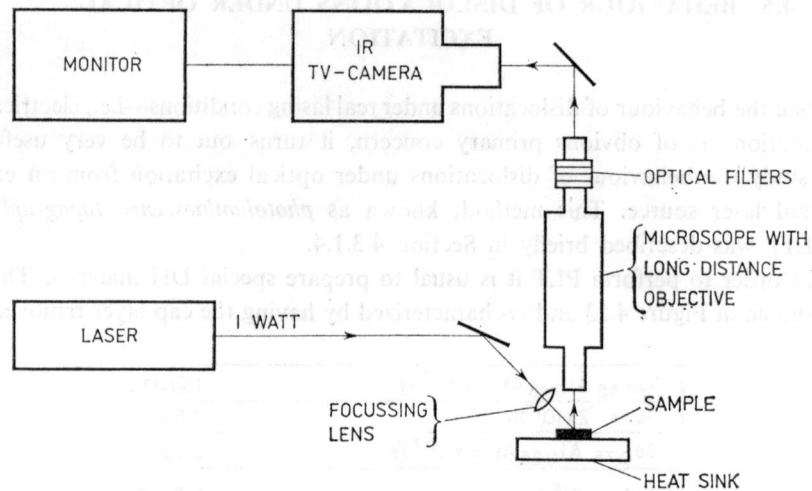

Figure 4.24 The experimental set-up used for photoluminescence topography (after Monemar (48))

In the sections which follow we will review the variety of phenomena which dislocations exhibit during PLT. For further details the interested reader is referred to the original publications: Monemar and Woolhouse (32), Monemar and Woolhouse (33), Woolhouse et al. (34), and Monemar et al. (35).

4.5.1 Dislocation glide

This is characterized by three features:

(a) It only takes place at freshly created dislocations (created by mechanical damage, rather than 'grown-in').
(b) It is usually very rapid—speeds up to 120 µm/s have been observed at excitations of 10^4 watts/cm^2.
(c) The dark lines created by glide are invariably ⟨110⟩ oriented.

Generation of Dislocations in Optoelectronic Materials 91

Figure 4.25 shows a PL topograph of a mechanically damaged region of a DH wafer before (a) and after (b) dislocation glide has taken place. Notice that the dark lines in (b) are without exception ⟨110⟩ oriented.

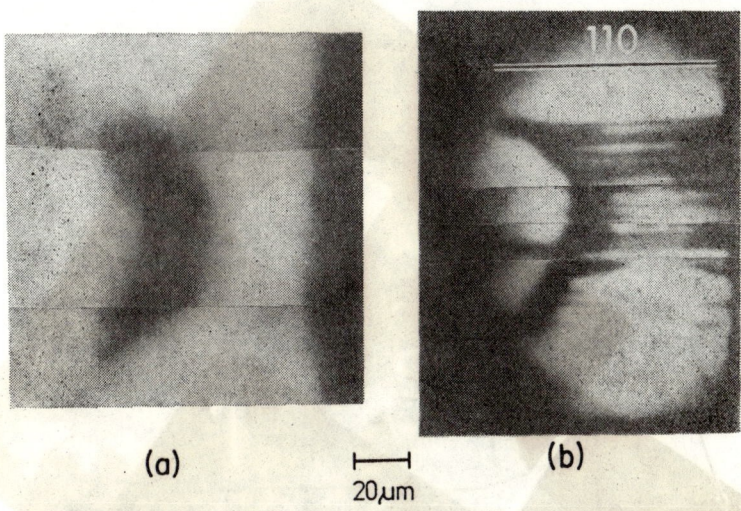

Figure 4.25 Photoluminescence topography of a mechanically damaged region of AlGaAs DH material before and after glide. See text for futher discussion. Reprinted from (32) by courtesy of the American Institute of Physics

Figure 4.26 shows a TEM picture of the region in which glide has taken place. The ⟨110⟩ oriented dark lines can be clearly seen as misfit dislocations which, for the most part, terminate in threading portions.

These observations suggest a model of the glide process which is shown in Figure 4.27. At the heart of the model lies the assumption that the region of mechanical damage has dislocation arcs associated with it similar to those depicted in Figure 4.21 and generated, for example, by a mechanism similar to that described in Figure 4.22. Such dislocation arcs in GaAs (or GaAlAs) are effectively 'frozen-in' at room temperature due to the very high friction stress of those materials, but under optical excitation the arcs can glide to the interface of the double heterostructure material creating lengths of misfit dislocation 'pulled' into position by the motion of the threading portion along its glide plane. TEM observations (Woolhouse (36)) have confirmed that the gliding dislocation does indeed have a unique glide plane associated

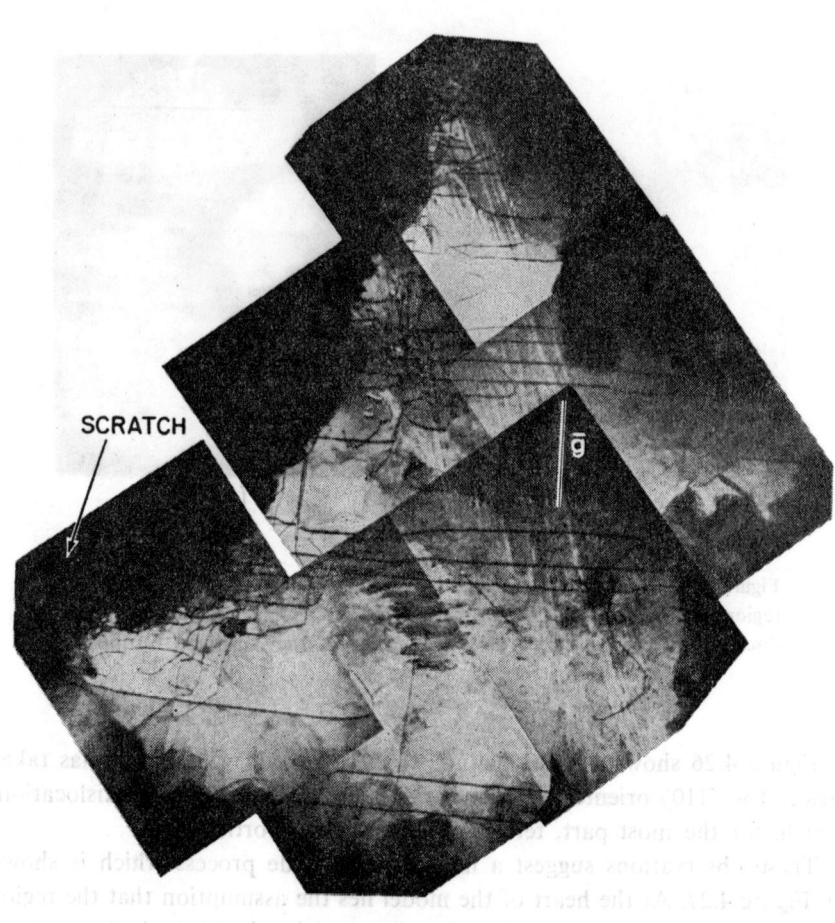

Figure 4.26 Transmission electron micrograph of the glided region of Figure 4.25 g = 220. Reprinted from (32) by courtesy of the American Institute of Physics

with it. It can thus be characterized as either α or β type. (The α dislocation has a plane of Ga atoms for slip plane, whereas the β dislocation has a plane of As atoms.) It is well known (for a review see Wolfson (37)) that α and β

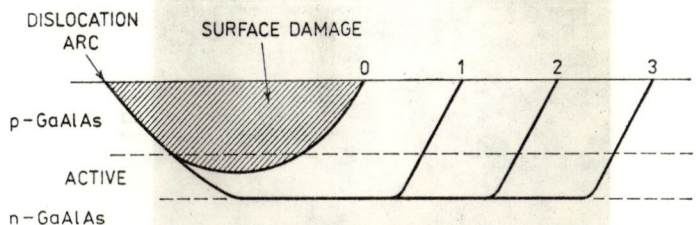

Figure 4.27 Model of the glide process. See text for further explanation

dislocations have different velocities under the same stress conditions. This behaviour was confirmed in studies of a movie of the glide process. The moving dark spots fell into two well-defined regimes of velocity.

4.5.2 Dislocation climb

This is also characterized by three features:

(a) It tends to take place at 'grown-in' dislocations. (See, however, Section 4.5.3.)
(b) It is many orders of magnitude slower than the glide—a typical climb speed is 0.5 μm/h at 10^4 watts/cm².
(c) The dark lines created by climb are usually oriented in directions other than $\langle 110 \rangle$.

A PL topograph of such a climb event is shown in Figure 4.28. The dark line has grown (b) from the dark spot (a) at an average speed of 0.5 μm/h in a $\langle 210 \rangle$ direction under 10^4 watts/cm² excitation.

Such optically-induced climb regions have been studied in the TEM by Petroff et al. (38) and by O'Hara et al. (39). Both sets of workers find highly convoluted giant dislocation dipoles of *interstitial* type.

4.5.3 Combined glide and secondary climb

In this experiment, described in greater detail by Woolhouse et al. (34), it was found that a dislocation could be optically excited from a source by glide in a $\langle 110 \rangle$ direction and then induced to climb in a $\langle 100 \rangle$ direction by

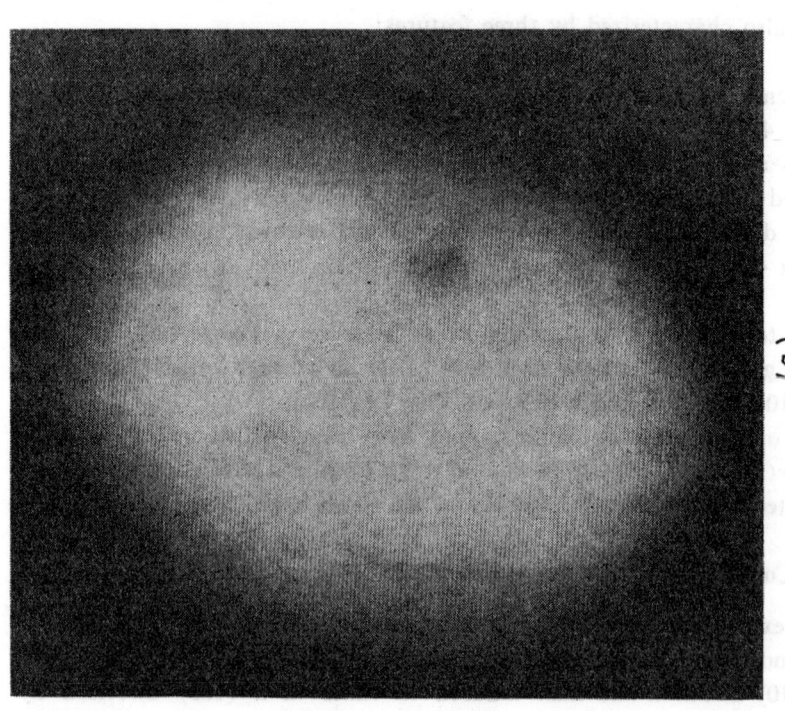
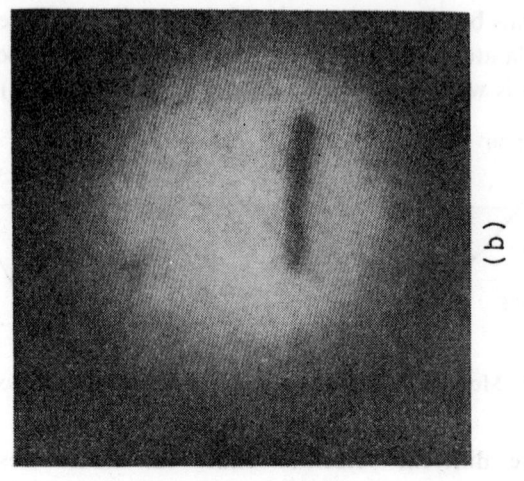

Figure 4.28 Photoluminescence topographs of the climb of a grown-in dislocation. See text for further explanation. Reprinted from (33) by courtesy of the Institute of Physics

excitation at a reduced level. The region of secondary climb (so designated to distinguish it from the primary climb of a 'grown-in' dislocation) was shown by a calibrated technique to be a giant convoluted dislocation dipole

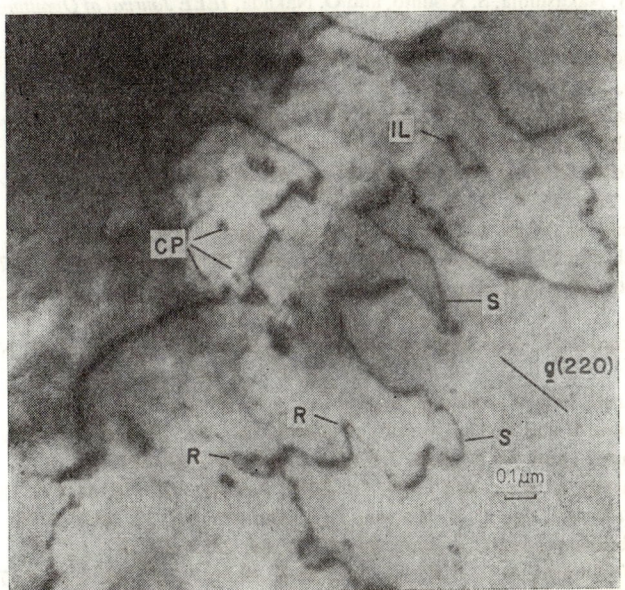

Figure 4.29 Transmission electron micrograph of a detail of the highly convoluted secondary climb region. Salient and re-entrant portions of the giant dislocation dipole are indicated 'S' and 'R' respectively 'CP' and 'IL' refer to coherent particles and internal dislocation loops

which had grown *by vacancy climb*. A detail from this highly convoluted region is shown in Figure 4.29.

While it seems that the glide process is fairly well accounted for by the model of Figure 4.27, the climb process is currently a subject of considerable controversy. There appear to be at least five different models for it extant in the literature (Matsui *et al.* (40), Van Vechten (41), Woolhouse (42), Petroff and Kimerling (43) and O'Hara *et al.* (44)). This does not seem to be the appropriate place to attempt to resolve this dilemma.

ACKNOWLEDGEMENTS

Thanks are due to M. H. Coden, H. F. Lockwood, F. W. Scholl and P. S. Zory for reading the manuscript and making useful comments on it.

REFERENCES

1. C. Kittel, *Introduction to Solid State Physics*, Wiley, New York, 1961, pp. 536, 557.
2. H. Kressel, and H. F. Lockwood, *Journal de Physique*, C3, **35**, C3-223 (1974).
3. R. Ito, H. Nakashima, S. Kishino, and O. Nakada, *IEEE Journal of Quantum Electronics*, QE-11, 551 (1975).
4. P. G. McMullin, *Scanning Electron Microscopy/1976*, Part IV, Chicago, III Research Institute, 543 (1976).
5. H. J. Leamy, L. C. Kimerling, and S. D. Ferris, *Scanning Electron Microscopy/1978*. Vol. I, Illinois: Scanning Electron Microscopy Inc., 717 (1978).
6. O. C. Wells, *Scanning Electron Microscopy*, McGraw-Hill Book Company, New York, 1974, p. 235.
7. A. L. Esquivel, W. N. Lin, and D. B. Wittry, *Applied Physics Letters*, **22**, 414 (1973).
8. A. Rasul, and S. M. Davidson, *Inst. Phys. Conf. Ser.*, No. 33A, 306 (1977).
9. J. M. Titmarsh, G. R. Booker, W. Harding, and D. R. Wright, *J. Mat. Sc.*, **12**, 341, (1977).
10. C. Werkhoven, C. Van Opdorp, and A. T. Vink, *Inst. Phys. Conf. Ser.* No. 33a, 317, (1977).
11. J. F. Petroff, and M. Sauvage, unpublished work, 1978.
12. J. K. Howard, and R. D. Dobrott, *J. Electrochem. Soc.*, **113**, 567 (1966).
13. W. J. Bartels, and W. Nijman, *J. Cryst. Growth*, **37**, 204 (1977).
14. W. J. Bartels, L. Blok, and C. W. Th. Bulle, *J. Cryst. Growth*, **34**, 181 (1976).
15. J. G. Grabmaier, and C. B. Watson, *Phys. Status Solidi*, **32**, K13 (1967).
16. M. S. Abrahams, and C. J. Buiocchi, *J.A.P.*, **36**, 2855 (1965).
17. G. R. Woolhouse, A. E. Blakeslee, and K. K. Shih, *J.A.P.*, **47**, 4349 (1976).
18. W. D. Johnston and B. I. Miller, *Appl. Phys. Lett.*, **23**, 192 (1973).
19. R. Ito, H. Nakashima, and O. Nakada, *Jap. J. Appl. Phys.*, **12**, 1272 (1973).
20. M. J. Whelan, and P. B. Hirsch, *Phil. Mag.*, **2**, 1121, 1303 (1957).
21. J. Friedel, *Dislocations*, Pergamon, 5 (1964).
22. J. W. Matthews, *Misfit Dislocations*, in 'Dislocations in Solids', Vol. II, F. R. N. Nabarro, Ed., Amsterdam, North Holland Pub. Co., 1978.
23. N. P. Sazhin, M. G. Milvidskii, V. B. Osvenskii, and O. G. Stolyarov, *Soviet Physics-Solid State*, **8**, 1223 (1966).
24. M. Ettenberg, and F. J. Paff, *J.A.P.*, **41**, 3926 (1970).
25. E. D. Pierron, D. L. Parker, and J. B. McNeely, *Acta, Cryst.*, **21**, 290 (1960).
26. G. A. Rozgonyi, P. M. Petroff, and M. B. Panish, *J. Cryst. Growth*, **27**, 106 (1974).
27. G. R. Woolhouse, *Inst. Phys. Conf. Ser.*, No. 33a, 17 (1977).
28. M. Ishii, H. Kan, W. Susaki, and Y. Ogata, *Appl. Phys. Lett.*, **29**, 375 (1976).
29. G. R. Woolhouse, *Phil. Mag.*, **36**, 597 (1977).
30. F. Seitz, *Phys. Rev.*, **79**, 723 (1950).
31. D. A. Jones, and J. W. Mitchell, *Phil. Mag.*, **3**, 1 (1958).
32. B. A. Monemar, and G. R. Woolhouse, *Appl. Phys. Lett.*, **29**, 605 (1976).
33. B. A. Monemar, and G. R. Woolhouse, *Inst. Phys. Conf. Ser.*, No. 33a, 400 (1977).
34. G. R. Woolhouse, B. A. Monemar, and C. M. Serrano, *Appl. Phys. Lett.*, **33**, 94 (1978).
35. B. A. Monemar, R. M. Potemski, M. B. Small, J. A. Van Vechten, and G. R. Woolhouse, *Phys. Rev. Lett.*, **41**, 260 (1978).
36. G. R. Woolhouse, unpublished work.

37. R. G. Wolfson, in *Treatise on Materials Science and Technology*, Vol. 6, R. J. Arsenault, Ed., Academic Press, New York, 1975, p. 366.
38. P. M. Petroff, W. D. Johnston, and R. L. Hartman, *Appl. Phys. Lett.*, **25**, 226 (1974).
39. S. O'Hara, P. W. Hutchinson, and P. S. Dobson, *Appl. Phys. Lett.*, **30**, 368 (1977).
40. J. Matsui, K. Ishida, and Y. Nannichi, *Japan. J. Appl. Phys.*, **14**, 1555 (1975).
41. J. A. Van Vechten, *J. Electrochem. Soc.*, **122**, 1556 (1970).
42. G. R. Woolhouse, IEEE, *J. Quant. Electron.*, **QE-11**, 556 (1975).
43. P. M. Petroff, and L. Kimerling, *Appl. Phys. Lett.*, **29**, 461 (1970).
44. S. O'Hara, P. W. Hutchinson, and P. S. Dobson, *Appl. Phys. Lett.*, **30**, 368 (1977).
45. K. K. Shih, G. R. Woolhouse, A. E. Blakeslee, and J. M. Blum, *Inst. Phys. Conf. Ser.*, No. **24**, 165 (1975).
46. M. B. Small, A. E. Blakeslee, K. K. Shih, and R. M. Potemski, *J. Cryst. Growth*, 30, 257 (1975).
47. G. R. Woolhouse, and S. J. Anderson, unpublished work.
48. B. A. Monemar, unpublished work.
49. Y. Ohno, S. Nakanura, T. Adachi, and T. Kuroda, *Surface Science*, **69**, 521 (1977).

Part II

Technological Problems

Chapter 5

Stoichiometric Crystallization Method of III-V Compounds for LEDs and Injection Lasers

J. Nishizawa and Y. Okuno

5.1 INTRODUCTION

The non-stoichiometry of compound materials has been neglected for many years, one of the main reasons being that crystals segregated from the liquid phase should follow the Gibbs' phase rule: $f = c - p + 2$, where c is the number of independent components, p is the number of phases and f is the number of freedom, and in the case of the Ga-As system, c is 2, p is 3, and then f becomes 1. It is understood that if we fix the vapour pressure, then the temperature is instantaneously determined without any doubt, and vice versa.

If vapour pressure of As is applied on the surface of molten Ga, As starts to dissolve into the molten Ga. If the vapour pressure is not high enough to overwhelm the equilibrium vapour pressure at the saturated situation, solubility balanced to the vapour pressure can be reversibly measured until the saturated solubility freedom f is 2, since in accordance with Gibbs' law the number of phases is 2, and both the temperature and the vapour pressure can be established.

When the vapour pressure of As is increased over the balanced pressure for saturated solubility, crystal growth segregated from molten Ga is started and continues until the vapour source As is evaporated out. The vapour pressure, on the other hand, decreases until the balanced pressure corresponds to saturated solubility and the segregation of GaAs attains equilibrium, or until the molten phase Ga is solidified out into solid GaAs and after the stoichiometry of the GaAs is started to change into balanced composition with the As vapour pressure. It has been believed for many years that the stoichiometry of segregated crystal from saturated liquid phase is perfect if the segregated crystal has no direct contact with the gas phase.

At the same time, however, a crystal with different deviation from stoichiometry should be balanced with molten phase at different solubility and then this solubility is easily understood to balance with different vapour pressure from that of saturated solubility.

Also, there is a possibility of a balanced state between the three phases, and this does not contradict Gibbs' law, because the composition can be changed to some extent; the so-called *non-stoichiometry*, and Gibbs' law, determine a certain pressure as a function of the deviation Δ from stoichiometric composition when the temperature is fixed. Then for each temperature, the balanced pressure can vary to some extent corresponding to each deviation Δ. At the same time, if we control the applied vapour pressure on the surface of the molten liquid phase, the composition of the segregated crystal can be controlled.

In this case, only the constancy of the saturation solubility for each temperature was omitted from consideration.

The main reason why this has been mentioned is that experiment gives many contradictory results of the constant stoichiometric crystal growth, regardless of the vapour pressure applied to the liquid surface.

5.2 EFFECT OF THE VAPOUR PRESSURE APPLIED IN THE LPE PROCESS

In the process of crystal growth by liquid phase epitaxy (LPE), crystal segregates as the temperature decreases due to a decrease of the solubility; however, the temperature change in the process of crystal growth usually changes the properties of the crystal. It has also been found that the changes in the growth temperature produce a large number of crystal defects and that defect-free layers can be obtained by growth under constant temperature (1). Furthermore, it has become clear that the crystallographic quality of epitaxial layers changes depending on the pressure of the phosphorus vapour applied to the solution surface (1, 2). Dislocation-free GaP crystals have been obtained by applying the optimum vapour pressure of phosphorus or arsenic.

Then, we applied a temperature difference to the molten zone and, in the higher temperature region, the material was dipped into the melt and there the solubility was rather higher because of the higher temperature, and in the lower temperature region the substrate crystal was dipped to come into contact with the melt, where the solubility was somewhat lower because of the lower temperature. As a result, the difference of the temperatures in each place produced a difference in the concentration, which promoted diffusion

Stoichiometric Crystallization Method of III–V Compounds 103

through the molten zone co-occurring with the thermal diffusion caused by the temperature difference.

In this manner, the crystal can be grown under constant temperature conditions which constitutes a convenient method of obtaining high quality crystals; however, the vapour pressure applied to the liquid surface has also been found to affect the morphology and the quality of the crystals. This will be described later on. In this chapter we report on the surface morphology and lattice constant variation of LPE layers of GaP grown by the temperature difference method (1) at constant temperature with an additional phosphorus pressure (2).

5.2.1 Experimental

A schematic diagram of the crucible for the purpose of explaining the growth principle of the temperature difference method (TDM) under controlled vapour pressure (CVP) is shown in Figure 5.1a. The apparatus used in this work is shown in Figure 5.1b. The vertical portion contains the solution

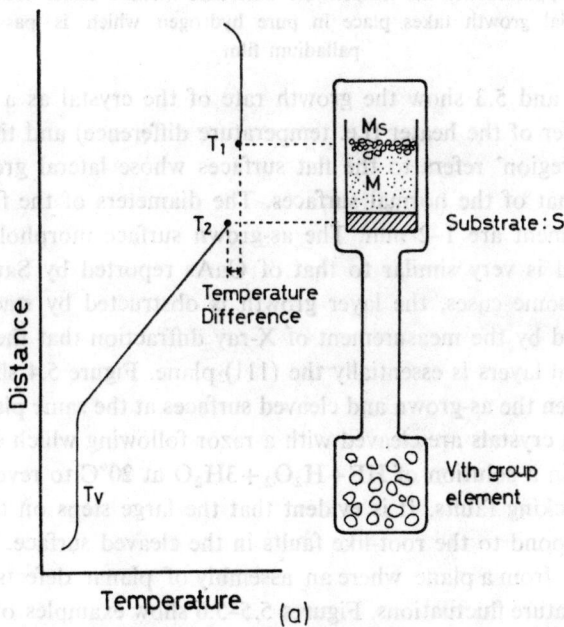

Figure 5.1(a) Schematic diagram of the apparatus for the temperature difference method (TDM) and controlled vapour pressure (CVP)

and excess GaP (floating on top). The solution is made hotter than the seed with an additional heater and there is provision for an additional source of phosphorus vapour. Two sets of experiments were made, one with additional phosphorus and one without. The first set of experiments was conducted without added phosphorus in the vapour phase.

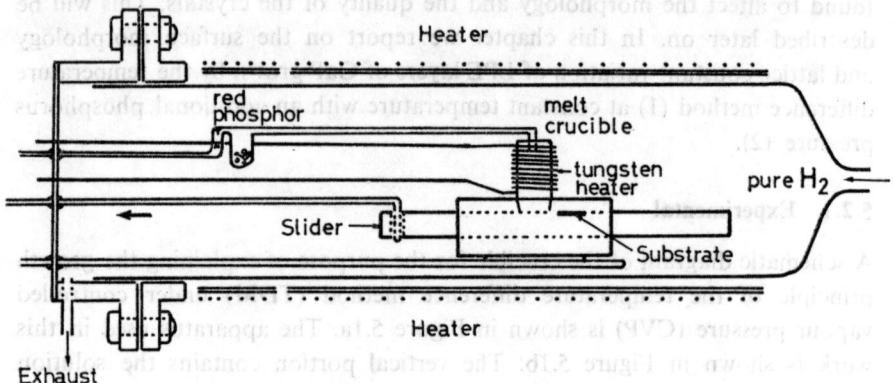

Figure 5.1 (b) Apparatus for the temperature difference method under controlled vapour pressure. Epitaxial growth takes place in pure hydrogen which is passed through a palladium film

Figures 5.2 and 5.3 show the growth rate of the crystal as a function of the input power of the heater (i.e. temperature difference) and time, respectively. 'Facet region' refers to the flat surfaces whose lateral growth rate is higher than that of the normal surfaces. The diameters of the facet regions in this experiment are 1–2 mm. The as-grown surface morphology of GaP is stepped and is very similar to that of GaAs reported by Saul et al. (3). However, in some cases, the layer growth is obstructed by stacking faults. We ascertained by the measurement of X-ray diffraction that the orientation of the epitaxial layers is essentially the (111) plane. Figure 5.4 shows a comparison between the as-grown and cleaved surfaces at the same place. The epitaxially grown crystals are cleaved with a razor following which stain etching is performed in a solution of $HF + H_2O_2 + 3H_2O$ at 20°C to reveal the rooty faults and stacking faults. It is evident that the large steps on the as-grown surface correspond to the root-like faults in the cleaved surface. These faults seem to arise from a plane where an assembly of planar defects is produced by the temperature fluctuations. Figures 5.5–5.6 show examples of the dependence of the cleaved surface morphologies of the crystal layers on the growth conditions shown by the accompanying schematic diagrams: Temperature control was carried out by the main heater wound on the reaction tube in

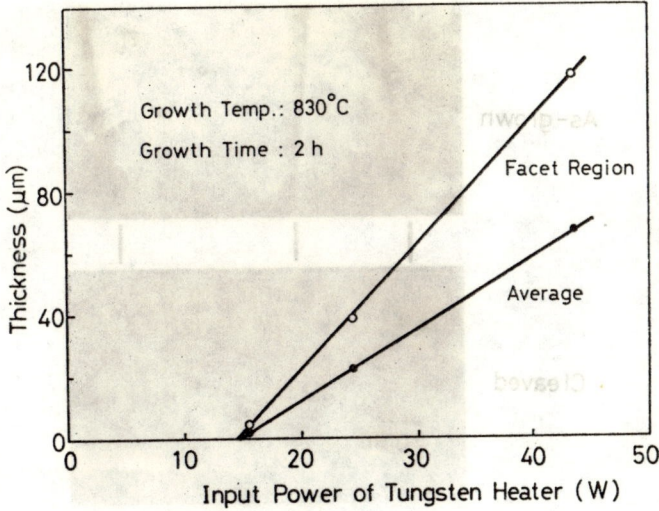

Figure 5.2 Relation between tungsten heater power and growth thickness. Growth temperature: 830° C; growth time: 2 h

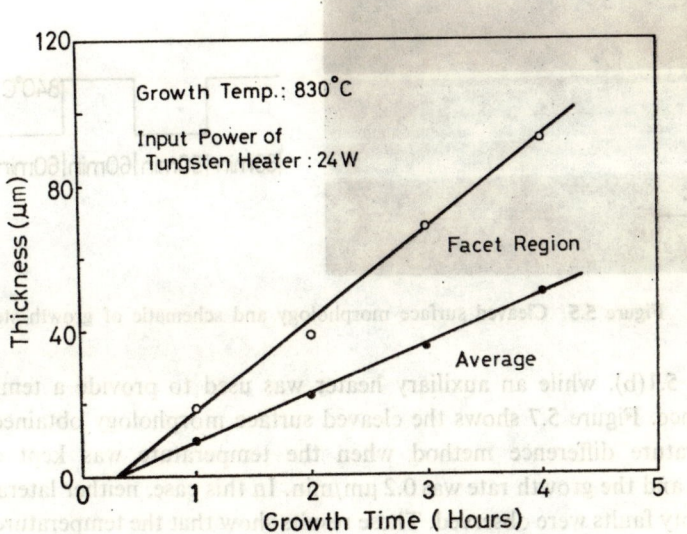

Figure 5.3 Relation between growth time and growth thickness. Growth temperature is 830° C and input power of tungsten heater to give temperature difference is 24 W which produces temperature difference of about 20° C in melt

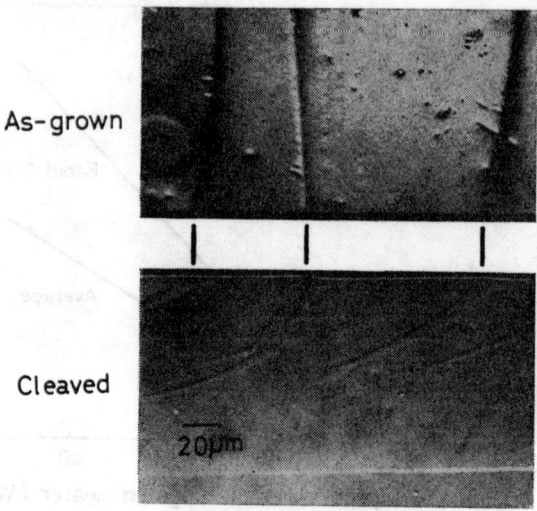

Figure 5.4 Relation of morphology of as-growth and cleaved surfaces in the same place

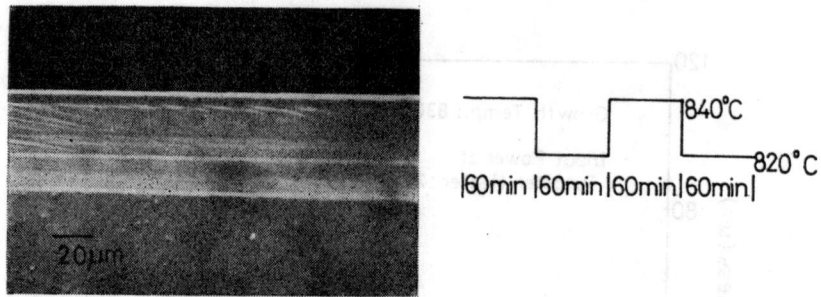

Figure 5.5 Cleaved surface morphology and schematic of growth steps

Figure 5.1(b), while an auxiliary heater was used to provide a temperature difference. Figure 5.7 shows the cleaved surface morphology obtained by the temperature difference method when the temperature was kept constant for 2 h and the growth rate was 0.2 µm/min. In this case, neither lateral stripes nor rooty faults were observed. These results show that the temperature change for the grown layers can generate stripes parallel to the interface and lateral growth.

In the second set of experiments, the effect of vapour pressure control was investigated. Figure 5.8 shows the cleaved surface of a layer grown under

Stoichiometric Crystallization Method of III–V Compounds

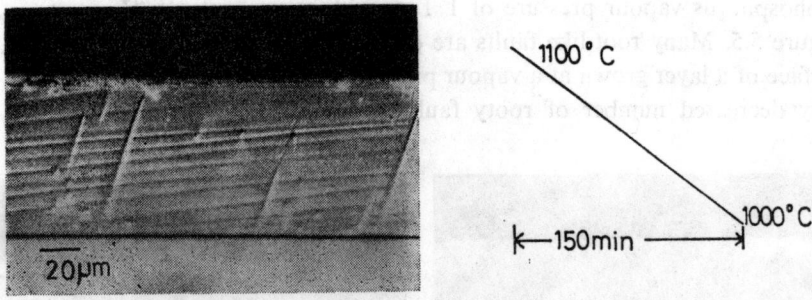

Figure 5.6 Cleaved surface morphology and schematic of temperature lowering process

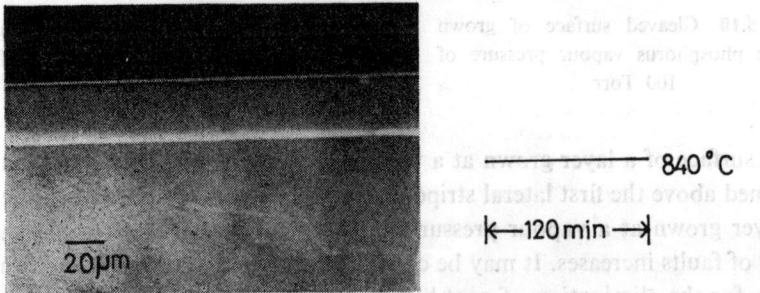

Figure 5.7 Cleaved surface morphology and schematic of crystal growth process at constant temperature

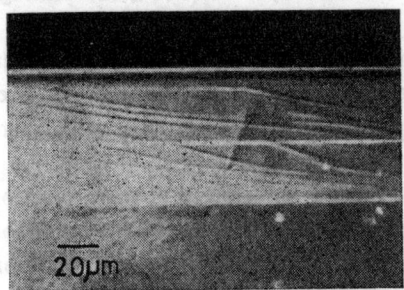 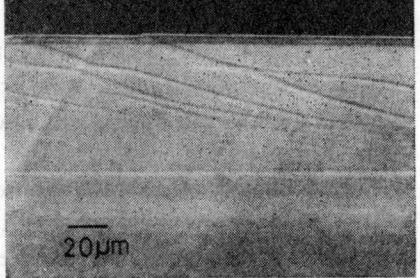

Figure 5.8 Cleaved surface of grown layer at phosphorus vapour pressure of 1 Torr

Figure 5.9 Cleaved surface of grown layer at phosphorus vapour pressure of 10 Torr

a phosphorus vapour pressure of 1 Torr with stepgrowth similar to that of Figure 5.5. Many root-like faults are observed. Figure 5.9 shows the cleaved surface of a layer grown at a vapour pressure of 10 Torr, in which a considerably decreased number of rooty faults is observed. Figure 5.10 shows the

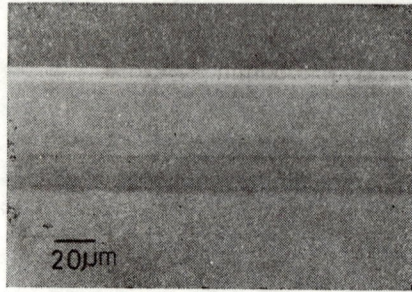

Figure 5.10 Cleaved surface of grown layer at phosphorus vapour pressure of 100 Torr

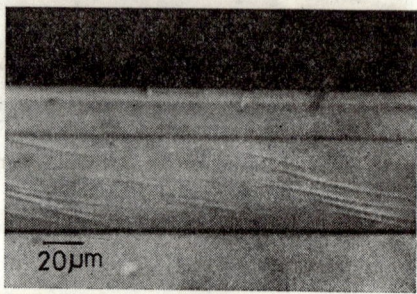

Figure 5.11 Cleaved surface of grown layer at phosphorus vapour pressure of 1000 Torr

cleaved surface of a layer grown at a vapour pressure of 100 Torr. No faults are formed above the first lateral stripe. Figure 5.11 shows the cleaved surface of a layer grown at a vapour pressure of 1000 Torr and it is seen that the number of faults increases. It may be concluded that an optimum phosphorus pressure for the elimination of root-like faults exists as a function of crystal growth temperature as shown in Figure 5.12, and also that it suppresses the growth of lateral stripes caused by the temperature change. Figure 5.13 shows the cleaved surface of a layer grown at only 750°C with a phosphorus vapour pressure of 42.5 Torr using the result of Figure 5.12. Although the crystal is grown at a very low temperature such as 750°C, it exhibits a morphology without any visible defects.

Figures 5.14(a)–(f) show X-ray topographs taken by the Berg–Barrett method. The epitaxial layers were grown under several phosphorus pressures for a constant temperature (800°C), temperature difference (20°C), and growth time (3 h).

A large number of defects are observed in the substrate, but the perfection of the crystals grown under controlled vapour pressure improved in comparison with the case of no vapour pressure control. Defect densities as a function of phosphorus pressures have not been measured quantitatively. In the crystal grown under $P_{GaP\,opt}$, relatively few defects (lattice defects and dislocations) exist over the whole surface (10 × 10 mm). In the crystals grown with phosphorus

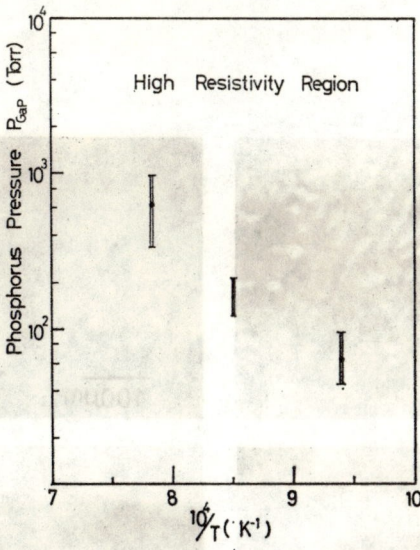

Figure 5.12 Relation between optimum phosphorus pressure to eliminate generation of rooty faults and crystal growth temperature. Black circle: phosphorus pressure to develop maximum lattice constant at each temperature (T_{GaP}) (cf. Reference 19)

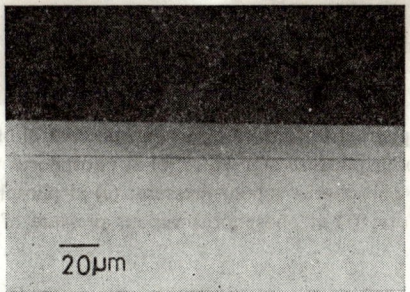

Figure 5.13 Cleaved surface of grown layer at 750° C at phosphorus pressure of 42.5 Torr

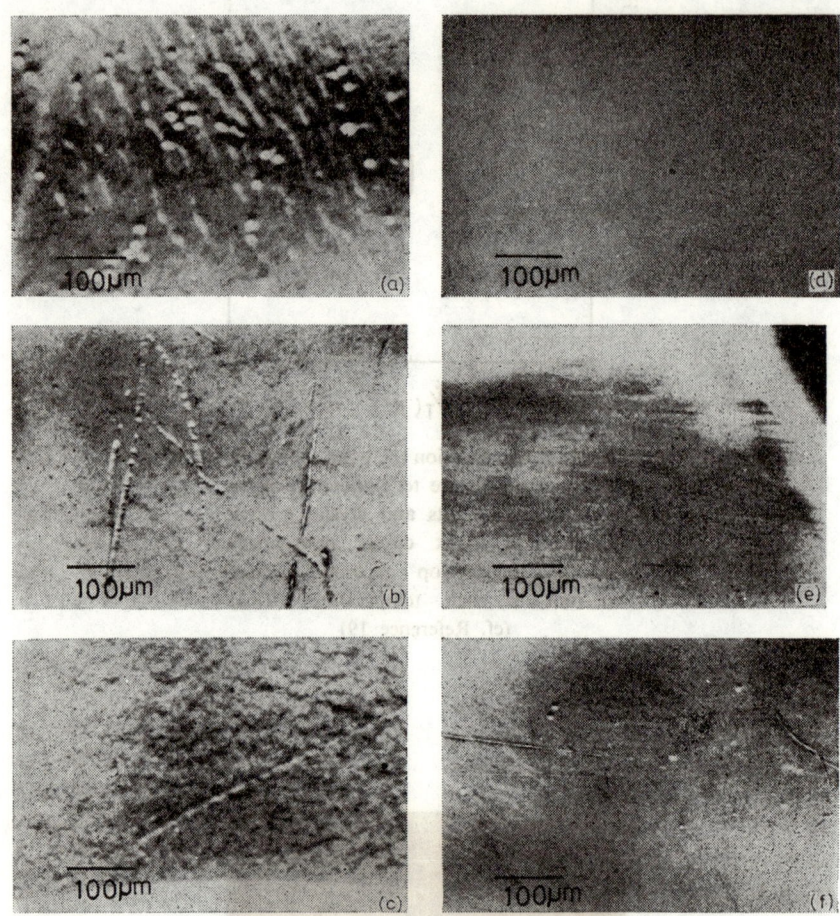

Figure 5.14 X-ray topographs of grown layers: (a) no control of the vapour pressure; (b) at phosphorus vapour pressure of 1 Torr; (c) at phosphorus vapour pressure of 10 Torr; (d) at optimum phosphorus vapour pressure; (e) at phosphorus vapour pressure of 300 Torr; (f) at phosphorus vapour pressure of 1000 Torr

Stoichiometric Crystallization Method of III–V Compounds

pressure lower or higher than $P_{GaP\,opt}$, a greater number of defects are generated and the habits of the crystals such as surface morphology are also affected.

5.2.2 Electrical properties of crystals grown by TDM under CVP

The dependence of carrier concentration of GaP epitaxial layers on phosphorus pressure was investigated and is shown in Figure 5.15. The Hall coefficients of the epitaxial layers grown on the high resistivity annealed at the optimum

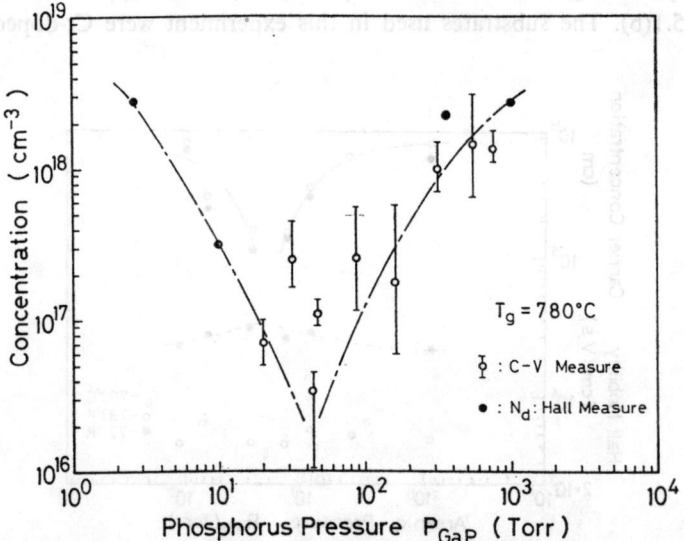

Figure 5.15 Dependence of electrical properties of GaP epitaxial layers on phosphorus pressure. ($\bar{\varphi}$) surface carrier concentration measured by C-V. (●) donor concentration established by Hall measurement

phosphorus pressure were measured by Van der Pauw's method in the temperature ranges from 77 to 400 K. The donor concentration N_d is calculated from the temperature dependence of carrier concentration. On the other hand, for the measurement of the surface concentration, thin Ni layers 0.5 mm in diameter were evaporated on the epitaxial layer grown on the n-type GaP substrates to form a Schottky barrier. The surface carrier concentration was measured over the whole surface. As shown in Figure 5.15, the donor concentration, N_d, becomes a minimum at about the same $P_{GaP\,opt}$ obtained in surface morphology and X-ray topography studies. The value $P_{GaP\,opt}$ is slightly lower because of the lower growth temperature (780°C). The random values in the C–V measurement are due to the fact that they were taken over the whole surface. The values in the centre part are almost equal. The phosphorus pressure $P_{GaP\,opt}$ at which N_d becomes a minimum carrier concentration shows a good coincidence with $P_{GaP\,opt}$ at which the annealed crystals convert to high resistivity. It can be inferred from these results that the stoichiometry of epitaxial layers can be controlled by phosphorus vapour pressure.

The epitaxial growth of GaAs is performed in the apparatus shown in Figure 5.1(b). The substrates used in this experiment were Cr-doped GaAs

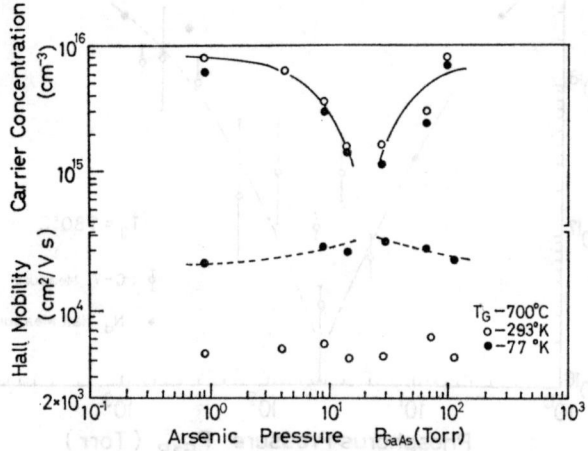

Figure 5.16 Relation between carrier concentration and Hall mobility and arsenic pressure

with surfaces parallel to the (100) plane. The electrical properties of epitaxial layers grown under several arsenic pressures at temperatures from 530 to 800°C, were measured and are shown in Figure 5.16 for $T_g = 700$°C. Carrier concentration are reduced to a minimum and the Hall mobility attains a maximum at a certain arsenic pressure. This arsenic pressure is defined as the optimum arsenic pressure ($P_{GaAs\,opt}$) and this value covers a narrow pressure range as shown in Figure 5.17. Figure 5.18 shows the optimum arsenic pressure versus crystal growth temperature along with the result of the annealing experiment of GaAs crystals (5). The optimum arsenic pressure $P_{GaAs\,opt}$

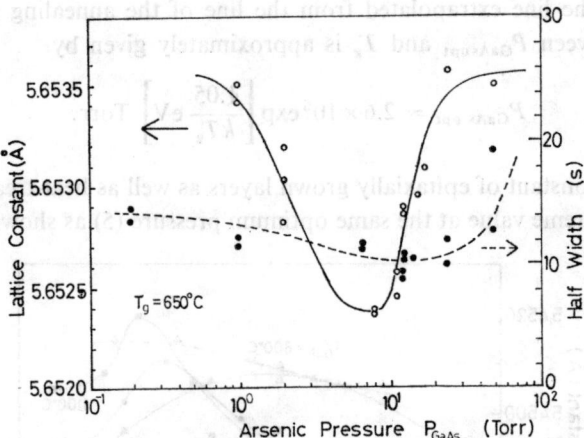

Figure 5.17 Relation between lattice constant of GaAs epitaxial layers and arsenic vapour pressure. The growth temperature is 650°C

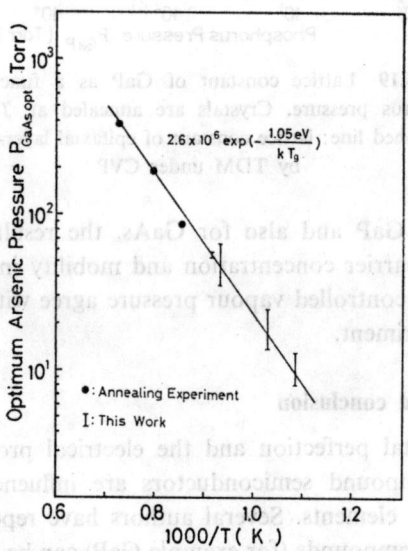

Figure 5.18 Relation between the optimum arsenic pressure and crystal growth temperature. Black circle: optimum pressure of annealing experiment of GaAs crystals

falls along the line extrapolated from the line of the annealing results. The relation between $P_{\text{GaAs opt}}$ and T_g is approximately given by

$$P_{\text{GaAs opt}} = 2.6 \times 10^6 \exp\left[\frac{1.05}{kT_g}\text{eV}\right] \text{Torr}.$$

The lattice constant of epitaxially grown layers as well as heat-treated crystals shows an extreme value at the same optimum pressure (5) as shown in Figures

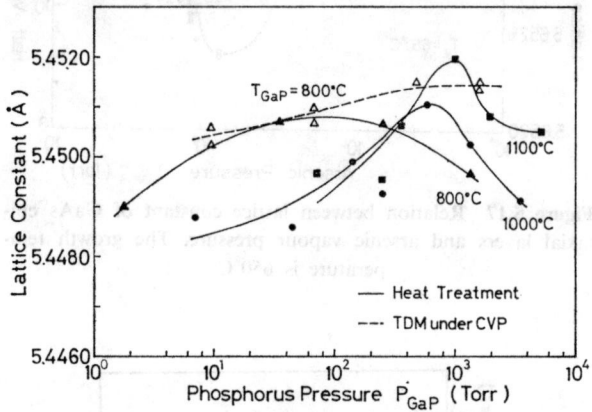

Figure 5.19 Lattice constant of GaP as a function of phosphorus pressure. Crystals are annealed at T_{GaP} for 50h. Dashed line: lattice constant of epitaxial layers grown by TDM under CVP

5.18 and 5.19. For GaP and also for GaAs, the results of measurements of lattice constant, carrier concentration and mobility in the grown crystals as a function of the controlled vapour pressure agree with the results of the heat treatment experiment.

5.2.3 Discussion and conclusion

It is clear that crystal perfection and the electrical properties of epitaxial layers of III–V compound semiconductors are influenced by the vapour pressure of group V elements. Several authors have reported that the stoichiometry of III–V compounds (for example GaP) can be controlled by group V elements (P) through the relation of solid (GaP)-vapour (P) (3). Furthermore, this experiment shows that the stoichiometry of III–V compounds can be controlled through the relationship of solid (GaP)-liquid (Ga)-vapour (P) by controlling the vapour pressure of group V elements.

There is an optimum vapour pressure which is a function of growth temperature in GaP and GaAs, and it is given in Torr by the equations

$$P_{\text{GaP opt}} = 4.67 \times 10^6 \exp\left[-\frac{1.01}{kT_g} \text{ eV}\right] \text{ Torr} \quad \text{(Reference 6)}$$

$$P_{\text{GaAs opt}} = 2.6 \times 10^6 \exp\left[-\frac{1.05}{kT_g} \text{ eV}\right] \text{ Torr}$$

That dependence is in good agreement with the relation of the annealing temperature and $P_{\text{GaP opt}}$, $P_{\text{GaAs opt}}$ in the annealing experiment. The X-ray topographs of epitaxial layers grown at optimum phosphorus or arsenic pressure show relatively perfect crystal quality with negligible defects.

In view of the fact that several characteristics of the heat treatment experiment and liquid phase epitaxial layers under controlled vapour pressure show good agreement we propose the following mechanisms to control the stoichiometry of GaP epitaxy: In epitaxy performed by this method, dissolved materials from the higher temperature region move to the GaP substrate through the Ga-solution by thermal and concentration diffusion. When the phosphorus vapour pressures are applied to the source material, the solubility of GaP (S_{GaP}) as a dissolved material in the Ga melt is different from the situation without phosphorus pressure. The S_{GaP}, in the case of increasing pressure, depends on the phosphorus vapour pressure. The stoichiometry of crystals deposited from the melt can be controlled by this phosphorus pressure. This phenomenon can be understood if excess or deficient phosphorus (or arsenic) is considered as an impurity.

It is usually assumed that an increase in the phosphorus pressure influences the growth rate of the epitaxial layer, but our studies with $\text{GaP}_y\text{Sb}_{1-y}$ growth from liquid show that y does not change with added phosphorus pressure over a wide range (7). As the lattice constant of epitaxial layers under controlled vapour pressure also shows a very large change, it plays an important role in lattice fitting (8).

5.2.4 Growth mechanisms of GaP and GaAs

As has already been made clear, the optimum vapour pressure applied on the top of the melt effectively improves the quality of the crystal segregated at the other end of the melt without any direct contact with the vapour pressure as is shown in Figures 5.8–5.16 (6, 9). Also, the density of dislocations included in the grown crystals is shown in Figure 5.20, (10) and the optimum vapour pressure applied for TDM method can grow nearly dislocation-free crystals.

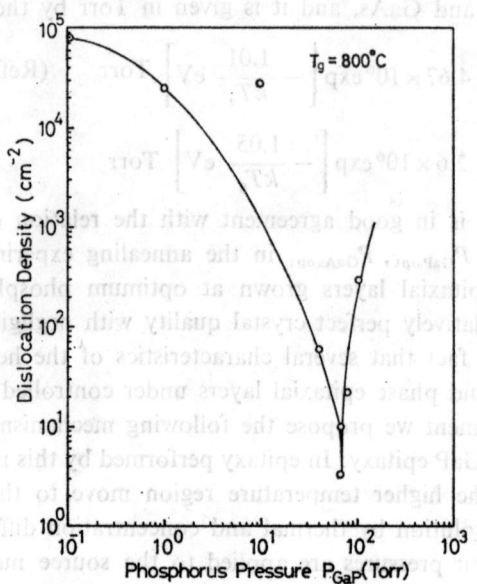

Figure 5.20 Phosphorus pressure dependence of dislocation density in the epitaxial layer. We have obtained nearly dislocation-free crystals by applying $P_{GaP} = 67$ Torr. The growth temperature was 800°C

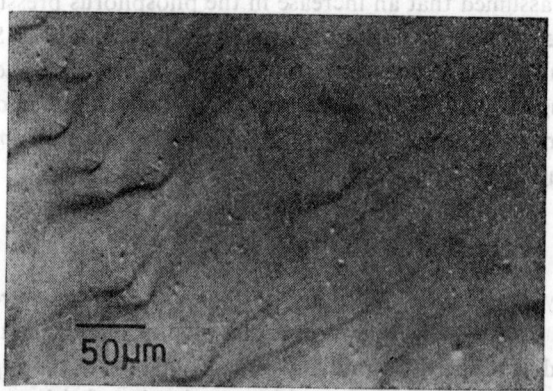

Figure 5.21 *Shadow* and stacking faults observed on the GaP epitaxial layer. The layer growth is obstructed by stacking faults

Stoichiometric Crystallization Method of III-V Compounds 117

At the same time, the growth mechanisms of GaP and of GaAs, and perhaps of all III-V compound semiconductors, have been found to follow the Kossel's model. Figure 5.21 (11) is very similar to those published by the author's group in the case of silicon (12) named as *shadow* and it can be un-

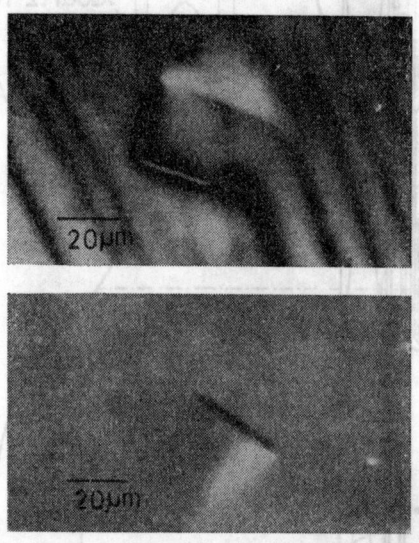

Figure 5.22 Morphology of stacking fault in the epitaxial layer on the (100) plane

derstood that the layer growth is obstructed by stacking faults resulting in the formation of a depressed region toward the fastest growth direction. Figure 5.22 shows the morphology of the stacking fault in the (100) plane (11).

5.3 EFFECT ON THE IMPURITIES SEGREGATED INTO CRYSTALS GROWN BY TDM UNDER CVP

The first experiment on tin-doped GaAs was reported many years ago (1). The experimental set-up which has a vertical closed structure is shown in Figure 5.23.

The photoluminescence spectrum depicted in Figure 5.24 shows that pressures of up to 1 Torr at 730°C decrease the relative emission intensity at 1.34 eV as shown in Figure 5.25, which is thought to be caused by deep Sn impurity level. It is understood that the number of Sn atoms incorporated in arsenic vacancy is decreased and this is also supported by the results of

118   J. Nishizawa and Y. Okuno

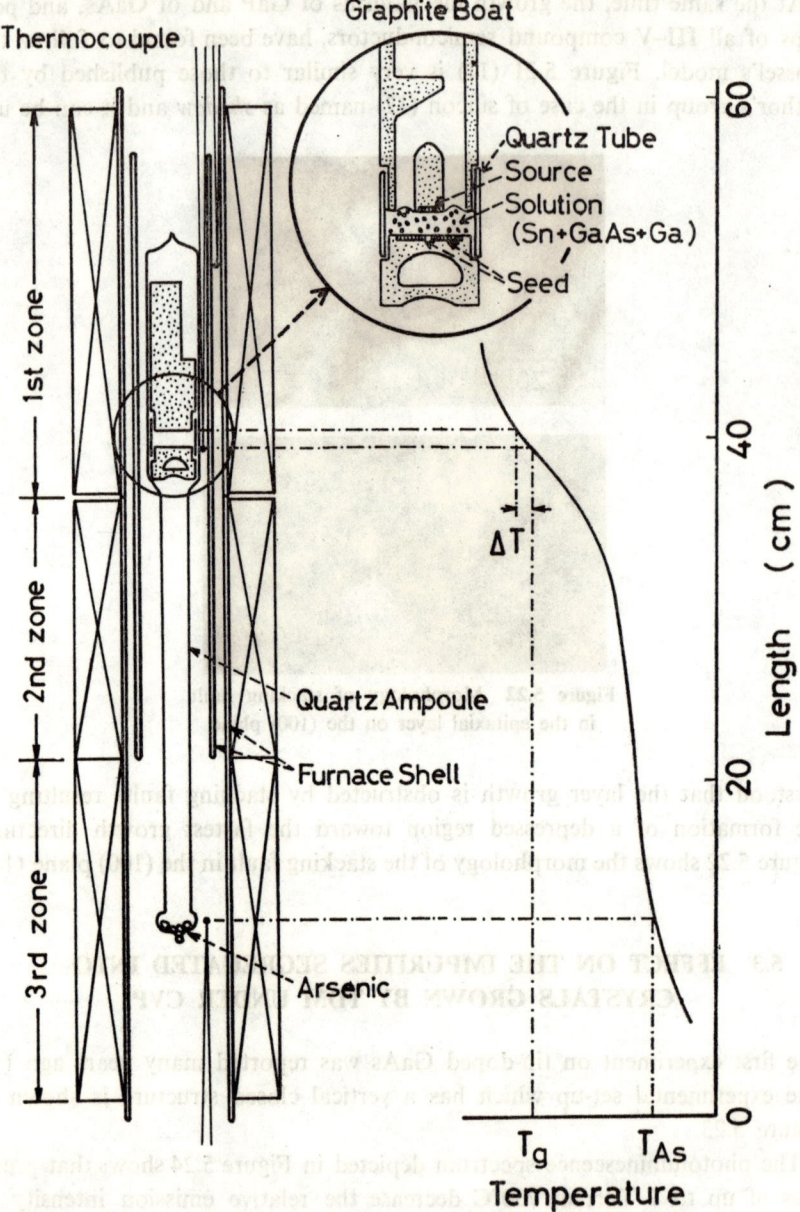

Figure 5.23 Structure of the furnace into which the graphite boat sealed in the quartz ampoule is placed, and the temperature distribution of the furnace. The structure of the boat is shown in the circle

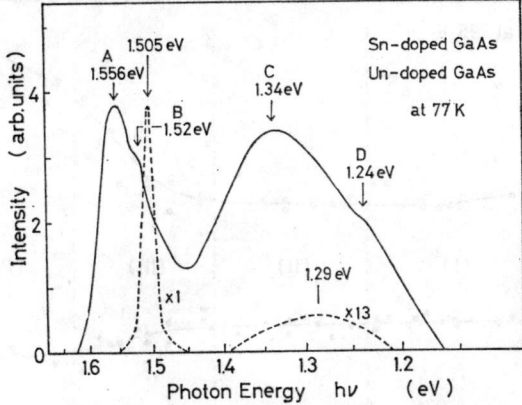

Figure 5.24 Typical photoluminescence spectrum of Sn-doped GaAs which has been grown in 8.4×10^{-6} Torr As pressure shown by the solid line. The dashed line shows the spectrum of undoped GaAs which has been grown from the Ga solution at 730°C in 1.3×10^{-2} Torr As pressure (No. G 1122). The spectra have been normalized to the same maximum value

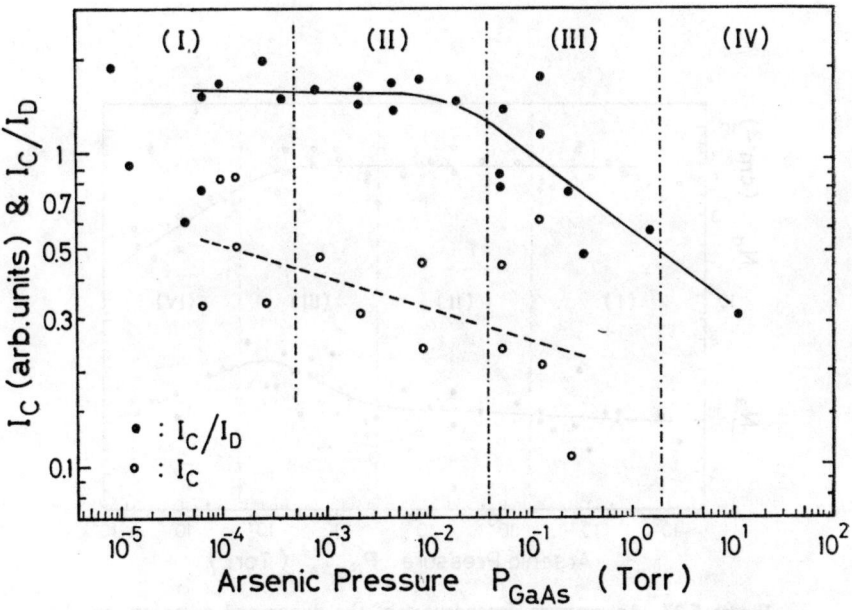

Figure 5.25 As pressure dependence of the intensity ratio of peak C to peak D (see Figure 5.24), and the As pressure dependence of the intensity of peak C

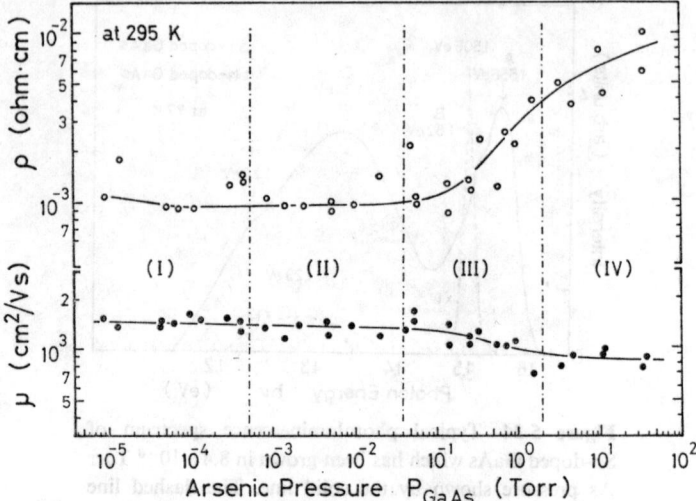

Figure 5.26 As pressure dependences of resistivity and Hall mobility for Sn-doped GaAs at 295 K

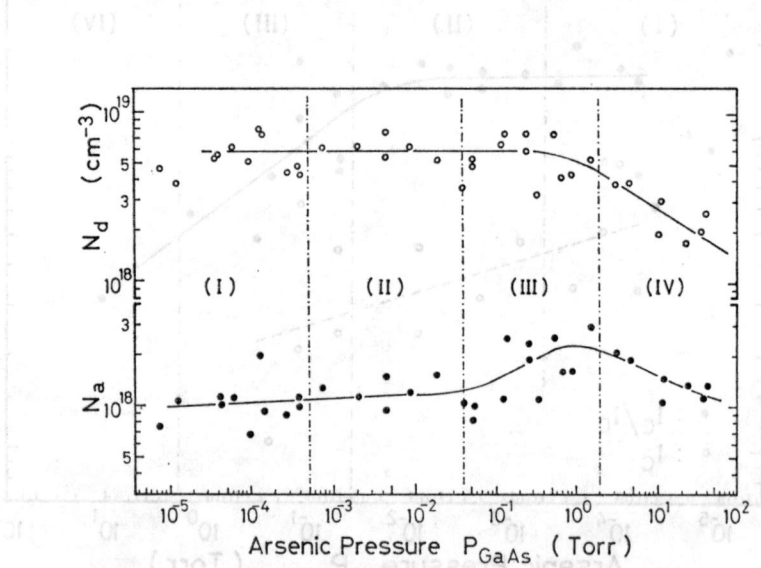

Figure 5.27 As pressure dependences of the donor and acceptor concentrations for Sn-doped GaAs at 295 K. N_d and N_a were calculated from Equation 5.4

the estimated deep donor concentrations, which also decreases with pressure, although at a pressure higher than 1 Torr, as shown in Figure 5.26 based on measured results of mobility and carrier concentrations shown in Figure 5.27.

These results can also be interpreted in terms of the possibility that the concentration of the Sn included in the crystals is controlled through the arsenic vacancy concentrations, which in this study was analysed on Ga vacancy. The concentrations of arsenic vacancies seem to be controlled by the arsenic vapour pressure through the liquid phase.

Another experiment on the properties of silicon atoms included in GaAs has been conducted by measuring the Hall effect and the resistivity of the grown crystals as a function of the As vapour pressure applied (13).

Crystals grown under lower pressure should have higher concentrations of arsenic vacancies and silicon impurity replaced for arsenic lattice point is expected to act as an acceptor and vice versa. Therefore, the silicon doped GaAs crystals should be converted from p-type to n-type with an increase of the vapour pressure under which the melt was settled to segregate the crystal at the other end.

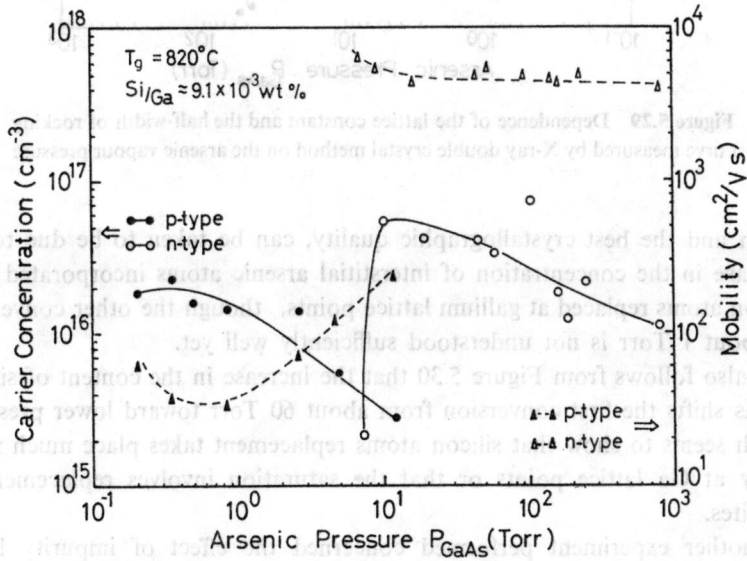

Figure 5.28 Dependency of carrier concentration and Hall mobility of the epitaxially grown Si-doped GaAs on the arsenic vapour pressure. Measurement are performed at room temperature. Si content in Ga melt is 9.1×10^{-3} wt % and growth temperature (T_g) is 820°C

Figure 5.28 gives the results of the measurements, which show conversion at about 60 Torr, which is in very good agreement with the results presented in this chapter. The increase of the half-width of the rocking curves shown in Figure 5.29 over this optimum pressure, which also gives the minimum half-

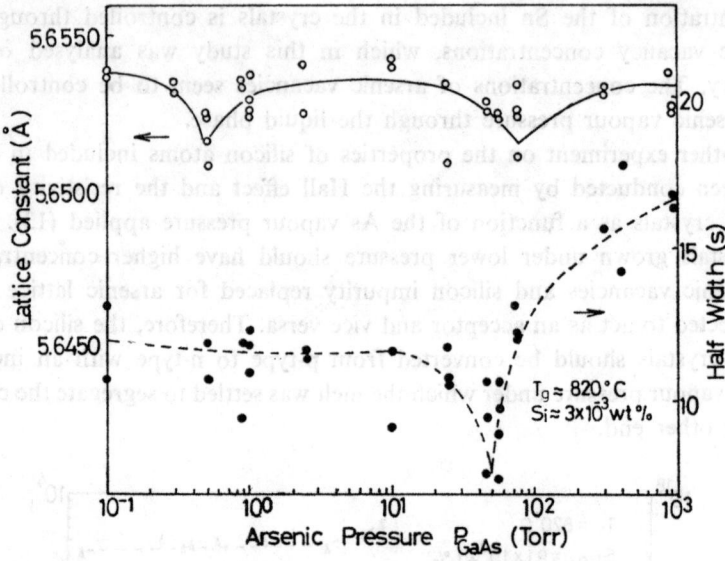

Figure 5.29 Dependence of the lattice constant and the half-width of rocking curve measured by X-ray double crystal method on the arsenic vapour pressure

width and the best crystallographic quality, can be taken to be due to the increase in the concentration of interstitial arsenic atoms incorporated with silicon atoms replaced at gallium lattice points, though the other conversion at about 1 Torr is not understood sufficiently well yet.

It also follows from Figure 5.30 that the increase in the content of silicon atoms shifts the first conversion from about 60 Torr toward lower pressure, which seems to show that silicon atoms replacement takes place much more easily at Ga lattice points or that the saturation involves replacement at As sites.

Another experiment performed concerned the effect of impurity levels in GaP as a function of phosphorus pressure also applied on the top surface of molten Ga to segregate GaP crystals at the other end (14).

Figures 5.31 and 5.32 give the measured results of the photocapacitance effect (15) and of the electroluminescence, respectively, on the crystals as

Stoichiometric Crystallization Method of III–V Compounds

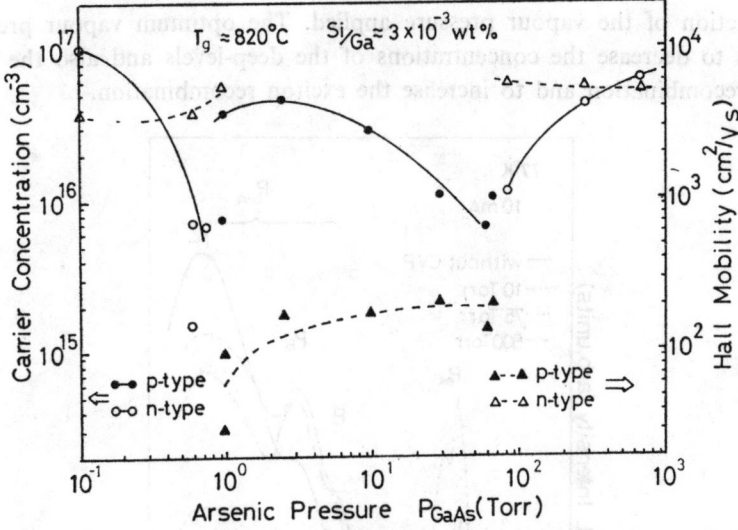

Figure 5.30 Dependence of mobility and carrier concentration on the applied arsenic pressures in the case of $Si/Ga \approx 3 \times 10^{-3}$ wt %

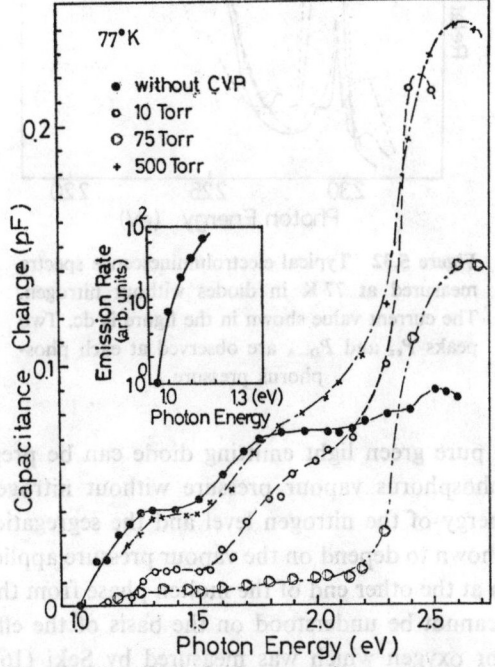

Figure 5.31 Typical photocapacitance spectra at each phosphorus pressure. The peaks in the lower energy region do not exist at $P_{GaP} = 75$ Torr

a function of the vapour pressure applied. The optimum vapour pressure seems to decrease the concentrations of the deep-levels and also the D-A pair recombination and to increase the exciton recombination.

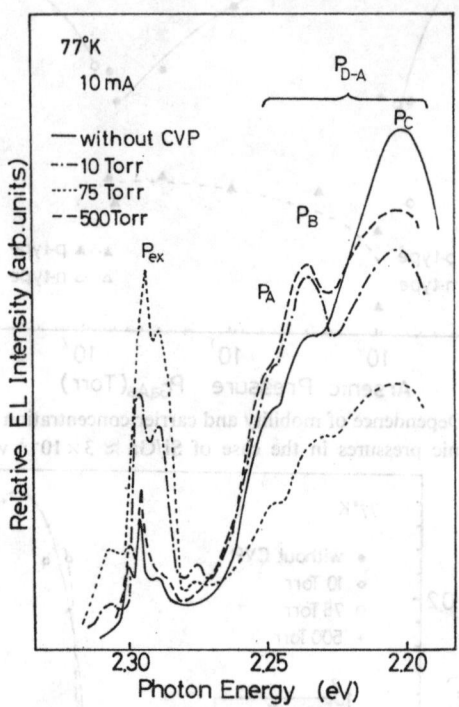

Figure 5.32 Typical electroluminescence spectra measured at 77 K in diodes without nitrogen. The current value shown in the figure is dc. Two peaks P_{ex} and P_{D-A} are observed at each phosphorus pressure

Consequently, pure green light emitting diode can be prepared by applying controlled phosphorus vapour pressure without nitrogen doping. Also the activation energy of the nitrogen level and the segregation constant into GaP have been shown to depend on the vapour pressure applied in the process of crystal growth at the other end of the molten phase from the substrate (14).

These results cannot be understood on the basis of the effect of dissolved silicon, carbon or oxygen which was measured by Seki (16) and shown to be of the order of $10^{13}/cm^3$, which is negligibly small compared with the results of the present experiment.

5.4 MEASUREMENT OF PHASE DIAGRAM

As has already been stated, there are many contradictory experimental results concerning the concepts under discussion, and this made us undertake measurements of the phase diagram in the Ga-As and Ga-P phase.

Some fairly successful experiments have already been carried out, e.g. by R. N. Hall (17), by J. van den Boomgaard and K. Schol (18), and others, but only the results of Boomgaard *et al.* concern the influence of vapour pressure.

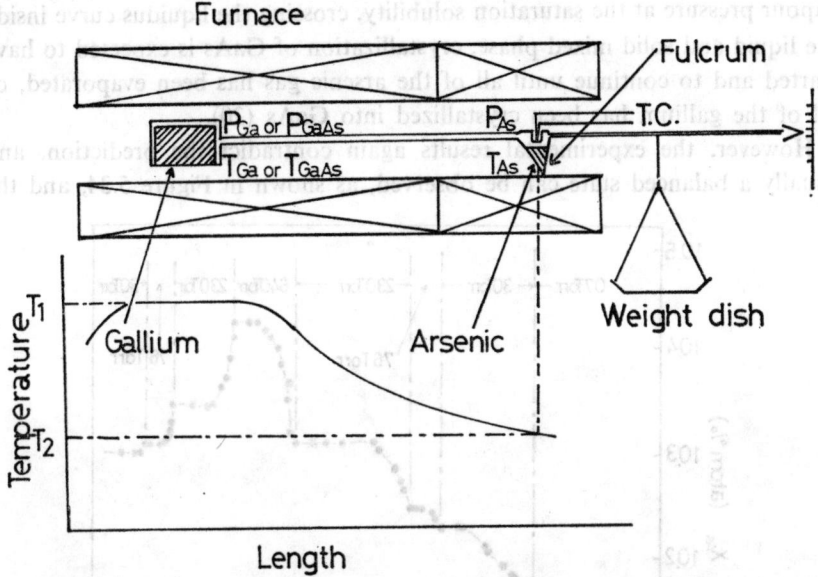

Figure 5.33 Schematic diagram of experimental set-up and temperature distribution of the furnace for determining the phase diagram

The experimental set-up is shown in Figure 5.33; it is based on the principles of balance to measure the weight of gas or the weight of solved materials, and vapour pressure control follows the formula

$$P_1 = \sqrt{T_1/T_{As}} \cdot P_{As}$$

which means that two vessels, one of which contains As at temperature T_{As}, and the other is vacant and set at temperature T_1, connected to each other by means of a fine tube, and the vapour pressure of the vessel with temperature T_1 can easily be derived from the published data of arsenic vapour pressure at temperature T_{As}.

The accuracy of this equipment and of the formula given above are high enough and will be published elsewhere (19).

When the vacant vessel is filled with Ga, the evaporated As starts to dissolve into Ga and the moment of inertia of the Ga vessel is increased, which can be estimated from the increased weight for the balance (20).

In this manner we can ascertain the reversible relationship between the weight of As dissolved into Ga outside the liquidus line. This is just the relation between solubility and the balanced pressure of arsenic (20).

When the arsenic vapour pressure has been raised over the equilibrium vapour pressure at the saturation solubility, crossing the liquidus curve inside the liquid and solid mixed phase, crystallization of GaAs is expected to have started and to continue until all of the arsenic gas has been evaporated, or all of the gallium has been crystallized into GaAs (20).

However, the experimental results again contradict the prediction, and usually a balanced state can be observed, as shown in Figure 5.34, and the

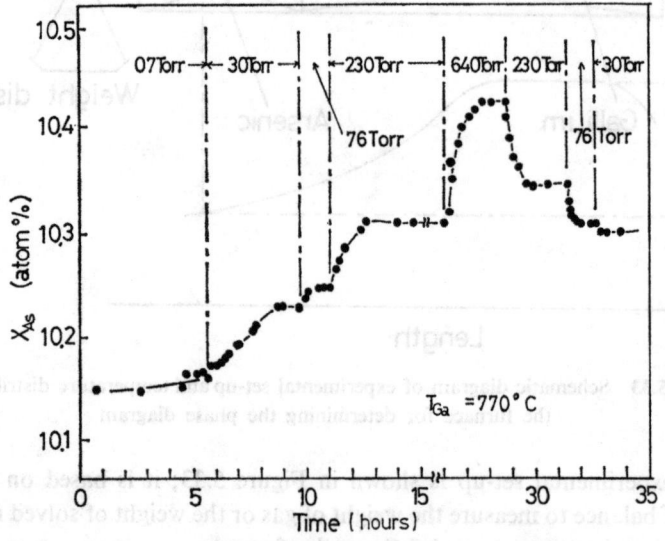

Figure 5.34 Dissolved quantity of As into Ga solution at different As vapour pressure

experiment carried out for over 150 hours did not show any monotonic deviation of the balance; it seems to be equilibrium rather than balance (20).

Also, Figure 5.34 shows the many repeats at 230 Torr and Figure 5.35 at 0.7 Torr after every repeat the weight of As dissolved in Ga; both forms of

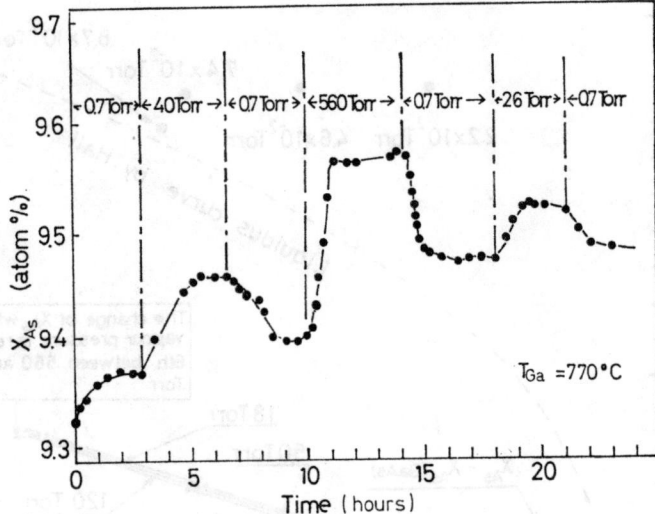

Figure 5.35 Dissolved quantity of As into Ga solution at different As vapour pressures. Cycling of As pressure is performed in the process of $P_{As} \rightarrow P_{As'} \rightarrow P_{As}$. ($P_{As}$ is the parameter from 26 Torr to 560 Torr; $P_{As'}$ is constant, 0.7 Torr)

As and GaAs showed hysteresis. This situation can be accounted for in terms of the three phase equilibrium mentioned earlier in this paper.

Those results are summarized in Figure 5.36. It is also surprising that the isobaric straight lines measured tend to level up at higher As content, these being measured accurately again. With repeats, these lines shift toward the right corresponding to the hysteresis as is shown by the small arrow in the figure (20).

Finally, we tried to check the direct influence of vapour applied at the surface through the liquid (20). The equipment used is shown in Figure 5.37. The vapour pressure of arsenic was controlled by the temperature T_2 of the vessel with the arsenic, where the vapour pressure P_2 of arsenic could be estimated following the published data. The arsenic gas was led to the gallium vessel, which had a manometer structure, and applied on top of the gallium melt. The vapour pressure of arsenic at the top of gallium P_4 could be estimated as

$$P_4 = \sqrt{T_4/T_2} \cdot P_2$$

and the other end of the manometer was connected to a fine quartz tube which led to the vacant vessel with the temperature T_1.

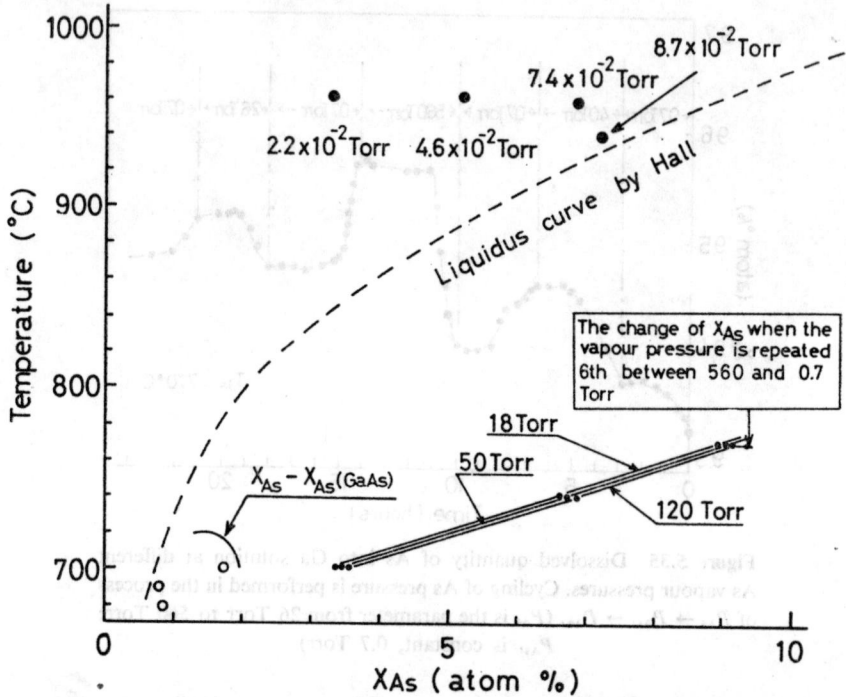

Figure 5.36 Phase diagram including the vapour pressure dependence in the Ga-As system. Isobaric curves are drawn with respective vapour pressures based on very simple calculation and the circles show the position of liquidus curve estimated from experiment by the deduction of weight of solid; however, the accuracy is not high

In this manner we measured the penetrated gas pressure at the other surface of the gallium melt in the manometer, in accordance with

$$P_{4'} = \sqrt{T_4/T_1} \cdot P_1$$

Some of the results obtained are given in Figure 5.38, where it can be seen that the gas pressure easily penetrates into the liquid phase and the diffusion constant of arsenic through the gallium melt is 5.3×10^{-3} cm^2/s at 800°C.

Even though the balanced solubility is not much changed by the vapour pressure, the vapour pressure can easily penetrate the liquid.

In this experiment, the Ga and As content in the equipment was 10 g and 2 g, respectively. Therefore, it was impossible for all gallium to be solidified, and in the experiment it is dissolving and segregating, even though the balance of vapour pressure through the liquid has already been established.

Stoichiometric Crystallization Method of III–V Compounds 129

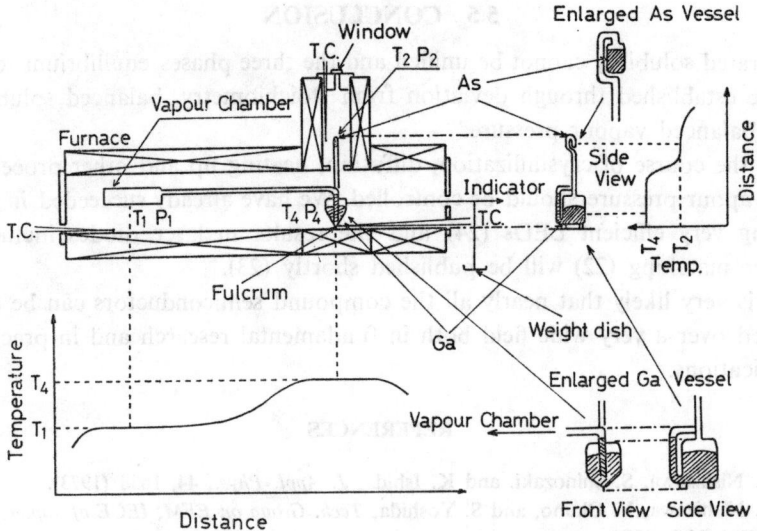

Figure 5.37 Schematic diagram of experimental set-up for the determination of the relation of arsenic vapour pressure in two vessels which were separated by Ga solution. Evaporated As vapour pressure applied from the Ga surface changes the solubility of As into Ga which changes the As vapour pressure of the vapour chamber in equilibrium with the other Ga surface

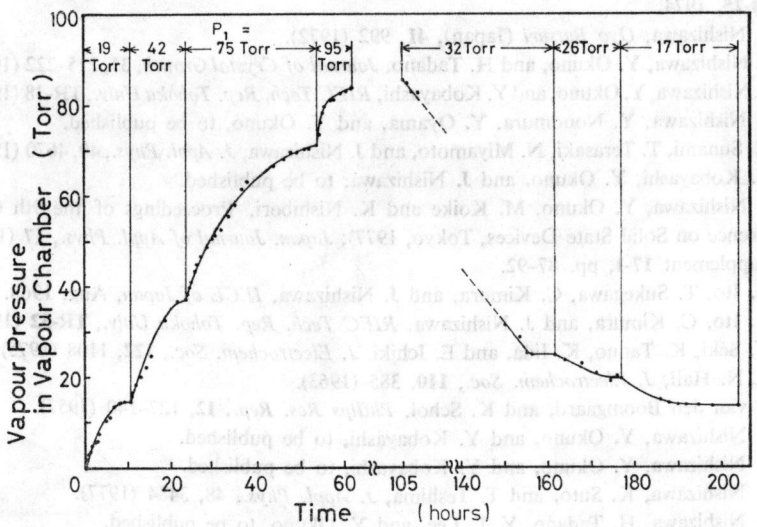

Figure 5.38 An example of results measured by the system of Figure 5.37

5.5 CONCLUSION

Saturated solubility cannot be unified and the three phases equilibrium seems to be established through deviation from stoichiometry, balanced solubility and balanced vapour pressure.

In the course of crystallization, diffusion, heating up and other processes, the vapour pressure should be controlled. We have already succeeded in preparing very efficient LEDs (24) and the results in laser diodes including lattice matching (22) will be published shortly (23).

It is very likely that nearly all the compound semiconductors can be controlled over a very wide field both in fundamental research and in practical applications.

REFERENCES

1. J. Nishizawa, S. Shinozaki, and K. Ishida, *J. Appl. Phys.*, **44**, 1638 (1973).
2. J. Nishizawa, Y. Okuno, and S. Yoshida, *Tech. Group on EFM, IECE of Japan, Rep.* EFM 74-3, 1974.
3. R. H. Saul and D. D. Roccasecca, *J. Appl. Phys.*, **44**, 1983 (1973).
4. J. Nishizawa, Y. Okuno, K. Suto, T. Sato, and S. Yamakoshi, *Solid State Commun.*, **14**, 889 (1974).
5. J. Nishizawa, H. Otsuka, S. Yamakoshi, and K. Ishida, *Japan. J. Appl. Phys.*, **13**, 46 (1974).
6. J. Nishizawa, Y. Okuno, and S. Yoshida, *Rept. Tech. Group on EFM, IECE of Japan*, No. EFM 74-3, 1974; to be published in *IEEE Trans. ED.*
7. J. Nishizawa and A. Koshino, *Rept. Tech. Group on SSD, IECE of Japan*, No. SSD 74-25, 1974.
8. J. Nishizawa, *Oyo Butsuri* (Japan), **41**, 992 (1972).
9. J. Nishizawa, Y. Okuno, and H. Tadano, *Journal of Crystal Growth*, **31**, 215–222 (1975).
10. J. Nishizawa, Y. Okuno, and Y. Kobayashi, *RIEC Tech. Rep. Tohoku Univ.*, TR-38 (1977).
11. J. Nishizawa, Y. Nonomura, Y. Oyama, and Y. Okuno, to be published.
12. H. Sunami, T. Terasaki, N. Miyamoto, and J. Nishizawa, *J. Appl. Phys.*, **40**, 4670 (1969).
13. Y. Kobayashi, Y. Okuno, and J. Nishizawa, to be published.
14. J. Nishizawa, Y. Okuno, M. Koike and K. Nishibori, 'Proceedings of the 9th Conference on Solid State Devices, Tokyo, 1977'; *Japan. Journal of Appl. Phys.*, **17** (1978) Supplement **17-1**, pp. 87–92.
15. A. Ito, T. Sukegawa, C. Kimura, and J. Nishizawa, *IECE of Japan*, Aug. 1968.
   A. Ito, C. Kimura, and J. Nishizawa, *RIEC Tech. Rep. Tohoku Univ.*, TR-32 (1969).
16. Y. Seki, K. Tanno, K. Iida, and E. Ichiki, *J. Electrochem. Soc.*, **122**, 1108 (1975).
17. R. N. Hall, *J. Electrochem. Soc.*, **110**, 385 (1963).
18. J. van den Boomgaard, and K. Schol, *Philips Res. Rep.*, **12**, 127–140 (1957).
19. J. Nishizawa, Y. Okuno, and Y. Kobayashi, to be published.
20. J. Nishizawa, Y. Okuno, and Y. Kobayashi, to be published.
21. J. Nishizawa, K. Suto, and T. Teshima, *J. Appl. Phys.*, **48**, 3484 (1977).
22. J. Nishizawa, H. Tadano, Y. T. Lee, and Y. Okuno, to be published.
23. J. Nishizawa, Y. T. Lee, T. Murakami, T. Yamada, and Y. Okuno, to be published.

Chapter 6

Interface Morphology and Inhomogeneity in Impurity Incorporation during Multilayer LPE Processes

T. Nishinaga

6.1 INTRODUCTION

The fact that LPE gives highly perfect crystals which contain fewer recombination centres compared with vapour growth makes it a very important technology, especially in the manufacture of optical devices. However, growth from solution cannot be free from the problem of segregation. This is due to the fact that the diffusion coefficient of the growing substance is much smaller in the case of LPE. This suggests that the impurity atoms or the growing solute are much easier to concentrate at one place of the growing surface. The segregation of the growing solute results in the geometrical irregularity of the growth front and that of impurity brings various electrical and optical inhomogeneities in LPE layer. Usually the above two phenomena are closely related. The former may cause the impurity inhomogeneity and vice versa.

The flatness of the interfaces between multilayer LPE is extremely important for the following two reasons. One is based on the fact that laser diode demands smooth interface. Otherwise, light beam may be scattered by local geometrical irregularity. The other is due to the fact that those morphological irregularities may bring about recombination centres which lower the light emission intensity.

In this chapter, we will first consider the results of experimental observation of the LPE morphology. These morphologies are thought to exist during the entire process of multilayer epitaxy. Hence, to obtain flat interface it is important to find the condition which gives a smooth LPE surface of a single layer. Secondly, meltback morphology is described. This technology is important not only in the preparation of clean and smooth substrate surface prior to the LPE growth but also from the point of view of determining the formation mechanism of ripples commonly found on the LPE crystal. Following that we will examine the relationship between the optical properties of

LPE layer and the growth morphology. Here experimental results on photoluminescence image are described. This clearly shows that the non-radiative centres introduced during LPE are closely connected with the growth morphology.

Finally, all experimental results are accounted for in terms of theory. The morphological stability theory which is a dynamic description of the constitutional supercooling has been applied to the present LPE case.

6.2 EXPERIMENTAL OBSERVATION OF LPE MORPHOLOGY

So far numerous types of morphology have been observed on LPE crystal surfaces, the most common of them being surface ripples. These ripples are roughly classified into two types. One has been studied by Small et al. (1) and is characterized by a series of thin lines running nearly perpendicular to the direction of liquid removal. The other type of ripples is distinguished from the previous meniscus lines by a terrace and/or wave morphology and has no relation to the liquid removal. The former type of ripple is formed by a 'stick-slip' motion of the edge of the liquid (1) and can be eliminated by, for instance, applying a pressure on the liquid. However, as for the second type of ripples, there is almost no common understanding concerning the mechanism of their formation and the method of their elimination. In this section we will discuss the experimental morphology usually observed in InP LPE. Nevertheless, exactly the same kind of morphology has been found in LPE of other III–V (2–5) or II–VI (6) compounds. Hence, the following results are commonly established for all crystals grown by LPE techniques.

Figure 6.1 shows a typical LPE morphology. Figure 6.1(a) represents beautiful wave morphology and Figure 6.1(b) shows rather terrace-like waves. The latter are quite frequently observed in the case of GaAs (3) and GaP (5) as well as InP. When the growth thickness becomes large it is often observed that the terrace is surrounded by a groove filled with metallic solution. A typical example is shown in Figure 6.2. This morphology is supposed to have a similar origin to that of cellular structure formed in the case of melt growth (7). All the above morphology is observed when the orientation of the substrate surface is close to the low index plane such as (111) or (100). On the other hand, on a highly misorientated substrate, one observes protuberance morphology as shown in Figure 6.3.

Among them, the most commonly observed morphology is the type shown in Figure 6.1(b). This is schematically illustrated in Figure 6.4. One finds that

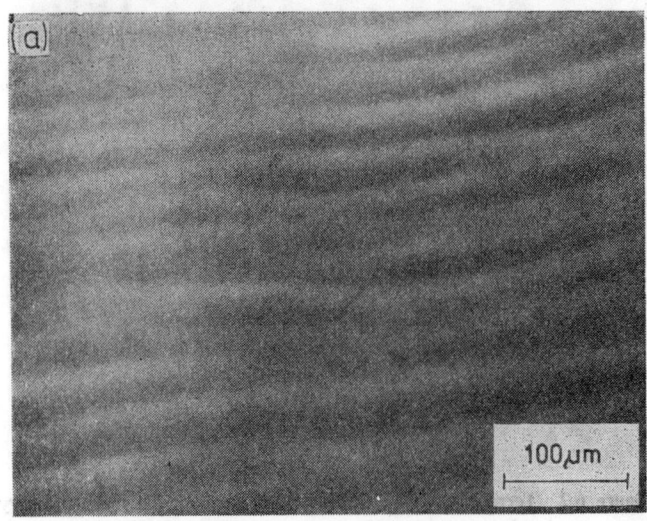

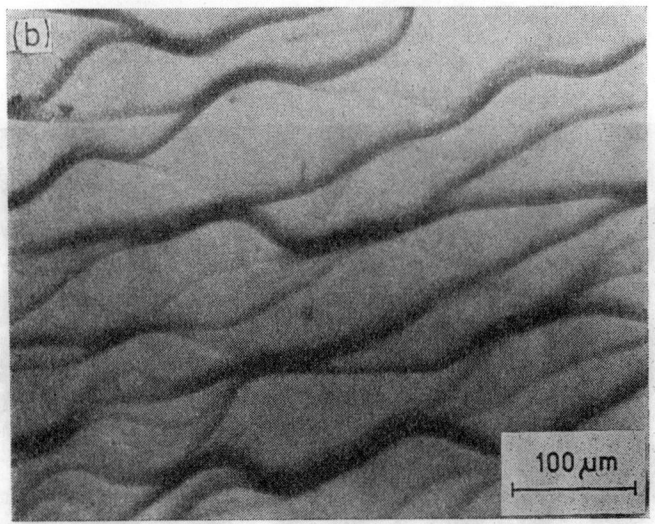

Figure 6.1 Two types of generally observed wave morphologies. (a) Wave with sinusoidal shape. (b) Terrace-like wave

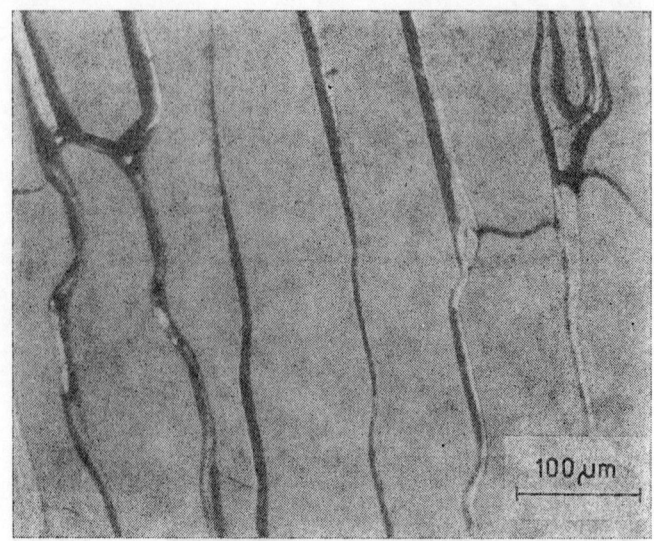

Figure 6.2 Terrace with an indium-filled groove. This morphology is observed when the growth thickness is large

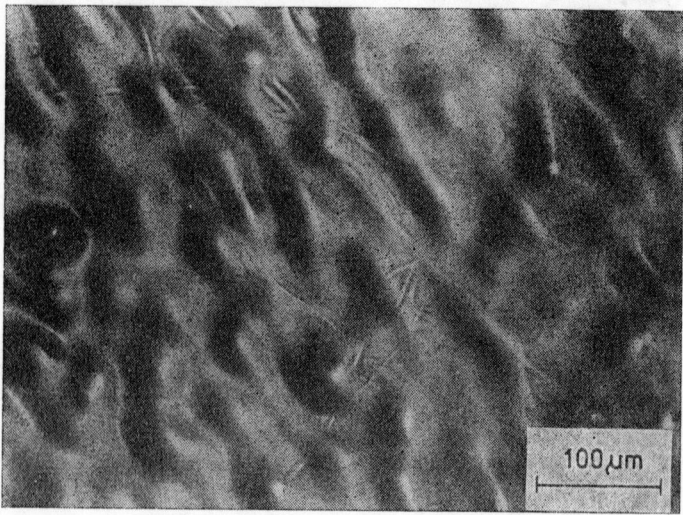

Figure 6.3 An example of protuberance morphology found on LPE layer grown with highly misoriented substrate

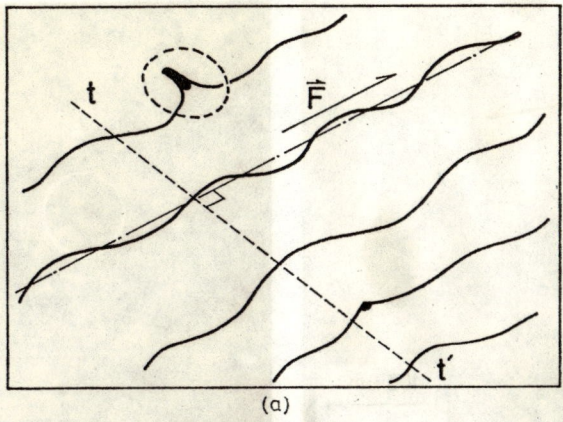

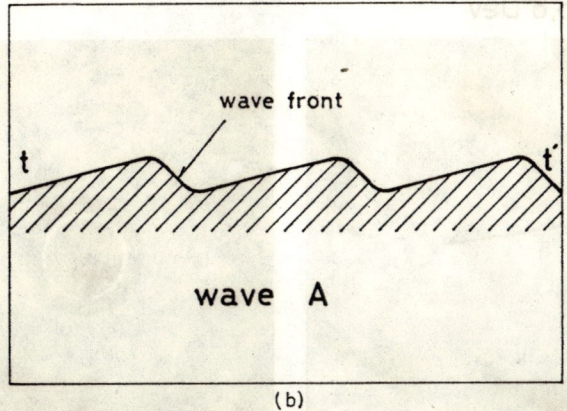

Figure 6.4 Schematic drawing of the LPE morphology. (a) A grown surface. $\vec{F}$ shows the average direction of the wave front. The wave B is defined as the wave of the wave front. Dotted circle illustrates the region where indium residue is found. (b) The cross section along the broken line in (a). This wave is named A

there are two different waves. Figure 6.4(b) shows a cross-sectional view of the morphology along the line (shown by a broken line in Figure 6.4(a)) perpendicular to the average wave front. We shall name this type of wave A. On the other hand, the wave front of A has a sinusoidal form as shown in Figure 6.4(a). We shall call this wave B. Sometimes this wave is found to be pinned by residue of the solvent metal which is schematically drawn in the same figure by a dotted circle. It has the appearance of a two-dimensional

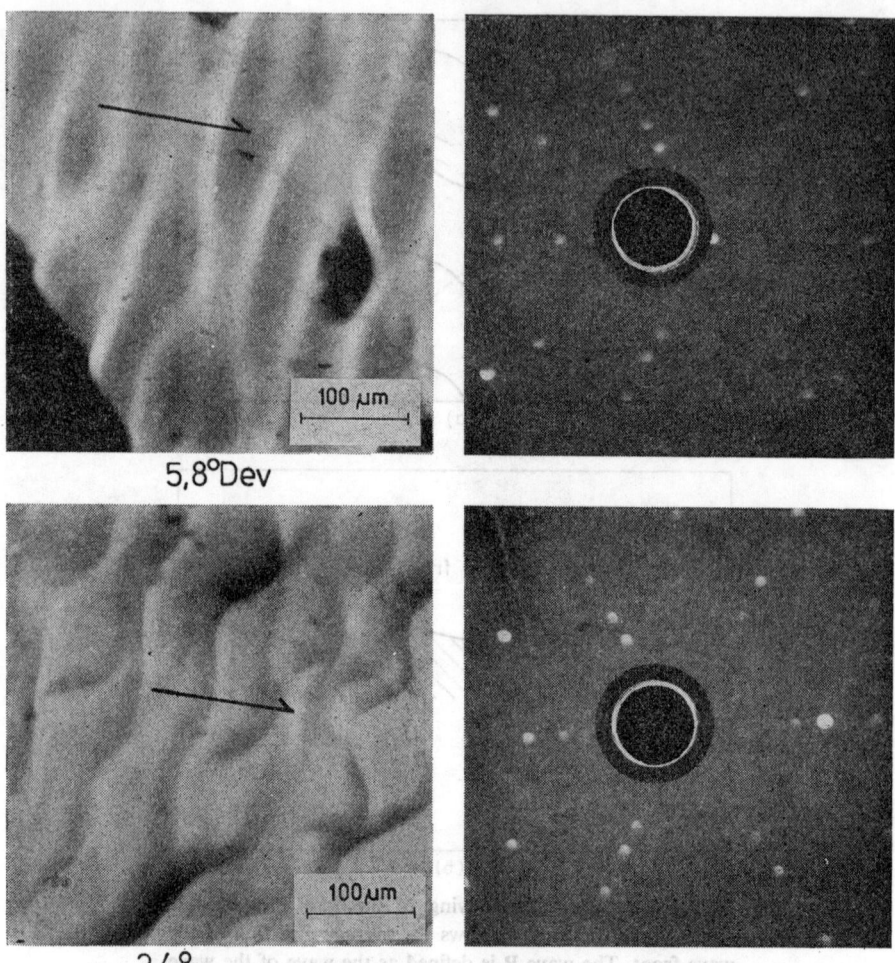

Figure 6.5 Relationship between wave morphology and X-ray back Laue photograph which was taken in the same position as the micrograph

cellular structure which has the same origin as that of Figure 6.2. We will not proceed with the discussion further since it is done elsewhere. In the following, the effect of growth parameters on the morphology of wave A will be described taking InP as an example.

6.2.1 Misorientation from (111) plane

Experimental results are as follows. Substrates with 3–6° off angle from (111)B were employed. Figure 6.5 shows the relationship between the X-ray back Laue photograph and the wave morphology. The figure showst hat the wave front is perpendicular to the gradient of the misorientated plane (shown by arrows in Figure 6.6). The direction of the gradient is defined in Figure 6.6.

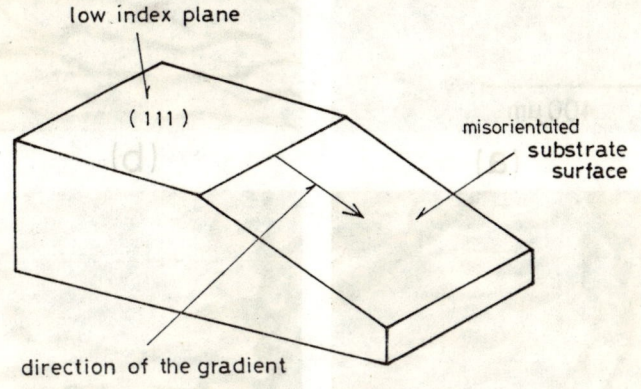

Figure 6.6 The definition of the vector showing the gradient of the misorientated plane

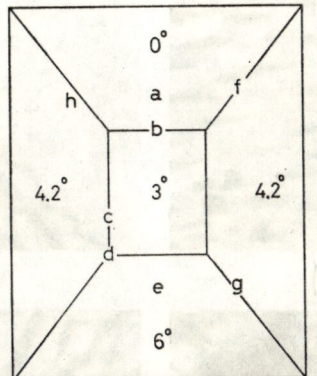

Figure 6.7 The arrangement of the misorientated surfaces on a single substrate

Figure 6.7 shows a single substrate with various misorientated surfaces. Side *a* has an exact (111) orientation (< 0.5°) and other surfaces have the misorientations indicated in the figure. The growth was made with an undercooling of 8°C for one hour. The resulting morphology is given in Figure 6.8. A striking phenomenon is the disappearance of the wave

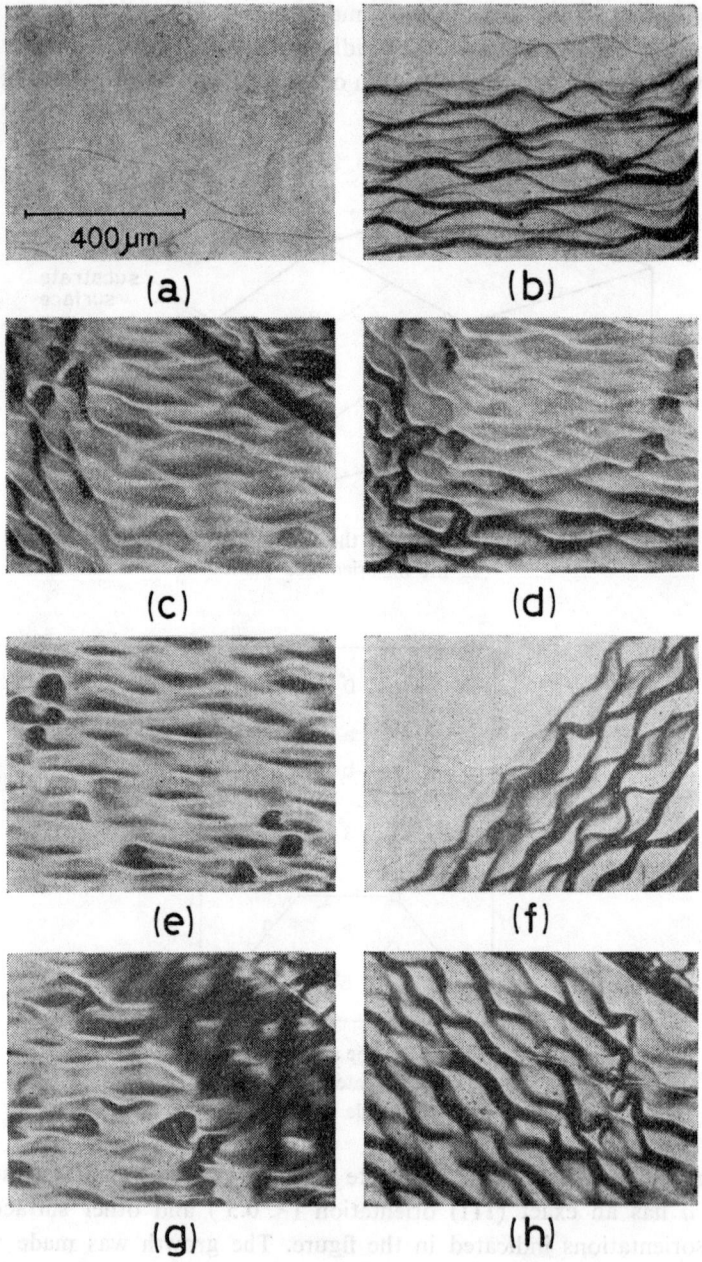

Figure 6.8 Micrographs of grown surfaces on the corresponding points in Figure 6.7

morphology on the exactly oriented surface. Of course, the growth was made there but still there were no waves. This phenomenon has also been observed in the case of GaP (5, 8) and GaAs (3). Nevertheless, the present results are important in that a single substrate with various misorientations was employed, which guaranteed the same growth conditions on all surfaces.

6.2.2 Supersaturation effect

The wavelength of wave A was found to vary with the degree of supersaturation. Table 6.1 shows the experimental results. It is clear that the wavelength becomes shorter with higher supersaturation. As was already seen in Figures 6.1(b) and 6.8, the shape of the wave is complicated because of the mixture of waves A and B. Consequently, the experimental wavelengths in Table 6.1

Table 6.1 The experimental (and theoretical) wavelengths for various supercoolings

Supercooling (°C)	Growth time (h)	λ_{exp} (μm)	λ_0 (μm)	λ_{max} (μm)
4	2.0	61	74	130
5	1.5	42	61	105
8	0.5	30	36	63
10	1.5	35	42	74
10	1.0	35	38	66

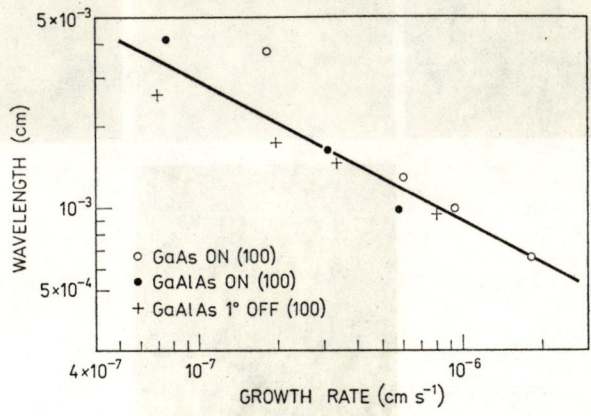

Figure 6.9 Wavelength as a function of growth rate for three cases (9)

140 T. Nishinaga

were obtained by averaging the numbers of waves in the arbitrary parts of the photograph. Nevertheless, a clear tendency is found in the table for the wavelength to decrease with higher supersaturation.

Small and Potemski (9) observed a similar effect. They studied the effect of the growth rate on the wavelength. Their results are given in Figure 6.9. The wavelength decreases as the growth rate increases. Since higher supersaturation results in the higher growth rate, their observation and ours concern the same phenomenon.

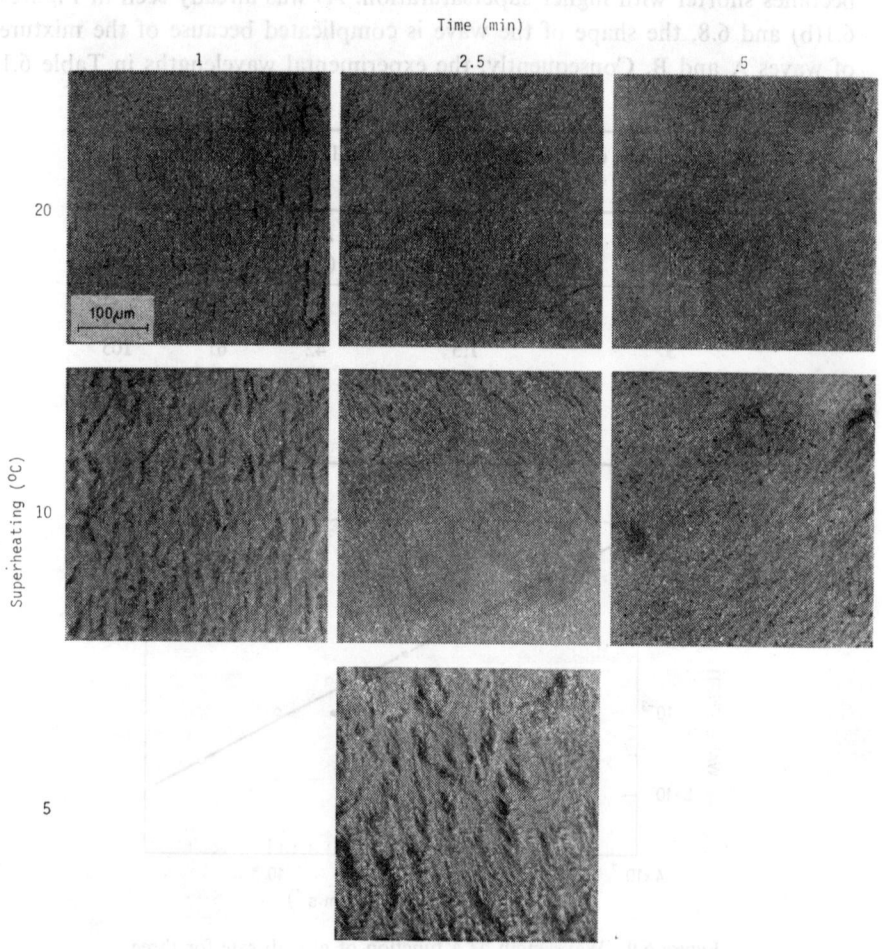

Figure 6.10 A matrix of photographs showing meltback morphologies

6.3 MELTBACK MORPHOLOGY

In this section, experimental meltback morphology is described taking InP as an example. InP samples were cut from (111)-oriented wafers. They were mechanically polished and chemically etched with a bromine-methanol solution. The (111)B surface was used for the meltback. For all samples, the meltback was carried out at 650°C. First, the melt was saturated with InP sources crystal for 40 min and then separated. Following that, the temperature was

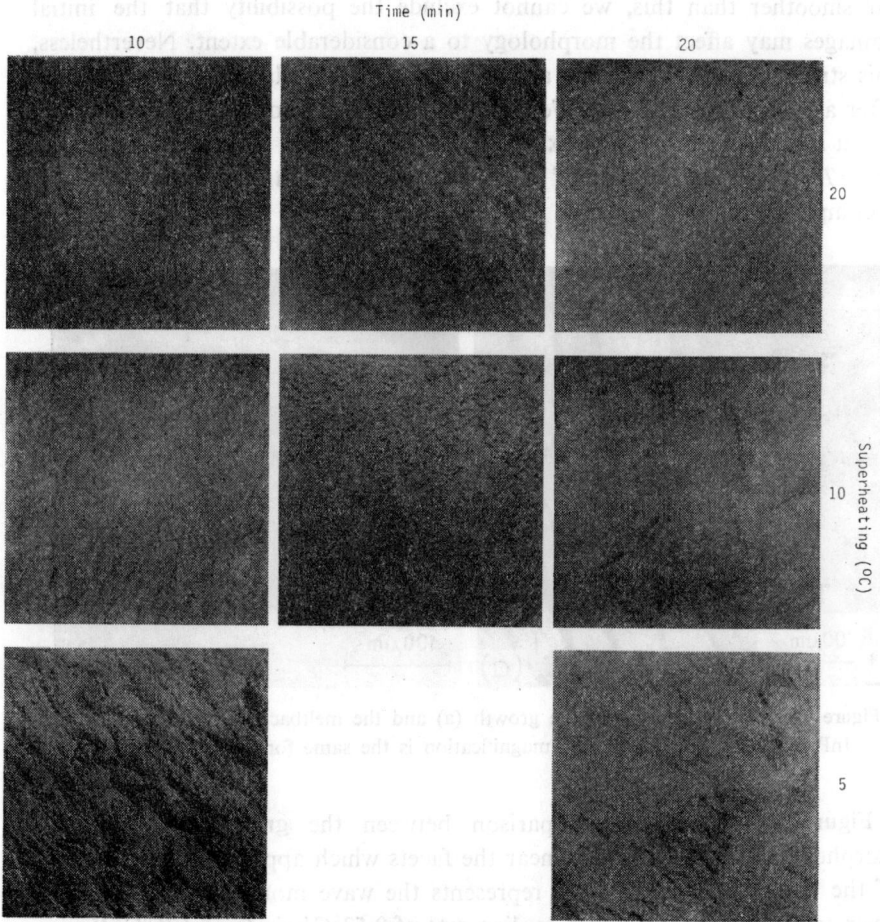

of InP for various superheatings and for different times

raised by ΔT so as to perform the meltback at 650°C, and then the melt was moved over the sample.

The resulting morphologies of the meltback are shown in Figure 6.10 for various superheatings and different times. In the figure, it is obvious that we can get smoother surfaces by increasing either the superheating, the time or both. For instance, in the case of superheating of 20°C, some structures are found on the surface at the beginning of the meltback. This structure seems to be related to the initial conditions of the surface, especially to the damages produced during the saturation period. Although the surface which was heat-treated at 650°C for one hour in the same atmosphere was far smoother than this, we cannot exclude the possibility that the initial damages may affect the morphology to a considerable extent. Nevertheless, this structure soon disappears and a macroscopically flat surface is obtained after a meltback of 2.5 min. For less superheating, a longer time is required to get a smooth surface. For example, at least a 10 min meltback is necessary for $\Delta T = 10°C$. In the case of $\Delta T = 5°C$, even 20 min is not enough to get a completely smooth surface.

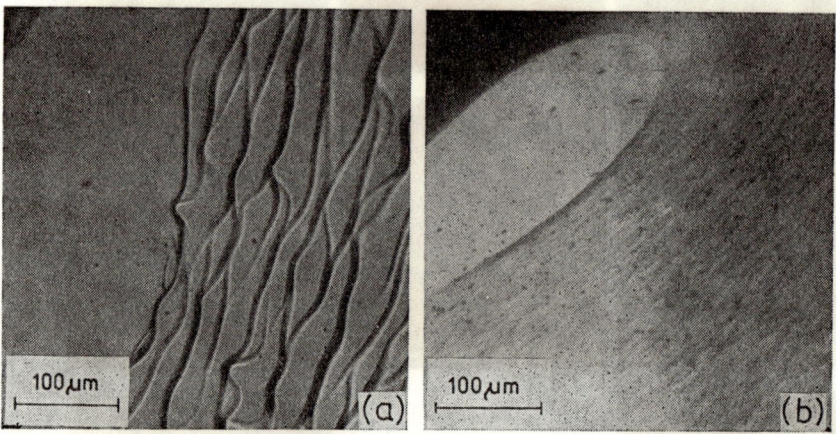

Figure 6.11 A comparison of the growth (a) and the meltback (b) morphologies of InP observed near facets. The magnification is the same for both photographs

Figure 6.11 shows a comparison between the growth and meltback morphologies. Both are taken near the facets which appeared in the corners of the samples. Figure 6.11(a) represents the wave morphology of an LPE layer grown for 15 min with a cooling rate of 0.53°C/min from 676°C. Figure 6.11(b) shows a photograph of the meltback surface which appeared in the case of 20°C superheating and 10 min meltback time. Since the magnification in the photographs is the same, it is clear that meltback and growth give quite different morphologies. The former gives a macroscopically very flat surface as compared with the latter.

6.4 INHOMOGENEITY OF IMPURITY INCORPORATION OBSERVED BY PHOTOLUMINESCENCE IMAGE

Kajimura, Aiki and Umeda (10) were the first to show clearly how non-radiative dark region is formed along the surface wave in GaP LPE. They used both photoluminescence and electroluminescence images to observe the light emission intensity distribution in LPE layer. We followed them exactly* and obtained the same results. Figure 6.12 shows the experimental

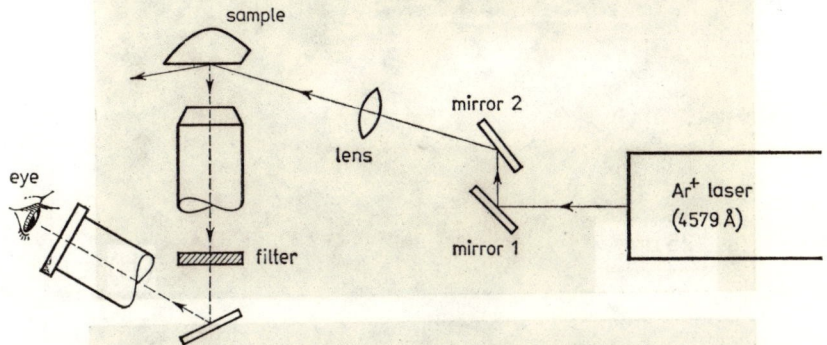

Figure 6.12 A system to observe photoluminescence image (11)

arrangements (11). Ar^+ laser beam (4579 Å) is focused on the cleaved cross section of GaP LPE layer. The luminescence is observed under an optical microscope through a yellow glass filter so as to cut out light other than the luminescence. Figure 6.13 shows the experimental results. Figure 6.13(a) is an optical micrograph of the cleaved surface and Figure 6.13(b) shows a photoluminescence image taken at the same position as (a); (c) is a magnified photograph of (b). There are two kinds of non-radiative lines (10); one has the dislocation as its centre but the other starts from the bottom of the wave. The latter is more clearly seen in Figure 6.14. Figure 6.14(a) shows the surface

---

* The author would like to acknowledge their kind assistance in repeating their experiment in our laboratory.

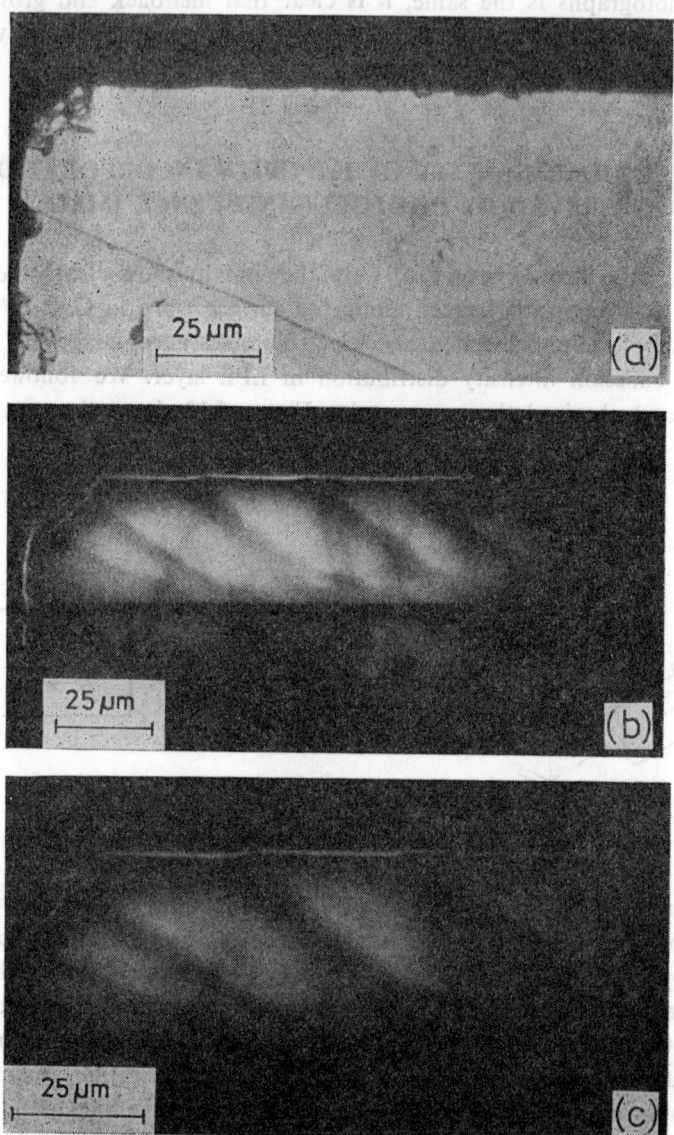

Figure 6.13 (a) Optical micrograph of cleaved cross section of LPE grown GaP layer. (b) Photoluminescence image of (a). (c) Magnified photograph of (b)

morphology with the waves and (b) shows the corresponding cross-sectional photoluminescence image. One-to-one correspondence between the wave bottom and the non-radiative line is observed. So, by some unknown mechanism the wave bottom creates recombination centres.

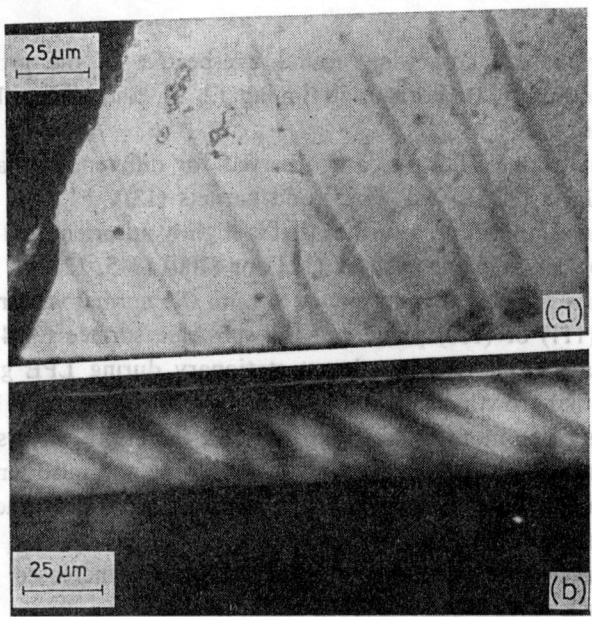

Figure 6.14 Positional relationship between surface morphology (a) and photoluminescence image (b)

The non-radiative line associated with the wave bottom indicates the trace of the position of the bottom during the LPE growth. This shows that the wave moves slightly in horizontal direction during the growth. If one imagines the waves to be the bunched steps and the growth to be performed by their horizontal movement, one must expect that many waves should pass at one place on the substrate. For instance, at least the pass of 40 waves is required to cause the growth of 40 μm since the usual wave has a height of less than 1 μm. However, as is seen in Figure 6.13, only a few waves have passed to cause the growth of approximately 40 μm. This means that the wave is not the usual bunched step whose horizontal movement causes the growth. Further examination of Figures 6.13(a) and (b) shows that the traces of wave bottoms sometimes join to form one trace. This means that the waves coalesce to form a wave with larger wavelength.

6.5 THEORETICAL CONSIDERATIONS

6.5.1 Growth morphology

The experimentally characterized features of the waves are summarized as follows:

  (i) The waves are commonly found irrespective of the growth system (e.g. vertical (12), horizontal tipping (2, 3), and sliding boat (4, 12) systems).
 (ii) The same kind of waves are observed for different crystals (e.g. InP (12), GaAs (2, 3), GaP (4, 5) and garnets (13)).
(iii) The waves appear on a surface that is slightly misorientated with respect to a low index plane such as (111) or (100) (3–5, 12).
 (iv) The wave front is at right angles to both the normal vectors of the low index (111) or (100) plane and the substrate surface (3, 4, 12).
  (v) The wave front remains almost stationary during LPE growth. This means that growth occurs not by advance of the wave fronts but, instead, it takes place more or less uniformly on the entire wave surface (10).
 (vi) The wavelength does not depend strongly on the misorientation (12).
(vii) By increasing the misorientation, the waves are transformed gradually to a cellular structure (5, 12).
(viii) The wavelength decreases as supersaturation is increased (9, 12).
 (ix) Meltback reveals quite smooth surfaces (14).

As for the origin of the waves, various models have been proposed which may be classified into the following five categories:

(1) constitutional supercooling (2, 13, 15, 16),
(2) crystallographic causes unrelated to constitutional supercooling (3, 4),
(3) equilibrium form (17),
(4) diffuse interface (18),
(5) cellular convection (19).

In this chapter the morphological stability theory is applied to explain the above experimental observations (i) to (ix).

6.5.1.1 Basic ideas

It is clear from (iii) and (iv) that the native steps supplied from the misorientated substrate play an essential role on the formation of the waves. The dimensions of the observed wave are several tens of micrometers in wavelength and 0.1–1µm high. This means that a single wave is composed of thousands of monoatomic steps. This fact, together with the discussion given in Section 6.4, dispels the impression that the waves are bunched steps which move in a lateral direction to cause growth. The usual crystals used for LPE substrates contain dislocations of the order of $10^5/\text{cm}^2$ and these also supply

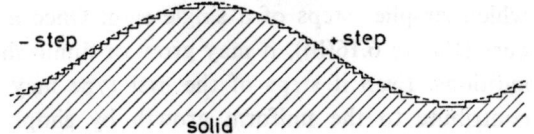

Figure 6.15 Steps supplied from the first type of step source

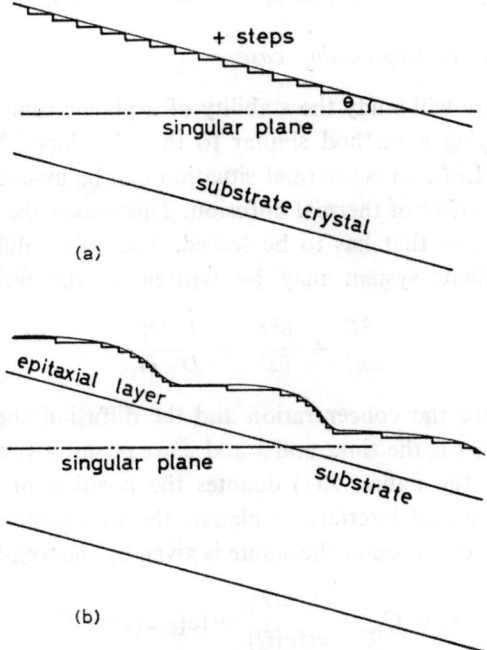

Figure 6.16 Mono-sign steps supplied from the second type of step source (a). Bunching of these steps, which acts as morphological perturbation of the solid–liquid interface (b)

steps on the growing interface. Under conditions of high supersaturation, two-dimensional nucleation supplies steps as well. Hence, all in all, we must consider three sources of steps. The steps supplied from dislocations and two-dimensional nuclei are schematically illustrated in Figure 6.15. We shall call these steps *sources of the first type* since they supply equal number of + and − steps. Two steps with opposite sign are annihilated when they meet. On a low index singular plane, only this type of sources exists and leads in growth to a macroscopically flat surface. Figure 6.16 shows the second type of source which supplies steps of a single sign. Once a morphological perturbation occurs (Figure 6.16(b)), it may grow or diminish under various experimental conditions. Even if steps of the first type may exist in superposition, they contribute to the uniform growth by simply covering and reproducing the original wave shape. Thus the wave of Figure 6.16(b) covered with the uniformly distributed first type of steps acts as an initial morphological perturbation. In the following, we will consider the stability of such surfaces by employing the morphological stability theory (20, 21).

6.5.1.2 Application to step-cooling case

In this section, we will study the stability of a planar solid-liquid interface in LPE by employing a method similar to that developed by Mullins and Sekerka (22). For LPE, an isothermal situation can be assumed (23). Hence, we can neglect the effect of thermal diffusion. This makes the solute diffusion equation the only one that has to be solved. The solute diffusion equation in the fixed coordinate system may be written in the well-known form,

$$\frac{\partial^2 C}{\partial x^2} + \frac{\partial^2 C}{\partial z^2} = \frac{1}{D}\frac{\partial C}{\partial t} \tag{6.1}$$

where C and D are the concentration and the diffusion coefficients of the solute, respectively, t is the time, and x and z are position variables as shown in Figure 6.17. In the figure, $Z(t)$ denotes the position of the solid-liquid interface. If the growing interface is planar, the first term in Equation 6.1 is zero and the concentration of the solute is given by the well-known formula,

$$C = C_\infty - \frac{C_\infty - C_0}{\mathrm{erfc}[Q]}\mathrm{erfc}[z/2(Dt)^{1/2}] \tag{6.2}$$

where Q is the growth constant determined through

$$\frac{C_\infty - C_0}{C_s - C_0} = \sqrt{\pi}Q\exp(Q^2)\mathrm{erfc}[Q] \tag{6.3}$$

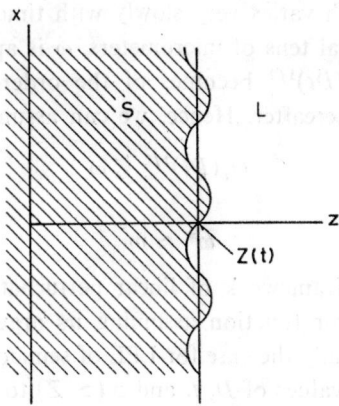

Figure 6.17 x and z axes in the fixed coordinate system. $Z(t)$ denotes the average position of solid-liquid interface

and C_s, C_0 and C_∞ are respectively the concentrations of the solute in the solid, at the growth front and at an infinite point in the liquid. The position of the interface, $Z(t)$, is simply given by

$$Z(t) = 2Q(Dt)^{1/2} \tag{6.4}$$

Next we introduce a sinusoidal morphological perturbation on the growing interface. The position of the interface is now given by

$$z_0 = Z + \delta \sin \omega x \tag{6.5}$$

where δ is the amplitude of the perturbation and is a function of time, ω is the angular frequency defined by $\omega = 2\pi/\lambda$, and λ is the wavelength. As usual, local equilibrium is assumed at the liquid-solid interface. The equilibrium solute concentration C at the perturbed boundary is written as

$$C_{0s}(z_0) = C_0(1+\Gamma_D K) = C_0(1+\Gamma_D \omega^2 \delta \sin \omega x) \tag{6.6}$$

where C_0 is the equilibrium concentration for planar interface, Γ_D is the capillary constant, and K is the curvature. With this perturbation, the solute concentration in the liquid may be written in the following form

$$C = C_\infty - \frac{C_\infty - C_0}{\text{erfc}[Q]} \text{erfc}[z/2(Dt)^{1/2}] + B \exp[-\omega^*(z-Z)]\sin \omega x \tag{6.7}$$

where ω^* is given by

$$\omega^* = \tfrac{1}{2}\{Q/(Dt)^{1/2} + [(Q^2/Dt) + 4\omega^2]^{1/2}\} \tag{6.8}$$

and B is a factor which varies very slowly with time. Since the wavelength is of the order of several tens of micrometers, ω is approximately 10^3 cm^{-1}. On the other hand, $Q/(Dt)^{1/2}$ becomes of the order of 1 cm^{-1} in several seconds and smaller thereafter. Hence, we can assume

$$Q/(Dt)^{1/2} \ll \omega \tag{6.9}$$

This leads to

$$\omega^* \approx \omega \tag{6.10}$$

To find B within the framework of linear perturbation theory, we expand the complementary error function assuming its argument to be much less than unity. This is actually the case for LPE of semiconductors. For instance, if one assumes typical values of D, t, and z ($\simeq Z$) to be 5×10^{-5} cm^2/s, 10 s and 10 μm, respectively, the argument becomes ~ 0.02 and is smaller as the growth time increases. Hence, the concentration may be approximated by

$$C = C_\infty - (C_\infty - C_0)\frac{1-(2Q/\sqrt{\pi}\bar{Z})z}{1-(2Q/\sqrt{\pi})} + B\exp[-\omega(z-Z)]\sin\omega x \tag{6.11}$$

in the region just near the interface where the waves are being formed. Then B is obtained by substituting (6.5) into (6.11) and by comparing it with (6.6) to obtain

$$B = \delta\left\{C_0\Gamma_D\omega^2 + (C_0 - C_\infty)\frac{2}{\sqrt{\pi}} \cdot \frac{Q}{1-(2Q/\sqrt{\pi})} \cdot \frac{1}{\bar{Z}}\right\} \tag{6.12}$$

Since δ and Z are functions of time, B actually depends on time, but the dependence is very weak (24). The mass balance equation at the interface is

$$D\left(\frac{\partial C}{\partial z}\right)_{z=z_0} = (C_s - C_{os})(\dot{Z} + \dot{\delta}\sin\omega x) \tag{6.13}$$

where C_s is the concentration of the solute in the solid. Equations 6.6, 6.11, 6.12 and 6.13 give

$$f(\omega) = \frac{\dot{\delta}}{\delta} = \frac{D\omega}{C_s - C_0}\left\{-C_0\Gamma_D\omega^2 \right.$$

$$\left. + (C_\infty - C_0)\frac{2}{\sqrt{\pi}} \cdot \frac{Q}{1-(2Q/\sqrt{\pi})} \cdot \frac{1}{\bar{Z}}\right\} \tag{6.14}$$

where we have used the inequality

$$\dot{Z}/D \ll \omega$$

which may be directly derived from (6.9). $f(\omega)$ is a third order equation which has a maximum at a certain positive ω_{max}. However, unlike the steady state case, $f(\omega)$ changes slowly with time. $f(\omega)$ is shown schematically in Figure 6.18. It is seen that $f(\omega)$ is positive in the region of $0 < \omega < \omega_0$ and negative for $\omega > \omega_0$. This means that the perturbations of the form described in 6.5.1.1

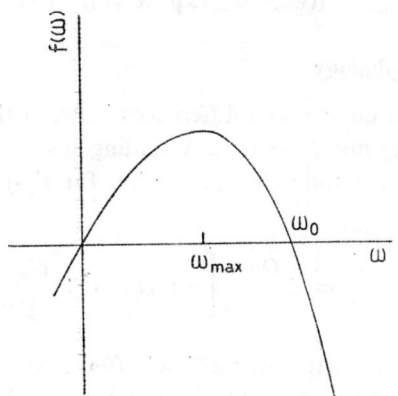

Figure 6.18 Schematic drawing of $f(\omega)$ versus ω

and with $\omega > \omega_0$ tend to disappear. In other words, coalesced waves with ω less than ω_0 will remain and grow during LPE growth. Since $f(\omega)$ is a function of t, the amplitude of such a wave is given by

$$\delta = \delta_0 \exp\left(\int_0^t f(\omega) dt'\right) \qquad (6.15)$$

where δ_0 is the initial amplitude of the perturbation. ω_{max}, which is the angular frequency of the fastest growing wave for any fixed time, is determined through

$$\frac{\partial f(\omega)}{\partial \omega} = 0 \qquad (6.16)$$

The resulting expressions for ω_0 and ω_{max} are

$$\omega_0 = [2(C_\infty - C_0)Q/\sqrt{\pi}C_0\Gamma_D(1-(2Q/\sqrt{\pi}))Z]^{1/2} \qquad (6.17)$$

and

$$\omega_{max} = \omega_0/\sqrt{3} \qquad (6.18)$$

The corresponding wavelength is

$$\lambda = 2\pi/\omega \qquad (6.19)$$

For usual semiconductor LPE, Q is assumed to be much less than unity which leads to

$$\omega_0 = [(C_\infty - C_0)/\sqrt{\pi} C_0 \Gamma_D \sqrt{Dt}]^{1/2} \tag{6.20}$$

and

$$\omega_{max} = [(C_\infty - C_0)/3\sqrt{\pi} C_0 \Gamma_D \sqrt{Dt}]^{1/2} \tag{6.21}$$

6.5.2 Meltback morphology

In principle, there are no essential differences between the stability problems of melting and growing interfaces (25). Assuming again that Q is much smaller than unity, one gets the following expression for $f(\omega)$ from Equations 6.4 and 6.14

$$f(\omega) = \frac{\dot{\delta}}{\delta} = \frac{D\omega}{C_s - C_0}\left[-C_0\Gamma_D\omega^2 + \frac{C_\infty - C_0}{\sqrt{\pi Dt}}\right] \tag{6.22}$$

Since $C_\infty - C_0$ is negative for the meltback, $f(\omega)$ is always negative. In other words, the melting interface is stable irrespective of the experimental conditions. On the other hand, the growth interface is unstable for a certain range of ω. Equation 6.22 is composed of two terms. The first term reflects the effect of the capillary force, while the second reflects the effect of the volume diffusion. In the case of meltback, the latter term stabilizes the interface as well as the former while in the growth the latter term causes instability.

6.6 COMPARISON BETWEEN THEORY AND EXPERIMENT

Typical morphologies of the experimentally found waves are drawn schematically in Figure 6.19. All three morphologies, Figures 6.19(a) to (c) correspond to Figures 6.1(a), (b) and 6.3, respectively. Figure 6.1(a) shows the simple sinusoidal wave. The most common form of the wave is given in (b) of the same figure. The wave front is modulated by another sinusoid. On the other hand, one finds a morphology such as shown in Figure 6.19(c) on a non-singular low index plane or on a highly misoriented surface. This morphology may correspond to the cellular structure popularly found in melt growth. Here, we compare with the theory only the waves of the type given in Figure 19(a) and wave A in 19(b).

Both the theoretical (calculated from Equations 6.20 and 6.21) and the experimental wavelengths are given in Table 6.1. In the calculation we employed

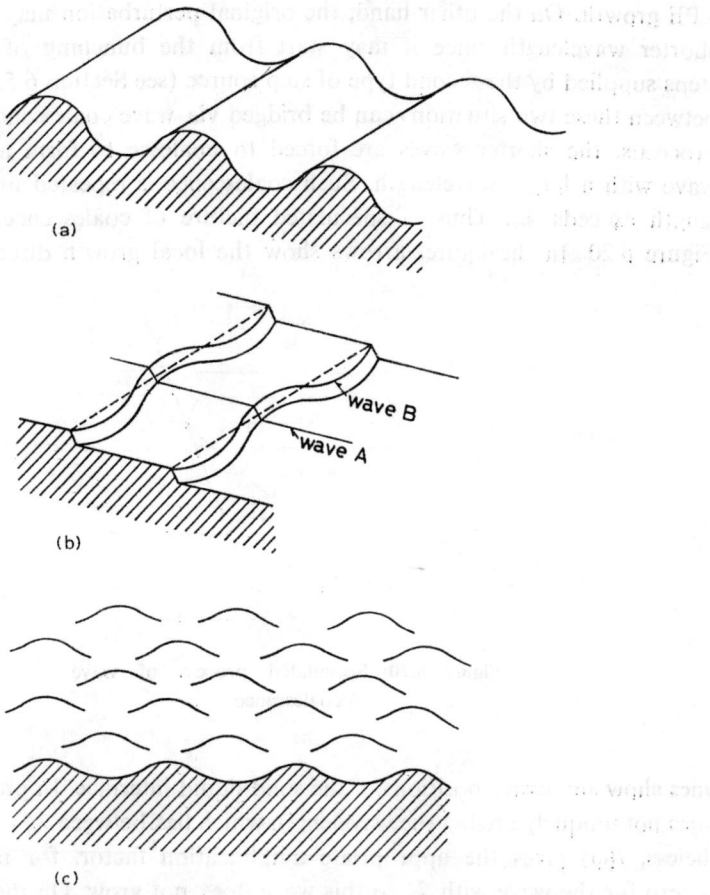

Figure 6.19 Three typical wave morphologies: (a) simple sinusoidal form, (b) waves with sinusoidal wave front, (c) multi-protuberance morphology

the following numerical values:

$$C(T) = 3.82 \times 10^{22} \times [\exp(12.9 - 1.24 \times 10^{-2}T) - 1]^{-1} \quad (26)$$
$$D = 5 \times 10^{-5} \text{ cm}^2/\text{s}$$
$$\Gamma_D = 7 \times 10^{-8} \text{ cm} \quad (27)$$

where $C(T)$ is the equilibrium concentration of phosphorus with InP in the indium liquid, and the same value of D as that of As in Ga solution is assumed. In principle, any wave with λ between λ_0 and ∞ can exist and grow during

LPE growth. On the other hand, the original perturbation may have a much shorter wavelength since it may start from the bunching of monoatomic steps supplied by the second type of step source (see Section 6.5.1.1). The gap between these two situations can be bridged via wave coalescence. As growth proceeds, the shorter waves are forced to coalesce to form a more stable wave with a larger wavelength. Such coalescence is repeated until the wavelength exceeds λ_0. Thus, a speculated picture of coalescence is shown in Figure 6.20. In the figure, arrows show the local growth direction and the

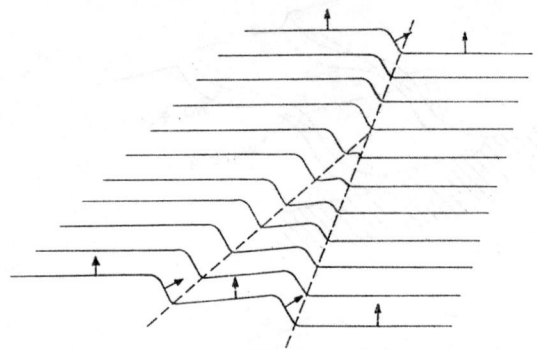

Figure 6.20 Speculated process of wave coalescence

lines show successive positions of the solid-liquid interface. In practice, theory does not uniquely predict the wavelength which lies between λ_0 and ∞. Nevertheless, $f(\omega)$ gives the appropriate amplification factor, for instance, $f(\omega)$ is zero for the wave with λ_0 so this wave does not grow. On the other hand, the wave with λ_{max} grows most rapidly and finally these waves may cover the whole surface. Actually, the wavelength has a value between λ_0 and λ_{max}. As is seen in Equations 6.20 and 6.21, λ_0 is smaller than λ_{max} by a factor of $\sqrt{3}$. This difference is quite small compared with the probable error to be found in parameters such as Γ_D. In view of the above considerations we can conclude that agreement between theory and experiment is fairly good. Furthermore, experiment clearly shows a tendency for the wavelength to decrease as the supersaturation increases. This is predicted theoretically by Equations 6.20 and 6.21. Another point which supports the present theory is the experimental fact that the meltback gives quite smooth surfaces. As was seen in Section 6.5.2, $f(\omega)$ is always negative. This means that the melting interface becomes smoother as the meltback time is increased. Furthermore,

$f(\omega)$ increases in negative value and hence the rate of smoothing increases as the superheating becomes large. This also agrees very well with the experimental results given in Section 6.3.

6.7 SUMMARY

To obtain flat interfaces between LPE multilayers it is essential to get flat solid-liquid interface during the growth. In this chapter, the most disadvantageous surface morphology which shows wavy structure has been considered from the point of view of theory and experiment. In the first part we were concerned with the way the wave morphology varies with the change of the various experimental conditions. It was found that the wave appears on a slightly misoriented surface and the wavelength decreases as the supersaturation is increased. Secondly, the meltback morphology was studied. Experiments show that the meltback surface becomes smoother as superheating or meltback time is increased. Thirdly, the optical properties of LPE layer were studied by photoluminescence image showing beyond doubt that there is a one-to-one correspondence between non-radiative region and the wave morphology. The non-radiative region accompanies the wave bottom and the region shows the trace of the bottom during LPE growth. It is concluded that the bottom creates recombination centres by some unknown mechanisms. Finally, the morphological stability theory was applied to the LPE growth. The theory gave a satisfactory explanation of the formation mechanism, of wave morphology, of growth and of the smoothing mechanism of the meltback.

REFERENCES

1. M. B. Small, A. E. Blakeslee, K. K. Shih, and R. M. Potemski, *J. Crystal Growth*, **30**, 257 (1975).
2. B. L. Mattes and R. K. Route, *J. Crystal Growth*, **16**, 219 (1972).
3. E. Bauser, M. Frik, K. S. Loechner, L. Schmidt, and R. Ulrich, *J. Crystal Growth*, **27**, 148 (1974).
4. R. H. Saul and D. D. Roccasecca, *J. Appl. Phys.*, **44**, 1983 (1973).
5. A. Mottram and A. R. Peaker, *J. Crystal Growth*, **27**, 193 (1974).
6. M. Yoshikawa, M. Itoh, H. Shinohara, R. Ueda and K. Kotani, *The Autumn Meeting of the Japan Soc. of Appl. Phys. Fukuoka*, November, 1975, 24a-K-9.
7. D. T. J. Hurle, *Crystal Growth, An Introduction*, ed. P. Hartman, 233, North-Holland, Amsterdam, 1973.
8. R. C. Peters, '1972 Proc. 4th Int. Symp. Gallium Arsenide and Related Compounds' 55, *Inst. Phys. Conf. Ser.*, No. 17, London, 1973.

9. M. B. Small and R. M. Potemski, *J. Crystal Growth*, **37**, 163 (1977).
10. T. Kajimura, K. Aiki, and J. Umeda, *Appl. Phys. Letters*, **30**, 526 (1977).
11. K. Aiki, T. Kajimura, and J. Umeda, *Report No. 367, Applied Electronics, Japan Soc. of Appl. Phys.*, **5**, 1976.
12. K. Pak, T. Nishinaga, and S. Uchiyama, *Japan. J. Appl. Phys.*, **16**, 949 (1977).
13. J. E. Davies and E. A. D. White, *J. Crystal Growth*, **27**, 261 (1974).
14. T. Nishinaga, K. Pak, and S. Uchiyama, *J. Crystal Growth*, **42**, 315 (1977).
15. I. Crossley and M. B. Small, *J. Crystal Growth*, **19**, 160 (1973).
16. J. T. Longo, J. S. Harris, Jr., E. G. Gertner, and J. C. Chu, *J. Crystal Growth*, **15**, 102 (1972).
17. D. L. Rode, *J. Crystal Growth*, **27**, 313 (1974).
18. B. L. Mattes and R. K. Route, *J. Crystal Growth*, **27**, 133 (1974).
19. J. A. Donahue and H. T. Minden, *J. Crystal Growth*, **7**, 221 (1970).
20. R. F. Sekerka, *Crystal Growth, An Introduction*, ed. P. Hartman, 403, North-Holland, Amsterdam, 1973.
21. R. L. Parker, 'Crystal Growth Mechanisms: Energetics, Kinetics and Transport', in *Solid State Physics*, vol. **25**, 151, Academic Press, New York and London, 1970.
22. W. W. Mullins and R. F. Sekerka, *J. Appl. Phys.*, **35**, 444 (1964).
23. D. L. Rode, *J. Crystal Growth*, **20**, 13 (1973).
24. T. Nishinaga, K. Pak, and S. Uchiyama, *J. Crystal Growth*, **43**, 85 (1978).
25. H. S. Chen and K. A. Jackson, *J. Crystal Growth*, **8**, 184 (1971).
26. R. N. Hall, *J. Electrochem. Soc.*, **110**, 384 (1963).
27. J. C. Brice, *The Growth of Crystals from Liquids*, 92, North-Holland, Amsterdam, 1973.

Chapter 7

Quaternary III–V Systems for Semiconductor Sources of Coherent and Non-Coherent Radiation

P. G. Eliseev

7.1 INTRODUCTION

The ability of many semiconductor crystals to form a variety of solid solutions has found a widespread optoelectronic application. This ability is characteristic of the so-called *isomorphic crystals*, i.e. crystals of the same lattice type. Change of the chemical content in such a mixed crystal (or alloy) is responsible for the variation of the energy band gap E_g, which determines the spectral position of the edge luminescence band, and the photosensitivity band of the material. The mixing of binaries does not deteriorate the luminescence and photoresponse capabilities of the solid solution. As a result, solid solutions can be used as the basis for the technology of optoelectronic devices operating in a wide spectral range.

At present the quaternary solid solutions play an increasingly important role in optoelectronic technology, especially in the preparation of a variety of heterojunction devices. The advantages of the heterojunction devices have been shown and discussed previously (1–3), mainly on the basis of investigations of AlGaAs/GaAs heterojunction systems. Such heterojunction optoelectronic devices compete successfully with other types in most applications at room and higher temperatures.

The AlGaAs alloy system covers a range of E_g between 1.43 and 2.17 eV, where a part of the range (namely, 1.43–1.93 eV) corresponds to the direct-band structure suitable for laser applications. The alloy system as well as the similar AlGaSb alloy system is based on Al–Ga substitution in III-group element sublattice which produces no significant effect on the lattice parameter of the sphalerite-type crystal lattice. This is due to the very close (but not perfect) coincidence of covalent radii of the elements. AlGaAs/GaAs heterojunctions can be prepared to be free of misfit defects at the heteroboundary. However, a small difference in the lattice parameter a_0 between the alloy and binary partners of the heterojunction produces an undesirable effect on the performance characteristics of the heterostructure, because of some tensile stress and defect instability during the operation of the device. AlGaSb/GaSb heterojunctions exhibit a larger lattice parameter difference (at given composition of the alloy) at room temperature.

The problem that arises now is how to improve the lattice matching at heteroboundaries, and how to widen the spectral range of heterojunction devices. In order to solve these problems, some new semiconductor materials should be used which satisfy the following requirements:

(1) their lattice parameters should be identical,
(2) there should be a difference in the energy band gap large enough to create electronic and optical barriers in heterojunctions.

The latter requirement excludes the use of dilute solid solutions. A minimum requirement for the band gap difference ΔE_g is to be more than several kT.

The above-mentioned problem can be solved by the use of quaternary and other multicomponent alloys. These mixed crystals possess a number (more than 1) of chemical degrees of freedom. Because of the continuous dependence of lattice parameters on two or more composition variables, it is easy to predict the existence of families of alloy compositions with the same lattice parameter (iso-lattice-parameter systems). The interrelation of E_g and the lattice parameter a for nine III–V binaries of the greatest importance is shown in Figure 7.1. The curves between the points show the ternary systems. The shaded areas between the curves correspond to quaternaries. The vertical straight lines along the shaded area correspond to the iso-lattice-parameter systems. Most important of such systems are those including binaries. Technologically, it is preferable to use binary crystals as a seed for further iso-lattice-parameter crystal growth of new mixtures. Quaternaries (as well as most ternaries) can be prepared by epitaxial procedure in the form of thin single-crystal layers. For the procedure (LPE, VPE and MBE) a single-crystal substrate is needed. Binary compounds are available in the form of high quality bulk crystals. Thus, it is easier to prepare iso-lattice-parameter systems related to any binary compound.

Recently, several quaternary III–V alloys have been obtained in this way for heterojunction devices, such as GaInPAs, AlGaAsP, AlGaAsSb, GaInAsSb, and some others. A number of new semiconductor sources of coherent and non-coherent radiation have been developed with quaternaries serving as the active medium. Most promising systems based on GaInPAs/InP double-heterostructure light-emitting devices (injection lasers and LEDs) were first obtained in 1974 (4). Such systems provide radiation sources for very

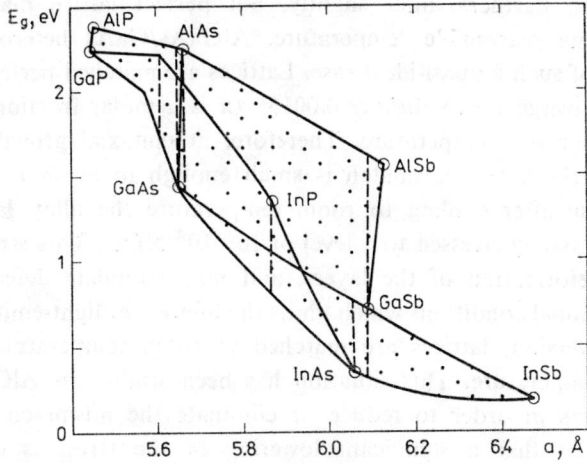

Figure 7.1 Interrelation of energy band gap E_g and the lattice parameter a for some III–V compounds and alloys. Iso-lattice parameter systems are shown by dashed vertical lines. The dotted area is the range of quaternary alloys

interesting spectral bands of high transparency and low material dispersion in optical fibres, and may be considered quite competitive in operational characteristics with AlGaAs/GaAs radiation sources. Remarkable success was achieved with AlGaAsSb/GaSb DH lasers and LEDs (5, 6), and with GaInAsSb/GaSb LEDs (6) in 1976. Very recently the latter system has been used for DH lasers at wavelengths about 2 μm (7).

The main results in the research and development of quaternary radiation sources will be the subject of this chapter. At first we shall consider the general approach to the construction of the iso-lattice-parameter alloy heterostructures and the method of predicting the properties of the alloys. Then we shall review the experimental data on the characteristics of LEDs and lasers on the basis of quaternary solid solutions.

7.2 ISO-LATTICE-PARAMETER MULTICOMPONENT SYSTEMS

7.2.1 Real structure of heteroboundaries

There are several degrees of lattice mismatch of heteroboundaries. The ideal case is a coincidence of lattice parameters at epitaxial growth temperature and room temperature. It can be obtained when thermal expansion coefficients α also coincide. An almost ideal lattice matching occurs when a of the two heterojunction partners differ slightly, but perfect lattice matching takes place at some reasonable temperature. AlGaAs/GaAs heteroboundary is an example of such a quasi-ideal case. Lattices are matched perfectly at about 930°C, but diverge up to $\Delta a/a \approx 0.0015x$ (x is a molar fraction of AlAs in the alloy) at room temperature. Therefore, at epitaxial growth conditions (700–900°C) the lattice mismatch is small enough to avoid a misfit defect formation, but after cooling to room temperature the alloy layers appear to be compressively stressed to a level of 10^7–10^8 N/m^2. This stress produces tetragonal deformation of the layers, and may stimulate defect formation under operational conditions which limits the lifetime of light-emitting devices. In another version, lattices are matched at room temperature, but differ at growth temperature. This situation has been studied in AlGaAsP/GaAs heterojunctions in order to reduce or eliminate the mismatch stresses (8). By this means that a significant lowering of the stress is obtained for $Al_{0.34}Ga_{0.68}As_{1-y}P_y$/GaAs heterojunction when $y \leqslant 0.015$. The lattice mismatch at the epitaxial growth temperature can lead to the formation of mismatch dislocation network. This is not, however, the case if the mismatch is small and the thickness of epitaxial layer is also small. There is a relationship between relative lattice misfit and the largest thickness t, at which mismatch dislocations do not yet appear. Incidentally, such a dislocation can appear at first by the transformation of a threading dislocation from the substrate into the misfit dislocation network (9). This provides a way of eliminating the threading dislocations in heteroepitaxial layers by controlled lattice parameter mismatch as has been demonstrated in the case of AlGaAsP/GaAs structures (10).

The next step to increase $\Delta a/a$ is the creation of a dislocation network. This is expected to take place at any reasonable thickness of the heterolayer ($t > 0.1$ μm) when $\Delta a/a > 10^{-3}$.

Early attempts to prepare new heterojunctions were made without lattice matching of the eventual multilayer structure to the substrate. Successful examples of such an approach are represented by DH lasers of GaPAs and GaAsSb active materials (11–13), and by a number of LEDs. In this way a preliminary preparation of intermediate grading layers is necessary. Gradual decrease and a step increase of the lattice parameter are preferred (13). Investigations of such grading layers and their interfaces show a typical defect structure, which appears in poorly matched heteroboundaries. Figure 7.2 is a TEM diffraction picture of GaAsSb/GaAs heterojunction with room-temperature mismatch about $\Delta a/a = 1.5 \times 10^{-3}$ (2 % of molar fraction of GaSb in the alloy) (14).

There is a (100)-plane of interface covered by two-dimensional network of misfit dislocations. The dislocations are mostly of the 60°-type with Burgers' vectors inclined at 45° to the interface plane extended mainly along the

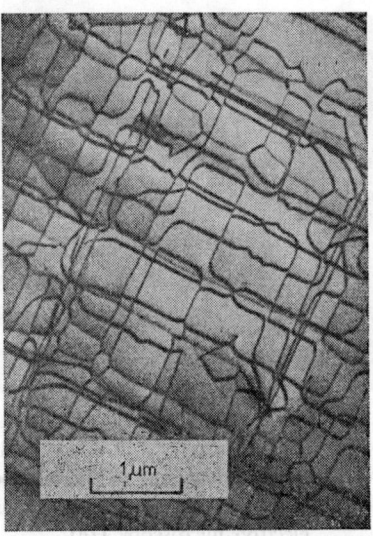

Figure 7.2 Two-dimensional misfit dislocation network at (100) interface of GaAsSb/GaAs heterojunction with molar fraction 2% of GaSb in the alloy. Calculated average spacing of dislocations in the network is about 0.37 μm. Transmission electron microscopy (14)

[011] and [0$\bar{1}$1] directions. Among them a number of edge dislocations (about 30% or less of the total) are observed in the network (14). Figure 7.3 shows a transverse section of two heterojunctions of the GaAsSb/GaAs type with dislocation networks. Such a micrography makes it possible to observe the formation of volume defected structure by initiation of inclined dislocations in the network.

The next degree of deviation from the ideal heteroepitaxy will occur when the crystallographic perfection of the layers grown is spoiled by threading dislocations, created at the imperfect interface. The data in (14) demonstrate the influence of the lattice misfit on volume dislocation density N_D in the heteroepitaxial layers of $GaAs_{1-x}Sb_x$ at GaAs-substrate. The density of the inclined dislocations appears to be significantly higher than in the substrate when $x \geqslant 0.02$ (i.e. $\Delta a/a \geqslant 1.5 \times 10^{-3}$); it increases from 10^6 cm^{-2} at $x \approx 0.02$ to 2×10^7 cm^{-2} at $x \approx 0.07$. The depth-variation of the dislocation density observed in 4–6 μm thick heteroepitaxial layer shows that there is

Figure 7.3 Transverse section of GaAsSb/GaAs heterojunction with molar fraction 9% of GaSb in the alloy. There are numerous dislocations in the heteroepitaxial layer nucleated at the heteroboundary. Transmission electron microscopy (14)

a remarkable annihilation of inclined dislocations, which leads to a 20–25 fold decrease of N_D when x is less than 0.07. At larger GaSb content (i.e. at $\Delta a/a \geqslant 0.005$) no such decrease is observed, while N_D can increase. Therefore, steps of $\Delta a/a \geqslant 0.005$ do not seem to be acceptable for preparation of grading layers, because of the irreversible deterioration of crystal perfection. Larger misfit produces heavily-defected structures and transformation to grain and polycrystal growth. In (14) it is also shown that the five-layer grading structure gives $N_D \approx 6 \times 10^5$ cm^{-2} when the eventual value of x = 0.135 (intermediate values are 0.015; 0.045; 0.075; 0.09; 0.13). By contrast, a two-layer structure (0.04; 0.12) gives $N_D \approx 2 \times 10^7$ cm^{-2} (whereas at substrate $N_D \approx 2 \times 10^4$ cm^{-2}).

To sum up, one can distinguish several cases of real heteroboundaries (quantitative estimations refer to GaAsSb/GaAs interface):

(1) Ideal matching of lattices, $\Delta \alpha/\alpha = 0$, $\Delta a/a = 0$.
(2) Quasi-ideal matching, $\Delta \alpha/\alpha \neq 0$, $\Delta a/a \leqslant 1 \times 10^{-3}$.
   (a) $\Delta a/a$ is equal to zero near the growth temperature, but not at room temperature, where it leads to arise of elastic stress.

(b) $\Delta a/a$ is equal to zero at room temperature, but not at the growth temperature. No stresses. In both versions the interface is free of misfit-dislocation networks.

(3) Poor matching, $1 \times 10^{-3} \leqslant \Delta a/a \leqslant 5 \times 10^{-3}$, there are misfit dislocation networks and a progressive increase of volume dislocation density in the heterolayers.

(4) Very poor matching, $\Delta a/a > 5 \times 10^{-3}$. High dislocation density in the heterolayers ($N_D \gtrsim 2 \times 10^7$ cm^{-2}), which are not of acceptable perfection even for buffer layers.

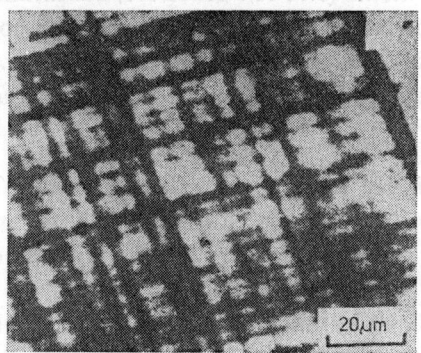

Figure 7.4 Cathodoluminescence topogram of the GaAs$_{1-x}$Sb$_x$ layer with the composition gradient about 50 cm^{-1} (the depth-variation of x is 0.005 per micrometer). Scanning electron microscopy (14)

Dislocations constitute a very effective non-radiative sink for excess electrons as can be seen in Figure 7.4, where a cathodoluminescence topogram of heteroepitaxial layer of GaAsSb with a composition gradient is presented. The dark lines correspond to dislocations just under the surface of the crystal.

7.2.2 Iso-lattice-parameter substitution concept

Let us now consider the condition of lattice parameter conservation at composition variation in multicomponent solid solution. In the substitution alloys this condition follows from the requirement of mutual compensation of specific effect of composition changes on the lattice parameter. Namely, if the lattice parameter of the alloy increases when atoms A are replaced partially by atoms A' and decreases when $B \to B'$, it is possible to determine both composition changes accurately to avoid variation of the lattice parameter.

For instance, starting from indium phosphide binary compound, one can obtain a quaternary iso-lattice-parameter alloy by dosing the addition of Ga and As atoms. Indeed, a partial replacement of In atoms by Ga atoms produces a decrease in the lattice parameter because of the smaller covalent radius of Ga (1.26 Å) as compared with In (1.46 Å). Actually, the decrease may be compensated for by a simultaneous partial substitution of In by any larger-size atoms, like Tl, from the IIIrd group of the Periodic Table. Another and easier way is the addition of As atoms with covalent radius 1.21 Å, which is smaller than that of In atoms. However, As atoms go to group V sublattice sites, and replace P atoms (1.10 Å). Therefore, substitution of P by As leads to an increase of the lattice parameter of the alloy in comparison with InP. The quaternary solid solution obtained is $Ga_xIn_{1-x}P_{1-y}As_y$, with atom fractions x and y, obeying the equation of iso-lattice-parameter substitution:

$$a(x, y) = a_0 \qquad (7.1)$$

where a_0 is the desirable value of the lattice parameter (in this case, one of InP). To solve (7.1) it is necessary to have an analytic expression of $a(x, y)$. The covalent radii of the elements involved usually conserve from compound to compound with high accuracy (but not perfectly). Due to this, $a(x, y)$ can easily be calculated by interpolation. In the generalized form, a hypothetical multi-component mixture of quasibinary type (15)

$$A^1_{x_1} A^2_{x_2} \ldots A^M_{x_M} B^1_{y_1} B^2_{y_2} \ldots B^N_{y_N} \qquad (7.2)$$

contains M kinds of atoms at A-sites of the lattice, and N kinds of atoms at B-sites. There are MN binary components and $M+N-2$ degrees of freedom (subtraction of 2 is due to stoichiometric requirements):

$$\sum_m^M x_m = \sum_n^N y_n = 1 \qquad (7.3)$$

(where x_m and y_n are the atomic fractions of elements at A- and B-sites, respectively). In the simplest approach using the covalent radii r_m, r_n of the elements involved, one can obtain

$$a(x_m, y_n) = (4/\sqrt{3})\left(\sum_m^M r_m x_m + \sum_n^N r_n y_n\right) \qquad (7.4)$$

In order to improve the approximation one may use the lattice parameters of binary components a_{mn}. The homogeneous mixture of (7.2) can be expressed as

$$(A^1 B^1)_{z_{11}} (A^1 B^2)_{z_{12}} \ldots (A^M B^N)_{z_{MN}} \qquad (7.5)$$

where the molar fractions of binary components z_{mn} obey the equation

$$\sum_{m}^{M}\sum_{n}^{N} z_{mn} = 1 \tag{7.6}$$

They may be obtained simply as

$$z_{mn} = x_m y_n \tag{7.7}$$

In this approximation,

$$a(x_m y_n) = \sum_{m}^{M}\sum_{n}^{N} a_{mn} x_m y_n \tag{7.8}$$

A higher step of interpolation can be reached by taking into account the mixing effects, as they exist in ternary (two-component) solid solutions. In the case of deviation from Vegard's rule, the lattice parameters of ternaries $A_{1-z}^1 A_z^2 B$ or $AB_{1-z}^1 B_z^2$ are presented approximately by the second power form:

$$a(z) = a(0) + [a(1) - a(0)]z + 4\beta z(1-z) \tag{7.9}$$

where β is the value of deviation $a(z)$ from its linearly interpolated values at $z = 0.5$. If (7.9) is valid for all the ternary systems confining the composition square $0 \leqslant x \leqslant 1$, $0 \leqslant y \leqslant 1$ of the quaternary alloy $A_x^1 A_{1-x}^2 B_{1-y}^2$, the following expression is obtained:

$$\begin{aligned}a(x, y) = {}& a_{00} + (a_{10} - a_{00} + 4\beta_{x0})x - 4\beta_{x0} x^2 + (a_{01} - a_{00} + 4\beta_{0y})y \\ & - 4\beta_{0y} y^2 + (a_{00} + a_{11} - a_{10} - a_{01} + 4\beta_{x1} + 4\beta_{1y} - 4\beta_{x0} - 4\beta_{0y})xy \\ & + 4(\beta_{x0} - \beta_{x1})x^2 y + 4(\beta_{0y} - \beta_{1y})xy^2 \end{aligned} \tag{7.10}$$

where

$$a_{00} = a(0, 0), \quad a_{10} = a(1, 0), \quad a_{01} = a(0, 1), \quad a_{11} = a(1, 1)$$

and where β_{x0}, β_{x1}, β_{0y} and β_{1y} are used for non-linearity coefficients of the ternaries at $y = 0$, $y = 1$, $x = 0$, and $x = 1$, respectively. Formula 7.10 does not contain the mixing effect specific of quaternary alloys. Such an effect can be approximately described by $16\beta_{xy} x(1-x)y(1-y)$, which vanishes on all sides of the composition square. The quantity β_{xy} may be found as a difference of real $a(x, y)$ from the calculated value by (7.10) at point $x = 0.5$, $y = 0.5$. These non-linearity corrections are probably quite small in comparison with the accuracy of experimental values. This holds for the lattice parameter, but the same interpolation scheme may be used to calculate the other alloy parameters. One such parameter is the thermal expansion coefficient α. The validity of linear interpolation for ternary alloys was noted previously

(13). The condition of ideal (temperature-independent) lattice matching can be expressed by the system of equations:

$$a(x_m, y_n)Z = a_0 \tag{7.11}$$

$$\alpha(x_m, y_n) = \alpha_0 \tag{7.12}$$

(a_0 and α_0 are the desirable binary values of a and α, respectively). In the frames of the quaternary alloys, the above equations give two almost straight lines in the plane. The only intersection of these lines is at the point corresponding to pure binary compound chosen as a substrate. Thus, the solution of (7.11) and (7.12) must be searched for among alloys with more components. For instance, the composition range of six-component alloys is confined inside a volume figure (a prism). In this case, Equations 7.11 and 7.12 give two surfaces. Their intersection is a line, which corresponds to the continuous variety of compositions for ideal matching if it is inside the above-noted volume figure: $0 \leqslant x_m \leqslant 1$, $0 \leqslant y_n \leqslant 1$.

7.2.3 Iso-lattice-parameter systems

Three types of quaternary alloys may by distinguished, two of them, $A^1_{1-x-y}A^2_x A^3_y B$ and $AB^1_{1-x-y}B^2_x B^3_y$, containing three binary components, and one, $A^1_{1-x}A^2_x B^1_{1-y}B^2_y$, four binary components. With six group III and V elements (Al, Ga, In, P, As, Sb) it is possible to form a number of quaternaries, namely, six three-component alloys and nine four-component ones. Four alloys of the latter type must be especially noted. They are GaInPAs, GaInAsSb, AlGaAsP and AlGaAsSb. Only the nearest neighbours in the Periodic Table meet in these alloys. This seems to be a very important technological point in view of the great difficulty involved in the mixing of elements which are rather far from each other in the Periodic Table. Indeed, little is known even about most of the ternary alloys formed by non-nearest neighbours like AlInSb or GaPSb.

The main quaternary systems of the III–V type are presented graphically in Figure 7.5.

Let us return to Figure 7.1 where all four of the quaternaries mentioned above are presented. A prominent tendency of E_g to decrease with an increase in a can be seen here. This is not, however, a definite dependence, and for this reason one can deal with quaternaries (otherwise the variation range of E_g reduces to zero).

Two kinds of iso-lattice-parameter systems can be distinguished: those containing any binary compound, i.e. downward directed systems (with the binary at the upper end of E range), and the upward directed systems (with the binary at the lower end). An interesting feature of systems of the latter kind is the transparency of the binary substrate to the edge radiation from alloy. LEDs of enhanced efficiency may readily be prepared with these systems. This is because external output efficiency of the planar LED may

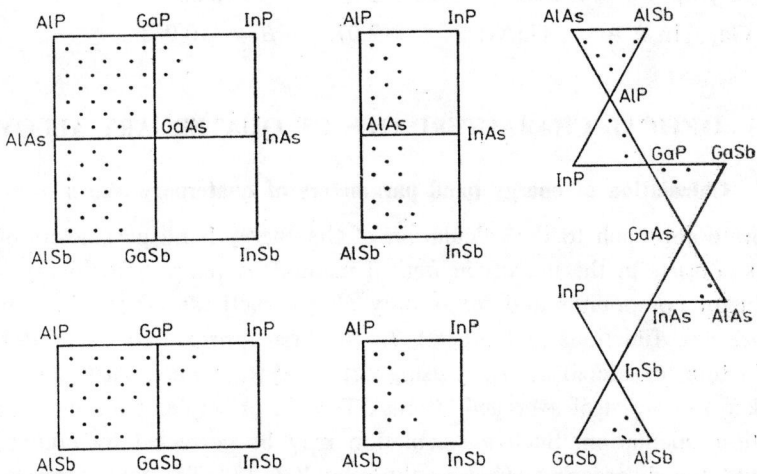

Figure 7.5 Fifteen quaternary alloys on the basis of six III–V-group elements (Al, Ga, In, P, As, Sb). The dotted areas are for indirect band-gap compositions

be increased by multifold reemission of radiation in the active layer if this radiation is not absorbed in the substrate. Similarly, the so-called 'photon recycling' in DH lasers may cause a lowering of the laser oscillations threshold.

Upward directed systems GaInPAs/InP, GaInAsSb/GaSb are of great interest for effective LEDs and lasers. They cover the spectral range from about 1 μm to more than 3 μm (at room temperature). Downward directed systems AlGaAsP/GaAs, AlGaAsSb/SaSb covering the spectral ranges 0.67–0.90 and 1.2–1.8 μm are also under intense investigation. Some excellent results have been obtained with quaternary/ternary heterojunctions, grown on a layered substrate which contained grading buffers. Examples of these heterojunctions are AlGaAsP/GaAsP, AlGaAsSb/GaAsSb.

Application of formulas developed in the previous section to iso-lattice-parameter systems gives a standard relationship of atomic fractions x and y:

$$y = \frac{Ax}{1+Bx} \quad (7.13)$$

with constants A and B calculated from binary lattice parameters using formula (7.8). They are as follows:

(1) $Al_xGa_{1-x}As_{1-y}P_y/GaAs$: $A = 0.069$, $B = -0.0128$;
(2) $Al_xGa_{1-x}As_ySb_{1-y}/GaSb$: $A = 0.0896$, $B = 0.058$;
(3) $Ga_xIn_{1-x}P_{1-y}As_y/InP$: $A = 2.2$, $B = 0.067$;
(4) $Ga_{1-x}In_xP_yAs_{1-y}/GaAs$: $A = 2.07$, $B = -0.063$.

7.3 OPTICAL CHARACTERISTICS OF QUATERNARY ALLOYS

7.3.1 Calculation of energy band parameters of quaternary alloys

Accurate approach to the calculation of the energy band parameters of the alloys consists in the use of theoretical methods developed for binary semiconductor compounds and for ternary alloys (methods of pseudopotential, of dielectric functions and others). In the first approximation mixed semiconductors considered in theory using virtual crystal model, where the crystal is taken to consist of averaged 'atoms'. This, in principle, is a kind of interpolation calculation. Such a calculation may be corrected by taking into account the disordering effect in the alloy (16–18). The final form of the computed energy band parameters of quaternary solid solution obeys the interpolation formula of the (7.10) type quite well. It can also be obtained by simple substitution in (7.10) of binary and ternary energy band parameters instead of lattice parameters. But the result of simple interpolation (as well as the result of accurate theoretical calculations) must be proved by comparison with experiment so that the influence of the mixing effect can be verified.

One of the predicted disordering effects in alloys is the broadening of the edges. On the whole, however, optical spectra of alloys are very similar to the spectra of binary compounds. This is due to the fact that the characteristic de Broglie wavelength of electrons and their free path lengths are longer than the distances of typical variation of the chaotic component of the crystalline potential. Consequently, the energy broadening parameter is rather small in comparison with the energy band width and does not influence the matrix elements of interband transitions in any significant way. The matrix elements as well as effective masses of current carriers can be simply interpolated for the alloys.

Calculations of the forbidden energy band E_g for quaternary alloys have been performed in several papers (15, 17, 19). Some of the results are presented

Quaternary III–V Systems

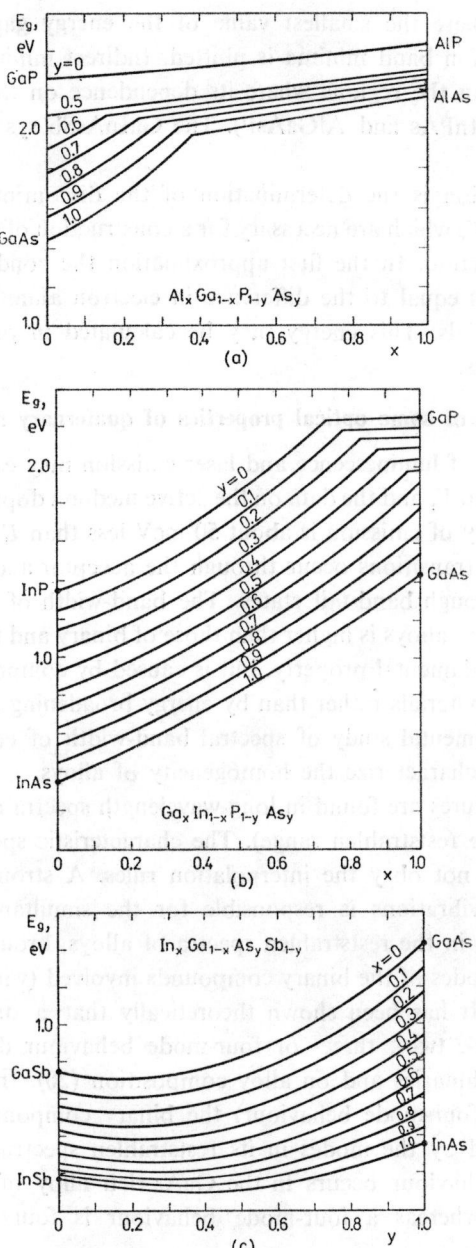

Figure 7.6 Calculated energy band-gap E_g of some quaternary alloys in the function of composition (15)

in Figure 7.6 where the smallest value of the energy gaps corresponding to three conduction band minima is plotted. Indirect minima determine the forbidden band in the regions where its dependence on the composition is rather weak (GaInPAs and AlGaAsP). The GaInAsSb system is an entirely direct one.

Another question is the determination of the discontinuity of the band edges ΔE_c and ΔE_v which are necessary for a construction of the band diagram of the heterojunction. In the first approximation the conduction band discontinuity ΔE_c is equal to the difference in electron affinity energies of the contacting materials. This energy may be calculated in principle by interpolation formulas.

7.3.2 Prediction of some optical properties of quaternary alloys

The wavelengths of luminescence and laser emission may easily be predicted using the band gap E_g and the data on the active medium doping level. Usually, the photon energy of emission is about 50 meV less than E_g. This is because optical radiative transitions occur through the acceptor and donor impurity levels and/or through band-tail states. The band-width of the luminescence band of quaternary alloys is higher than those of binary and ternary materials. This is not a fundamental property, but is caused by compositional inhomogeneities in the materials rather than by energy broadening specific of alloys. Therefore, experimental study of spectral band-width of edge luminescence may be used to characterize the homogeneity of alloys.

Interesting features are found in long-wavelength spectra of ir-active lattice vibrations (in the reststrahlen range). The characteristic spectral parameters in the range do not obey the interpolation rules. A strong localization of the short-wave vibrations is responsible for the simultaneous appearance of several modes in the reststrahlen spectra of alloys, brought about by the corresponding modes of the binary compounds involved (with some deformation and shift). It has been shown theoretically that a quaternary system may display one-, two-, three- or four-mode behaviour depending on the property of the binaries and on alloy composition (20). In other words, in the latter case (four-mode behaviour) the binary components of the alloy can be numbered by the modes in its reststrahlen spectrum. For example, a three-mode behaviour occurs in the GaAs-rich alloy of AlGaAsP quaternary system, whereas a four-mode behaviour is found in a GaP-rich alloy (20).

Data on the refractive index in the vicinity of the edge luminescence peak are important for the construction of laser heterostructures. Low threshold of oscillations in DH lasers is ensured by optical confinement, i.e. by the light-guiding effect in the active heterostructure layer. To obtain the optical confinement effect larger refractive index of the narrow-band active layer is preferred over the wide-band passive layers. Standard dispersion curves in semiconductors make it possible for this condition to be fulfilled in a number of heterostructures. Refractive index in quaternary alloys has not yet been measured in detail. Ternary AlGaAs is an example of the systems under investigation. At fixed photon energy, 1.38 eV, corresponding to the laser wavelength of GaAs, the refractive index n varies with alloy composition from 3.59 in GaAs to 2.971 in AlAs with a slight deviation from linear dependence, namely, $n(x)$ is smaller than the value obtained by a linear approximation between the values for GaAs by 1.5% at $x = 0.5$ (21). Using linear extrapolation, one can predict the existence of optical confinement in most of the iso-lattice-parameter systems except GaInAsSb/GaSb, where anti-guiding effect rather than optical confinement may be expected.

7.4 HETEROSTRUCTURES AND LIGHT-EMITTING DEVICES BASED ON QUATERNARY ALLOYS

7.4.1 GaInPAs/InP

The range of E_g which is covered by this iso-lattice-parameter system is from 1.35 eV in InP to 0.75 eV in $Ga_{0.47}In_{0.53}As$ (Figure 7.7). Effective double heterostructure can be prepared for the wavelength from about 1 μm to 1.68 μm at room temperature. Injection lasers were first reported in Reference 4,

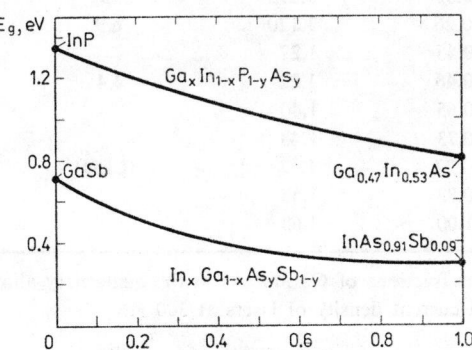

Figure 7.7 Calculated energy band-gap E_g of iso-lattice-parameter systems GaInPAs/InP and GaInAsSb/GaSb (15)

where lasing at about 1.06 μm had been obtained at 77 K in DH diodes prepared by the LPE method on the (111)-substrate of n-InP. The formula of the double heterostructure was pInP: Zn/pGa$_{0.1}$In$_{0.9}$P$_{0.78}$As$_{0.22}$: Zn/nInP: Te. Following that, room temperature laser action and high efficiency luminescence were described (22). A number of papers have appeared recently concerning this system for the entire spectral range (see Table 7.1). Laser action is obtained over the spectral range 1.06–1.31 μm. The properties of DH lasers at 1.10–1.12 μm are discussed in Reference 25. The lowest value of threshold current density at room temperature obtained was 2.8 kA/cm^2 at optimal thickness of the active layer of about 0.4–0.5 μm.

Table 7.1 GaInPAs/InP double heterostructure lasers and LEDs

x	y	λ, μm	j, kA/cm^2	References
0.1	0.2	1.06–1.10	10.4	(22) 1975
—	—	1.145	8.7	(23) 1976
0.17	0.34	1.10	(LED)	(24) 1976
0.12	0.23	1.12	2.8	(25) 1976
0.11	0.30	1.133	(LED)	(26) 1977
0.16	0.40	1.147		
0.16	0.50	1.279		
0.20	0.48	1.246	4.4–8.0	(27) 1977
0.22	0.47	1.208		
0.16	0.39	1.154	—	(28) 1977
0.25	0.55	1.20	(LED)	(29) 1977
0.28	0.58	1.26		
0.28	0.58	1.299	3–5	(30) 1977
0.12	0.26	1.120	6.9	(31) 1977
0.24	0.45	1.27	—	(32) 1977
0.20	0.46	1.22	4.4	(33) 1977
0.29	0.65	1.40		
0.32	0.73	1.48		
0.35	0.77	1.52	(LED)	(34) 1977
0.37	0.83	1.58		
0.47	1.00	1.68		

x and y are molar fractions of Ga and As in the quaternary alloy,
j is the threshold current density of lasers at 300 K

Our interpolation estimations of GaInPAs refraction index along the InP iso-lattice-parameter line shows that relative discontinuity $\Delta n/n$ at wavelength 1.1 μm is about 1.8%. This implies the smallest thickness for first order transversal mode of about 0.8 μm. Experimentally, the first order mode appeared at thickness d exceeding 1.1 μm, whereas at smaller d values only zero-order mode was observed. Estimation of the optimal active layer thickness leads to a value of 0.45 μm which is in fair agreement with the experimental value. Hence, we are dealing with weaker waveguiding than in conventional AlGaAs DH diodes, and one may expect to obtain more power in zero-order mode of the DH waveguide. In accordance with our observations such zero-order mode operation of laser diodes in the 1.06–1.12 μm range has been observed at $d \leqslant 1$μm, and sometimes at $d \leqslant 1.5$μm, Angular divergence of the laser emission plane, which is perpendicular to the waveguide, corresponds to about 40° (half-magnitude full-width) when $d \approx 1$ μm.

An enhancement of the waveguiding effect is anticipated when laser wavelength increases. Steps $\Delta n/n$ of about 4.5% at 1.27 μm and of about 9% at 1.68 μm may be expected.

Luminescent and laser properties of GaInPAs/InP structures have been described in a number of papers (22–40). Efficient DH LEDs of GaInPAs/InP have also been reported (22, 24). External efficiencies as high as 8.4% have been obtained at room temperature in planar-geometry diodes (22). The spectral

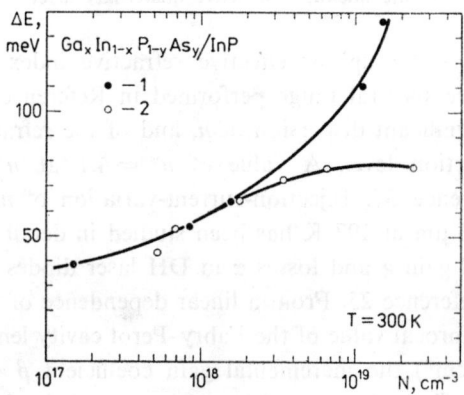

Figure 7.8 Spectral bandwidth ΔE of non-coherent photoluminescence of GaInPAs alloy at InP substrate near $1.08 \pm 0,01$ μm in the function of impurity concentration of (1) donors (Sn) and (2) acceptors (Zn)

bandwidth ΔE of spontaneous photoluminescence of single GaInPAs layers lattice-matched to InP substrate in 1.08 ± 0.01 μm wavelength range at 300 K is plotted in Figure 7.8 as a function of impurity concentration. A regular increase of ΔE is observed when the concentration increases, which is in agreement with the generally accepted concept of the influence of impurities on radiation spectral band. The absolute value of ΔE at the low concentration limit of Figure 7.8 approaches those of binary InP at the same doping level.

Variation of the laser threshold current density with an active layer thickness for 1.08 ± 0.01 μm DH lasers of GaInPAs/InP is shown in Figure 7.9, where a straight line is plotted for the linear dependence with a slope of 7.5 kA/cm²μm.

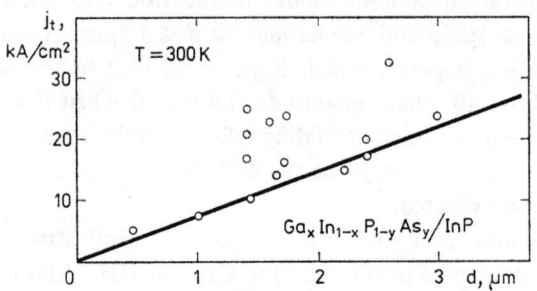

Figure 7.9 Dependence of the threshold current density j_t in DH lasers of GaInPAs/InP at room temperature on the thickness of active quaternary layer

Determinations of 'group' or effective refractive index $n^* = n - \lambda \mathrm{d}n/\mathrm{d}\lambda$ in the luminescence spectral range performed in References 22, 31, 35 give evidence of the significant dispersion of n, and of the refractive index sensitivity to the injection level. A value of $n^* = 4.1$ at $n \approx 3.3$ has been reported in Reference 35. Injection-current-variation of n^* in the spectral range of 1.08–1.12 μm at 197 K has been studied in detail in Reference 31.

Data on optical gain g and losses α in DH laser diodes of GaInPAs/InP are reported in Reference 25. From a linear dependence of threshold current density on the reciprocal value of the Fabry–Perot cavity length (in the range from 3 to 6 kA/cm²) the incremental gain coefficient $\beta = \mathrm{d}g/\mathrm{d}j$ is found to be about 30 cm/kA whereas the loss parameter α is about 68 cm⁻¹.

We have recently obtained some data on the temperature dependence of threshold current density j of GaInPAs/InP lasers. We found the rate of temperature increase of j_t to depend on the energy-band-gap step in the heterojunction. We have measured $T^* = (\mathrm{d}\ln j/\mathrm{d}T)^{-1}$ in the vicinity of 290 K in DH diodes operating at wavelengths 1.125 μm, 1.230 μm and 1.270 μm. The value of T^* appeared to increase with the wavelength and was equal to 46, 61 and 106 K, respectively. This may be understood to have resulted from closer carrier confinement in DH active layer due to higher potential barriers. The corresponding values of the energy-band-gap step were 163, 257 and 289 meV. However, more detailed temperature investigation of j_t (see Figure 7.10) shows significant variations of T^* with temperature.

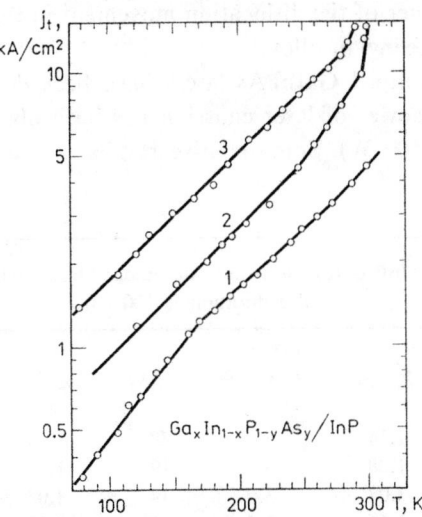

Figure 7.10 Temperature dependence of the threshold current density of DH lasers of GaInPAs/InP emitting at room temperature at wavelength of 1.27 μm (1), 1.23 μm (2), and 1.125 μm (3)

The most important question is the reliability of GaInPAs/InP laser diodes. Operation lifetime testing for CW lasers at room temperature has been performed (40). No deterioration of characteristics was observed during 1500 h CW operation at wavelength 1.15 μm (39). A ca. 3000 h degradation-free performance has also been reported. More recently, a 500 h operational lifetime of GaInPAs/InP CW lasers has been reported (at wavelength 1.31 μm) (40). During this time a 30% rise of threshold current has been noticed. It is interesting to compare these results with those of AlGaAs DH lasers. It was pointed out in (39) that rather long operational lifetime (more than 10^3 h) was obtained despite the high dislocation density N_D in the InP substrate (more than 10^5 cm^{-2}). If this value of N_D were detected in GaAs-substrates, operational lifetime of DH lasers would appear to be limited to about 10 h. This suggests that the physical mechanism which is responsible for the high rate of degradation in AlGaAs lasers, and which is very sensitive to the dislocation density does not appear to be very important in the quaternary GaInPAs lasers (39). This difference in the behaviour of AlGaAs and GaInPAs lasers may be caused by smaller electron-hole pair energy (i.e. by smaller band gap E_g in quaternary lasers). Another cause of the difference may be in the manner of the dislocation movement in the crystals during the multicomponent mixing in alloy.

Electron-beam-pumped GaInPAs lasers have been described also in (41). Rather high peak power of laser emission has been observed in these lasers (up to more than 100 W). Some relative results of the study are presented in Table 7.2.

Table 7.2 Epitaxial GaInPAs/InP electron-beam-pumped lasers (electron energy is 50 keV, pulse duration is 100 ns)

N	DOPANT	300 K			80 K		
		λ, μm	j, A/cm²	P, W	λ, μm	j, A/cm²	P, W
1	Ge	1.14	3.5	50	1.06	0.4	75
2	Te, Zn	1.08	3.0	19	1.03	0.3	42
3	Zn	1.08	3.0	14	1.02	0.3	66
4	Te	1.15	3.5	22	1.06	0.3	110
5	Sn	1.15	3.5	6.2	1.09	0.3	46
6	Sn	—	—	—	1.03	0.3	80
7	Te, Zn	1.13	3.5	70	—	—	—
8	Sn	1.09	2.5	44	—	—	—

j threshold current density, P maximum output power (41)

Electrical and photoelectrical properties of GaInPAs and of heterojunction GaInPAs/InP have been investigated, though not in detail. Efficient photodiodes for 1.06 μm emission have been developed with the sensitivity exceeding 0.6 A/W. Some suggestions have been made on separate determination of steps in energy band edges in GaInPAs/InP heterojunctions. When considering the photosensitivity and electroluminescence spectra it was supposed that most of the energy-band-gap discontinuity corresponds to the conduction band edge step ΔE_c (36). This conclusion applies to $Ga_xIn_{1-x}P_{1-y}As_y$ quaternary alloy-InP heterojunctions with the composition range $0.1 \leqslant x \leqslant 0.15$. Some material characterization data may be found in Ref. 38, where quaternary alloy layers of relatively high purity are described. At $x = 0.12$, $y = 0.25$ the epitaxial layer was characterized by a differential impurity concentration N of about 5.5×10^{-15} cm⁻³ and by electron mobility 4030 and 14,460 cm²/Vs at 300 and 77 K, respectively. At

$x = 0.2$, $y = 0.44$ and $N \approx 2.3 \times 10^{15}$ cm^{-3} mobilities of 6600 and 16830 cm²/Vs were obtained for identical temperatures (38).

All these data show that the system of GaInPAs/InP lattice-matched heterojunctions may be considered to be a very promising basis for various optoelectronic devices, especially for injection lasers. In this connection it is necessary to mention that the system covers a spectral range where the characteristics of optical fibre transmission are most attractive.

7.4.2 GaInPAs/GaAs and GaInPAs/GaPAs

The significant progress in developing visible light emitting sources is connected with these heterojunction systems (42–48). As can be seen in Figure 7.6(b), the largest value of direct band gap corresponds to the GaInP composition and is equal to $E_g \approx 2.10$ eV at room temperature (about 2.16 eV at 77 K). This composition may be matched by lattices to GaP$_z$As$_{1-z}$ substrates of $Z \approx 0.38$. As for the GaInPAs/GaAs iso-lattice-parameter system, it covers the E_g-range of 1.43–1.90 eV; therefore, this latter system overlaps with the AlGaAs-systems.

Single heterostructures of $p\text{Ga}_x\text{In}_{1-x}\text{P}_{1-y}\text{As}_y/n\text{GaP}_z\text{As}_{1-z}$ have been developed and were described in Reference 42. Laser action was observed for wavelength range near 0.63 µm at 77 K when $z = 0.40$, $x = 0.70$, $y = 0.01$.

The direct-indirect transition region of the quaternary system has been investigated and found to correspond to the following relationships (43):

$$300 \text{ K}: \quad x_c + 0.526 y_c = 0.726 \tag{7.14}$$

$$77 \text{ K}: \quad x_c + 0.516 y_c = 0.719 \tag{7.15}$$

where x_c and y_c are atomic fractions of Ga and As in the alloy. Iso-lattice parameter lines may be represented in terms of the expression (43, 46)

$$a = 6.058 - 0.405 x - 0.189 y - 0.015 xy.$$

When a substrate of $z = 0.42$ is used, the direct-indirect transition occurs between 77 and 300 K ($x = 0.72$, $y = 0.01$).

The DH of $\text{Ga}_{x'}\text{In}_{1-x'}\text{P}_{1-y'}\text{As}_{y'}/\text{Ga}_x\text{In}_{1-x}\text{P}_{1-y}\text{As}_y/\text{Ga}_{x'}\text{In}_{1-x'}\text{P}_{1-x'}\text{As}_{y'}$ have been developed for injection lasers with yellow and green-yellow light emission (44–48). Photoluminescence and laser electroluminescence data are presented in Table 7.3 as an example of DH diode characteristics (44, 45). There, $y' = 0$ was used and x and y varied near the direct-indirect transition region. The shortest laser wavelength obtained was about 0.575–0.585 µm at 77 K and about 0.637 µm at room temperature (45–48).

Table 7.3 Some characteristics of GaInPAs double heterojunction lasers in visible (spectral) range (44, 45)

| | | $\hbar\omega_{PL}$, eV | | 300 K | | 77 K | |
	GaPAs Substrate	GaInP Wide band layers	GaInPAs Active layer	$\hbar\omega$, eV	j, kA/cm²	$\hbar\omega$, eV	j, kA/cm²
1	1.400	1.905	1.850	—	—	1.910	3.2
2	1.400	1.905	1.780	1.735	15	1.815	0.6
3	1.710	2.090	1.920	1.880	17	1.950	0.52
4	1.770	2.120	1.950	1.900	32	1.985	4.6
5	1.840	2.145	1.960	1.914	129	2.026	3.7
6	1.860	—	1.980	1.947	250	2.055	3.6
7	1.870	2.160	2.090	—	—	2.156	13.0

$\hbar\omega_{PL}$ photoluminescence peak photon energy at 300 K,
$\hbar\omega$ photon energy of injection laser emission,
j threshold current density

Investigations of DH lasers suggest that the effects of optical and electron confinement are present in this system, but optical and electrical parameters of the materials involved are poorly known. Interesting proof of the electron confinement effect in DH laser of the GaInPAs/GaPAs type is found in the temperature dependence of the threshold current density which is expressed in an approximate way by the exponential function:

$$j_t \sim \exp(T/T^*) \qquad (7.16)$$

where T^* is a characteristic parameter. It has been found that the value of T^* varies from 27 K when there is no energy-band-gap step ΔE_g (i.e. when heterojunction laser structure is used) to 51 K ($\Delta E_g = 63$ meV), 71 K ($\Delta E_g = 90$ meV), 74 K ($\Delta E_g = 137$ meV) and 109 K ($\Delta E_g = 274$ meV).

7.4.3 AlGaAsP/GaAs and AlGaAsP/GaAsP

Heterostructures of this type were first reported in References 11, 49, where DH laser of visible (red) emission was also described. Some details of the preparation of AlGaAsP/GaAsP are presented in (50). The AlGaAsP/GaAs system was also examined from the point of view of fine lattice matching for improvement of conventional AlGaAs/GaAs lasers (8, 10).

Addition of a small amount of phosphides to AlGaAs grown on a GaAs substrate makes it possible to obtain perfect lattice matching at room perature (instead of 930°C in the AlGaAs/GaAs case). When such perfect matching occurs no residual stress arises in the heterostructure; this fact may be useful if one wants to reduce the laser threshold and to improve the reliability characteristics. As was said earlier, mismatch dislocation may also be avoided despite the small lattice mismatch at growth temperature. Phosphorus concentration in the $Al_xGa_{1-x}As_{1-y}P_y$ alloy required for the elimination of residual stress is found to be about $y = 0.006$–0.008 when $x = 0.24$; because of strong segregation of phosphorus in the melt and its escape by evaporation, efficient stress elimination may be obtained in rather thin (2 μm or less) quaternary layers (51, 52).

As was mentioned above, AlGaAsP-system was the first quaternary system of the III–V type used for heterojunction light-emitting devices. In (11) DH lasers are described as prepared on GaP_zAs_{1-z} substrates ($z = 0.067$, 0.138 and 0.40); laser action at 300 K is obtained at threshold as high as 66 kA/cm². Lower thresholds (~ 17 kA/cm²) have been reported in (49) with quaternary alloy as the active medium of the laser.

7.4.4 AlGaAsSb/GaAsSb and AlGaAsSb/GaSb

Substrate preparation for AlGaAsSb/GaAsSb heterojunction includes the growing of intermediate layers to change the lattice parameter at the substrate surface, if the desired value is to be obtained. Some impressive results have been obtained in this way in the 1 μm spectral range (12, 4, 53, 54, 56). New

Table 7.4 Room temperature AlGaAsSb DH lasers

Substrate	Active layer		λ, μm	j, kA/cm²	References
	x	y			
GaAs	—	0.1	0.98	8.5	(12)
GaAs	0.1	0.1	0.95	10 (77 K)	(4)
GaAs	0.07	0.124	1.01	2.1	(53)
GaAs	—	0.124	1.0	1.9	(54)
GaAs	—	0.1	1.0	1.2	(56)
GaSb	—	1.0	1.70–1.78	5	(5)

x and y molar fraction of Al and Sb in quaternary alloy, respectively
j threshold current density

lattice-matched DH lasers for the 1.5–1.8 μm spectral range have been developed on AlGaAsSb/GaSb system (5); these results are summarized in Table 7.3. The first DH formula was pAlGaAsSb/pGaAsSb/nAlGaAsSb/

GaAsSb ... /GaAs (the latter material is the substrate) (12). To prepare efficient laser and light-emitting diodes it is necessary to prepare several layers of $GaAs_{1-y}Sb_y$ of intermediate lattice parameters. For instance, in (53) three preliminary steps of antimonide concentration were used ($y = 0.025$, 0.058 and 0.093) before the final composition with $y = 0.124$ was grown. The diode lasers obtained emit at near 1 μm spectral band in a CW mode of operation. The threshold current density normalized to active layer thickness (in micrometers) appears to be slightly less than those of AlGaAs lasers and equals about 4–4.2 kA/cm^2 μm at room temperature. Incremental efficiencies as high as 23% in the pulse regime and 11% in the continuous-wave regime of operation have been obtained. CW threshold current is found to be in the 100–200 mA range (54).

Longer-wavelength diode laser operating at room temperature (1.78 μm) on the base of $pAl_xGa_{1-x}As_ySb_{1-y}/pGaSb/nAl_xGa_{1-x}As_ySb_{1-y}/GaSb$ heterostructure is reported in (5). The normalized threshold current density of about 5 kA/cm^2 μm was obtained for a laser diode of this type, and pulse operation efficiency up to 37% was reported (5, 55).

Little information is available about the properties of AlGaAsSb alloys. Some investigations of this quaternary alloy are reported in (57) for the composition range of $0.05 \leqslant y \leqslant 0.15$ (y is mole fraction of antimonides). External efficiency of DH diode electroluminescence of about 1% is reported in the spectral range near 1.00–1.02 μm. Electroluminescent spectra containing only narrow-band material emission are observed in pAlGaAsSb/nGaAsSb heterojunction; otherwise, spectra containing some signs of wide-band material emission are observed in nAlGaAsSb/pGaAsSb heterojunctions (57). This suggests that the condition-band-edge step in the heterojunction should be close to ΔE_g (with an accuracy to kT) (37).

7.4.5 GaInAsSb/GaSb and some others

A wide spectral range from 1.8 to 3.5 μm (300 K), is covered by GaInAsSb/GaSb iso-lattice-parameter system. Systematic investigation of the GaInAsSb alloy has been commenced only recently.

Luminescence and laser action under electron beam pumping is reported in (58, 59). An efficiency of spontaneous electroluminescence of about 1% at 2 μm wavelength has been obtained (59). Room temperature laser action at electron beam pumping has also been obtained at wavelengths up to 2.2 μm (59). A DH injection laser operating at wavelength 1.85–1.90 μm was first reported in (7). The design was based on the following structures: pGaSb : Zn/pGaInSbAs : Zn/nGaSb : Te, and a quaternary active layer thickness of about 1 μm was used. Liquid-nitrogen temperature luminescence was detected in the spectral range of 1.8–1.9 μm. Spectral narrowing was observed at 77 K when pumping current density approached the value of about 1 kA/cm² as shown in Figure 7.11.

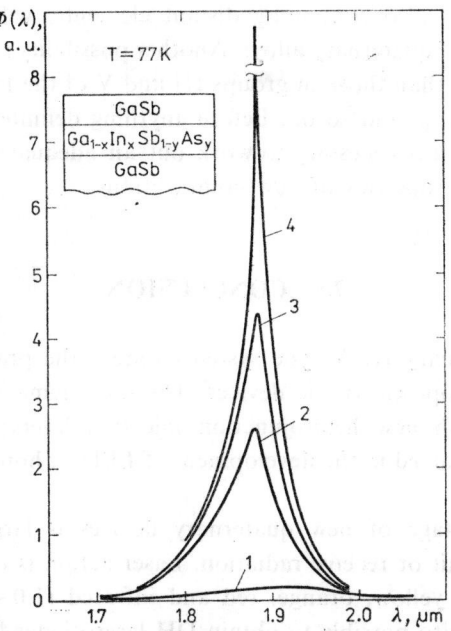

Figure 7.11 Spectral narrowing and laser action in the GaInSbAs/GaSb DH laser at 77 K. Current densities are equal to 50 A/cm² (1), 400 A/cm² (2), 600 A/cm² (3), 900 A/cm² (4)

A possible feature of this heterostructure system is a refractive index discontinuity at $Ga_{1-x}In_xSb_{1-y}As_y$/GaSb interface. The interpolation approach to this quantity leads to an 'anti-waveguide' effect rather than the usual waveguiding. This follows from lower refractive index of InAs than GaSb at some fixed wavelength. When $x \approx y \approx 0.1$ and the emission wavelength is about 2 μm one may estimate the relative step $\Delta n/n$ to be -0.6%. This quantity attains a larger negative value when x and y increase (possibly to about -9%). However, this speculative estimation has to be verified experimentally. If it turns out to be true, this system will be a candidate for longitudinal injection lasers.

A number of other quaternary systems have been proposed for experimental testing. There is a lot of interest recently in systems which involve distant (not the nearest) neighbouring elements of the Periodic Table. 'Triangular' AlGaInAs and GaPAsSb systems may be pointed out as examples. Realization of this iso-lattice-parameter heterojunction (as well as of some 'square' quaternary systems also containing distant elements) will be the next step in the technology of quaternary alloys. Another possibility are systems containing elements other than those in groups III and V of the Periodic Table, such as $(GaAs)_x(ZnSe)_{1-x}$, and so on. Before anything definite can be said about these perspectives, it is necessary to work out an adequate approach to predicting the basic properties of new compositions.

7.5 CONCLUSION

Use of the quaternary III–V type systems widens the practical applicability of heterojunction optoelectronic devices. The most remarkable achievements are connected with new heterojunction injection lasers; important results have also been obtained in the development of LEDs, photodiodes and photocathodes.

The first advantage of new quaternary devices is larger spectral range where they can emit or receive radiation. Laser action is obtained in yellow-green (0.575 μm), yellow, orange, red and infrared (1.0–1.78 μm) emission bands. Now it is also possible to obtain DH laser diodes for communication links operating at one of three optimal transparency bands of optical fibres (0.8–0.85 μm, 1.0–1.1 μm, 1.6–1.7 μm), or at a wavelength close to material dispersion minimum in optical fibre ($\sim$ 1.27 μm). Some of these DH laser diodes are operating continuously at room temperature; all of them may be readily modulated at frequencies as high as some hundreds of megacycles per second.

LEDs of GaInPAs/InP and GaInAsSb/GaSb are interesting because of substrate transparency to quaternary alloy emission. It is convenient for optimal construction of the body of emitting crystal. In planar-geometry LED, this body does not include any absorbing region except an active one. Because of this, photons emitted there and reflected by a surface of crystal inside can be absorbed in active layer and then re-emitted. This recycling of photons leads to an increase of external output efficiency because at every such cycle some portion of radiation is contributed into external emission flow. As an example, DH diodes of GaInPAs/InP are found to be able to emit 1.08–1.10 μm radiation with external efficiency as large as 8.4% at room temperature (in contrast with 1.4% value predicted by single-pass model) (22). This testifies the existence of re-emission effect in such DH diodes. It is also possible to decrease the laser threshold due to re-absorption of spontaneous emission in the active layer exclusively.

From the point of view of technology, the use of quaternary iso-lattice-parameter systems can simplify the preparation procedure of DH diodes by avoiding grading layers. This leads also to improvement of crystallographic perfection and of reliability of devices. For example, CW laser operation in GaInPAs/InP DH diodes as long as three thousand hours has been demonstrated (27, 28) in contrast to about 20 hours operation of AlGaAsSb/GaAsSb DH diodes prepared with grading GaAsSb multilayer structure at GaAs substrate. Note that double heterostructure of GaInPAs/InP preparation may include the growing of only two layers at the substrate; for some integrated-optical purposes waveguiding structure can be prepared by growing of one layer of GaInPAs at InP substrate.

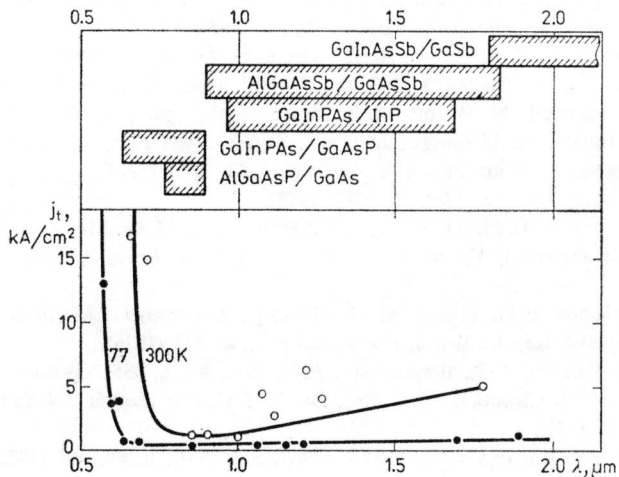

Figure 7.12 Spectral ranges of some iso-lattice-parameter quaternary III–V systems and typical values of threshold current density over all spectral ranges (compiled)

Spectral ranges of the quaternary III–V type alloy system under investigation are presented in Figure 7.12, and laser threshold current densities are plotted over all these ranges.

The author is very grateful to his colleagues L. M. Dolginov, L. V. Druzhinina, E. G. Shevchenko, M. G. Milvidski, L. M. Margulis, B. N. Sverdlov for ready and fruitful cooperation.

REFERENCES

1. Zh. I. Alferov, *Vestnik AN SSSR*, 7, 28 (1976), *Thin Solid Films*, 36, 2, 441 (1976).
2. M. B. Panish, I. Hayashi, *Appl. Solid State Sci.*, 4, 235 (1974).
   M. B. Panish, *Proc. IEEE*, 64, 10, 1540 (1976).
3. P. G. Eliseev, *Kvantovaya Elektronika*, 6 (12), 3 (1972).
4. A. P. Bogatov, L. M. Dolginov, L. V. Druzhinina, P. G. Eliseev, B. N. Sverdlov, E. G. Shevchenko, *Kvantovaya Elektronika*, 1, 10, 2294 (1974); English transl.: *Sov. J. Quant. Electron.*, 4, 1281 (1975).
5. L. M. Dolginov, L. V. Druzhinina, P. G. Eliseev, M. G. Milvidski, B. N. Sverdlov, *Kvantovaya Elektronika*, 3, 2, 465 (1976); English transl.: *Sov. J. Quant. Electron.*, 6, 257 (1976).
6. L. M. Dolginov, L. V. Druzhinina, P. G. Eliseev, M. G. Milvidski, B. N. Sverdlov E. G. Shevchenko, *Kratkie soobshcheniya po fizike, FIAN, M.*, 8, 29 (1976).
7. L. M. Dolginov, L. V. Druzhinina, P. G. Eliseev, N. A. Lapshin, M. G. Milvidski, B. N. Sverdlov, *Kvantovaya Elektronika*, 5, 3, 720 (1978).
8. G. A. Rozgonyi, M. B. Panish, *Appl. Phys. Lett.*, 23, 10, 533 (1973).
9. J. W. Matthews, S. Mader, T. B. Light, *J. Appl. Phys.*, 41, 3800 (1970).
10. G. A. Rozgonyi, P. M. Petroff, M. B. Panish, *Appl. Phys. Lett.*, 24, 6, 251 (1974).
11. R. D. Burnham, M. Holonyak, Jr., D. R. Scifres, *Appl. Phys. Lett.*, 17, 455 (1970).
12. K. Sugiyama, H. Saito, *Japan. J. Appl. Phys.*, 11, 1057 (1972).
13. H. Kressel, *J. Electron. Mat.*, 4, 1081 (1975).
14. A. V. Govorkov, L. M. Dolginov, L. V. Druzhinina, M. G. Milvidski, L. M. Margulis, V. B. Osvenski, Yu. M. Fishman, T. G. Yugova, *Kristallographia*, 22, 5, 1060, (1977).
15. L. M. Dolginov, P. G. Eliseev, M. G. Milvidski, *Kvantovaya Elektronika*, 3, 7, 1381 (1976); English transl.: *Sov. J. Quant. Electron.*, 6, 747 (1976).
16. J. A. Van Vechten, T. K. Bergstrasser, *Phys. Rev.*, B1, 8, 3351 (1970).
17. A. Onton, P. J. Chicotka, *Phys. Rev.*, B4, 1847 (1971); *Izvestia AN SSSR, ser. fiz.*, 37, 3, 560 (1973).
18. A. Pikhtin, V. Razbegaev, D. Yaskov, *Phys. Stat. Sol.*, B, 50, 717 (1972).
19. G. A. Antypas, R. L. Moon, *J. Electrochem. Soc.*, 120, 1574 (1973).
20. P. N. Sen, G. Lucovsky, *Phys. Rev.*, B12, 8, 2998 (1975).
21. H. C. Casey, Jr., D. D. Sell, M. B. Panish, *Applied Phys. Lett.*, 24, 2, 63 (1974).
22. A. P. Bogatov, L. M. Dolginov, P. G. Eliseev, M. G. Milvidski, B. N. Sverdlov, E. G. Shevchenko, *FTP*, 9, 1956 (1975).
23. K. Oe, K. Sugiyama, *Japan. J. Appl. Phys.*, 15, 10, 2003 (1976).
24. T. P. Pearsall, B. I. Miller, R. J. Capik, K. J. Bachmann, *Appl. Phys. Lett.*, 28, 9, 499 (1976).
25. J. J. Hsieh, *Appl. Phys. Lett.*, 28, 5, 283 (1976).

26. J. J. Hsieh, M. C. Finn, J. A. Rossi, 'Gallium Arsenide and Related Compounds', *Proc. Sixth Int. Symp.*, *Cont. ser.*, **336**, p. 37 (1976).
27. J. J. Hsieh, C. C. Shen, *Appl. Phys. Lett.*, **30**, 8, 429 (1977).
28. C. C. Shen, J. J. Hsieh, T. A. Lind, *Appl. Phys. Lett.*, **30**, 7, 353 (1977).
29. K. Oe, S. Ando, K. Sugiyama, *Japan. J. Appl. Phys.*, **16**, 9, 1693 (1977).
30. K. Oe, S. Ando, K. Sugiyama, *Japan. J. Appl. Phys.*, **16**, 7, 1273 (1977).
31. P. D. Wright, E. A. Rezek, M. J. Ludowise, N. Holonyak, Jr., *J. Appl. Phys.*, **48**, 5 2091 (1977).
32. I. Yamamoto, K. Sakai, S. Akiba, Y. Suematsu, *Electron. Lett.*, **13**, 5, 142 (1977).
33. K. Wakao, K. Moriki, T. Kambayashi, K. Iga, *Japan. J. Appl. Phys.*, **16**, 11, 2073) (1977).
34. H. Nagai, *Noguchi 100C-77*, **B2-2**, 201 (1977).
35. J. J. Hsieh, J. A. Rossi, J. P. Donnelly, *Appl. Phys. Lett.*, **28**, 12, 709 (1976).
36. L. M. Dolginov, N. Ibrakhimov, V. Yu. Rogulin, E. G. Shevchenko, *FTP*, **10**, 6, 1224 (1976).
37. L. M. Dolginov, P. G. Eliseev, M. G. Milvidski, B. N. Sverdlov, E. G. Shevchenko, *FIAN, M.*, **8**, 38 (1976).
38. R. Sankaran, G. A. Antypas, R. L. Moon, J. S. Escher, L. W. James, *J. Vac. Sci. Technol.*, **13**, 4, 932 (1976).
39. C. C. Shen, J. J. Hsieh, T. A. Lind, *Appl. Phys. Lett.*, **30**, 7, 353 (1977).
40. T. Yamamoto et al., *Japan. J. Appl. Phys.*, **16**, 9, 1699 (1977).
41. L. M. Dolginov, L. V. Druzhinina, E. M. Krasavina, I. V. Kryukova, E. B. Matveenko, Yu. V. Petrushenko, S. P. Prokofieva, V. P. Tsiganov, E. G. Shevchenko, *Kvantovaya Elektronika*, **3**, 11, 2490 (1976).
42. J. J. Coleman, N. Holonyak, Jr., M. J. Ludowise, P. D. Wright, W. O. Groves, D. L. Keune, *IEEE J. Quant. Electron.*, **QE-11**, 7, 471 (1975).
43. R. J. Nelson, N. Holonyak, Jr., W. R. Hitchens, D. Lazarus, M. Altarelli, *Solid State Commun.*, **18**, 321 (1976).
44. Zh. I. Alferov, I. N. Arsent'ev, D. Z. Garbuzov, S. G. Konnikov, V. D. Rumyantsev, *Pisma ZTF*, **1**, 305 (1975).
    Zh. I. Alferov, I. N. Arsent'ev, D. Z. Garbuzov, V. D. Rumyantsev, *Pisma ZTF*, **1**, 406 (1975).
45. Zh. I. Alferov, I. N. Arsent'ev, D. Z. Garbuzov, V. D. Rumyantsev, V. P. Ulin, *Pisma ZTF*, **2**, 241 (1976).
    Zh. I. Alferov, I. N. Arsent'ev, D. Z. Garbuzov, V. D. Rumyantsev, V. P. Ulin, *Pisma ZTF*, **2**, 481 (1976).
46. J. J. Coleman, N. Holonyak, Jr., M. J. Ludowise, P. D. Wright, *J. Appl. Phys.*, **47**, 5, 2015 (1976).
    P. D. Wright, J. J. Coleman, N. Holonyak, Jr., M. J. Ludowise, G. E. Stillman, *J. Appl. Phys.*, **47**, 3580 (1976).
47. N. Holonyak, Jr., R. Chin, J. J. Coleman, D. L. Keune, W. O. Groves, *J. Appl. Phys.*, **48**, 635 (1977).
48. R. Chin, N. Holonyak, Jr., H. Shichijo, W. O. Growes, D. L. Keune, J. A. Rossi, *J. Appl. Phys.*, **48**, 3991 (1977).
49. R. D. Burnham, N. Holonyak, Jr., H. W. Korb, H. M. Macksey, D. R. Scifres, J. B. Woodhouse, Zh. I. Alferov, *Appl. Phys. Lett.*, **19**, 25 (1971).

50. Zh. I. Alferov, V. M. Andreev, T. B. Godlinnik, S. G. Konnikov, V. R. Larionov, *Kristall und Technik*, **10**, 633 (1975).
51. R. L. Brown, R. G. Sobers, *J. Appl. Phys.*, **45**, 4735 (1974).
52. M. A. Afromowitz, D. L. Rode, *J. Appl. Phys.*, **45**, 4738 (1974).
53. R. E. Nahory, M. A. Pollack, *Appl. Phys. Lett.*, **27**, 562 (1975).
54. R. E. Nahory, M. A. Pollack, E. D. Beebe, J. C. DeWinter, R. W. Dixon, *Appl. Phys. Lett.*, **28**, 19 (1976).
55. L. M. Dolginov, L. V. Druzhinina, P. G. Eliseev, I. V. Kryukova, V. I. Leskovich, M. G. Milvidski, B. N. Sverdlov, E. G. Shevchenko, *IEEE J. Quant. Electron.*, **QE-13**, 609 (1977).
56. R. E. Nahory, M. A. Pollack, J. R. Abrokwah, *J. Appl. Phys.*, **48**, 3988 (1977).
57. L. M. Dolginov, L. V. Druzhinina, N. Ibrakhimov, V. Yu. Rogulin, *FTP*, **10**, 847 (1976).
58. S. A. Bondar, N. A. Borisov, D. N. Galchenkov, E. M. Lavrushin, V. V. Lebedev, S. S. Strelchenko, *Kvantovaya Elektronika*, **3**, 94 (1976).
59. L. M. Dolginov, L. V. Druzhinina, P. G. Eliseev, V. I. Kryukova, V. I. Leskovich. M. G. Milvidski, B. N. Sverdlov, V. A. Chapnin, *Kvantovaya Elektronika*, **3**, 932 (1976).
60. R. E. Nahory, M. A. Pollack, J. C. DeWinter, *J. Appl. Phys.*, **48**, 320 (1977).

Chapter 8

Electroepitaxy of $A^{III} B^{V}$ Compounds and its Application to Optoelectronic Devices Technology

T. Bryśkiewicz

8.1 INTRODUCTION

Most types of the electronic device structures of $A^{III}B^{V}$ compounds are based on epitaxial layers grown from high temperature solutions (1). During liquid phase epitaxial growth the substrate is thermodynamically equilibrated with A^{III}-rich solution (2), and then growth takes place upon supersaturation of the solution. It can be achieved in the following ways (Figure 8.1):

— by slow cooling of the solution (2),
— in a steady-state temperature gradient with the use of the source wafer (1),
— by passing direct current across the substrate-solution interface at constant temperature (3–20).

Current-controlled LPE growth method has been successfully developed in recent years (3–20). Schematic diagram of the electroepitaxial growth system is shown in Figure 8.2. It utilizes the standard LPE equipment modified to permit passage of electric current through the seed-melt interface (3). Two stainless steel electrodes are threaded into the slider and the solution holder respectively, and both parts of the graphite boat are electrically insulated with boron nitride. Thus, electric current can flow only when the seed is brought into contact with the melt. A uniform electric contact between the bottom part of the substrate and the graphite slider can be established with a thin gallium layer, a Ta foil (a few microns thick) and a second thin layer of gallium (5, 18). In a typical growth experiment the A^{III}-rich melt is heated to 30–50° C above the growth process T_0 for at least an hour to ensure adequate mixing of all the constituents (A^{III}-solvent, AB-compound or compounds, and dopants (2)). The furnace is then allowed to cool to the temperature T_0 and the source wafer preceeds the substrate into the melt for an hour

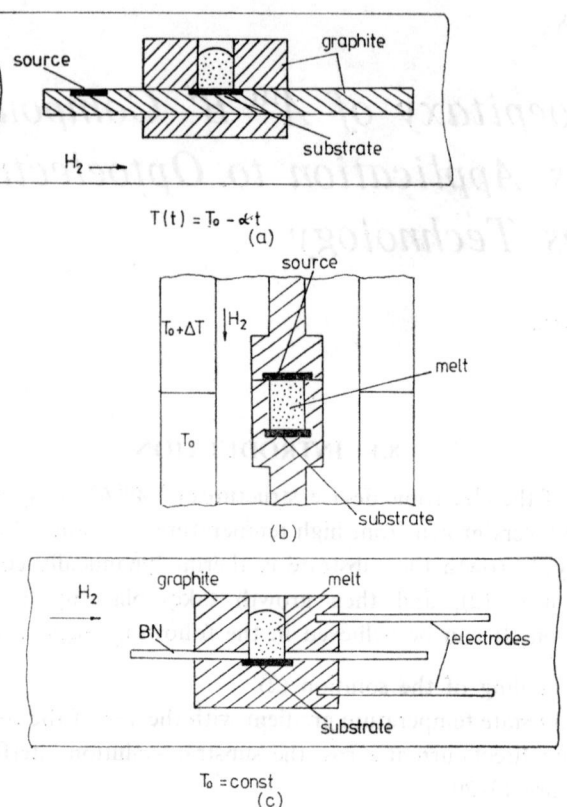

Figure 8.1 Schematic cross section of the boat used for LPE growth: (a) growth by linear ramp cooling of the melt, (b) growth in the steady-state, imposed-temperature-gradient, (c) growth produced by an electric current flowing across the substrate-melt interface

to complete its saturation (5). Finally, the substrate is brought into contact with the melt and electric current is passed through the interface at a density 0.3–60 A/cm^2 (3–20). The substrate is usually n-type and has a positive polarity with respect to the melt (3–20). Growth is terminated by switching the current off and moving the slider. Epitaxial layers range in thickness from 0.3 to 150 μm and the growth time—from a few seconds (16) to a few hours (5, 18). Electroepitaxy* has since been employed for the growth of InSb (3), GaAs (5, 13–20), InP (11), GaAlAs (6, 7) as well as GaAs-GaAlAs DH lasing wafers (16)

---

* This term has been proposed by J. Łagowski (18).

Electroepitaxy of $A^{III}B^V$ Compounds

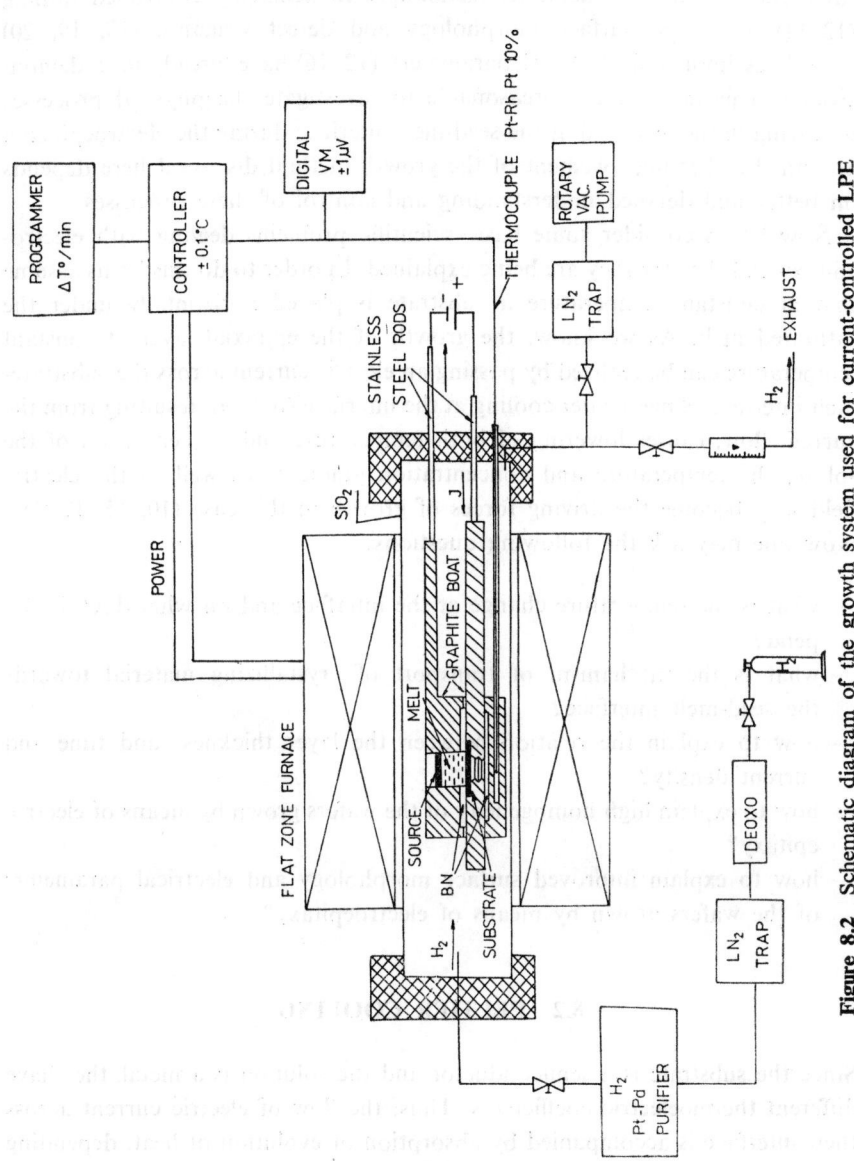

Figure 8.2 Schematic diagram of the growth system used for current-controlled LPE

and p–n junctions (12). Although the growth method discussed here is in the early stage of development, its advantages in achieving controlled doping (12–14), improved surface morphology and defect structure (13, 19, 20) as well as improved electrical parameters (12, 16) have already been demonstrated. This fact makes it reasonable to investigate the physical processes occurring in the melt and at the seed-melt interface during the electroepitaxial growth. Further improvement of the growth method discussed here depends on better and detailed understanding and control of these processes.

Now let us consider some basic scientific problems dealing with electroepitaxy and the way they are being explained. In order to do this let us assume that at constant temperature a substrate is placed horizontally under the saturated melt. As we know, the growth of the epitaxial layer at constant temperature can be realized by passing an electric current across the substrate-melt interface. Since Peltier cooling at the interface (6, 8, 9), resulting from the current flow, causes lowering of both temperature and concentration of the solute, the temperature and concentration gradients as well as the electric field may become the driving forces of growth in this case (10, 15, 17, 18). Now one may ask the following questions:

— what is the temperature change at the interface and on what does it depend?
— what is the mechanism of transport of crystallizing material towards the seed-melt interface?
— how to explain the relation between the layer thickness and time and current density?
— how to explain high homogeneity of the wafers grown by means of electroepitaxy?
— how to explain improved surface morphology and electrical parameters of the wafers grown by means of electroepitaxy?

8.2 PELTIER COOLING

Since the substrate is a semiconductor and the solution is a metal, they have different thermoelectric coefficients. Thus, the flow of electric current across their interface is accompanied by absorption or evolution of heat, depending on the current direction. The magnitude of this heat Q is equal to (8)

$$Q = \pi \cdot J \qquad (8.1)$$
$$|\pi| = E_c - E_f + 2kT \qquad (8.2)$$

where π is the difference in Peltier coefficients of the melt and the substrate, J is the current density, E_c and E_f are energies of the bottom of the conduction band and of the Fermi level.

In order to answer the question 'What is the temperature change at the interface and on what does it depend?' Daniele and Michel as well as E. K. Stefanokos et al. (6, 8, 9) used a simple thermal conduction model. They assumed that heat flows solely by conduction that there is no thermal contact resistance between materials, and that the Thompson effect is negligible (see Figure 8.3). When an electric current is passing through the system, heat is

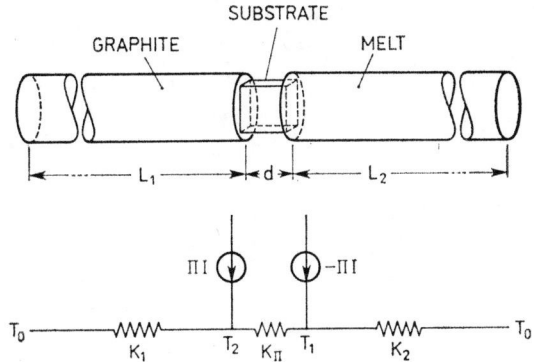

Figure 8.3 (8) Schematic diagram of graphite-substrate-melt region and the corresponding electrical analog circuit

removed by the Peltier effect from the melt-seed junction and accumulated at the seed-graphite junction with a rate equal to $\pi \cdot J$. The heat flow in each region is defined as follows:

$$Q_1 = (T_0 - T_1) \cdot K_1 \quad (8.3)$$

$$Q_2 = (T_2 - T_0) \cdot K_2 \quad (8.4)$$

$$Q = \pi \cdot J - (T_2 - T_1) \cdot K_\pi \quad (8.5)$$

and

$$K_1 = K_L \cdot \left(\frac{A}{L_1}\right), \quad K_2 = K_G \cdot \left(\frac{A}{L_2}\right), \quad K_\pi = K_S \cdot \left(\frac{A}{R}\right) \quad (8.6)$$

where K_1, K_2, and K_π are thermal conductances; K_L, K_S and K_G are thermal conductivities of the melt, substrate, and graphite pedestal, respectively; A is

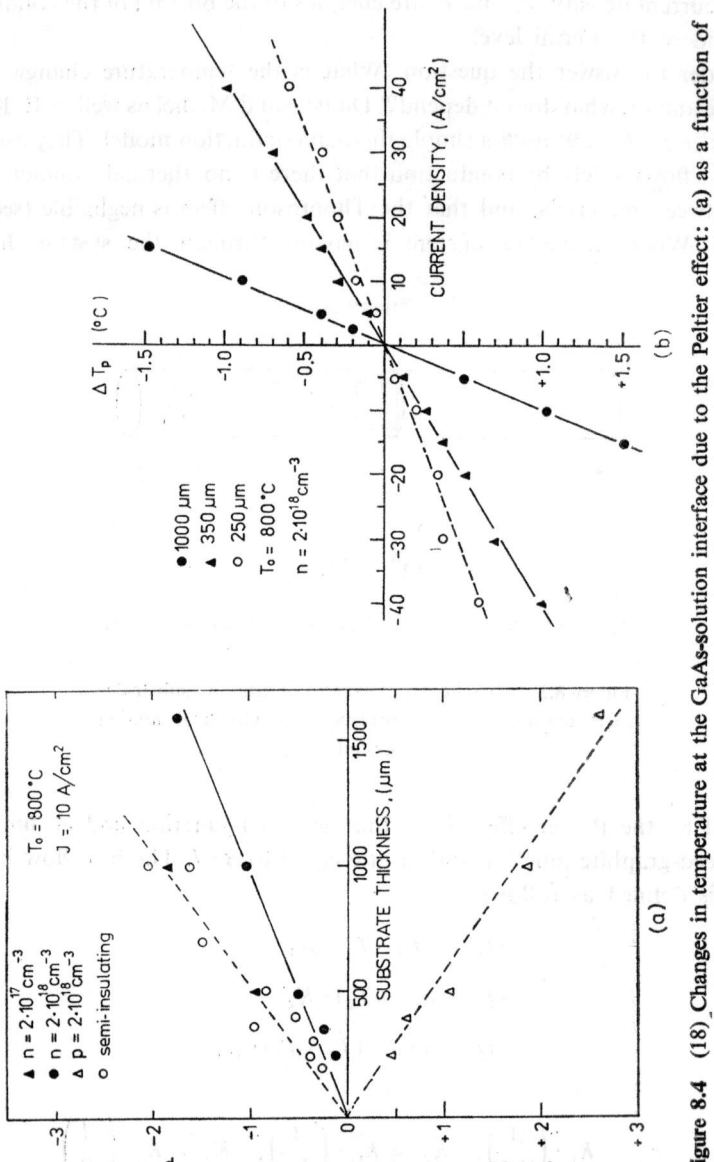

Figure 8.4 (18) Changes in temperature at the GaAs-solution interface due to the Peltier effect: (a) as a function of substrate thickness, (b) as a function of current density

the cross-section area; R is the thickness of the substrate; L_1 and L_2 are the heights of the melt and the graphite pedestal; T_0, T_1, and T_2 are furnace temperature, temperature of the upper and bottom parts of the substrate. Since the heat flow is continuous, we may write

$$Q_1 = Q_2 = Q \qquad (8.7)$$

and the following equations can be obtained

$$T_0 - T_1 = \frac{\pi \cdot J}{(1 + K_1/K_2) \cdot K_\pi + K_1} \qquad (8.8)$$

$$T_2 - T_0 = \frac{\pi \cdot J}{(1 + K_2/K_1) \cdot K_\pi + K_2} \qquad (8.9)$$

It can be seen that the temperature changes $T_1 - T_0$ and $T_2 - T_0$ depend on the electrical parameters of the substrate (π), current density J, thickness R of the substrate and the growth cell geometry. Stefanokos et al. (8, 9) as well as L. Jastrzebski et al. (18) measured Peltier cooling at the GaAs substrate-Ga rich solution and the InP substrate-In rich solution interfaces vs. current density J, growth temperature T_0, sample thickness R and carrier concentration of p- and n-type substrates. The results are shown in Figure 8.4. It can be seen that in electroepitaxy the parameter ΔT can be established by applying a given carrier concentration and thickness of the substrate as well as the proper direction and density of the electric current.

8.3 GROWTH KINETICS

Let us assume that at a temperature T_0 a substrate is placed horizontally under the saturated melt of concentration $C_L(T_0)$ and thickness L (Figure 8.5). Let a source wafer be in turn situated on the surface of the melt. As we know, epitaxial growth at constant temperature can be achieved by passing an electric current across the substrate-melt interface (3). Since Peltier-cooling at the interface, produced by the current, causes a lowering of the temperature and concentration of the solute down to T_1 and $C_L(T_1)$, respectively (Figure 8.6), the solute concentration and temperature gradients as well as the electric field may in this case become the driving forces of growth. Thus, quantitative analysis of electroepitaxial growth kinetics requires taking into consideration the following transport mechanisms of the crystallizing material:

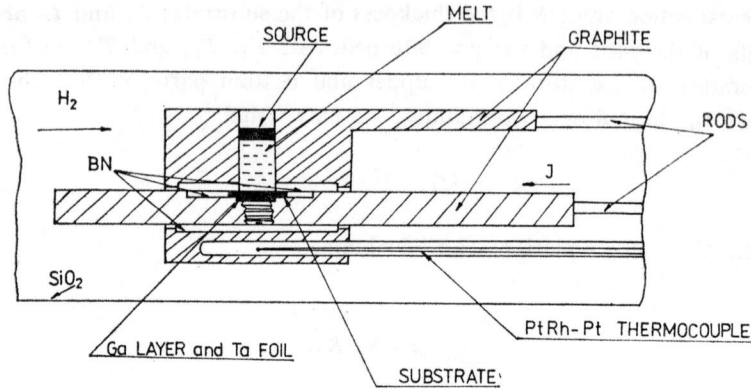

Figure 8.5 Schematic cross section of the graphite boat used for current-induced LPE

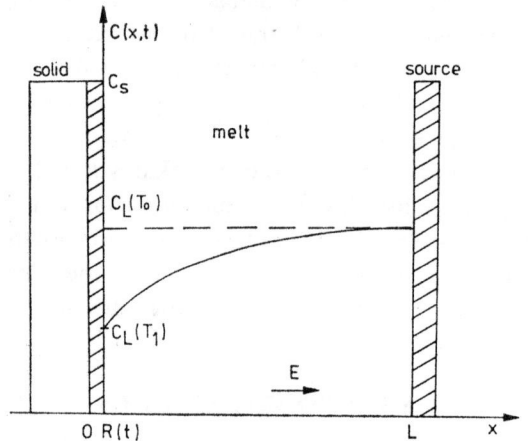

Figure 8.6 Distribution of the solute concentration $C(x, t)$ in the melt and solid: C_L—equilibrium composition of the melt, R—thickness of the epitaxial layer, L—thickness of the melt, t—time, E—electric field intensity

— diffusion due to the gradient of the solute concentration. The current of the diffusing particles can be expressed by (22)

$$J_D = -D \frac{\partial C}{\partial x} \tag{8.10}$$

where D is the diffusion coefficient, and C is the solute concentration;

— convective flow of material due to differences of the solution density induced by a horizontal temperature gradient (22)

$$J_V = V \cdot C \tag{8.11}$$

where V is the volume of the melt passing through the unit plane in one second;
— electromigration due to the direct interaction of the electrostatic field $E = J/\sigma$ with ionized particles (22)

$$J_E = u_0 \cdot F_E \cdot C \tag{8.12}$$

$$F_E = Z \cdot E, \quad u_0 = \frac{F \cdot D}{R \cdot T} \tag{8.13}$$

where u_0, F, and σ are the mobility, Faraday's number, and conductivity of the melt, and F and Z are the electrostatic force and valency of solute ions:
— electrotransport due to collisions (momentum exchange) between electrons flowing across the metallic solution and solute particles

$$J_{el} = u_0 \cdot F_{el} \cdot C \tag{8.14}$$

$$F_{el} Z' \cdot E, \quad Z' = e \cdot n \cdot l \cdot \sigma^* \tag{8.15}$$

where F_{el} is called the *drag, friction force* or *electron wind*; e and n are elementary charge and concentration of free electrons in the melt; l and σ^* are free path of electrons on Fermi-level and solute ions cross section due to electron scattering.

Finally, the total current of solute particles flowing towards the interface is equal to

$$J_{tot} = J_D + J_V + J_E + J_{el} = -D \cdot \frac{\partial C}{\partial x} + u_0 \cdot Z_{eff} E \cdot C + V \cdot C \tag{8.16}$$

$$Z_{eff} = Z + e \cdot n \cdot l \cdot \sigma^* \tag{8.17}$$

where Z_{eff} is the effective charge (valency) of solute ions. Hence, distribution of solute particles in the melt during electroepitaxy can be expressed as follows:

$$\frac{\partial C}{\partial t} + \mathrm{div} J_{tot} = 0 \tag{8.18}$$

$$\frac{\partial C}{\partial t} = D \cdot \frac{\partial^2 C}{\partial x^2} - u_0 \cdot Z_{eff} \cdot E \cdot \frac{\partial C}{\partial x} - V \cdot \frac{\partial C}{\partial x} \tag{8.19}$$

Attempts at a quantitative analysis of the current-induced growth kinetics of epitaxial layers have been, thus far, limited in scope. It was originally proposed (3) that Peltier cooling at the growth interface is responsible for supersaturation of the melt and epitaxial growth. Subsequently, it was demonstrated that under certain conditions electromigration (10, 18), or rather electrotransport (17), constitutes the dominant contribution to supersaturation, while in other studies it was shown that solute convection due to temperature gradients induced by the electric current dominates electroepitaxy (15).

Current-controlled LPE layers of GaAs prepared for growth kinetics analysis by L. Jastrzebski et al. (10) were grown in a commonly employed apparatus shown in Figure 8.2. n-Type GaAs layers obtained in 90 min at 850°C had a thickness of 60 µm, while a total temperature decrease of only 0.2°C was measured with a thermocouple at a distance of about 200 µm from the growth interface. In a conventional cooling experiment using the same experimental arrangement, an epitaxial layer of about 0.5 µm was obtained upon a temperature decrease of 0.2°C. These results indicate that Peltier cooling is not the primary driving force in current-controlled LPE; more likely, migration of particles in the electric field causes supersaturation of the melt at the interface and growth of the epitaxial layers at a virtually constant temperature.

To demonstrate the electromigration of As and Ga, L. Jastrzebski et al. (10) performed the following experiment. A p-type Zn-doped GaAs wafer ($2 \times 10^{18}/cm^3$) was brought into contact with Ga-As melt, in equilibrium at 850°C, contained in two wells as shown schematically in Figure 8.7. The

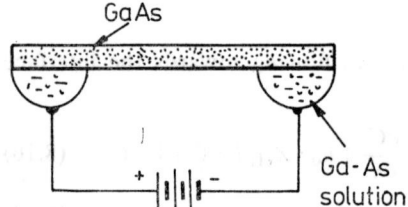

Figure 8.7 (10) Schematic diagram of experimental arrangement for observing electromigration in solutions of III-V compounds

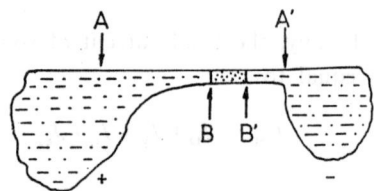

Figure 8.8 (10) Electromigration of Ga and As (see text). A and A' are starting points of electromigration of Ga and As, respectively. At B and B' electromigration was terminated

result obtained after passing the electric current (0.5 A/cm² for 40 s) is shown in Figure 8.8. The solidified melts are attached to the ends of the wafer. It is seen that migration of the solution took place along the surface of the wafer up to regions B and B' in Figure 8.8. Electron microprobe analysis showed that the concentration of As in the solution on the wafer near B' was more than 50% higher than on the surface near B. Apparently electromigration of charged As and Ga species caused the observed migration of the solution; no solution migration was obtained in the absence of an electric current. The distance of migration on the wafer was found to increase linearly with time and with electric field. Accordingly, effective mobility for Ga μ_{Ga} migrating to the negative end of the wafer and for As μ_{As} migrating to the positive end were calculated to be 2×10^{-2} and 7×10^{-3} cm^2/Vs, respectively. These mobility values are believe to be smaller than in the LPE process, since surface tension is acting against solution migration in the experimental arrangement of Figure 8.7.

Since the contribution of diffusion and the movement of the growth interface can be neglected, the flux of As towards the interface is (10)

$$J_{As} = C_{As} \cdot C_{Ga} \cdot \overline{V}_{Ga} \cdot U \tag{8.20}$$

where C_{As}, C_{Ga}, and $\overline{V}_{Ga}$ are concentration of As, concentration of Ga and the partial molar volume of Ga ($C_{Ga} \cdot \overline{V}_{Ga} \cong 1$), and U is the difference in mobilities $\mu_{Ga} - \mu_{Aa}$.

On the basis of the above relationship, and assuming that $U \approx 2.7 \times \times 10^{-2}$ cm^2/Vs, $E \approx 10^{-3}$ V/cm, and that there is no nucleation of GaAs in the melt, a growth rate of ~ 0.8 μm/min was obtained. This value is in good agreement with the experimentally determined values of 0.4–0.7 μm/min.

The above experiment were also carried out employing In-Sb solutions and InSb substrates at 450°C (10). The results were consistent with those obtained for the GaAs system.

A more precise analysis of the current-controlled growth kinetics of LPE GaAs has been made by Bryśkiewicz (17). However, his analytical treatment applies only to such a substrate-melt arrangement in which the natural convection can be neglected. In order to find an analytical formula expressing the rate $V(t)$ of the current-induced LPE growth he made a general balance of material in the melt and epitaxial layer at moment t (Figure 8.6)

$$\varrho^l C_L(T_0) \cdot L = \varrho^s \cdot C_s \cdot R(t) + \varrho^l \int_R^L C(x, t) \mathrm{d}x \tag{8.21}$$

$$\frac{\partial C}{\partial t} = D \frac{\partial^2 C}{\partial x^2} - Z_{eff} \cdot u_0 \cdot E \cdot \frac{\partial C}{\partial x} \tag{8.22}$$

where ϱ^s and ϱ^l are densities of the solid and melt, C_s is the composition of the solid, and L and $R(t)$ are thicknesses of the melt and epitaxial layer.

Differentiation of Equation 8.21 with respect to t and substitution of Equation 8.22 into 8.21 yields

$$0 = \varrho^s \cdot C_s \cdot V(t) + \varrho^l \cdot D \int_R^L \frac{\partial^2 C}{\partial x^2} dx -$$

$$- Z_{eff} \varrho^l \cdot u_0 \cdot E \int_R^L \frac{\partial C}{\partial x} dx - \varrho^l \cdot C(R, t) \cdot V(t)$$

Taking into account the fact that $\left.\frac{\partial C}{\partial x}\right|_{x=L} = 0$, $C(R, t) = C_L(T_1)$ and $E(L) = 0$, one obtains the following formula:

$$V(t) = \frac{D \left.\frac{\partial C}{\partial x}\right|_{x=R} - Z_{eff} \cdot u_0 \cdot E \cdot C_L(T_1)}{C'_s - C_L(T_1)} \qquad (8.24)$$

where $C'_s = C_s \frac{\varrho^s}{\varrho^l}$. Distribution of the solute particles near the growth interface can be estimated from Equation 8.22 with the proper boundary condition. L. Jastrzebski et al. (18) obtained the following analytical formula

$$V(t) = \frac{\Delta T}{C'_s - C_L(T_1)} \cdot \frac{dC_L}{dT} \left(\frac{D}{\pi t}\right)^{1/2}$$

$$\times \frac{\exp\left[-\frac{(Z_{eff} u_0 E t)^2}{4Dt}\right]}{\operatorname{erfc}\left(-\frac{Z_{eff} u_0 E t}{2\sqrt{Dt}}\right)} - Z_{eff} \cdot u_0 \cdot E \cdot \frac{C_L(T_1)}{C'_s - C_L(T_1)} \qquad (8.24')$$

valid for a very thick solution layer ($L \to \infty$).

In the case of electroepitaxy from A^{III}-rich solutions the condition

$$Z_{eff} u_0 E \sqrt{t} \ll 2 \sqrt{D} \qquad (8.25)$$

for the time periods of less than about 10 hours is fulfilled (18). Hence

$$V(t) = \frac{\Delta T \frac{dC_L}{dT} \left(\frac{D}{\pi t}\right)^{1/2} - Z_{eff} u_0 E C_L(T_1)}{C'_s - C_L(T_1)} \qquad (8.24'')$$

The first term in Equation 8.24″ represents the contribution of the Peltier effect, whereas the second term represents the contribution of migration to the growth rate; both terms depend linearly on the current density, through ΔT and E, respectively. It can easily be seen that the growth rate calculations require the dependences: $D(T)$, $C_L(T)$, $\sigma(C)$, $\Delta T = T_1 - T_0$ as well as the Z_{eff} parameter to be known. To the best of our knowledge only $C_L(T)$

and $D(T)$ are known (21), $\sigma(C)$ and Z_{eff} have not been measured yet, while $\Delta T = f(J, T)$ may depend on the graphite boat geometry (8, 9) and is required to be estimated separately for each technological equipment. Since an exact measurement of the ΔT parameter by means of a thermocouple is very difficult, Bryśkiewicz estimated it indirectly taking into account the fact that the growth of epitaxial layer without the source on the solution surface occurs until the following condition is fulfilled

$$J_{\text{tot}} = -D\frac{\partial C}{\partial x} + Z_{\text{eff}} u_0 EC = 0 \qquad (8.26)$$

which means that the diffusion, electromigration and electrotransport streams balance one another. Solving (8.26) and substituting into (8.21) gave the following formula

$$C_L(T_1) = \frac{C_L(T_0) - C' R_{\max}/L}{1 - (u_0 L/2D) Z_{\text{eff}} E} \qquad (8.27)$$

where $R_{\max}$ is the maximum thickness of the layer.

Equation 8.27 and an exact knowledge of the liquidus curve $C_L(T)$ (1) made it possible to estimate the parameter ΔT. Unfortunately, neither the electric conductivity σ of Ga solution nor the parameter Z_{eff} is known. Thus, the product EZ_{eff}, linearly dependent on current density J, has to be treated as the parameter which fits the theoretical dependences $R = f(t, J)$ with the experimental data. The result of the numerical solution of the set of Equations 8.27, 8.22 and 8.24 with the boundary conditions

$$C(x, 0) = C_L(T_0) \qquad (8.28)$$

$$C(0, t) = C_L(T_1) \qquad (8.29)$$

$$C(L, t) = C_L(T_0), \qquad \frac{\partial C}{\partial x}\bigg|_{x=L} = 0 \qquad (8.30)$$

as well as the fitting procedure made by Bryśkiewicz (17) is shown in Figure 8.9. As a criterion of fitting the following condition was used

$$\frac{J}{Z_{\text{eff}} E} = \text{const.} \qquad (8.31)$$

This means that the fitting parameter $Z_{\text{eff}} E$ has to be proportional to the current density J. One may prove that an agreement within $\pm 20\%$ between the calculated dependences $R = f(t, J)$ and the measured ones can easily be achieved.

As we see now, the values of the $Z_{\text{eff}} E$ parameter obtained from the fitting procedure permit estimation of the electric conductivity $\sigma(C)$, the effective charge Z_{eff} as well as the electric field intensity E. It follows from Ficks' theory that the effective charge of solute particles migrating in a dilute metallic solution can be expressed by the following formula (22)

$$Z_{eff} = Z - Z_0 \frac{\sigma^{-1}(C) - \sigma^{-1}(0)}{C\sigma^{-1}(0)} \tag{8.32}$$

where Z and Z_0 are valencies of solute and solvent ions, and $\sigma(C)$ and $\sigma(0)$ are conductivities of the solution and pure solvent.

Hence

$$Z_{eff}E = \left[Z - Z_0 \frac{\sigma^{-1}(C) - \sigma^{-1}(0)}{C\sigma^{-1}(0)}\right] J\sigma^{-1}(C) \tag{8.33}$$

and

$$\sigma^{-1}(C) = \sigma^{-1}(0) \cdot \left(1 + \frac{Z}{Z_0}C\right) \frac{1 + \sqrt{1-B}}{2} \tag{8.34}$$

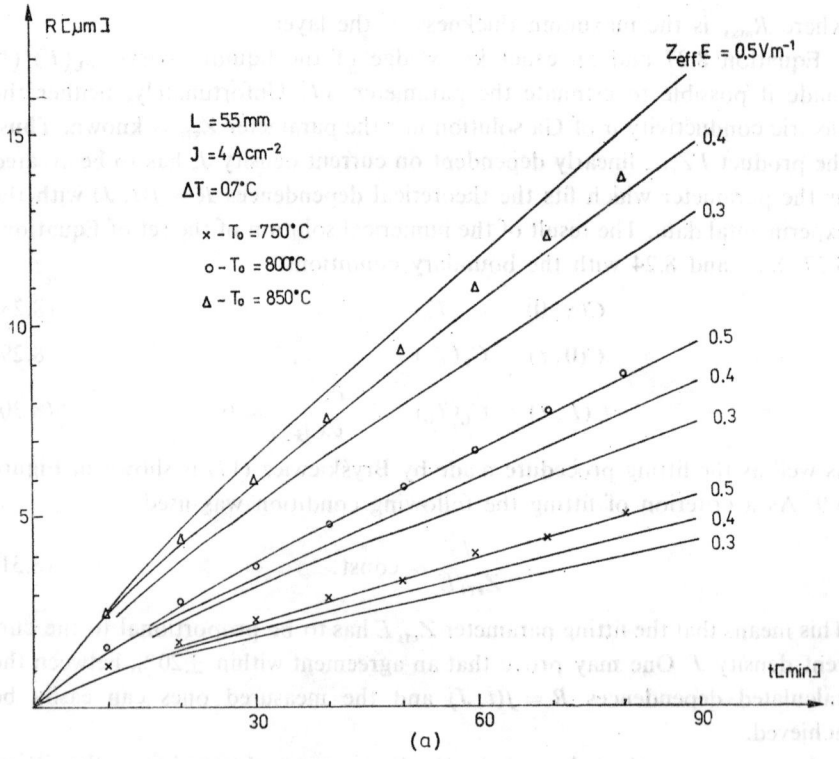

Figure 8.9 (17) Layer thickness R versus time t and current density J calculated ($\times$, $\circ$, $\triangle$) with the use of

and
$$Z_{eff} = Z - Z_0 \frac{\left(1+\dfrac{Z}{Z_0}C\right)\left(1+\sqrt{1-B}\right)-2}{2C} \qquad (8.35)$$
where
$$B = \frac{4Z_{eff}E \cdot C}{Z_0 \sigma^{-1}(0) J \left(1+\dfrac{Z}{Z_0}C\right)^2} \qquad (8.36)$$

Bryśkiewicz assumed (17) that for the GaAs system $Z(As^{---}) = -3$, $Z_0(Ga^{+++}) = +3$ and $\sigma^{-1}(0) \sim 4.5 \times 10^{-5}\Omega$ cm at 750–850°C. Taking into account the product $Z_{eff}E$ obtained from the fitting procedure he estimated

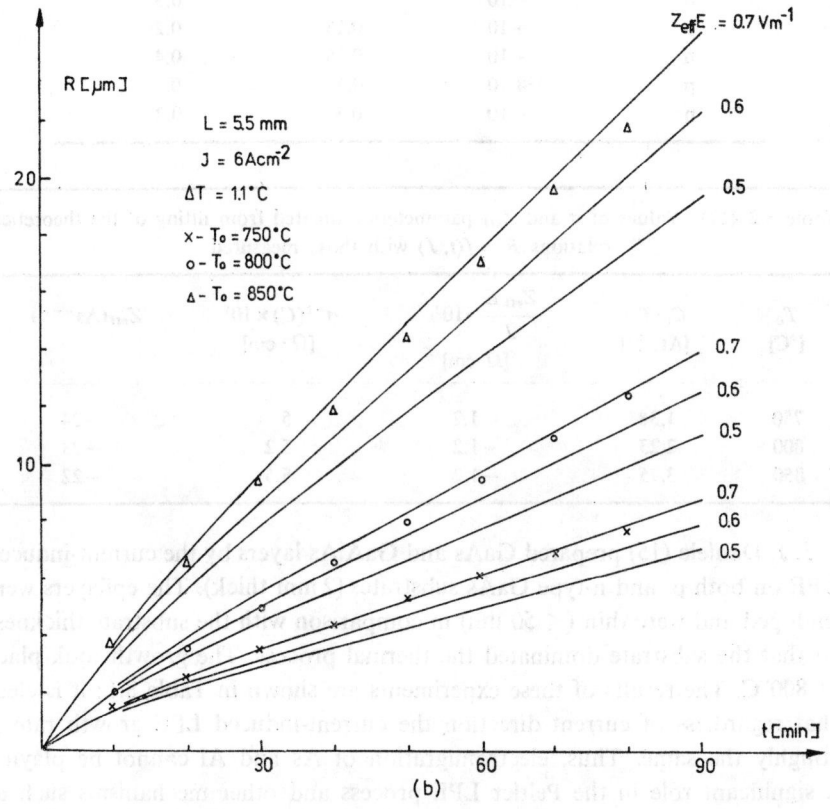

(solid line) from the set of Equations 8.27, 8.22, 8.24 and 8.28–8.30 and measured current marks of time (3)

the conductivity of Ga-rich melt as well as the effective charge of arsenic ions. The results are seen in Table 8.2. The high value of the effective charge $Z_{eff} \approx -22$ justifies our conclusion that the arsenic transport in the melt is realized due to collisions with electrons flowing across the melt rather than due to direct interaction with the electrostatic field.

Table 8.1 Experimental results (15)

Substrate type	Normalized current [A/cm^2]	X_{Al} (solid)	Average growth rate (v_p), [μm/min]
p	+10	0	0.5
n	−10	0	0.5
p	+10	0.15	0.25
n	−10	0.15	0.4
p	+10	0.3	0.3
n	−10	0.3	0.3

Table 8.2 (17) Values of σ and Z_{eff} parameters estimated from fitting of the theoretical relations $R = f(t, J)$ with those measured

T_0 [°C]	$C_L(T_0)$ [At. %]	$\dfrac{Z_{eff} E}{J} \cdot 10^3$ [Ω·cm]	$\sigma^{-1}(C) \times 10^5$ [Ω·cm]	$Z_{eff}(\text{As}^{---})$
750	1.38	−1.2	5	−24
800	2.23	−1.2	5.2	−23
850	3.75	−1.2	5.5	−22

J. J. Daniele (15) prepared GaAs and GaAlAs layers by the current-induced LPE on both p- and n-type GaAs substrates (2 mm thick). The epilayers were undoped and were thin ($\leqslant$ 50 μm) in comparison with the substrate thickness so that the substrate dominated the thermal process. The growth took place at 800°C. The results of these experiments are shown in Table 8.1; it is clear that regardless of current direction the current-induced LPE growth rate is roughly the same. Thus, electromigration of As and Al cannot be playing a significant role in the Peltier LPE process and other mechanisms such as convection and boundary layer formation should be more closely investigated. Moreover, preliminary experiments showed (15) that the Peltier LPE growth rate of GaAs and GaAlAs decreases with diminishing melt height until a critical height of 2 mm was reached. These experiments indicate that convective stirring and formation of a boundary layer may be present and can account for the observed elevated growth rate.

L. Jastrzebski *et al.* (18) identify convective flow of material with the horizontal temperature gradients resulting in the solution from Joule heating which was found by them to be more pronounced at the contact of the current carrying rods and the graphite segments of the boat. As a criterion of convective flow the magnitude of the thermal Grashof number has to be considered

$$\mathrm{Gr} = \frac{\alpha g}{v^2} \cdot \frac{L^4}{W} \Delta T \tag{8.37}$$

where α, v, and g are the coefficient of thermal expansion, the kinematic viscosity and the gravitational constant, and W is the width of the cell.

On the basis of the results obtained by L. Jastrzebski *et al.* (18) it appears that the current-induced LPE growth of GaAs is not affected by convective flow until Gr exceeds a value of the order of 10^4. It is not understood, at this time, why such a relatively high value of Gr is necessary before significant convective flow is present in the solution to affect the growth process. It is possible that a stabilizing solute boundary layer is present. However, no quantitative data are available at this time to determine the thickness of such a stabilizing solute layer.

In order to find an analytical formula expressing the rate of the current-induced LPE growth one has to consider a solution of both (8.19) and the Navier–Stockes equations (22), a very difficult task in this case. Assuming a static boundary diffusion layer in a moving solution immediately adjacent to the growth interface (Nernst's approximation (22)), L. Jastrzebski *et al.* (18) obtained the following analytical formula

$$V(t) = \frac{\Delta T \dfrac{dC_L}{dT} \cdot \dfrac{D}{\delta} - Z_{\mathrm{eff}} u_0 E C_L(T_1)}{C'_s - C_L(T_1)} \tag{8.38}$$

where δ is thickness of the Nernst diffusion layer (22).

8.4 SURFACE MORPHOLOGY

Lawrence and Eastman have proved (13) that the undesirable surface features most commonly observed on layers grown by electric current controlled LPE are identical to those found on epitaxial layers grown by conventional techniques, namely terraces and cusps. In addition, there are long-range thickness variations across the epitaxial layers. These variations are typically $\pm 15\%$ for layers grown with 5 or 10 A/cm², and often $\pm 20\%$ for layers grown with 15 or 20 A/cm², neglecting excess edge growth, which sometimes equals the average layer thickness. The thickness variations usually occur over lateral distances of about 0.4 cm and the excess edge growth extends approximately 0.05 cm inward the edge of the epitaxial layer. The experiments of Lawrence and Eastman indicate that in the case of electric current controlled LPE, the severity of the terracing depends strongly upon the electric current density used in the growth process. The formation of severe terraces and cusps is favoured when one attemps to force growth directly upon the substrate with a current density of 15 or 20 A/cm². For current densities of 5 or 10 A/cm² the terracing is milder and in several cases large areas of the layers are nearly featureless. In addition, the terracing is less severe when growth with 15 or 20 A/cm² is carried out upon growth with 5 or 10 A/cm² rather than directly upon the substrate. Another observation made by Lawrence and Eastman is that, in all cases, for a given growth temperature and current density, a layer grown on a semi-insulating substrate exhibits worse terracing than one grown on an n^+ substrate. These observations, together with the larger long-range thickness variations observed in epitaxial layers grown with higher current densities, suggest that all of these features depend largely upon the non-uniformity of the current density across the substrate. There are three possible causes of the non-uniformity: non-uniform substrate conductivity, non-uniform electrical contact to the back of the substrate, and non-simultaneous initial wetting of the substrate by the melt. The latter two cases are believed to be the most important in the experiments reported here.

The uniformity of the layer thickness in conjunction with various growth conditions such as current density, extent of Joule heating, solution height and the nature of the electric contact to the substrate was studied by L. Jastrzebski et al. (19). They observed two types of epilayer thickness variations: random fluctuations and systematic variations. The random fluctuations were found to be associated with non-uniform current density at the growth interface resulting from random defects in the electric contact to the substrate. Random fluctuations were also found to result from fluctuations in the resistance across the substrate, which in turn were caused by fluctuations in the depth of substrate dissolution by the Ga contact layer. For relatively thin substrates the dissolution depth represented a significant fraction of their thickness since a relatively thick layer of Ga (~ 150 μm) is necessary for a satisfactory spread. These fluctuations were eliminated by placing a tantalum foil in contact with the gallium layer (5, 19). This foil dissolves part of the gallium when the assembly is heated to establish electric contact. Thus, the Ta foil essentially decreases the effective thickness of the gallium layer and thus the extent of dissolution of the substrate by the gallium.

Systematic fluctuations of the layer thickness were found by L. Jastrzebski et al. (19) to be associated with temperature gradients or convective flow in the solution. In the presence of convection in the solution the thickness profile of the layers was found to reflect the convective flow pattern in the solution. The primary cause of convection was identified with the horizontal temperature gradients resulting from Joule heating which was found to be more pronounced at the contacts of the current carrying rods and the graphite segments of the boat. Convective flow affecting the growth was observed for thermal Grashof numbers greater than about 10^4. Convective interference was eliminated by minimizing the temperature gradient across the solution through reduction of the system's resistance and by positioning the solution well in the furnace so that the horizontal gradient in the solution prior to current flow was of the same magnitude but opposite sign to that induced by Joule heating. The thickness profiles of the layers grown without convective interference were found to reflect the gradient of the current density across the growth interface associated with the geometry of the growth system and a very small horizontal temperature gradient due to Joule heating. These two effects compensated each other at high current densities, and thus epilayers with the most uniform thickness were obtained at current densities of 50–60 A/cm^2. Although L. Jastrzebski et al. (19) minimized, or essentially eliminated, random thickness fluctuations and convective interference, attempts are being made to obtain epilayers with perfectly uniform thickness.

A comparison of the surface morphology of GaAs layers grown by electroepitaxy and thermal LPE, performed by Y. Imamura et al. (20), was based on layers of the same thickness, obtained in the same apparatus, with the same average growth rates, on substrates having the same orientation and electrical properties, and from solutions equilibrated at the same temperature. It was found that over a broad spectrum of growth conditions GaAs layers grown by electromigration—controlled electroepitaxy were flat and exhibited no terracing or other morphological defects. The occasional appearance of defects such as dishes or hillocks were invariably related to defects in the electrical contact of the substrate. Y. Imamura et al. (20) suggest that in view of the above results, the pronounced differences in the surface morphology of the layers grown by electroepitaxy and of those grown by thermal LPE cannot be attributed to specific growth conditions. Rather they must be associated with fundamental differences in the growth mechanisms involved in the two growth methods. In electroepitaxy, supersaturation takes place in the immediate vicinity of the growth interface and growth is controlled by the rate of transfer of solute, under an electric field, towards the interface. Under these conditions growth must take place under nearly equilibrium (isothermal) conditions and interface instabilities due to constitutional supercooling cannot occur. On the other hand, in thermal LPE supersaturation is induced throughout the solution and nucleation takes place not only on the substrate but also on the walls of the solution container. Unless special precautions were taken, random nucleation, fluctuations in solute concentration and/or temperature gradients and constitutional supercooling were potential causes of non-uniform growth leading to terraced surface morphology.

8.5 DOPING MODULATION AND COMPOSITION STABILIZATION

It has been proved (5, 12–14) that modulation of the dopant concentration can be achieved in the growing layers by a corresponding variation of the current density. L. Jastrzebski and H. C. Gatos (12) have analysed this modulation in GaAs epitaxial layers highly doped with Te, Sn and Si. The dependence of the carrier concentration in the Te-doped layers grown at 850°C from 0.03 at.% doped solution on the current density is shown in Figure 8.10. It is interesting to note that the electron concentration in the layers increases about 40% as the current density changes from 0.5 to 40 A/cm². This change in the current density corresponds to a change in the growth velocity from 0.04 to 3.5 µm/min. To distinguish between the effects of the electric current and growth velocity on the impurity segregation observed during current growth, L. Jastrzebski and H. C. Gatos prepared epilayers thermally with the same constant velocity from the same solution by a 10°C temperature decrease. The measured concentration is also presented in Figure 8.10. No significant change of the carrier concentration is evident in the thermally grown layers for various growth velocities. The same trend was found for Sn-doped layers grown at 950°C from 2.5 at.% Sn-doped solution (Figure 8.10). On the other hand, the change in the carrier concentration caused by increased current density, in the case of Sn, is about 35%. It is important to note that changes in the concentration caused by the electric current are very similar for both impurities despite a large difference in the value of the segregation coefficient ($K_{Te} = 0.7$ at 850°C, $K_{Sn} = 0.01$ at 950°C (12)). The discussed changes in

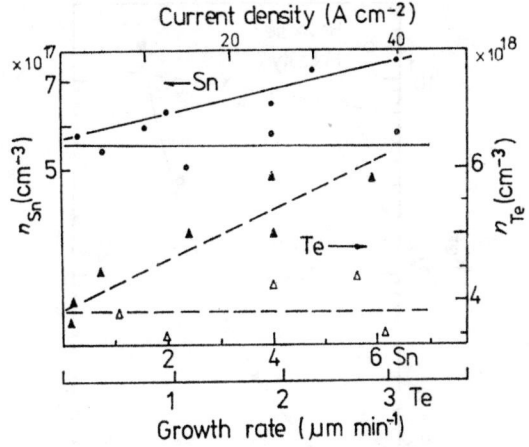

Figure 8.10 (12) Electron concentration at room temperature in Te-doped layers grown at 850°C (▲) and Sn-doped layers grown at 950°C (●) as a function of current density and growth velocity. Currier concentration as a function of growth velocity in Te- and Sn-doped layers grown by a temperature decrease from 850 to 830°C (△) and from 950 to 940°C (○)

the concentration caused by electric current are similar to those reported by J. Lawrence and L. F. Eastman (5, 13) for low levels of doping (10^{15}–10^{16} cm^{-3} at 700–800°C) although no saturation of impurity concentration as a function of current density is observed in Figure 8.10.

L. Jastrzebski and H. C. Gatos (14) have demonstrated that growth of GaAs epitaxial layers from 2% Si-doped solution by current-controlled LPE exhibits unusual kinetics and segregation characteristics. The growth rate, as a function of current density at 900°C, was found to be significantly smaller than that from undoped melts. At 950°C the growth rate was smaller below a certain current density and greater above this current density than that from undoped melts. Similarly, the growth rate at a constant current density was found to be smaller below 950°C and greater above this temperature than the growth rate from undoped solutions. For a constant current density the Si distribution coefficient increased by two orders of magnitude in going from 825 to 975°C (Figure 8.11); changes of this magnitude in the distribution coefficient were not observed in LPE by thermal growth. The Si distribution coefficient in current-controlled LPE increased with current density, whereas in LPE by temperature decrease it remained constant as a function

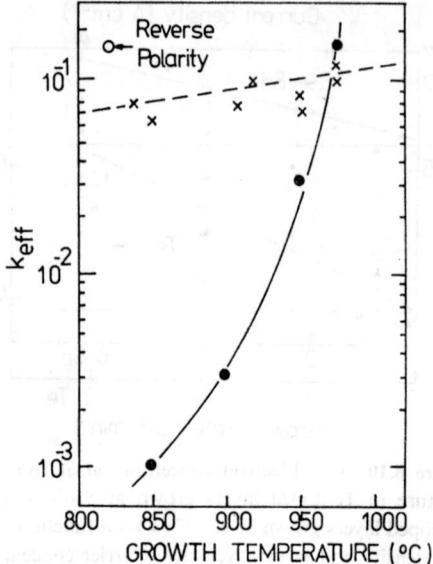

Figure 8.11 (14) Dependence of Si segregation coefficient on growth temperature for layers grown at a constant current density of 10 A/cm² (solid line) or by temperature-controlled LPE (dashed line) from 2% Si-doped solution

of the growth rate. Layers grown at 900°C exhibited p- or n-type conduction depending on the current density. Such amphoteric behaviour of Si was observed in thermally grown GaAs from 2% Si solutions when the growth temperature decreased from above 925°C to values below this temperature. L. Jastrzebski and H. C. Gatos account for their findings in terms of a qualitative model based on the presence of positively charged Ga-Si-As and negatively charged Si-As complexes in the solution and their electromigration during current flow. From Ga-Si-As type complex Si is incorporated into As lattice positions (acceptor) and from Si-As is incorporated into Ga positions (donor).

The results discussed in this section prove that the impurity concentration in the epitaxial layers depend significantly on their electromigration (or electrotransport) characteristics in the solution during current growth rather than on their distribution coefficient or the microscopic rate of growth.

Very interesting results can be obtained when solid-solutions of $A^{III}B^{V}$ compounds are grown by means of electroepitaxy. J. I. Daniele has proved

(7) that in addition to electrical control over the growth rate, the current has a strong stabilizing influence on the composition of the growing layer. Figure 8.12 shows that the aluminium concentration in $Ga_{1-x}Al_xAs$ layers

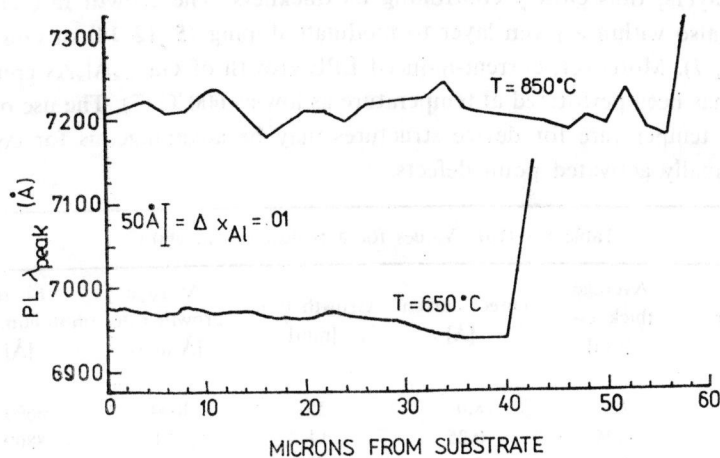

Figure 8.12 (7) Plot of peak photoluminescence wavelength versus distance from the substrate for GaAlAs layers grown at 850 and 650°C. At 850°C, currents of 20/30/20/20/30 A/cm² were applied for 15 min each, and at 650°C a current of 24 A/cm² was applied for 5 h. The sharp decrease of X_{Al} at the end of the layer resulted from overgrowth that occured due to incomplete melt wiping

grown at 850°C as well as at 650°C is constant within ±0.005 even though the current and growth rate were varied by a factor of 1.5 (7).

J. I. Daniele observed this uniformity of composition in all layers grown by current-induced LPE. It becomes highly significant in low-temperature growth where melt depletion usually limits the thickness of the layer and very sharply grades its composition profile. This is especially important for $Ga_{1-x}Al_xAs$, which has an aluminium segregation coefficient of about 260 at 700°C.

8.6 FABRICATION OF DEVICES

The current-controlled LPE growth method has unique advantages for the growth of p–n junctions (12), DH lasing wafers (16) as well as other complex multilayer structures. Since the rate of electroepitaxial growth is roughly proportional to the applied current over a wide range, each layer in a layer structure can be grown at virtually any desired rate, independent of the growth rate of the other layers. For instance, in DH lasing wafers growth (16), the very thin active layer can be grown much more slowly than the other layers, thus closely controlling its thickness. The growth rate can be varied also within a given layer to modulate doping (5, 12–14) or composition (6, 7). Moreover, current-induced LPE growth of $Ga_{1-x}Al_xAs$ epitaxial layers has been performed at temperature as low as 600°C (7). The use of low growth temperature for device structures may be advantageous for control of thermally activated point defects.

Table 8.3 (16) Values for a typical run at 800°C

Layer	Average thickness [μm]	Direct current [A]	Growth time [min]	Average growth rate [Å/min]	El. or photolum. peak [Å]
1	7.5	8.0	25	3000	6690
2	0.24	0.25	14	171	8800
3	1.4	8.0	30	467	6840
4	2.6	3.0	10	2600	8720

$GaAs$-$Ga_{1-x}Al_xAs$ DH lasing wafers grown by means of electroepitaxy were prepared for the first time by J. J. Daniele (16). Table 8.3 contains the the currents, growth rates, growth times and resulting layer thicknesses of the wafer grown at 800°C. The dopants used in the structure were Sn for melt No. 1 and Ge for melts Nos. 2, 3 and 4. The substrates were (100)-oriented GaAs doped with Si ($n \approx 3 \times 10^{18}$ cm^{-3}). The growth area was 1.0×1.0 cm, and the surface of the prepared epilayers was smooth and mirror-like. Mesa stripe geometry (17 μm stripe width) lasers fabricated from those wafers lased cw at room temperature. Other devices (60-μm mesas from several wafers) showed I_{th}/d of 5 kA/cm² (I_{th} the threshold current, d the active layer thickness) and device characteristics were comparable to devices produced by normal LPE. Thus, J. J. Daniele has shown for the first time that current-induced LPE is a viable technique for device growth.

8.7 CONCLUSIONS

On the basis of the current state of knowledge of the physical processes which occur during electroepitaxy, one can answer the questions given in the Introduction as follows:

- The temperature change due to Peltier-cooling at the growth interface depends on thickness and electrical properties of the substrate, direction and density of the current as well as the growth cell geometry.
- During electroepitaxy the solute concentration (diffusion) and temperature gradients (convection) as well as the electric field (electromigration and electrotransport) become the driving forces of growth. It has been proved that for low horizontal temperature gradients in the melt resulting from Joule heating (Gr $< 10^4$) convective flow of the solute particles is negligible and electromigration or rather electrotransport dominates electroepitaxy. In the case of high temperature gradients (Gr $> 10^4$) solute convection in the melt may be present and can account for the observed elevated growth rate.
- It has been demonstrated that the rate of current-induced LPE growth is roughly proportional to the applied current. The relation between the growth rate and current density can be explained theoretically taking into consideration the diffusion, electromigration and electrotransport of ionized particles in the melt as well as solute convection. However, attempts at a quantitative analysis of the growth kinetics have been, thus for, limited in scope.
- High homogeneity of the solid-solutions prepared by means of electroepitaxy would probably be explained with high effective charges of the solute particle migrating toward the growth interface or with convective mixing of the melt.
- It has been found that epitaxial layers grown by migration-controlled electroepitaxy are flat and exhibit no terracing or other morphological defects. The occasional appearance of defects such as dishes or hillocks are invariably related to defects in the electrical contact of the substrate. The improved surface morphology of epilayers can be attributed to the fact that during electroepitaxy supersaturation of the melt takes place in the immediate vicinity of the growth interface and growth is controlled by the rate of transfer of solute due to migration. Under these conditions growth must take place under nearly equilibrium (isothermal) conditions and interface instabilities due to constitutional supercooling cannot occur.

The author hopes that further analysis of the current-induced LPE growth process will result in revealing sone new interesting phenomena which, studied and explained, will pave the way for improvements in the technology of optoelectronic device structures.

ACKNOWLEDGEMENTS

The author is deeply indebted to Professor H. C. Gatos, Professor A. F. Witt, Dr. J. Lagowski, Dr. L. Jastrzebski and Dr. Y. Imamura of MIT for sending manuscripts of their papers submitted for publication.

REFERENCES

1. M. A. Herman, Ed., 'Semiconductor Sources of E-M Radiation', *Proceedings of the International Autumn School on Semiconductor Optoelectronic Cetniewo 1975*, p. 231, PWN, Warszawa, 1976.
2. H. Nelson, *RCA Rev.*, **24,4**, 603 (1963).
3. M. Kumagawa, A. F. Witt, M. Lichtensteiger, H. C. Gatos, *J. Electrochem. Soc.*, **120,4**, 583 (1973).
4. G. M. Blom, J. J. Daniele, T. Kyros, A. F. Witt, *J. Electrochem. Soc.*, **122,11**, 1541 (1975).
5. D. J. Lawrence, L. F. Eastman, *J. Cryst. Growth*, **30**, 267 (1975).
6. J. D. Daniele, C. Michel, 'Gallium Arsenide and Related Compounds, Deauville 1974', *Inst. Phys. Conf. Ser.*, **24**, p. 155.
7. J. J. Daniele, *Appl. Phys. Lett.*, **27**, 373 (1975).
8. E. K. Stefanokos, A. Abul-Fadl, M. D. Workman, *J. Appl. Phys.*, **46,7**, 3002 (1975).
9. A. Abul-Fadl, E. K. Stefanokos, *J. Appl. Phys.*, **47, 10**, 4627 (1976).
10. L. Jastrzebski, H. C. Gatos, A. F. Witt, *J. Electrochem. Soc.*, **123,7**, 1121 (1976).
11. A. Abul-Fadl, E. K. Stefanokos, *J. Cryst. Growth*, **39**, 341 (1977).
12. L. Jastrzebski, H. C. Gatos, 'Gallium Arsenide and Related Compounds, St. Louis 1976', *Inst. Phys. Conf. Ser.* **33b**, p. 88.
13. L. Lawrence, L. F. Eastman, *J. Electric Mat.*, **6,1**, 1 (1977).
14. L. Jastrzebski, H. C. Gatos, *J. Cryst. Growth*, **42**, 309 (1977).
15. J. J. Daniele, *J. Electrochem. Soc.*, **124,7**, 1143 (1977).
16. J. J. Daniele, D. A. Cammack, P. M. Asbeck, *J. Appl. Phys.*, **48,3**, 914 (1977).
17. T. Bryśkiewicz, *J. Cryst. Growth*, **43**, 567 (1978).
18. L. Jastrzebski, J. Lagowski, H. C. Gatos, A. F. Witt, *J. Appl. Phys.*, **49**, 5909 (1978).
19. L. Jastrzebski, Y. Imamura, H. C. Gatos, *J. Electrochem. Soc.*, **125**, 1140 (1978).
20. Y. Imamura, L. Jastrzebski, H. C. Gatos, submitted to *J. Electrochem. Soc.*
21. T. Bryśkiewicz, *J. Cryst. Growth*, **43**, 101 (1978).
22. V. G. Levich, *Physicochemical Hydrodynamics*, Prentice-Hall, Inc., 1962, pp. 1–138.
23. J. N. Pratt, R. G. R. Sellors, *Electrotransport in Metals and Alloys*, Switzerland 1973, pp. 2–38.

Part III
Optoelectronic Devices

Chapter 9

Heterostructures in Semiconductor Electronics*

Zh. I. Alferov

The problem of contact between semiconductors with different forbidden gaps has long attracted the attention of specialists. A. F. Ioffe and A. V. Ioffe from the Leningrad Physico-Technical Institute studied contact of two different semiconductors in the pre-war period. It was also then that they studied selenium rectifiers in which rectification is determined by a heterojunction. However, unsurpassable difficulties arose with the treatment of experimental data caused by the inability to control the purity of the surface or by the polycrystalline nature of the objects under study.

Creation of pure and doped Ge- and Si-monocrystals and development of methods of obtaining p–n junctions in them have stimulated investigations on semiconductor physics and technology. This has resulted in a revolution in radio engineering, electronics, and electrical engineering. The significance of this revolution is apparently no less than that of the discovery of nuclear energy for power engineering and military purposes.

In the first two post-war decades, the development of semiconductor electronics was associated mostly with application of p–n junctions (p–n homojunctions) and was based on the possibility of a controlled doping of semiconductor crystals. The desired structure was obtained by means of changing the doping type and concentration.

Development of methods of epitaxial semiconductor crystal growth led in the 60's to systematic studies of monocrystalline heterojunctions in semiconductors (1), i.e. of graded and abrupt contacts of semiconductors with different chemical composition realized in a single monocrystal. Such structures generally not only undergo a change in the forbidden gap width, but

---

* This chapter has also been published (with some abbreviations) in *Europhysics News*, **9**, no. 12 (1978), on the occassion of granting to Zh. I. Alferov the Hewlett-Pachard award of the European Physical Society 1978 for his outstanding works in heterojunction physics research.

other fundamental properties are changed too. They are—zone structure, effective masses of current carriers, their mobility, etc. The possibility of controlling these properties, as well as the forbidden gap width, has opened new prospects for physical investigations and practical application of semiconductors in devices and crystals.

The experimental realization of graded (2) and abrupt (3) AlAs-GaAs heterostructures with properties close to those of ideal models and the discovery of effective injection and of the superinjection effect in these structures have served as a basis for the creation of a large number of devices with a heterostructure as the main active element. Of greatest interest has been the application of heterojunctions in optoelectronic semiconductor devices such as injection lasers, light emitting diodes, solar-cell batteries and all kinds of photocells, modulators, amplifiers, and light converters. Although the use of heterojunctions for conventional injection devices has not yet given sufficiently effective results, there is no doubt that heterostructures will become the main element in such widely used devices as rectifying diodes, transistors, dynistors, and thyristors.

A most interesting and promising feature of the research is the development of complex monolithic integrated systems, based on heterostructures, which could determine the future course of integrated optics.

In this chapter a short review is given of the physical investigation of heterojunctions and their application in optoelectronics, together with some new results in this field, obtained mainly at the Laboratory of Contact Phenomena in Semiconductors of the A. F. Ioffe Physico-Technical Institute of the U.S.S.R. Academy of Sciences.

9.1 SOME PECULARITIES OF ELECTRICAL AND OPTICAL PHENOMENA IN HETEROSTRUCTURES

In p–n heterostructures with properties close to ideal, i.e. in heterostructures containing sufficiently small amounts of interface states in regions of changing chemical composition and forbidden bandwidth (in which phenomena take place simultaneously), mainly one-sided effective injection occurs from the wide gap into the narrow gap region, independently of the doping levels of the p and n regions. An extremely valuable property of p–n heterostructures is the possibility of injecting practically any concentration of non-equilibrium carriers from the wide gap material into narrow gap material, thus considerably exceeding their equilibrium concentration in the emitter. In graded p–n heterojunctions (p–n structures with a variable forbidden gap), owing to the influence of built-in quasi-electric fields, a marked increase or decrease in the effective diffusion length of the injected carriers also occurs (4).

Thus heterostructures appear to have a powerful means of controlling carrier flow, i.e. carrier concentration, diffusion rate and recombination region dimensions. Moreover, heterostructures also permit effective control of light flow in crystals, both introduced from outside and generated due to radiative recombination inside.

The so-called *wide gap window effect*, whose possible use in solar batteries has long been recognized, permits light radiation of appropriate spectral composition to be introduced into the required structural region without losses. A multilayer heterostructure is a dielectric thin film waveguide with effective light wave guiding in a narrow gap material. This waveguide effect permits controlled propagation of light flow in crystals; its application in heterolasers was pointed out at the beginning of the 1960s (5). The above peculiarities of injection and optical heterojunction properties were first experimentally discovered and studied in AlAs–GaAs heterostructures (1, 2, 6, 7).

A heterojunction is a 'pair' and the choice of an 'ideal pair' is naturally a complicated problem. It requires the fulfilment of numerous 'compatibility' conditions as regards mechanical, crystallochemical and thermal properties as well as crystalline and energetic structure of the materials in contact. Gallium arsenide has been widely used in different semiconductor devices owing to the successful combination of a small effective mass and a large forbidden gap, effective radiative recombination and a sharp optical absorption edge due to a 'direct' zone structure, high mobility in the absolute minimum of the conduction band and its abrupt decrease in the nearest minimum at point $\langle 100 \rangle$. Considering the fact that the use of a heterojunction between a semiconductor which serves as an active material and a material with a wider forbidden gap results in a maximum effect, AlAs–GaAs, GaP–GaAs and AlP–GaAs heterojunctions on the basis of GaAs may be expected to be promising. Since coincidence of lattice parameters is the first and the most important criterion of 'compatibility' of the materials of a 'pair', AlAs–GaAs heterojunctions are preferable.

A simple technological solution involving the application of an easily realizable LPE method, made it possible to master the production of multilayer AlAs–GaAs heterostructures on an industrial scale.

9.2 HETEROSTRUCTURES IN DISCRETE OPTOELECTRONIC DEVICES

The above peculiarities of the electrical and optical properties of heterostructures have been successfully applied in fundamental improvement of the parameters of a large number of semiconductor devices and in the development of new devices which could not be created on the basis of homostructures.

9.2.1 Heterostructure lasers

The injection laser has been one of the best applied achievements of solid state physics during the last decade. A theoretical analysis of the possibilities of creating lasers on p–n junctions, performed at the Physical Institute of the U.S.S.R. Academy of Sciences (8) and the discovery of stimulated radiation in GaAs p–n junctions (the A. F. Ioffe Physico-Technical Institute (9)), lead to the construction of semiconductor injection lasers (10, 11). High efficiency, the possibility of modulation up to frequencies of hundreds of MHz, small dimensions, long life-time and small consumed power—all these properties were expected to guarantee a wide application of this device. The study of injection laser properties, particularly of those fabricated on GaAs, which are considered to be the highest quality and practically the most valuable ones, has attracted the attention of physicists in numerous laboratories the world over. Hundreds of papers were published in 1963–1966.

Unfortunately, the prospects soon turned out to be much less bright than had been hoped for, and the sphere of application of injection lasers began to decrease rapidly. The extremely high value of the threshold current required to start generation in a device and its strong temperature dependence made it necessary to use complicated cooling systems and short-pulse feeding. High efficiency appeared to be realizable only in the generation regime. External quantum efficiency equals several parts of 1 per cent at a current density below the threshold, and increases sharply only after the generation has started. As a result of the high threshold current density ($j_{thr} = 25 \times 10^3$ A/cm^2 being the highest value for GaAs lasers at room temperature), the operation region of currents cannot considerably exceed the threshold. Therefore, with a high differential quantum efficiency the summary quantum efficiency did not exceed 2–3% at 300 K and there seemed to be no way to increase it. For the same reason the life-time of the devices appeared to be short; after several hours of operation degradation of the working parameters was observed.

The physical cause of these defects can easily be seen if we consider the simplest scheme of a device in operation. To obtain inverse population it is necessary that a divergence of quasi-Fermi levels exceed the energy of a quantum of generated radiation $h\nu_{gen}$ (in injection lasers this value is very close to the semiconductor forbidden gap width, i.e. $h\nu_{gen} \simeq E_g$). Thus, a laser based on an ordinary p–n junction implies that a 'degenerated' junction has been used.

The concentration of p–n junction injected electrons and holes falls gradually on both sides of the junction. Therefore, the recombination region and that of inverse population do not coincide. This means that additional recombination losses take place in passive regions.

The optical inhomogeneity of the medium is associated only with impurity distribution and with injected carriers concentration. Thus, of importance are the diffraction losses of light radiation from the active region into the passive one. To compensate for the considerable losses and to achieve generation, a high amplification coefficient is needed, which means that the threshold current density should be sufficiently high. Losses unavoidable in principle make the external quantum efficiency much lower than the internal one and, therefore, create a principal obstacle in achieving quantum efficiency close to 100% (even at low temperatures).

In double heterostructure lasers (7) the regions of recombination, light radiation and population inversion coincide completely and are confined to the middle layer. Owing to the heterojunction potential barriers, recombination losses in passive regions are negligible and the electron-hole plasma is confined to a potential box in the middle layer. Owing to the marked difference in refractive indices the middle layer serves as a high-quality waveguide and light radiation losses in passive regions are negligible. Population inversion is obtained by double injection and neither a high doping level nor even degeneracy are needed. All this leads to a considerable reduction in losses and in threshold current density. In such a structure both external and internal quantum efficiencies almost coincide and there seems to be a strong possibility that a quantum efficiency close to 100% will be obtained.

Injection heterolasers with threshold current density of less than 1000 A/cm^2 were created in 1970 (12). External differential quantum efficiency and full efficiency reached 70% and 25%, respectively. The low threshold current density permitted CW operation at room temperature (12, 13), which is impossible with p–n homojunctions. For AlAs-GaAs heterostructures the energy of coherent radiation is determined by the composition of the $Al_xGa_{1-x}As$ solid solution in the middle layer and may be found for wavelengths in the 6900–9000 Å range. With the possibility of obtaining uniform generation in the middle layer, the light beam divergence in plane parallel to the junction can be less than 0.5°.

During the past 8 to 9 years, since the creation of low threshold heterolasers and the realization of CW operation at room temperature, scientists in laboratories all over the world have achieved considerable success in the optimization of heterostructure technology and parameters, the best results having been achieved in 1969–70. In a number of papers certain double-structure modifications were suggested and studied, but they resulted in little improvement in the parameters. Of considerable interest has been the physico-technological study of defects occurring in heterostructures after long-term operation. As a result DHLs with a 10,000 h life of continuous operation were created and in the author's opinion the operational life will reach 100,000 h (14). This work has made the construction of optical communication systems based on light guides with low losses and on DHLs a real possibility.

The most promising avenues for future development of DHLs would appear to be the pursuit of a further decrease in the threshold current density and an increase in efficiency and the creation of DHLs with coherence and beam divergence parameters comparable with those of solid state and gas lasers.

A decrease in threshold current in DHLs with a consequent increase in efficiency may be achieved to some degree by the method proposed in the first papers on DHLs and discussed in more detail in the theoretical work which followed, i.e. by the optimization of doping levels and DHL layer thicknesses. It is obvious that even a considerable decrease in the amplitude of the light power and the achievement of an active region of several hundred Å by, for instance, molecular beam epitaxy would hardly result in a significant decrease in the threshold value. A decrease in threshold without a reduction in power may be realized in heterostructures with the substrate removed and with light-reflecting contacts. For the shortwave part of spontaneous radiation multiple emission of photons is accompanied by their self-absorption and re-radiation in the active region. This leads to an increase in the effective lifetime of non-equilibrium carriers in the active region compared with their ordinary spontaneous radiative lifetime τ_r and, accordingly, to an increase in their concentration Δn at a given rate I: $\Delta n = I\tau_{\text{eff}}$.

With the possibility of obtaining a 100% internal quantum efficiency of spontaneous recombination in double heterostructures (DHs), as demonstrated in Reference 15, it seems that this means of reducing the threshold current in DHLs may be extremely effective. In fact, optical pumping of such structures has brought about a reduction in the threshold excitation level by a factor of 2.5 from the effect of multiple spontaneous radiation absorption (16). Important and fruitful lines of study appear to be the creation of DHSs based on other semiconductor materials and also the operation of DHLs in a wide spectral region.

Whereas an attempt to enter the IR region, using heterostructures containing a solid solution of Pb chalcogenides (17), has been quite successful, mastering the visible range has proved more difficult. Heterostructures based on quaternary A_3B_5 compounds may be considered the most promising for creating injection DHLs for the yellow and green spectral regions. After the first AlGaAsP DHs (18) were grown—of theoretical rather than practical importance—attention was concentrated on the system InGa–AsP, in which DHs and red DHLs operating at room temperature (19) have recently been obtained. This system is the most realizable for the yellow spectral region (20). For the creation of green DHLs a knowledge of the technology of obtaining an AlGa–InP system will be necessary; the metallurgy of the system is extremely complex. The development of multichannel optical communication systems increases the importance of injection coherent radiation sources in the 1–1.5 μm range. Encouraging results have recently been obtained with double heterostructures in the InGaAsSb and AlGaAsSb systems (21–23).

Laser-dynistors on p–n–p–n heterostructures (24) promise to be useful for different spheres of application. These devices combine a low generation threshold with sufficiently high on-voltages, which considerably simplifies the schemes of feeding and control of lasers.

9.2.2 Heterostructure light-emitting diodes (LEDs)

Development of injection spontaneous radiation sources, i.e. light-emitting diodes (25, 26), coincided with the creation of semiconductor lasers. Initially these devices were fabricated only on the base of p–n homostructures. However, heterostructures have a number of indisputable advantages for LEDs as well. These advantages arise from the possibility of coupling the radiation out from the recombination region to the crystal surface without self-absorption and also from the use of the superinjection effect to choose optimum doping levels and either to restrict or expand recombination regions, since both abrupt and graded heterojunctions have built-in potential barriers.

Since their fabrication is fairly simple, LEDs with graded heterojunctions (2, 27) were developed first (i.e. p–n structures in crystals with a variable band gap). In the most efficacious constructions (28, 29) LEDs were obtained by sequential growth of two layers with different gaps when the p–n junction position either coincided with, or was slightly removed from, the metallurgical interface. The recombination region is 'pressed' to the wide gap window and recombinative radiation is coupled out without absorption as a result of a small internal electric field gradient. LEDs obtained on the basis of such structures in an AlAs-GaAs system have yielded an external quantum efficiency of 1.0 to 1.5% at 300 K in the near infrared and red (up to $\lambda = 0.65$ μm) spectral region for a flat construction without anti-reflection coverings. The response of light sources is less than 10^{-8} s. However, DHs without absorbing layers are more promising for LEDs (28); the external efficiency (30) obtainable at present for flat LEDs is 30–40% at 100% internal quantum efficiency, and further development will undoubtedly produce an efficiency of up to 50–80%. Homo p–n structures had efficiency values close to those quoted above (28%) only in a semi-spherical construction of Si-doped GaAs (31). Such high efficiencies of LEDs have guaranteed the possibility of their application as powerful light sources and for pumping of solid state lasers on the basis of YAG : Nd^{3+} (32).

Extension of the range, as with lasers, is connected with the use of new materials for the heterostructures.

9.2.3 Photoelectric cells and infrared-to-visible converters

The main advantages of the use of heterojunctions for photoelectric devices arise from the possibility of using the wide gap part of a heterojunction as a transparent 'window' for light absorbed in its narrow part. Electron-hole generation takes place immediately in the space charge region of a p–n junction which, in principle, permits high response to be obtained, eliminating recombination losses in the illuminated surface and making recombination losses in the narrow gap material negligible. In the photon energy region restricted by forbidden gap widths of the narrow and wide gap materials constant spectral sensitivity may be obtained.

However, the above advantages can be realized only when the heterojunction band diagram is such that non-equilibrium carriers are divided by the junction contact field. p-$Al_xGa_{1-x}As$–n–GaAs heterojunctions fully meet this demand, because the discontinuity in the valence band is equal to zero and the interface is perfect, so that it is possible to obtain almost complete separation of non-equilibrium carriers (33). Such devices as position-sensitive photoelements (34), solar-cell batteries (35), and selective photoelements (36) have been created on the basis of p–n heterojunctions in the AlAs-GaAs system.

When a light beam enters the centre of a position-sensitive photoelement, $i_1 = i_2$, and the photovoltage V between the edge contacts is zero. When the light beam moves to the left, V is positive and when it moves to the right, V is negative. When the region near a contact is illuminated, $V = V_{max}$. Owing to the low level of leakage through the interface, an increase in positional sensitivity can be obtained by increasing the resistance of the wide gap region; light absorption takes place only in the active region of the device. A technology for fabricating high ohmic p–$Al_xGa_{1-x}As$ layers by Ge doping has been developed. It allows the sensitivity of photoelements to be raised to 2000 mV/mm^{-3}mW^{-1} with a base of 0.5 cm, which exceeds the sensitivity of similar devices using p–n homojunctions by two orders of magnitude.

For the creation of high efficiency solar-cell batteries the reverse problem was solved, i.e. the doping conditions of a wide gap material for providing minimum spread resistance were determined. In this case Zn was used as the doping impurity. The full efficiency of solar-cell batteries at room temperature is at present about 20–22% (37). Recently, in the USA, using solar energy concentrators at 19.1% efficiency and light concentration of 1735 'suns' (38), the power of 24 Watt/cm^2 has been obtained for an AlAs-GaAs heterophotoelement. Another advantage of heterophotoconverters is the possibility of using them at high temperatures; at 200°C their efficiency decreases by 15–20% only. These results prove solar converters on heterostructures to be promising for terrestrial power generation. The use of concentrators can make the construction of solar electric power plants economically profitable, it being, moreover, the only absolutely 'pure' trend in power generation as regards its effect on the environment, and causing no changes in the energetic balance of the planet. A further increase in the efficiency of photoelectric converters may be achieved with the use of graded heterostructures. In this case, the presence of a volume photoeffect in a crystal with a variable forbidden gap, ensures a more complete utilization of the energy of absorbed photons, which results in a decrease in Stocks losses. It is estimated that a 50% efficiency of photocells is possible (56). For a long time no noticeable e.m.f. had been observed in crystals with a variable forbidden gap. Recently, in experiments with AlGaAs crystals ($\Delta E_g \simeq 0.5$ eV) illuminated by an argon laser equivalent to solar light concentration of ~ 1000 suns, a volume photo e.m.f. of 0.3 V (57) has been observed. This means that at present there is a real possibility to obtain a further considerable increase of the efficiency of solar photoconverters on heterostructures. Controlling the composition of heterojunction components permits the heterophotoelement spectrum sensitivity region to be changed and photocells sensitive in a narrow predetermined region, i.e. selective photoelements, can be obtained.

Assuming different values for E_{gr} and E_g selective photoelements may be obtained by the use of heterojunctions in the AlAs-GaAs system within the whole range of $Al_xGa_{1-x}As$ solid solution compositions with a direct band structure, i.e. within the energy region 1.4–2.0 eV. Thus, application of heterojunctions for photoelements results not only in high efficiency solar-cell batteries and position-sensitive photocells, but also in easy control of the spectral sensitivity of these devices. This also applies to photocells with internal amplification, for instance transistors and photocells with multiplication at the avalanche breakdown (39). Selective photocells for other spectral regions have also been obtained by now on the basis of InGaAsP, and AlGaAsSb, and GaInAsSb heterostructures.

A number of experimental techniques of military concern (for instance 'night vision') require the conversion of infrared radiation into visible radiation. The technical solution of the problem has so far been provided on the basis of vacuum optoelectronic converters. The application of structures with abrupt heterojunctions or of crystals with a variable forbidden gap, i.e. of graded heterojunctions, offers a new approach to creating solid state infrared-to-visible converters. When a graded heterojunction is illuminated from the narrow gap side by infrared light $h\nu_1$, electron-hole generation is observed near the illuminated surface. The external electric field 'pulls' non-equilibrium carriers (holes) into the wide gap region where they recombine radiating quanta of visible light $h\nu_2$, $h\nu_3$ and so on (6). In an abrupt heterojunction which is directly illuminated it is possible to increase the current and the recombination in the wide gap region slightly when a narrow gap region is being illuminated. The efficiency of this kind of converter is not great, because it is implicit in the mode of operation that it lacks a light-amplifying mechanism.

In a four-layer $n-Al_xGa_{1-x}As-p-Al_xGa_{1-x}As-SiGaAs-p-Al_xGa_{1-x}As$ structure the use of a wide gap emitter junction results in a solid state light converter with a photosensitive Si region (semi-insulating GaAs) which operates as an infrared photoconductor (40). When a positive bias is applied (with + directed to the edge p region) the voltage is divided between the edge junctions switched-in in a forward direction, the middle junction biased in a reverse direction and the high ohmic semi-insulating region. As with an ordinary p-n-p-n structure, when the direct biases are sufficiently great the middle junction is 'overflowing' with injected carriers and becomes conductive. With a voltage close to the point of such a 'switching' this transition (from one state to another) may occur when the middle regions are illuminated.

Under certain conditions the use of semi-insulating GaAs doped with Cr in one of these bases results in a considerable increase in the diffusion length, and thus in the sensitivity for controlling a transition to the conductive state in the structure. Illuminating semi-insulating GaAs (from the side or through wide gap emitters) with infrared radiation absorbed in Si–GaAs may then bring the device to the conductive state and convert the radiation into the visible owing to non-equilibrium carriers recombining in the wide gap region.

The study of the spectral characteristics of such devices has shown that the long wave conversion edge is equal to 0.75 eV, or 1.65 µm. The application of semi-insulating GaAs doped with other impurities (e.g. with Fe) will apparently permit values of 3–3.5 µm to be obtained. A great advantage of these devices is their high amplifying coefficient (10^2–10^3) during conversion. The threshold sensitivity even at present is equal to 10^{-8} W.

A poor agreement of radiation spectra of gas-discharge pumping source and of the active medium absorption spectra is known to be the main reason of the low efficiency of solid state lasers. Thus, for instance, not more than 7–8% of the light power generated by the pumping lamp is absorbed in the rod of pulse lasers based on YAG:Nd^{3+}. The rest of the radiation is left beyond the absorption bands and, eventually, changes into heat losses. This gave rise to the idea that semiconductor LEDs with a comparatively narrow radiation band, the maximum of which may coincide with one of the active medium absorption bands of a laser, could be used as pumping sources. However, creation of large square LEDs for laser pumping is technologically difficult. To assemble pumping systems with a great number of LEDs is also a fairly difficult task.

In view of the above a hybrid type of pumping has been developed in which semiconductor structures, similar to hetero-LED ones, convert the wide discharge lamp radiation spectrum into a narrow radiation band coinciding with one of the absorption bands YAG:Nd^{3+} ($\lambda = 0.805$ µm) (41). The active layer of wide gap p–$Al_xGa_{1-x}As$ with $E_g = 1.545$ eV and 2–3 µm thick, is placed between the two layers of a more wide gap solid solution. Radiation from the lamp reaches the converter on the side of a wide gap window with the energy band width for direct junctions of about 2.8 eV. Photons having passed this layer are absorbed in the narrow gap region causing luminescence with $\lambda \simeq 0.8$ µm.

The luminescence light, after multiple reflections and re-radiation, leaves the converter through the layers of wide gap materials. The external quantum efficiency of such conversion is now 40–50% which makes it possible to double the efficiency of the pumping systems and to reduce the pumping threshold power by the same factor. Fabrication of converters with a unitary square of several cm² does not cause any technological difficulties. Combination of such converters with a p-n junction in a GaAs substrate makes it possible to create high-efficiency photocells for the operation at solar light concentrations of 2000 suns and even more (37).

9.2.4 Powerful diodes, transistors, and thyristors

Further progress in powerful semiconductor technology is associated with the application of A^3B^5 wide gap materials, and especially of GaAs and GaAs–AlAs solutions. For a long time the use of these materials was thought to be impossible because of the difficulties in obtaining their pure crystals and achieving large diffusion lengths of non-equilibrium carrier required for the base region conductivity modulation of a diode or thyristor with the thickness of 100 μm, which is necessary for obtaining high reverse voltages. These difficulties may be overcome due to the increase in the diffusion length of injected current carriers caused by internal electric fields in graded AlAs–GaAs heterostructures (2) and by the re-radiation mechanism at 100% internal efficiency. This makes it possible to create powerful diodes on the base of GaAs (42) with the working current density of 10^3 A/cm² and reverse voltage of 1000 V, as well as thyristors (43) with the voltage of 100–200 V and working current density of 200 A/cm². An advantage of these devices is that they can operate at increased frequencies (several MHz) and high limit temperatures (300°–400°C). The possibility of controlling light fluxes in heterostructures plays an important role in powerful devices as well (conversion of a certain amount of power into light (5, 42), thyristors with optical control) since the control power is converted into light and is absorbed without losses at the collector junction (43).

Although the original suggestion was to use a wide gap emitter in a transistor (44), and attempts to create such a transistor were carried out a fairly long time ago (45), the first high efficiency transistors on heterostructures have been obtained just recently (46). At present, a transistor with heterojunctions on the basis of GaAs–AlAs solid solutions is the best kind of a bipolar transistor for high frequencies and increased working temperatures. It enables the amplification coefficient to be independent on current practically over the entire range of current densities available.

9.3 HETEROSTRUCTURES IN INTEGRATED OPTICS

The science of integrated optics is based on the possibility of generating and controlling light waves located in a thin film dielectric waveguide. The main purpose of the investigations is the application of thin film technology and integrated techniques for the creation of new optical and optoelectronic systems. The parallel problem of creating discrete thin film optical devices is being solved (the devices are usually fabricated on the basis of perfect volume crystals). While many scientists aim at perfection of volume crystals and optical devices, it would appear that a number of practical problems will be solved with the help of thin film analogies.

Although investigations of optical waveguide structures based on various organic and inorganic materials are being carried out at present, it is quite evident that multifunctional integrated optical schemes will be based on single-crystal heterojunctions. This is not only because the optical properties of semiconductor crystals can be effectively controlled but also because hetero-epitaxial structures permit immediate and highly effective conversion of electric signals into optical effects and vice versa.

Since the final aim of integrated optics is to fabricate all optoelectronic elements on a single semiconductor heterojunction platelet (an integrated scheme being made from a silicon platelet) it is necessary to choose from a number of heterostructures the one which combines all the necessary passive and active elements of an integrated scheme. Just as in microelectronics the transistor is an essential element, so in integrated optics lasers are of paramount importance.

Of the known heterostructures DHLs meet the requirements best of all. It should be emphasized that this structure permits planar combination in a single platelet of all the necessary active and passive elements for integrated optics such as lasers, LEDs, photocells, modulators, deflectors and light-beam input and output systems.

9.3.1 Distributed-feedback DHLs

The possibility of replacing the Fabry–Perot type of resonator or similar ones in lasers by a periodic structure has been shown in (47). The authors described a laser with an active medium in which the refractive index changes periodically from layer to layer. If an inhomogeneity period equals an integral number of half wavelengths, then owing to the optical inhomogeneity the wave is Bragg reflected in a similar way to the reflection of light from the Fabry–Perot resonator mirrors. As the reflection originates in the active medium itself the lasers were named distributed feedback lasers (DFBLs). The same authors later developed a theory of these lasers (48). At the same time a new type of semiconductor laser with light output through a diffraction grating on the surface of a heterostructure active layer was proposed at the A.F. Ioffe Physico-Technical Institute of the USSR Academy of Sciences, the idea being to reduce the radiation divergence and to replace the Fabry–Perot resonator by a distributed-feedback structure formed by the interaction between the waveguide modes and the surface diffraction grating; also a detailed theoretical analysis of diffraction grating laser operation was made (49).

The possibility of creating low divergence semiconductor lasers, with radiation output through a plane parallel to the active layer due to the interaction between a waveguide mode and the diffraction grating on the waveguide surface, was first demonstrated experimentally in the A. F. Ioffe Physico-Technical Institute (50).

A diffraction grating with period a equal to 0.22 µm was embedded at the waveguide surface by interference photoetching (51). Radiation divergence in a plane perpendicular to the waveguide layer and the grating grooves measured in the most intensive beam was 30' and was determined by the width of the generation line. The threshold density slightly exceeded the threshold in specimens without an uncovered passive region and was 3–5 kA/cm^{-2}, which confirms the low losses caused by this region and the wide perspectives of the method used for obtaining the gratings (interference photoetching). Simultaneous generation in a laser with a distributed-feedback resonator with a radiation output through the diffraction grating on the surface of a heterostructure waveguide layer has been obtained quite recently (52). n-$Al_xGa_{1-x}As$–p–GaAs heteroepitaxial structures grown on n$^+$-GaAs substrates were used as experimental specimens. A diffraction grating up to 900 Å deep was obtained by interference photoetching on the surface of the waveguide layer.

An important peculiarity of lasers with radiation output through the diffraction grating is the almost 100% polarization of the coupled out radiation with vector E parallel to the grating lines. In the case of distributed feedback, radiation divergence in a plane perpendicular to the active layer and the grating grooves is considerably less than in the case where radiation generated in a Fabry–Perot resonator is coupled out by the grating. Near the generation threshold the divergence does not exceed 0.1° and is close to the aperture divergence.

A valuable property of DFB lasers is the rather weak temperature dependence of the generation line position in the spontaneous radiation spectrum within a small range of working temperature changes (several dozens of degrees). This dependence is determined not by the E_g temperature dependence, but by the temperature changes of the grating period.

Recently, a continuous operation regime at room temperature has been realized for both DFB lasers (the grating being in the active laser region, which is injection pumped) (53), and DBR lasers (54) (in which the grating is separated from the active region to improve the degradation characteristics).

The first operating monolithic integrated optical scheme was obtained in Japan (55). It was an AlAs–GaAs heterostructure with six DFB lasers, the generation wavelength of each laser being shifted by 20 Å. Radiation from the lasers was coupled out into a common waveguide.

Further development of DFB lasers probably will result in the creation of better longlife coherent radiation sources with such advantages of semiconductor lasers as coherence, small radiation divergence, frequency stability with temperature changes, the possibility of obtaining single-mode and single-frequency regimes. Their spectral range broadening is connected with the development of multicomponent semiconductor solid solutions technology.

At present, fabrication of both discrete and integrated devices without the application of heterostructures leads to a decrease in their efficiency and to a restriction as to their application. Optical communications, cosmic and terrestrial energetics, information processing systems, coloured projection television—these are the perspectives for the application of heterojunction devices.

The p–n homostructure which has served semiconductor electronics for about 30 years is yielding its position to the more complex but considerably more effective heterostructure.

REFERENCES

1. R. L. Anderson, *IBM J. Res. Develop.*, **4**, 283, (1960).
2. Zh. I. Alferov, V. M. Andreyev, V. I. Korol'kov, D. N. Tret'yakov, *FTP*, **1**, 1579 (1967).
3. Zh. I. Alferov, V. M. Andreyev, V. I. Korol'kov, E. L. Portnoy, D. N. Tret'yakov, *FTP*, **2**, 1016 (1968).
4. H. Kroemer, *RCA Rev.*, **18**, 332 (1957).
5. Zh. I. Alferov, *FTP*, **1**, 430 (1967).
6. Zh. I. Alferov, V. M. Andreyev, V. I. Korol'kov, E. L. Portnoy, A. A. Yakovenko, *FTP*, **3**, 541 (1969).
7. Zh. I. Alferov, V. M. Andreyev, V. I. Korol'kov, E. L. Portnoy, D. N. Tret'yakov, *FTP*, **2**, 1545 (1968);
   Zh. I. Alferov, V. M. Andreyev, E. L. Portnoy, M. K. Trukan, *FTP*, **3**, 1328 (1969).
8. N. G. Basov, O. N. Krokhin, Yu. M. Popov, *ZhETF*, **40**, 1879 (1961).

9. D. N. Nasledov, A. A. Rogachev, S. M. Ryvkin, B. V. Tsarenkov, *FTT*, **4**, 1062 (1962).
10. R. N. Hall, G. E. Fenner, J. D. Kingsley, T. J. Soltys, R. O. Carlson, *Phys. Rev. Lett.*, **9**, 366, (1962);
    M. I. Nathan, W. P. Dumke, G. Burns, F. H. Dill, G. L. Lasher, *Appl. Phys. Lett.*, **1**, 62, (1962);
    N. Holonyak, S. F. Bevaequa, *Appl. Phys. Lett.*, **I**, 82 (1962).
11. V. S. Bagayev, N. G. Basov, B. M. Vul, B. D. Kopylovsky, O. N. Krokhin, Yu. M. Popov, E. P. Markin, A. N. Khvoschev, A. P. Shotov, *DAN SSSR*, **150**, 275 (1963).
12. Zh. I. Alferov, V. M. Andreyev, D. Z. Garbuzov, Yu. V. Zhilayev, E. P. Morozov, E. L. Portnoy, V. G. Trofim, *FTP*, **4**, 1826 (1970).
13. I. Hayashi, M. B. Panish, P. W. Foy, S. Sumski, *Appl. Phys. Lett.*, **17**, 109 (1970).
14. R. L. Hartman, R. W. Dixon, *Appl. Phys. Lett.*, **26**, 239 (1975).
15. Zh. I. Alferov, V. M. Andreyev, D. Z. Garbuzov, M. K. Trukan, *FTP*, **8**, 561 (1974).
16. Zh. I. Alferov, V. M. Andreyev, D. Z. Garbuzov, V. R. Larionov, V. D. Rumyantzev, *Pis'ma ZhTF*, **1**, 401 (1975).
17. L. R. Tomasetta, G. G. Foustad, *Appl. Phys. Lett.*, **25**, 440 (1974).
18. R. Burnham, N. Holonyak, H. Korb, H. M. Macksey, D. R. Scifres, J. B. Woodhouse, Zh. I. Alferov, *Appl. Phys. Lett.*, **19**, 25 (1971);
    R. D. Burnham, N. Holonyak, H. Korb, H. M. Macksey, D. R. Scifres, J. B. Woodhouse, Zh. I. Alferov, *FTP*, **6**, 97 (1972).
19. Zh. I. Alferov, I. N. Arsent'yev, D. Z. Garbuzov, V. D. Rumyantsev, *Pis'ma ZhTF*, **1**, 406 (1975);
    Zh. I. Alferov, I. N. Arsent'yev, D. Z. Garbuzov, V. D. Rumyantsev, V. P. Ulin, *Pis'ma ZhTF*, **2**, 241 (1976);
    J. J. Coleman, N. Holonyak, M. J. Ludowize, P. D. Wright, R. Chin, W. O. Groves, D. L. Keune, *Appl. Phys. Lett.*, **29**, 167 (1976).
20. W. R. Hitchens, N. Holonyak, P. D. Wright, J. J. Coleman, *Appl. Phys. Lett.*, **27**, 245 (1975);
    Zh. I. Alferov, I. N. Arsent'yev, D. Z. Garbuzov, V. D. Rumyantsev, V. P. Ulin, *Pis'ma ZhTF*, **2**, 481 (1976).
21. A. P. Bogatov, L. M. Dolginov, L. V. Druzhinina, P. G. Yeliseyev, B. N. Sverdlov, E. L. Shevchenko, *Quantum Electronics*, **1**, 2294 (1974).
22. K. Sugiyama, H. Saito, *Japan J. Appl. Phys.*, **11**, 1057 (1972).
23. R. E. Nahory, M. A. Pollack, *Appl. Phys. Lett.*, **27**, 562 (1975).
24. Zh. I. Alferov, V. M. Andreyev, V. I. Korol'kov, V. G. Nikitin, E. L. Portnoy, A. A. Yakovenko, *FTP*, **6**, 739 (1972).
25. N. Holonyak, S. F. Bevaequa, *Appl. Phys. Lett.*, **1**, 82 (1962);
    N. Holonyak, S. F. Bevaequa, C. V. Bielon, S. J. Lubowski, *Appl. Phys. Lett.*, **3**, 47 (1963).
26. A. A. Gutkin, A. A. Rogachev, V. E. Sedov, B. V. Tsarenkov, *PTE*, **4**, 187 (1963).
27. H. S. Rupprecht, I. M. Woodall, G. D. Pettit, *Appl. Phys. Lett.*, **11**, 88 (1967).
28. Zh. I. Alferov, V. M. Andreyev, V. I. Korol'kov, E. L. Portnoy, A. A. Yakovenko, *FPT*, **3**, 930 (1969).
29. Zh. I. Alferov, R. I. Chikovani, R. A. Charmakadze, G. M. Mirianashvili, N. K. Zosimov, N. A. Grigoryan, *FTP*, **6**, 2289 (1972);
    Zh. I. Alferov, D. Z. Garbuzov. R. A. Charmakadze *et al.*, *ZhTF*, **43**, 2413 (1973).

30. Zh. I. Alferov, V. M. Andreyev, D. Z. Garbuzov, N. Yu. Davidyuk, V. R. Larionov, L. T. Chichua, *Pis'ma ZhTF*, **2**, 1066 (1976); Zh. I. Alferov, V. M. Andreyev, D. Z. Garbuzov, N. Yu. Davidyuk, B. V. Pushny, L. T. Chichua, *Pis'ma ZhTF*, **3**, 657 (1977).
31. N. S. Dubrovskaya, R. I. Krivosheyeva, N. F. Nedel'ski, V. N. Ravich, V. I. Sobolev, B. V. Tsarenkov, L. A. Chicherin, *FTP*, **3**, 1815 (1969).
32. Zh. I. Alferov, V. M. Andreyev, D. Z. Garbuzov, N. Yu. Davidyuk, V. R. Larionov, P. P. Pashinin, A. M. Prokhorov, V. D. Rumyantsev, V. M. Tuchkevich, M. M. Khaleyev, *ZhTF*, **45**, 368 (1975).
33. Zh. I. Alferov, V. M. Andreyev, N. S. Zimogorova, D. N. Tret'yakov, *FTP*, **3**, 1633 (1969).
34. Zh. I. Alferov, V. M. Andreyev, E. L. Portnoy, I. I. Protasov, *FTP*, **3**, 1324 (1969).
35. Zh. I. Alferov, V. M. Andreyev, M. B. Kagan, I. I. Protasov, V. G. Trofim, *FTP*, **4**, 2378 (1970).
36. Zh. I. Alferov, O. A. Ninua, I. I. Protasov, V. G. Trofim, *FTP*, **5**, 988 (1971).
37. V. M. Andreyev, M. B. Kagan, G. L. Lyubashevskaya, T. A. Nuller, D. N. Tret'yakov, *FTP*, **8**, 1328 (1974); Zh. I. Alferov, V. M. Andreyev, G. S. Donetski, M. B. Kagan, N. S. Lidorenko, V. M. Tuchkevich, *Report on WELC*, Moscow (1977); J. A. Hutchby, *Appl. Phys. Lett.*, **26**, 457 (1975).
38. L. W. James, R. L. Moon, *Appl. Phys. Lett.*, **26**, 467 (1975).
39. Zh. I. Alferov, F. A. Akhmedov, V. I. Korol'kov, V. G. Nikitin, *FTP*, **7**, 1159 (1973); Zh. I. Alferov, F. A. Akhmedov, V. I. Korol'kov, A. A. Yakovenko, *FTP*, **8**, 1741 (1974); F. A. Akhmedov, V. I. Korol'kov, Yu. M. Makushenko, *FTP*, **8**, 1032 (1974).
40. Zh. I. Alferov, V. I. Korol'kov, V. G. Nikitin, D. N. Tret'yakov, *FTP*, **5**, 1503 (1971).
41. Zh. I. Alferov, V. M. Andreyev, D. Z. Garbuzov, N. Yu. Davidyuk, V. R. Larionov, P. P. Pashinin, A. M. Prokhorov, V. M. Tuchkevich, *Quantum Electronics*, **3**, 1349 (1976).
42. Zh. I. Alferov, V. I. Korol'kov, V. G. Nikitin, M. N. Stepanova, D. N. Tret'yakov, *Pis'ma ZhTF*, **2**, 201 (1976).
43. V. I. Korol'kov, V. G. Nikitin, N. Rakhimov, *Pis'ma ZhTF*, **2**, 941 (1976).
44. W. Shockley, *U. S. Patent*, **2.569.347** (1951).
45. H. J. Hovel, A. G. Milnes, *IEEE Trans. Electron. Dev.*, **ED-16**, 766 (1969).
46. Zh. I. Alferov, F. A. Akhmedov, V. I. Korol'kov, V. G. Nikitin, *FTP*, **7**, 1159 (1973).
47. H. Kogelnik, C. V. Shank, *Appl. Phys. Lett.*, **18**, 152 (1971).
48. H. Kogelnik, C. V. Shank, *J. Appl. Phys.*, **43**, 2327 (1972).
49. R. F. Kazarinov, R. A. Suris, *FTP*, **6**, 1359 (1972).
50. Zh. I. Alferov, S. A. Gurevich, R. F. Kazarinov, M. N. Mizerov, E. L. Portnoy, R. P. Seysyan, R. A. Suris, *FTP*, **8**, 832 (1974); Zh. I. Alferov, S. A. Gurevich, R. F. Kazarinov, V. R. Larionov, M. N. Mizerov, E. L. Portnoy, *FTP*, **8**, 2031 (1974).
51. L. V. Belyakov, D. N. Goryachev, M. N. Mizerov, E. L. Portnoy, *ZhTP*, **44**, 1331 (1974).
52. Zh. I. Alferov, S. A. Gurevich, N. V. Klepikova, V. I. Kuchinski, M. N. Mizerov, E. L. Portnoy, *Pis'ma ZhTF*, **2**, 245 (1976).

53. M. Nakamura, K. Aiki, J. Umeda, A. Yariv, *Appl. Phys. Lett.*, **27**, 403 (1975).
54. Zh. I. Alferov, S. A. Gurevich, M. N. Mizerov, E. L. Portnoy, *Pis'ma ZhTF*, **3**, 983 (1977).
55. K. Aiki, M. Nakamura, J. Umeda, *Int. Conf. Semicond. Laser*, Japan (1976).
56. V. M. Yevdokimov, A. F. Milovanov, D. S. Strebkov, *FTP*, II, 2224 (1977).
57. Zh. I. Alferov, V. M. Andreyev, Yu. M. Zadiranov, V. I. Korol'kov, N. Rakhimov, T. S. Tabarov, *Pis'ma ZhTF*, **4**, No. 7 (1978).

Chapter 10

Processes in Heavily Doped Rare-Earth Materials and their Applications to Optoelectronic Devices

F. E. Auzel

10.1 INTRODUCTION

As you may have noticed, the subject of this chapter is somewhat different from others in the volume. In fact, the materials I am going to discuss here, are not usually semiconductors and one may wonder why I am going to talk about properties of rare-earth doped insulators at a school wholly devoted to semiconductor optoelectronics.

The reason is that two types of optoelectronic devices involving heavily doped rare-earth materials closely associated with semiconductor light sources have been of interest recently; these are displays and light sources with up-conversion phosphors (1) and 'mini-lasers' with stoichiometric materials (2). Both types of devices are meant to be pumped by tightly coupled GaAs incoherent or coherent diodes.

In this chapter, I plan to explain what happens to energy excitations in materials with trivalent rare-earth content higher than 10^{21} ions/cm^3 and to give a brief account of the state of the art in applying such materials to light sources and lasers.

In the first part, I am going to review the effect of concentration upon relaxation, energy transfer, and radiative and non-radiative decay.

In the second part, I will consider up-conversion by energy transfer from the point of view of the principles as well as applications with emphasis upon displays and LEDs involving Yb^{3+}, Er^{3+} and Tm^{3+} ions.

In the third part, self-activated (stoichiometric) Nd^{3+} laser materials will be considered with a discussion of the compromise between high gain per unit length and self-quenching due to ion–ion interaction. Some current results as well as some new ones will be presented.

Finally, common problems which are yet to be solved will be considered in view of the effective useful applications of both up-conversion and laser devices.

10.2 DESCRIPTION OF BASIC PHENOMENA INVOLVED IN THE KINETICS OF HIGHLY DOPED RARE-EARTH MATERIALS

Besides the phenomena intrinsically arising only at high concentration, I shall first recall one which is roughly concentration independent but which can be clearly investigated at low concentration; this is the multiphonon non-radiative decay, which governs the concentration independent part of emission quantum yields.

10.2.1 The multiphonon non-radiative decay

Once an energy level has been excited, whatever the means, energy can relax down non-radiatively through the level ladder, from one level to the next nearest one as shown schematically in Figure 10.1(a). It has been found experimentally that the quantum efficiency of an excited state can be lower than expected of one phonon interaction even at low concentrations, that is, without any possibility for energy transfer towards sinks to take place. A well-known example of this situation is provided by the relation between quantum yields of levels and their difference relative to the next lower level (3, 4). Monochromatic excitation has shown (5) that decay of energy proceeds effectively by cascade as shown in Figure 10.1(b). This is related to the fact that the non-radiative transition probability is well described by an exponential law with respect to energy differences ΔE between two consecutive levels (6) (Figure 10.1(b)).

$$W_{NR}(\Delta E) = W_{NR}(0)\exp(-\alpha_{NR}\Delta E) \qquad (10.1)$$

Here, ΔE is larger than the highest phonon frequency of the matrix and so the non-radiative decay has to involve several simultaneous phonons: α_{NR} can be theoretically given as (7)

$$\alpha_{NR} = \alpha_S - (\hbar\omega_m)^{-1}\frac{2}{\bar{N}}\ln\frac{N}{S_0} \qquad (10.2)$$

with $\hbar\omega_m$ being the highest phonon energy of the matrix, S_0 the Pekar–Huang–Rhys electron-phonon coupling constant, and $\bar{N} = E/\hbar\omega_m$ the average number

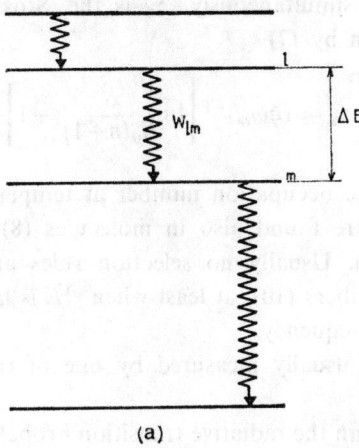

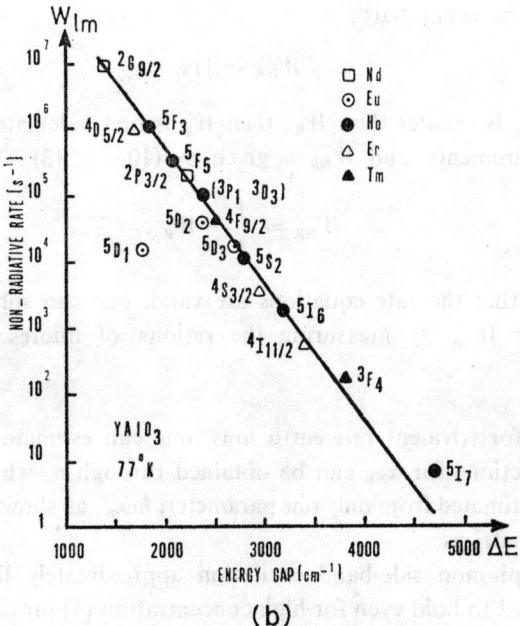

Figure 10.1 The multiphonon non-radiative decay:
(a) a sequential, 'ladder' type process; (b) an example
of the exponential law for the non-radiative decay rate
versus energy gap. From Reference 13

of phonons involved simultaneously. α_S is the Stokes multiphonon excitation parameter given by (7)

$$\alpha_S = (\hbar\omega_m)^{-1}\left[\ln\frac{\bar{N}}{S_0(n+1)} - 1\right] \qquad (10.3)$$

where $\bar{N}$ is the average occupation number at temperature T.

Analogous results are found also in molecules (8) and for deep centres in semiconductors (9). Usually no selection rules are found with respect to levels quantum numbers (10), at least when ΔE is larger than the phonon or vibration highest frequency.

Practically, W_{NR} is usually measured by one of the following methods:

(a) If W_{NR} is larger than the radiative transition probability W_R, then a direct measurement of W_{NR} is obtained by measuring the lifetime of the considered level (11) from a sample with dilute concentration effects which could not be separated:

$$W_{NR} = 1/\tau \qquad (10.4)$$

(b) When W_{NR} is smaller than W_R, then W_R is first calculated from absorption measurements, and W_{NR} is given by (10, 12, 13)

$$W_{NR} = \frac{1}{\tau} - W_R \qquad (10.5)$$

(c) Assuming that the rate equations are valid, one can solve the equation system for W_{NR} by measuring the rations of fluorescence intensities (14, 15).

Practically, for trivalent rare-earth ions, one can estimate that $S_0 \simeq 0.05$ (7), and predictions for α_{NR} can be obtained through α_S which can be approximately estimated from only one parameter, $\hbar\omega_m$, as shown in Figure 10.2 for different matrices.

From multiphonon side-band excitation approximately the same values can be estimated to hold even for high concentration (7) provided the phonon frequencies are not modified too much by the doping ions. This implies the use of a matrix which already contains an 'optically neutral' ion (Y, La, Gd, Lu) as a constituent, thus permitting dilution of the active ion concentration without changes in the crystal structure.

Now let us consider the effects arising primarily at high concentrations.

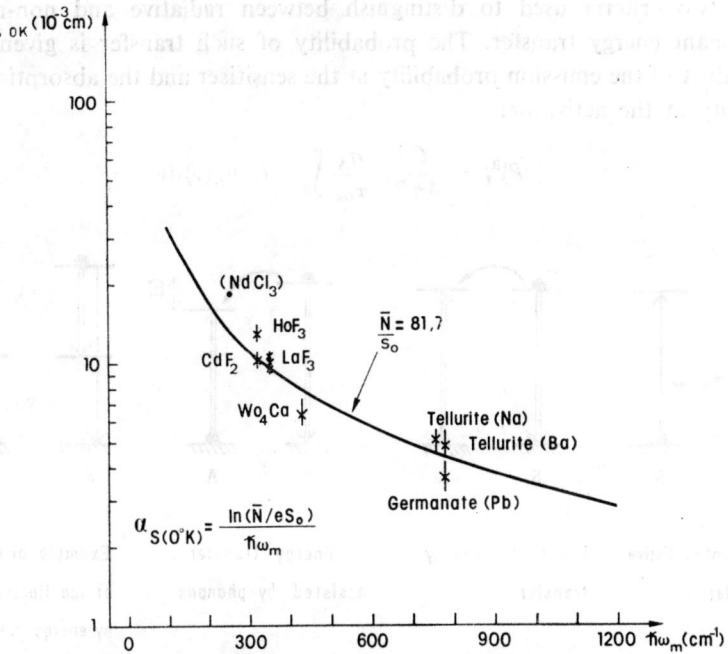

Figure 10.2 The multiphonon Stokes excitation exponential parameter versus effective phonon frequency for different matrices. Tellurite and germanate are glasses

10.2.2 Different types of energy transfer encountered in rare-earth spectroscopy

Energy transfers occur in luminescence when absorption and emission do not take place within the same centre. They may occur without charge transport. It is then possible to distinguish between radiative and non-radiative resonant energy transfer and phonon-assisted non-radiative energy transfer (Figure 10.3).

10.2.2.1 The resonant radiative energy transfer

When the transfer is radiative, the photons emitted by 'sensitiser ions' are absorbed by any 'activator ions' within the photon travel distance, so that transfer depends upon the shape of the sample. Moreover, depending on the degree of overlap between the emission spectrum of the sensitiser (S) and the absorption spectrum of the activator (A), the structure of the emission spectra of the sensitiser will change with activator concentration. These are the two criteria used to distinguish between radiative and non-radiative resonant energy transfer. The probability of such transfer is given by the product of the emission probability at the sensitiser and the absorption probability at the activator:

$$P_{SA}^{(R)} = \frac{C^2}{4\pi R^2} \frac{\sigma_A}{\tau_{0s}} \int_0^\infty g_s(\nu) g_A(\nu) \, d\nu \qquad (10.6)$$

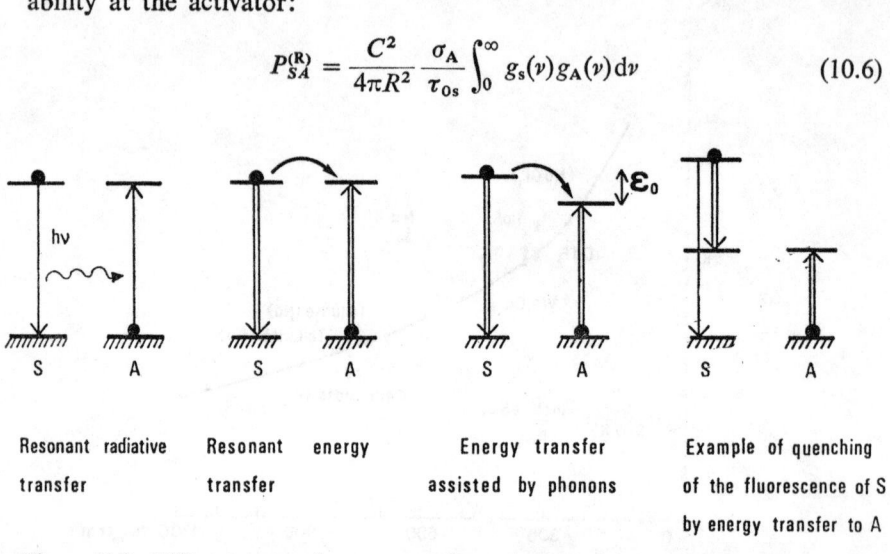

Resonant radiative transfer | Resonant energy transfer | Energy transfer assisted by phonons | Example of quenching of the fluorescence of S by energy transfer to A

Figure 10.3 Different types of energy transfers towards an activator in its ground state

the integral being the spectral overlap integral between $g_A(\nu)$, the absorption shape function of the activator, and $g_s(\nu)$, the emission shape function of the sensitiser; R is the $S-A$ distance; τ_{0s} is the radiative lifetime of S; σ_A is the integrated cross section of A. Such resonant transfer may permit long-range energy diffusion between identical ions.

10.2.2.2 The resonant non-radiative energy transfer

Let us consider a simple case of two ions, each with one excitable electronic state separated from its electronic ground state by nearly equal energy. With suitable interaction between the two electronic systems, the excitation will jump from one ion to the other before it is able to emit a quantum of fluorescence. The mutual interactions are the Coulomb interactions of the Van der Waals type between the two ions. Förster (16), who first treated such a case theoretically in terms of the quantum mechanical theory, considered the dipole–dipole interactions. He assumed that the interaction is the strongest if for both transitions electric dipole transitions are permitted (17). The interaction energy is then proportional to the inverse of the third power of the interionic distance and the transfer probability is given by:

$$P_{SA} = \frac{2\pi}{\hbar} |\langle S^e A^0 | H_{SA} | S^0 A^e \rangle|^2 \varrho_E \qquad (10.7)$$

where H_{SA} is the interaction Hamiltonian, and ϱ_E is the density of states provided by the vibrational motion contributing to the line broadening of the transition, which is proportional to the inverse sixth power of that distance. The wavefunctions to be considered for the matrix element describe the initial state of the system with the sensitiser in its excited state and the activator in its ground state, the final state having the sensitiser in its ground state and the activator in its excited state.

Therefore, the transfer probability can be written as

$$P_{SA} = \frac{1}{\tau_s} \left(\frac{R_0}{R}\right)^6 \qquad (10.8)$$

where τ_s is the actual lifetime of the sensitiser excited state, including multiphonon non-radiative decay, with the critical transfer distance for which excitation transfer and spontaneous deactivation of the sensitiser having equal probability. R_0 can be written as (18)

$$R_0^6 = \frac{3}{64\pi^6} \cdot \frac{1}{\varepsilon^2} \cdot \frac{\sigma_A \eta_s^0}{\bar{\nu}^4} \int_0^\infty g_S(\nu) g_A(\nu) d\nu \qquad (10.9)$$

where ν is the wavenumber, $g_A(\nu)$, $g_S(\nu)$ normalized shape functions, η_s^0 the quantum efficiency of the sensitiser in the absence of the activator, ε the refraction index, $\bar{\nu}$ the average wavenumber of the transition, and the integral represents the energy overlap between the absorption in the sensitiser and the activator emission. However, Dexter pointed out (19) that this theory should be extended to include higher multipole and exchange interactions. In fact, for an isolated atom, one can consider the transition probability as decreasing as $(a_0/\lambda)^{2n}$, where a_0 is the Bohr radius, λ the wavelength, and n an integer. By contrast, in an energy transfer process with a dependence on near-zone interactions, the transition probabilities drop off as $(a_0/\varrho)^{2n}$, where ϱ is the separation of the interacting ions. ϱ can be as much as three orders of magnitude smaller than λ, so that the energy transfer effect tends to be more pronounced in systems with forbidden transitions (19). This holds true for the rare-earth ions which we shall discuss later.

The energy transfer probability for electric multipolar interactions can be more generally written as (19)

$$P_{SA}(R) = \frac{(R_0/R)^s}{\tau_s} \qquad (10.10)$$

where s is a positive integer taking the following values:

$$s = \begin{cases} 6 & \text{for dipole-dipole interactions} \\ 8 & \text{for dipole-quadrupole interactions} \\ 10 & \text{for quadrupole-quadrupole interactions} \end{cases}$$

It should be noted that for dipole–dipole interaction differences between radiative and non-radiative resonant transfer consist essentially in the fact that for radiative transfer there is no critical R_0 depending only upon concentration, and the sensitiser lifetime does not depend upon this distance either.

10.2.2.3 The phonon-assisted non-radiative energy transfer

If we now consider two ions with excited states of different energies, the probability of energy transfer should drop to zero where the overlap integral $\int g_S(\nu) g_A(\nu) d\nu$ vanishes. However, it has been found experimentally that energy transfer can take place without electronic overlap provided that the overall energy conservation is maintained by production or annihilation of phonons with energies approaching $k\theta_d$ or kT, where θ_d is the Debye temperature of the host matrix (20). Then energy transfer assisted by one phonon can take place (21). However, in energy transfer between rare-earth ions energy mismatch as high as several thousand reciprocal centimetres are encountered (22, 23). This is much higher than the Debye cutoff frequency generally found in normally encountered hosts, so that multiphonon phenomena have to be considered here.

This was done by Miyakawa and Dexter in a general theory of multiphonon processes which arose from the necessity of obtaining a theoretical estimation of the probability of energy transfer between Yb^{3+} and Tm^{3+}, Er^{3+} (24) processes which are the object of this chapter. They showed that it is also legitimate to write the probability of energy transfer in the form

$$P_{SA} = \frac{2\pi}{\hbar} |H_{SA}|^2 S_{SA} \tag{10.11}$$

where S_{SA} is the overlap of the line-shape functions for absorption of ion S and emission of ion A, but in this case the phonon sidebands must be included in the line shape in order to evaluate S_{SA}. It is necessary to consider each partial overlap between the m-phonon emission line shape of ion A and the n-phonon absorption line shape of ion S. This mathematical assumption has been given a physical meaning recently by demonstration of the existence of multiphonon side-bands for trivalent rare-earth ions even in the case of very small electron-phonon coupling (7).

S_{SA} can be expressed as follows:

$$S_{SA} = \sum_N e^{-(S_{0S}+S_{0A})} \left[\frac{(S_{0S}+S_{0A})N}{N!}\right] \sigma_{SA}(0,0;E)\delta(N, \Delta E/\hbar\omega) \quad (10.12)$$

where S_{0S} and S_{0A} are the respective lattice coupling constants for the ions S and A, N is the order of the multiphonon process with $N = \Delta E/\hbar\omega_m$, ΔE the energy mismatch between the two ions, and $\hbar\omega_m$ the phonon cutoff frequency. $\sigma_{SA}(0,0;E)$ is the zero-phonon overlap integral between S and A.

The expression for S_{SA} with an energy mismatch of ΔE for small S_0 constant and for an occupation number $\bar{n} = (\exp(\hbar\omega/kT)-1)^{-1}$ not exceeding unity at the operating temperature, can be approximated by

$$S_{SA}(\Delta E) = S_{SA}(0)e^{-\beta \Delta E} \quad (10.13)$$

where $S_{SA}(0)$ is the zero-phonon overlap between S and A in the case where there is no energy mismatch between the two ions. β is given by

$$\beta = (\hbar\omega)^{-1}\left\{\log(N/S_0(\bar{n}+1)-1) - \log\left(1+\frac{S_{0A}}{S_{0B}}\right)\right\} \equiv \alpha_s - \gamma \quad (10.14)$$

This exponential dependence on energy mismatch agrees well with experiment (25).

This expression is particularly useful since we have seen the multiphonon relaxation rate to be also given by an exponential function with respect to energy (26, 27). Thus the energy transfer probability in a given host can be estimated from relaxation probability measurements for the individual ions (24, 12).

10.2.2.4 The macroscopic case of energy transfer in real samples

Up to this point we have been dealing with the microscopic case of two ions interacting with one another. To discuss the case of real macroscopic samples with many ions and to establish a link with experimental facts, a statistical analysis of the energy transfers is necessary. Supposing only a sensitiser-activator interaction, an average transfer efficiency can be calculated (17, 19). This was studied in some detail by Inokuti and Hirayama (28). They considered the number of activators situated at random in a sphere around a sensitiser such that the activator concentration is constant when the volume of the sphere and the number of activator ions considered goes to infinity:

then the average probability of transfer from one sensitiser to any acceptor is

$$W_{SA} = N_A \int_{R_{min}}^{\infty} P_{SA}(R) 2\pi R^2 dR \tag{10.15}$$

Introducing into (10.15) the expression for the intensity emitted by all sensitisers each with different activator neighbourhood, they obtained the following relation for the intensity decay of the emission of the sensitiser surrounded by many activators

$$I(t) = \exp\left[-\frac{t}{\tau_{S0}} - \Gamma\left(1 - \frac{3}{s}\right)\frac{C}{C_0}\left(\frac{t}{\tau_{S0}}\right)^{3/s}\right] \tag{10.16}$$

where τ_{S0} is the decay constant of the sensitiser in the absence of activator, C is the activator concentration, C_0 is the critical activator concentration, and s is the parameter of the multipolar interaction.

The comparison between experimental decay and this theoretical expression has been widely used to determine the index of the multipolar interaction involved (12, 29, 30). This theory is valid only when there is no sensitiser-to-sensitiser transfer or activator-to-activator transfer, i.e. this formulation has to be modified for high sensitiser and activator concentrations. Then rapid energy migration between sensitisers or between activators is possible, because of the perfect resonance conditions. The general result is complicated (31) but, for large t, $I(t)$ decays exponentially (32):

$$I(t) = \exp\left(-\frac{t}{\tau_s} - \frac{t}{\tau_D}\right) \tag{10.17}$$

Two cases can then be distinguished:

(i) *The diffusion limited case* for which spontaneous decay of excited sensitiser, diffusion among sensitisers and energy transfer between sensitisers and activators are of about the same order.

For sufficiently long times and dipole–dipole interactions one has (32, 33, 34):

$$\frac{1}{\tau_D} = 4\pi D N_A \varrho \tag{10.18}$$

where ϱ is a length defined by $\varrho = 0.68(C/D)^{1/4}$, D being the diffusion constant, N_A the activator concentration, and C is the sensitiser-activator energy transfer constant such that $C = R_0^s/\tau$, R_0 being the critical transfer distance of Equation 10.10.

For the diffusion limited case, the product $D\varrho$ is found to depend linearly upon sensitiser concentration (35), which has been experimentally verified (36); one has

$$\tau_D^{-1} = V N_S N_A \tag{10.19}$$

with $V = 8C^{1/4}C_{SS}^{3/4}$.

(ii) *The fast diffusion case.* For high sensitiser concentration, the diffusion rate can be faster than spontaneous sensitiser decay or sensitiser-activator energy transfer. The limiting step is no longer diffusion and D appears to saturate with increasing donor concentration; each activator has the same excited sensitiser neighbourhood.

$D\varrho$ is now a constant (35). We have

$$\tau_D^{-1} = U N_A \tag{10.20}$$

where U is a constant depending on the type of interactions (37).

Another approach to the macroscopic case is the use of the well-known rate equations which deal with population of ions in a given state. This was used as a phenomenological approach in studies of lasers. In the same manner, rate equations were derived for the energy transfer between Yb^{3+}–Er^{3+} (38, 39, 1). The applicability of those equations in relation to the Inokuti and Hirayama statistics has been discussed by Grant (40). On the basis of the first principles of quantum statistics, Grant shows that the average transition probability which enters into the rate equation, provided diffusion is sufficiently fast, depends on the powers of the concentration reflecting the number of types of coupled particles.

The basic Grant's result is that energy transfer probability is proportional to the activator concentration:

$$W_t = U N_A \tag{10.21}$$

This is the same result as the one obtained by fast diffusion studies (Equation 10.20).

The practical interest in considering diffusion is that decay is again exponential since when ions do not interact, the use of rate equations is justified. Further, the constant U can be estimated from Dexter's theory (19) and recently we gave a practical single form regardless of the type of multipolar or exchange interaction (41) generalizing expressions given by Kushida (42). A comparison of the calculated values for U makes it possible to predict the type of multipolar interactions involved (37).

A summary of the different cases for energy transfer is given in Table 10.1.

Table 10.1 Summary of the different cases of energy transfer processes involved in rare-earth ions interactions

Microscopic Case (theoretical one sensitiser–one activator case)		Macroscopic Case (experimental all sensitisers–all activators case)
Resonant **Radiative:** $P_{SA}^{(R)} = \dfrac{C_0^2 \sigma_A}{4\pi R^2 \tau_S} \int g_A(\nu) g_B(\nu) d\nu$ **Non-radiative:** $P_{SA}^{(R)} = \dfrac{(R_0/R)^s}{\tau_S}$; $(s = 6, 8, 10)$ with: $R_0^s \propto \int g_A(\nu) g_B(\nu) d\nu$	Intermediate statistics: average probability for one sensitiser → all activators	**Without diffusion among sensitisers:** non-exponential decay: $I(t) = \exp\left[\dfrac{-t}{\tau_S} \right.$ $\left. -\Gamma\left(1 - \dfrac{3}{s}\right)\dfrac{C}{C_0}\left(\dfrac{t}{\tau_S}\right)\right]^{3/s}$
Non-resonant **Non-radiative** $P_{SA}^{(R)} = \dfrac{(R_0/R)^s}{\tau_S}$; $(s = 6, 8, 10)$ with: $R_0^s \propto e^{-\beta \Delta E}$	W_{SA} $= N_A \int P_{SA}(R) \times$ $\times 4\pi R^2 dR$	**With diffusion among sensitisers:** (long time approximation) exponential decay: $I(t) = \exp\left(\dfrac{-t}{\tau_S} - \dfrac{t}{\tau_D}\right)$ Diffusion limited: $\tau_D^{-1} = VN_S N_A$ Fast diffusion: $\tau_D^{-1} = UN_A$ (rate equation case)

10.2.2.5 Energy transfer between ions in the excited states

Until the mid-1960's all energy transfers considered between rare-earth ions were of the type shown in Figure 10.3, where the sensitiser ion is in one of the excited states while the activator is in the ground state. This type of energy transfer explains fluorescence sensitisations as well as concentration quenching. In 1966, I reported evidence (22) for the energy transfer via the different processes shown in Figure 10.4. In these cases both the activator and sensitiser ions are in the excited states prior to energy transfer.

The probability calculation described previously applies to energy transfers between excited states (1) and rate equations are particularly useful (1, 38, 39). This new type of energy transfer paved the way for the up-conversion processes I am going to discuss now.

10.3 UP-CONVERSION BY ENERGY TRANSFER

In this section I am going to describe the use of energy transfers between excited states to obtain incoherent energy up-conversion (anti-Stokes luminescence). I will not, however, discuss other kinds of up-conversion sometimes experimentally indistinguishable but conceptually different, i.e. the cooperative sensitisation effect of Ovsyankin and Feofilov (45) and the cooperative luminescence of Nakazawa and Shionoya (46). The reason is that we are interested here in the most efficient effects.

Table 10.2 Comparison between different multiphoton upconverting processes (after (1))

Mechanisms	Efficiency (cm²/W)	Material
Two successive transfers	$\sim 10^{-3}$	$YF_3:Yb:Er$
Two-step absorption	$\sim 10^{-5}$	$SrF_2:Er$
Cooperative sensitisation	$\sim 10^{-6}$	$YF_3:Yb:Tb$
Second harmonic generation	$\sim 10^{-11}$	KDP
Two-photon absorption	$\sim 10^{-13}$	$CaF_2:Eu^{2+}$

A comparison of different two-photon up-converting processes is given in Table 10.2 showing why I have chosen to talk about the multiple sequential energy transfer.

10.3.1 *The APTE principle; the role played by high concentration*

Since the process described in Figure 10.4 is possible for any excited state provided its lifetime is larger than the duration of an energy transfer, there is no reason not to generalize it to the several excited levels of an activator A. For such a system, for instance, three incoming photons are absorbed in sequence by one sensitiser; at each step the coupled activator receives the sensitiser energy by energy transfer. Between each transfer this energy is kept by the intermediate excited states of the activator. Finally ion A reaches the nth excited state emitting a photon $h\nu_2 \leqslant nh\nu_1$ the energy of n incoming photons. This is basically the APTE (*addition de photons par transferts d'energie*) process.

To optimize the APTE efficiency, the activator concentration N_A is set at the highest level compatible with the small activator-activator cross relaxation which is known to produce self-quenching of the type described in

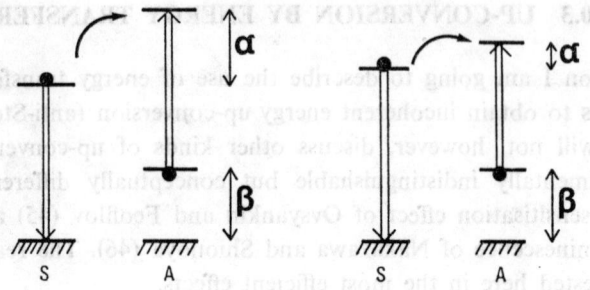

Resonant case   Phonon assisted case

$\beta - \alpha = 0$ $\beta - \alpha = \varepsilon_0$

Figure 10.4 Energy transfer towards an activator in one of its excited states in a resonant and in a non-resonant case

Figure 10.3. Yet N_A has to be sufficient in order to assure a strong energy transfer (Equation 10.21).

N_S, the sensitiser concentration, has to be high for two reasons: (i) to absorb the maximum of pumping photons, (ii) to assure a rapid migration of this energy towards as many activator ions as possible. Practically a limitation for N_S is found in the back-transfer from ion A to ions S if N_S is too large.

Among the different pairs of rare-earth ions which have shown such APTE effect and which I have already reviewed up to 1973 (1), I would like to report some recent results dealing with the following pairs: $Yb^{3+}-Er^{3+}$ and $Yb^{3+}-Tm^{3+}$ in view of their application in display devices.

10.3.2 Efficiencies and merit factor for APTE materials
10.3.2.1 Efficiencies

As can be understood from their principle, APTE materials are basically non-linear since they are characterized by multiphoton processes. Therefore the emitted intensity is proportional to a power of the excitation equal to the number of absorbed photons in the absence of saturation. Such behaviour can, of course, be predicted from rate equations.

Conversion efficiency is generally defined for continuous (CW) excitation as

$$\eta = \frac{I_{\text{emitted}}}{I_{\text{absorbed}}}$$

but in our case η has to be defined for each excitation intensity since I_emitted depends upon I_incident. In this way two types of efficiencies have been defined:

(i) a 'thick coating' efficiency (47) given by

$$\mu_n = \frac{P_e/\text{Area}}{(P_i/\text{Area})^n} \quad \text{in} \quad \text{cm}^{2(n-1)}\,\text{mW}^{-(n-1)} \tag{10.22}$$

where P_e and P_i are the output and incident power, respectively, n being the order of the multiphoton process. In such a coating nearly the whole incident power is absorbed thus justifying the 'thick coating' name. Practically, the difficulty in measuring μ_n is related to the fact that it is difficult to measure P_e in a thick sample due to diffusion and reabsorption.

(ii) a 'thin coating' efficiency (48) given by

$$\eta_{Nn} = \frac{P_e/\text{Area}}{(P_i/\text{Area})^{n-1}(\varepsilon P_i/\text{Area})} \quad \text{in} \quad \text{cm}^{-2(n-1)}\,\text{mW}^{-(n-1)}, \tag{10.23}$$

where ε is a function of P_i absorbed inside the sample. ε is sufficiently small to be homogeneous within the sample thickness, whence the 'thin coating' name.

A simple relation can be found between both efficiencies (49):

$$\eta_{Nn} = n\mu_n \tag{10.24}$$

Physically this arises from the fact that the absorbed power inside a coating of thickness x varies as $[1-\exp(-\sigma_s N_s x)]$; whereas the emitted power varies as $[1-\exp(-n\sigma_s N_s x)]$; σ_s is the sensitiser absorption cross section.

We have been considering η_{Nn}, because its definition is closer to what is effectively realised in a light emitting device, as will become obvious later.

10.3.2.2 Factor of merit

Up to now I have discussed continuous excitation efficiencies. Yet, in non-linear processes, three types of 'concentration' are useful for devices which have to be finally coupled to a time integrating receiver such as the human eye:

(a) spatial concentration represented by $(\text{Area})^{n-1}$ in (10.23);
(b) spectral concentration represented by ε in (10.23);
(c) time concentration which means pulsed excitation.

Pulsed excitation leads us to deal with time averaged efficiencies which can provide a gain g over CW efficiencies. The lower limit of g is fixed by the delay time of the eye response, the upper one depends on the longer time constant of the APTE (practically the rise-time constant). Moreover, in optoelectronics, one must often make a time multiplexing which necessarily leads to a pulsed-mode. The brightness increase obtained for pulsed excitation may be written as

$$g = \left(\frac{\overline{I_f(t)}}{I_{fCW}}\right) \quad \text{for} \quad \overline{I_{ex}(t)} = I_{exCW} \quad (10.25)$$

where $\overline{I_f(t)}$ is the average of $I_f(t)$ over the time, $I_f(t)$ is the fluorescence signal at time t under pulsed excitation, I_{fCW} is the fluorescence signal under CW excitation, $I_{ex}(t)$ is the infrared excitation in the pulsed mode, and I_{exCW} is the CW infrared excitation. The condition $\overline{I_{ex}(t)} = I_{exCW}$ is considered, because excitations are generally limited by heat dissipation ability which fixes a maximum average value for the available power. This implies that the thermal time-constant of the infrared excitation device is larger than the optical time-constants, which is generally the case. The intrinsic efficiency (taking into account the excitation concentration in time, space and spectral range) is then

$$H = \eta_{Nn} g \quad (10.26)$$

and is explained as follows.

Let us consider infrared excitation pulses, with a duration θ and a period T; if $\tau_1 \ll \theta < T$, τ_1 being the APTE rise time, we may write

$$I_{fCW} = \eta_{Nn} I_{exCW}^n; \quad I_S(t) = \eta_{Nn} I_{ex}^n(t)$$

hence

$$\overline{I_f(t)} = \frac{\theta}{T} I_{fmax}(t) = \frac{\theta}{T} \eta_{Nn} I_{exmax}^n(t) = \frac{\theta}{T} \eta_{Nn} \frac{T^n}{\theta^n} I_{exCW}^n$$

The condition $\overline{I_{ex}(t)} = I_{exCW}$ being written

$$\frac{\theta}{T} I_{exmax}(t) = I_{exCW}$$

hence

$$\overline{I_f(t)} = \left(\frac{T}{\theta}\right)^{n-1} I_{SCW}$$

and

$$g = \left(\frac{I}{\theta}\right)^{n-1} \ll \left(\frac{T}{\tau_1}\right)^{n-1}$$

The limit factor is then: $g_1 = (T/\tau_1)^{n-1} = [T \cdot 2\pi f_c]^{n-1}$, where $1/\tau_1$ is equivalent to a cut-off pulsation. Hence, the limit intrinsic efficiency is

$$H_1 = \eta_{Nn}(f_c T)^{n-1} = MT^{n-1} \quad \text{with} \quad M = \eta_{Nn}\left(\frac{1}{2\pi\tau_i}\right)^{n-1} \quad (10.27)$$

The period T depends solely on the investigator's eye, and according to the admitted degree of blinking, T ranges between 20 and 40 ms. Thus, the coefficient M is an intrinsic characteristic of the materials used; it is called the *figure of merit* and, for $n = 2$, it is analogous to the gain-bandpass product usually defined for current electronic components.

10.3.3 The Yb^{3+}–Er^{3+} pair

In the case of the rare-earth ions Er^{3+} and Yb^{3+}, the sensitiser ion is Yb^{3+} and the activator is Er^{3+}. Either a green, red, or blue UV output is obtained when the ion system is irradiated with near IR photons. Here we will confine ourselves to the green emission only.

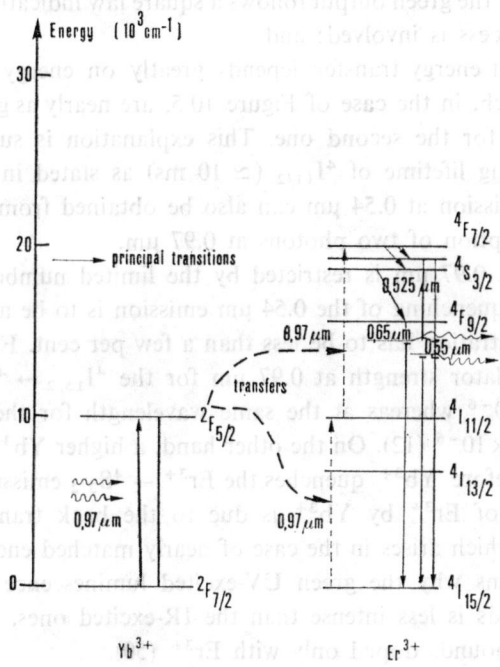

Figure 10.5 Actual scheme for summation of two photons in the Yb^{3+}–Er^{3+} couple

Figure 10.5 gives the energy paths of the phenomena which were first observed in a germanate glass, then in $NaYb(WO_4)_2$: Er (1, 43, 50). For the sake of clarity, only the pertinent levels are given. The green emission at 0.52 and 0.55 μm corresponds to the transition $^2H_{11/2} \to {}^4I_{15/2}$ and $^4S_{3/2} \to {}^4I_{15/2}$ of Er^{3+} and is the same as the emission found under UV excitation, $^2H_{11/2}$ being mainly thermally populated from $^4S_{3/2}$ (12). The excitation of $^4F_{7/2}$, $^2H_{11/2}$, and $^4S_{3/2}$ is obtained first through absorption by Yb^{3+} of photons at 0.97 μm in a $^2F_{7/2} \to {}^2F_{5/2}$ transition then by resonant energy transfer; as described in Section 10.2.2.2, the excitation of Er^{3+} in the $I_{11/2}$ state is obtained. Finally, either the same Yb^{3+} ion absorbs a second photon at 0.97 μm, or another nearby Yb^{3+} ion, being still in a $^2F_{5/2}$ state, transfers its energy to the same Er^{3+}. The Er^{3+} reaches the $^4F_{7/2}$, $^2H_{11/12}$ and $^4S_{3/2}$ states by the phenomena described in Section 10.3.1 respectively, with and without phonon assistance. This double successive energy transfer explanation is supported by:

(i) the fact that the excitation spectrum shows the same features as the reflectance spectra of the Yb^{3+} matrix;
(ii) the fact that the green output follows a square law indicating that a double photon process is involved; and
(iii) the fact that energy transfer depends greatly on energy matching conditions which, in the case of Figure 10.5, are nearly as good for the first transfer as for the second one. This explanation is supported by the required long lifetime of $^4I_{11/2}$ ($\simeq 10$ ms) as stated in Section 10.3.1.

The green emission at 0.54 μm can also be obtained from Er^{3+} alone by successive absorption of two photons at 0.97 μm.

Absorption at 0.97 μm is restricted by the limited number of Er^{3+} ions available if self-quenching of the 0.54 μm emission is to be avoided; in fact, the Er^{3+} concentration has to be less than a few per cent. Furthermore, the absorption oscillator strength at 0.97 μm for the $^4I_{15/2} \to {}^4I_{11/2}$ transition is about 0.4×10^{-6} whereas at the same wavelength for the $^2F_{7/2} \to {}^2F_{5/2}$ of Yb^{3+}, it is 4×10^{-6} (12). On the other hand, a higher Yb^{3+} concentration may be used before Yb^{3+} quenches the $Er^{3+} \to {}^4S_{3/2}$ emission (50, 51, 52). The quenching of Er^{3+} by Yb^{3+} is due to the back transfer Er^{3+} $^4F_{7/2}$ $\to Yb^{3+}$ $^2F_{5/2}$ which arises in the case of nearly matched energy differences. This also explains why the green UV-excited luminescence of Yb^{3+}–Er^{3+} doped compounds is less intense than the IR-excited ones, but the reverse is true for compounds doped only with Er^{3+} (50).

Those processes have been analysed systematically assuming the applicability of rate equations (38, 44, 49).

In particular, introducing the photon cutoff frequency $\hbar\omega_m$ as a parameter in those equations and considering all the different relaxation processes discussed in Section 10.2 the curve of Figure 10.6 has been obtained in par-

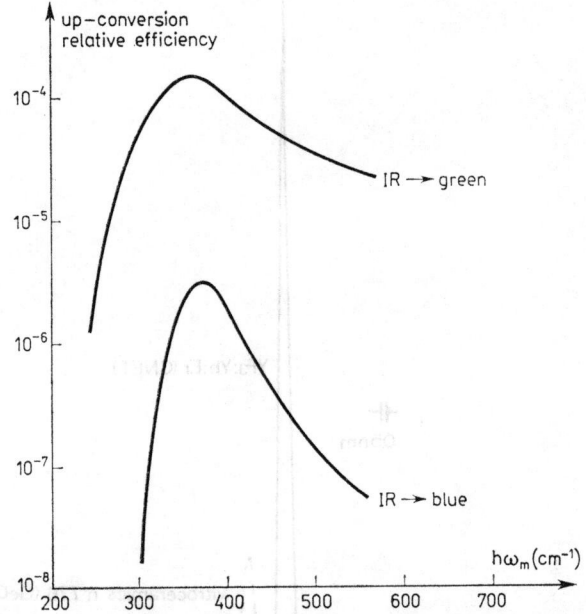

Figure 10.6 Theoretical optimization with respect to phonon cutoff frequency for up-conversion IR → green (Yb^{3+}–Er^{3+}) and IR → blue (Yb^{3+}–Tm^{3+})

ticular for Yb^{3+}–Er^{3+} showing that an optimized material should have $\hbar\omega_m = 360$ cm^{-1} (53) which is the case for fluoride.

Experimentally it is found that to our knowledge the best materials are YF_3:Yb:Er; $NaYF_4$:Yb:Er; and vitroceramics with fluoride constituent as shown in Table 10.3: Rare-earth contents have been optimized for CW. This confirms the theoretical predictions, as for concentration and $\hbar\omega_m$, and indicates that improved materials could be obtained by selecting more covalent hosts in order to increase the transition probability but keeping the same cutoff frequency (53).

Figure 10.7 gives the type of emission spectra obtained for YF_3 and vitroceramics.

Figure 10.8 presents a comparison of the excitation spectra of the best materials.

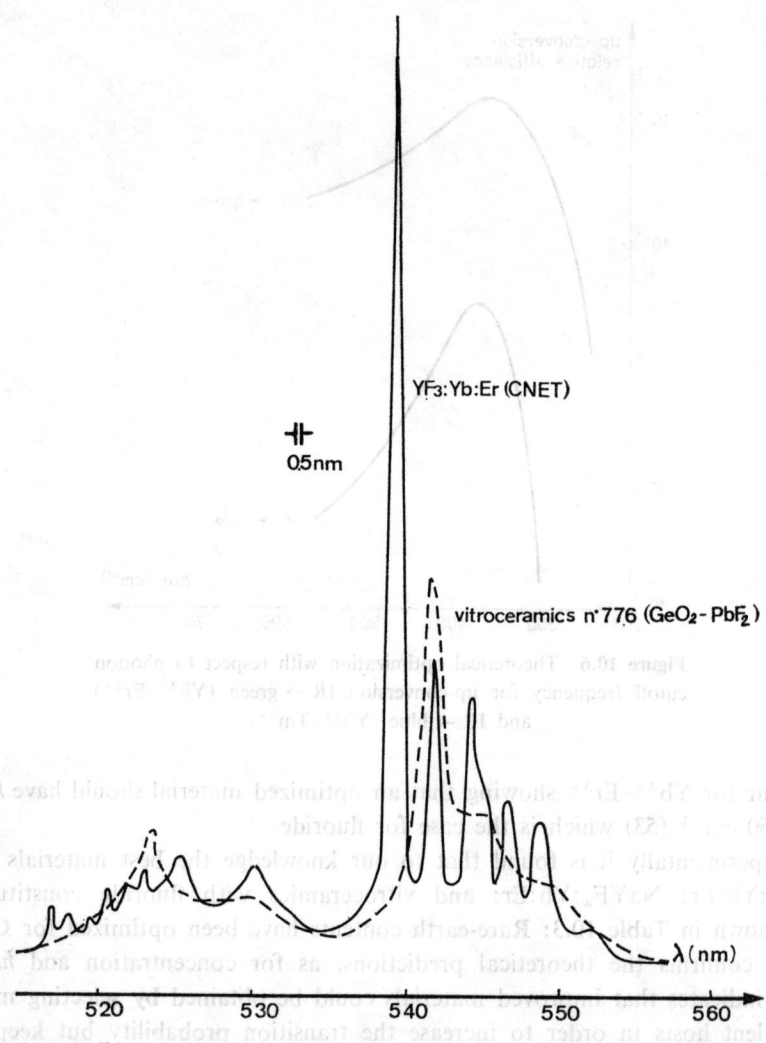

Figure 10.7 Emission spectra in up-conversion for Yb^{3+}–Er^{3+} in YF_3 and a vitroceramics

Table 10.3 Comparison of efficiency and transient times for Yb^3-Er^3-doped materials (IR → 0.54 μm)

Sample chemical formula powdered samples (Ø 50 μm)	Infrared → 0.54 μ conversion			Figure of merit ($\times 10^3$ Hz)
	Fall time, τ_1 (ms)	Fall time, τ_2 (ms)	Efficiency $\eta_{N2}(\times 10^6)$	
Vitroceramics No. 776 (PbF_2-GeO_2):Yb-Er PbF_2 78%, GeO_2 22%, +10% Yb_2O_3+0.75% Er_2O_3 (per cent by weight)	2	0.65	285	22.8
YF_3:Yb:Er (CNET 1972)* $Y_{0.80}Yb_{0.19}Er_{0.01}F_3$	3.2	1.2	285	15
YF_3:Yb:Er (BTL 1970)* $Y_{0.84}Yb_{0.15}Er_{0.01}F_3$	3.5	1.3	185	8.8
LaF_3:Yb:Er (GE 1970)* $La_{0.86}Yb_{0.12}Er_{0.02}F_3$	4.25	1.6	165	7

* The samples are discussed in Table X of Reference 1 where a comparison between different samples from different origins at one point in their development is reported.

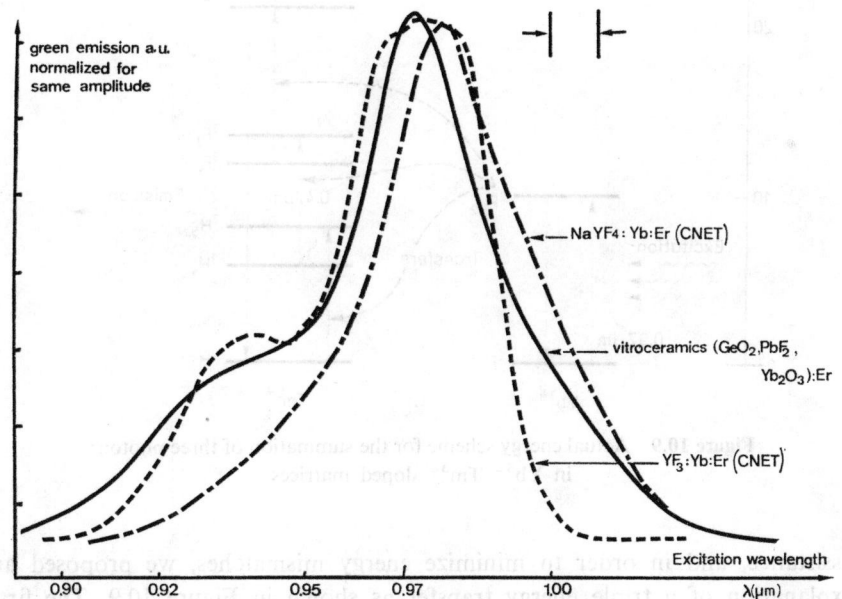

Figure 10.8 Comparison for excitation spectra of different materials doped with $Yb^{3+}-Er^{3+}$

10.3.4 The $Yb^{3+}-Tm^{3+}$ pair

Shortly after discovering the summation of photons by energy transfer in $Yb^{3+}-Er^{3+}$, we showed that a similar process occured in matrices doped with $Yb^{3+}-Tm^{3+}$ (22). By irradiating with 0.97 µm photons a germanate glass doped with these ions, we could get a faint blue light. Using the same ion pair in $NaYb(WO_4)_2$ in order to enlarge the phenomena, a strong blue emission was found having an excitation spectrum identical to the one of the $Yb^{3+}-Er^{3+}$ couple and corresponding to the diffused absorption spectra of Yb^{3+} (22, 51), the main difference being that the power law of the emission with respect to the IR excitation was nearly cubic instead of quadratic. This showed that a three-photon process was taking place. With our hypothesis of energy transfer between ions in their excited states (Section 10.2.2.5) and phonon

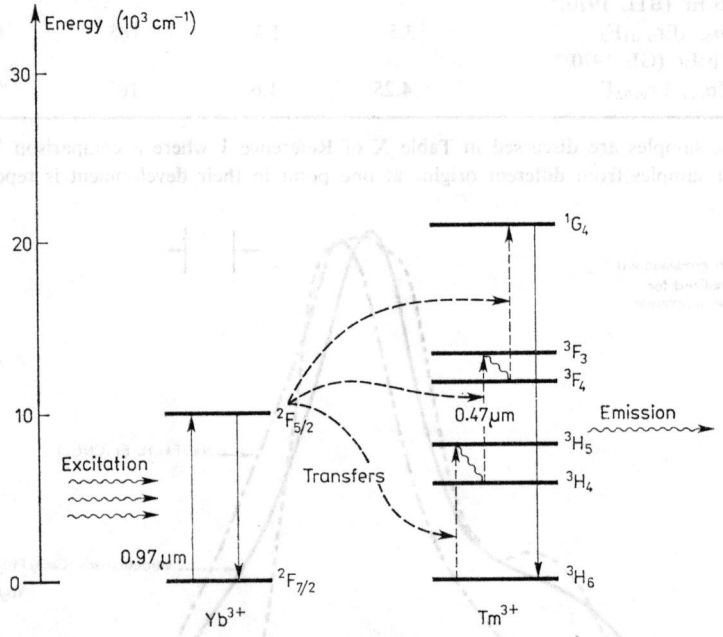

Figure 10.9 Actual energy scheme for the summation of three photons in $Yb^{3+}-Tm^{3+}$ doped matrices

assistance, and in order to minimize energy mismatches, we proposed an explanation of a triple energy transfer as shown in Figure 10.9. The first transfer with phonon assistance initiated in Tm^{3+} a $^3H_6 \rightarrow {}^3H_5$ transition

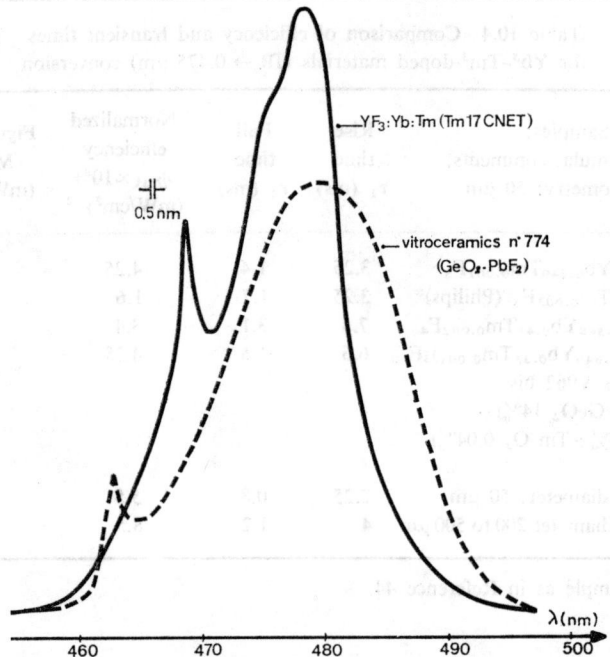

Figure 10.10 Emission spectra as obtained by the summation of three IR photons in YF$_3$ and in a vitroceramics

followed by a phonon relaxation to $^3H_4$; then a second phonon-assisted energy transfer initiated the $^3H_4 \rightarrow {}^3F_3, {}^3F_2$ transition followed by phonon relaxation to $^3F_4$, from which the $^1G_4$ state is reached by a third phonon-assisted transfer. This level then emits a photon at 0.47 μm corresponding to the three IR photons absorbed by Yb^{3+} ions which have successively transferred their energy to the Tm^{3+} ion. Figure 10.10 gives the emission spectra in a YF$_3$ matrix and in a vitroceramics by summation of three photons at 1.06 μm.

Using rate equations, a theoretical optimization can be performed (53) as for the Yb^{3+}–Er^{3+} pair giving the result of Figure 10.7. Here again $\hbar\omega_m$ should be around 360 cm^{-1}, but this figure is somewhat more critical. This can be traced back to the need for three multiphonon-assisted energy transfers (49). This result is experimentally confirmed since the optimized hosts are YF$_3$ and vitroceramics with fluoride surroundings for the rare-earth ions (54). The best results are summarized in Table 10.4. Of course, peak positions in the excitation spectra are about the same for both pairs Yb^{3+}–Fr^{3+}

Table 10.4 Comparison of efficiency and transient times for Yb^3–Tm^3-doped materials (IR → 0.475 μm) conversion

Samples: No., formula, comments, granulometry: 50 μm	Rise time τ_1 (ms)	Fall time τ_2 (ms)	Normalized efficiency $\eta_{N3}(\times 10^8)$ $(mW/cm^2)^{-2}$	Figure of merit $M (\times 10^{10})$ $(mW/cm^2)^{-2}s^{-2}$
Tm17, $Y_{0.65}Yb_{0.349}Tm_{0.001}F_3$	3.25	1.4	4.25	102
$Y_{0.647}Yb_{0.35}Tm_{0.003}F_3$ (Philips)*	3.25	1.7	1.6	38
Na22, $NaY_{0.548}Yb_{0.45}Tm_{0.002}F_4$	7.5	3.1	3.4	15
KTm7, $K(Y_{0.649}Yb_{0.35}Tm_{0.001})_3F_{10}$	6.6	1.5	4.25	24
Vitroceramics V962 bis $(PbF_2\ 86\% + GeO_2\ 14\%) +$ $+ (Yb_2O_3\ 25\% + Tm_2O_3\ 0.04\%)$ by weight:				
grain diameter: 50 μm	2.25	0.8	5.5	276
grain diameter 200 to 500 μm	4	1.2	8.5	135

* Same sample as in Reference 44.

Table 10.5 Results and comparison of green 7-segment displays with APTE and GaP technology under continuous excitation

Current intensity in one segment	APTE (green) optimized for diode, and up-converter thickness		GaP (Commercially available)					
			Monsanto MAN 5 yellow		X citon yellow		X citon green	
	V	I	V	I	V	I	V	I
10 mA	1.17 volts	32 μCd	1.96 volts	18 μCd	2 volts	90 μCd	2.04 volts	19.5 μCd
50 mA	1.64 volts	1850 μCd	2.9 volts	380 μCd	3.5 volts	430 μCd	3.07 volts	84 μCd

and $Yb^{3+}-Tm^{3+}$ and Figure 10.8 is also valid for $Yb^{3+}-Tm^{3+}$. The results are presented in Table 10.4.

It can be noticed in Tables 10.4 and 10.5 that the efficiencies for vitroceramics increase with the grain diameter. This increase saturates for a grain larger than $\simeq 300$ μm. This is due to the resonant radiative energy transfer between Yb^{3+} ions, as shown by the simultaneous increase of time constants (44, 55).

10.3.5 Application of APTE to LED and displays

10.3.5.1 The IR pumping diode and its coupling

Back in 1966, we failed to detect IR emission from Zn-alloyed GaAs diodes made at CNET. It was not until Galginaitis and Fenner (56) used Si-doped GaAs that IR detection was possible. In addition to being more efficient Si-doped GaAs diodes have somewhat longer emission wavelength than GaAs diodes. Such an IR source made it possible to obtain visible light-emitting diode of practical significance (56).

(a) *Description of phosphor-coated diodes and displays*

The principal scheme of the diodes is outlined in Figure 10.11. An IR-emitting diode is coated with phosphor giving the desired colour. For technological applications one must consider the phosphor intensity of IR energy input in space, time, and spectral domain. The use of dome-shaped diodes (57) (Figure 10.11(b)), a structural form in which most of the internal IR energy is extracted, does not seem to be a practical solution unless some external intensity concentration is used.

One method of obtaining space concentration is to have the phosphor as close to the emitting junction as possible and with the smallest volume compatible with diode reliability, for which 200 A/cm^2 is a convenient limiting threshold. Such confinement of the phosphor on the diode may be realized by cutting a moat into a planar diode structure. Filled with phosphor, seven such moat-diode segments on a GaAs chip make a single alphanumeric character (58).

An increase in the spatial IR density is also obtained by index matching between the semiconductor bulk ($n \simeq 4$) and the phosphor powder ($n \simeq 1.4$). This can be done by using an intermediate index gel or epoxy resin, being careful to exclude air between the powder particles and the semiconductor bulk. The higher the index of the epoxy, the better is the result. Since the sexcitation is made through the phosphor coating, a small thickness is

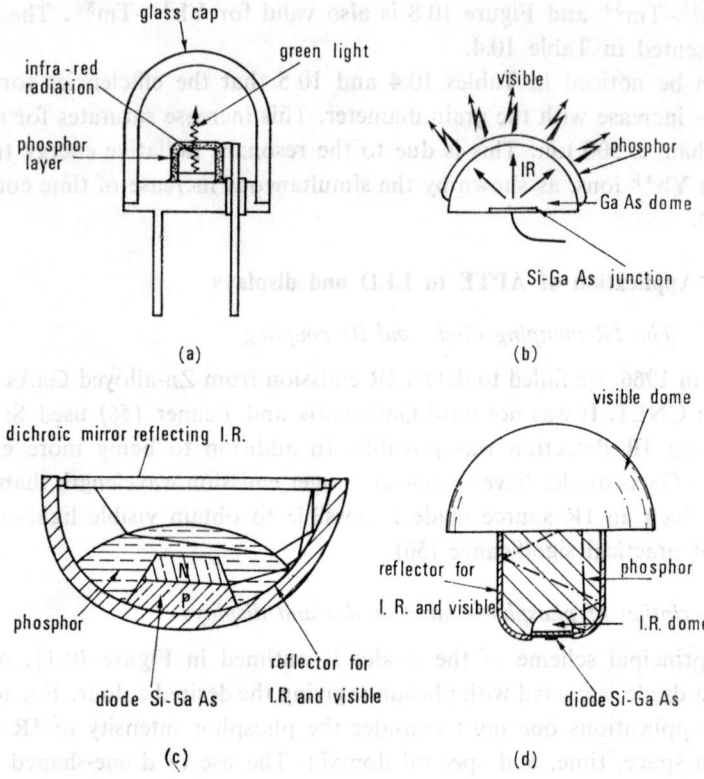

Figure 10.11 Different configurations for phosphor coated diodes after (1). (a) simple coating and planar diode; (b) domed diode and coating; (c) irregular shaped planar coated diode with dichroic mirror; (d) double dome configuration

required. An optimum seems to be about 70 mg/cm² for a YF$_3$ host, for grains diameter of 20 to 60 μm; for larger grains the optimum thickness tends also to be larger (59). However, with such a small thickness, only a small fraction, typically 1 to 2 per cent, of the IR emission is absorbed (60). One way to avoid this loss in IR power is to enclose the phosphor inside a dielectric coating transparent to the visible light and reflecting the IR emission onto the phosphor (61) thereby creating an IR cavity (Figure 10.11(c)). If reabsorption of visible light is small, a thick layer of phosphor may be used. An example is the double Weierstrass geometry presented in Figure 10.11(d), where one hemisphere extracts the maximum IR energy and the other hemisphere extracts the maximum visible light (62). The IR would then be completely absorbed by the

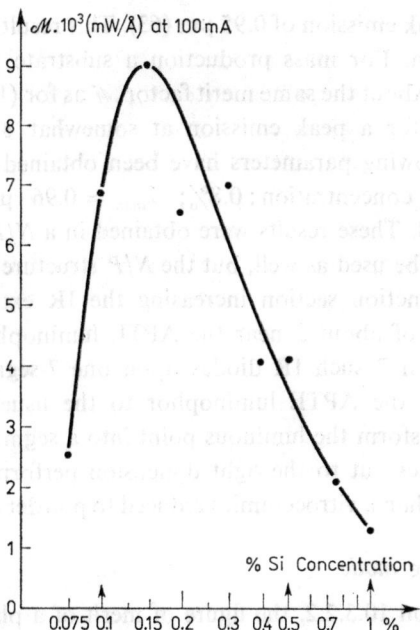

Figure 10.12 Merit factor for the IR GaAs diode versus Si source content of liquid phase epitaxy

phosphor. However, both the pump light and the phosphor emission are scattered preferentially in the backward rather than the forward direction (57, 60).

To increase the IR spectral power density one has to match the GaAs diode emission to the absorption spectra of Yb^{3+}, taking into acount the thermal shift of the diode emission at the operating temperature. This is obtained by varying the Si content of the Si-doped GaAs diode.

The efficiency is improved because the semiconductor is then transparent to its own emission. However, there are some drawbacks. The higher the Si content, the lower the efficiency becomes and the emitted IR spectra become broader. To optimize the Si concentration for a given current and a given temperature of the diode, one can define the coefficient of merit: $\mathcal{M} = P_{total} \times E_{0.97}/\Delta\lambda \times E_{max}$, where P_{total} in milliwatts is the integrated emitted IR power at given current, $E_{0.97}$ is the ordinate of the emitted spectra at 0.97 μm, E_{max} is the ordinate at the peak of the spectra, and $\Delta\lambda$ is the spectral width in angstroms at half power; Figure 10.12 shows that the optimum Si content in the source of the liquid-phase epitaxy process is 0.15% which corresponds to a peak emission of 0.95 µm (63). This result is valid for a (111) substrate orientation. For mass production a substrate (100) is more convenient for cutting. About the same merit factor $\mathcal{M}$ as for (111) can be obtained for (100) (64) but for a peak emission at somewhat shorter wavelength. Practically, the following parameters have been obtained and retained (64): substrate (100): Si concentration: 0.8%; λ_{max} = 0.96 µm; IR efficiency: : 30 mW/A: $\mathcal{M}$ = 8. These results were obtained in a N/P structure, though a P/N structure can be used as well, but the N/P structure permits a chemical reduction of the junction section increasing the IR extraction and power density by a factor of about 3 near the APTE luminophor (64). A display can be obtained with 7 such IR diodes upon one 7-segment display comb and by substituting the APTE luminophor to the usual dispersive Al_2O_3 powder used to transform the luminous point into a segment. Also a massive piece of vitroceramics cut to the right dimension performs equally well, but it is less convenient than a vitroceramics reduced to powder and put in a binder.

(b) *The pulse driving mode*

As noted in Section 10.3.2.2, the figure of merit of a phosphor defines the theoretical limit in the improvement obtainable under pulsed operation.

Figure 10.13 presents (48) the factor g measured for pulses with lengths both smaller and larger than τ_1. Plotted on the same diagram is the

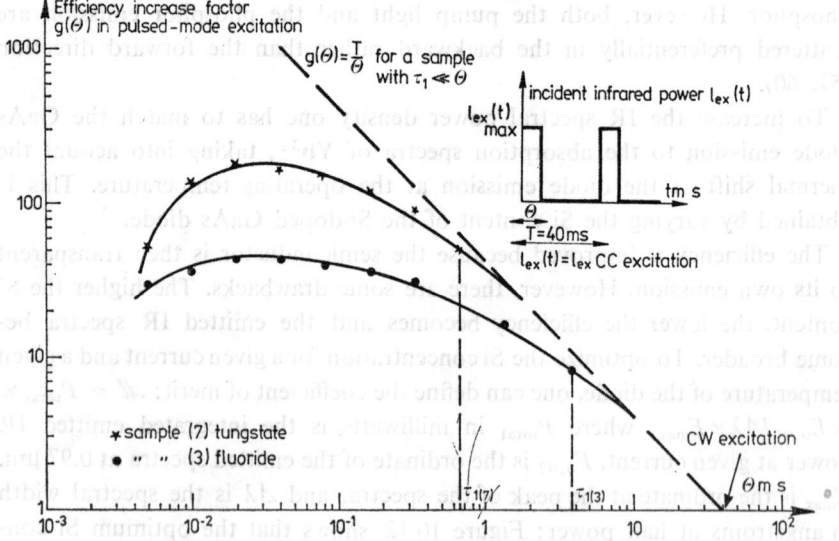

Figure 10.13 Pulsed excitation improvement g in a two-photons case (IR → green)

theoretical factor $g = T/\theta$, which is valid over the range $\tau_1 < \theta < T$ and which diverges from the experimental curve as θ becomes lower than τ_1. The experimental curves reach their maximum for $\theta < \tau_1$, because the emission increases after the end of excitation. These results indicate an increase in the practical efficiency with pulsed excitation by two orders of magnitude.

Analogous results have been obtained by Barnett and Henmann (58) for fluoride hosts. For practical purposes the increase in efficiency by the pulse driving mode is limited by the accepted degree of flicker. This limits T for the optimum θ.

(c) *Possibility of colour modulation*

Since the output of double-pumped phosphors may vary with the number of photons engaged in the process, the phosphors have an inherent tuning ability. For instance, a mixture of Yb^{3+}–Er^{3+} and Yb^{3+}–Tm^{3+} doped tungstates varies from green to blue depending on the IR excitation intensity (65). This is due to the fact that the green, produced from a two-photon process, follows a quadratic excitation law and the blue, produced from a three-photon process, follows a cubic excitation law. Similar tunability was observed in Yb^{3+}–Er^{3+} systems where the route for the red emission is through a three-photon process and the green through a two-photon process (66). A green-to-red switching is then obtained by electric means. It should be noted that phosphor mixtures should produce more subtle control than a tunable composition, because in this way each colour component can be optimized independently.

In addition to being sensitive to excitation intensity, the emitted colour may be changed by pulse modulation of the excitation. A red-to-green modulation is obtained by a pulsewidth change (67). Tuning is accomplished because the rise time of the red light is longer than the rise time of the green light (66, 58). These colour modulation effects have to be considered when the pulse-driving mode is used, which in turn may affect the choice of the host. By carefully selecting the host and ions (Tm^{3+}, Er^{3+}) 'primary colours' can be obtained. Since the white point lies inside the triangle of the primaries, mixing of these colours can produce white light (1).

10.3.5.2 Results and comparison with large-gap semiconductor diodes and displays

In addition to the variety of colours obtainable from phosphor-coated Si–GaAs diode, their efficiency is comparable with large-gap semiconductor diode. Since the time the first review was published (1) a number of technological improvements for the green emission in phosphors and for the coupling to the diode have been made, yielding the results of Figure 10.14 for LED, and of Table 10.5 for 7-segment displays. As for the blue, improvements

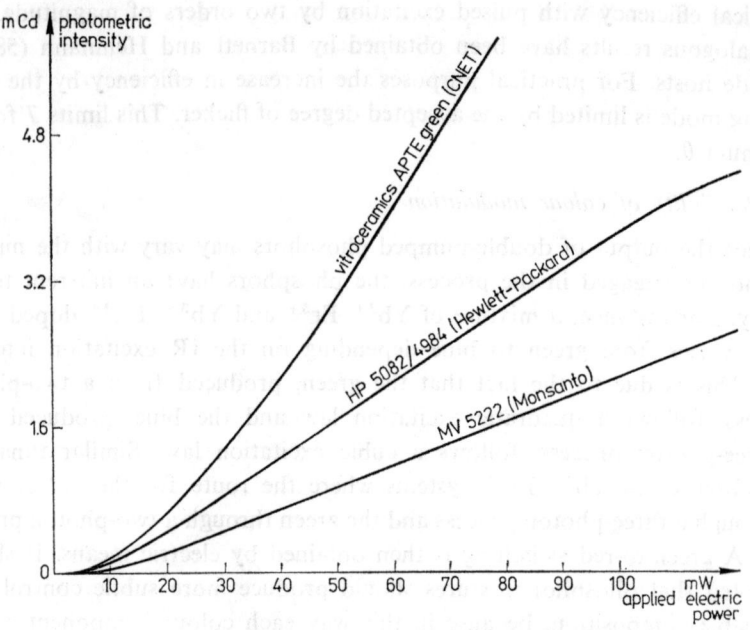

Figure 10.14 Comparison for photometric green emission intensity versus input electrical power for an APTE vitroceramic coated diode and two commercial GaP diodes with built in diffuser

depend on the powers of the IR excitation and are thus more difficult to make; in particular, for 7-segment displays, homogeneity of a segment can be questioned if only one diode is used. However, since much less work has been done in this area than for the green emission, one cannot settle the point. From the point of view of reliability, experiments indicate (68) that if good encapsulation is used, the lifetime obtained should depend on the GaAs:Si and binder; good reliability should be obtained at least for the GaP diodes (68). As for overall efficiency, only one order of magnitude is available since it depends on the input level. For the green, results indicate 5×10^{-4} and for the blue 5×10^{-5} at $\simeq$ 100 mA. Since output is nearly monochromatic, comparison with large-gap sources is questionable (69) and in fact it depends upon particular applications.

(a) 4-Nd^{3+} *Self-activated minilaser materials*

Now that low attenuation optical fibres in the 1.06 μm region are available, some new interest arises in Nd^{3+} laser materials (70). However, the conditions are somewhat different from the well developed Nd^{3+}-doped YAG lasers; for a Nd^{3+} laser sources to compete with semiconductor laser the following characteristics are necessary:

(i) overall length should be less than one centimeter;
(ii) laser threshold should be obtained for a transverse pumping from semiconductor diodes (71, 72, 73).

From such specifications, we are going to discuss the conditions imposed upon the material in relation to the physical processes discussed in Section 10.2 and we shall see analogies and differences with APTE materials.

Compactness requirement as well as transverse pumping impose a high concentration for Nd^{3+}. This is related to the fact that compactness means a high gain per unit length, i.e. a high number of active ions per unit volume. As for transverse pumping, easily manufactured planar LED can provide high power density only when close coupling with index matching is used, just as for APTE materials, the difference here being that the polycrystal coating is replaced by a single crystal. The active region in the single crystal has to be of about the same dimension as the high brilliance region of the pumping diode. This means that absorption of pumped light has to take place within less than 100 μm. This condition also imposes a high concentration of the absorbing ions.

A difference in comparison with APTE materials is that ions of the same type Nd^{3+} contribute both to absorption and emission, so that a separate optimization for absorption and emission is no longer possible. What one does it try to reduce the constraint upon the emission optimization by using materials with small self-quenching (74).

10.4.1 Self-quenching—Different hypothesis

To obtain a high Nd^{3+} concentration ($> 10^{21}$ cm^{-3}) Nd^{3+} has to be one of the matrix constituents. Such materials are usually called *stoichiometric* (2) or more appropriately *self-activated* (75). Unfortunately, however, for any luminescence system a high activator concentration generally leads to poor quantum efficiency for emission; this is the pervading effect of self-quenching due to ion-ion interactions. The levels of Nd^{3+} involved in pumping, emission and self-quenching are recalled in Figure 10.15; pumping is planned to be made by emission of $Al_xGa_{1-x}As$ diodes in absorption bands of $^4I_{9/2} \rightarrow \, ^2H_{9/2}$, $^4S_{5/2}$ transition at 0.8 μm (70, 76, 77); laser transitions take place at $\simeq$ 1.06 μm corresponding to $^4F_{3/2} \rightarrow \, ^4I_{11/2}$, or at 1.35 μm for $^4F_{3/2} \rightarrow \, ^4I_{13/2}$. Self-quenching is usually studied through $^4F_{3/2}$ lifetime variations with respect

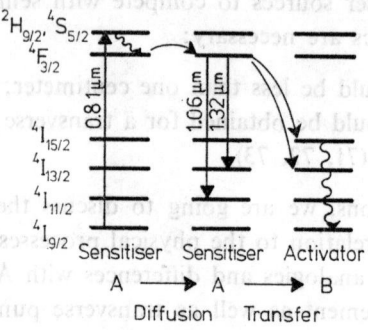

Figure 10.15 Energy levels of Nd^{3+} ions involved in diode excited laser emission and quenching

to Nd^{3+} concentration increase, because lifetime is an easily obtained parameter related to quantum efficiency for all emissions from a given level j by

$$\eta_j = \frac{\tau}{\tau_{0j}} \tag{10.28}$$

where τ is the lifetime effectively measured for $F_{3/2}$; τ_{0j} is the purely radiative lifetime of $^4F_{3/2}$ which can be linked to transition probabilities by:

$$\tau_{0j}^{-1} = \sum_{i=1}^{4} A_{ij} \qquad (i \equiv \, ^4I_{9/2}; \, ^4I_{11/2}; \, ^4I_{13/2}; \, ^4I_{15/2}) \tag{10.29}$$

where A_i is the Einstein coefficient for spontaneous emission for the transition j from $^4F_{3/2}$.

If one assumes τ_0 to be independent of the Nd^{3+} concentration (C_{Nd}) the variations of $\tau \, (C_{Nd})$ are a good representation of those of $\eta \, (C_{Nd})$. This point may be questioned (2) because for high Nd^{3+} concentration there can be some overlapping between the 4f configuration of one ion and the 5d configuration of its next neighbour. Considering this, Danielmeyer (78) proposed that the lifetime decrease with concentration could be the indication of some transition probability enhancement. As is well known, transitions within the 4f configuration of trivalent rare-earth ions are forbidden in the first order by the Laporte rule. The transitions are the result of parity mixing between 4f and 5d configurations coming from the odd terms of the crystal field. By analogy with this effect the above-mentioned enhancement would come from crystal field overlap mixing of pairs (CFOM) (78). By oscillator strength measurement with respect to concentration in $La_{1-x}Nd_xP_5O_{14}$ (NdLaUp) we failed to show any probability increase beyond measurement errors (79). One can thus safely assume that CFOM is not operative at least in NdLaUp and that a lifetime decrease follows a quantum yield decrease. The problem now is to understand why, in YAG for instance, concentration quenching is important even at $\simeq 10^{20}$ cm^{-3}, whereas in some other materials such as NdLaUp it is not. The strong self-quenching of Nd^{3+} in $Na_{0.5}Gd_{0.5}WO_4$:Nd (80), LaF_3:Nd (81), YAG:Nd (82) or glass (83) has been well known for a long time (80) to be due to the following cross relaxation energy transfer

$$Nd_A^{3+}(^4F_{3/2}) + Nd_B^{3+}(^4I_{9/2}) \rightarrow 2Nd^{3+}(^4I_{15/2}) \qquad (10.30)$$

as depicted in Figure 10.15. Such transfer is of the non-radiative resonant type (see 10.2.2.2). When energy matching for resonance is too bad, then phonon-assisted energy transfers can take place (84) in addition to Equation 10.30 according to

$$Nd_A^{3+}(^4F_{3/2}) + Nd_B^{3+}(^4I_{9/2}) \rightarrow 2Nd^{3+}(^4I_{13/2}) + N_1\hbar\omega_m \qquad (10.31)$$

$$Nd_A^{3+}(^4F_{3/2}) + Nd_B^{3+}(^4I_{9/2}) \rightarrow Nd_A^{2+}(^4I_{13/2}) + Nd_B^{3+}(^4I_{15/2}) + N_2\hbar\omega_m \qquad (10.32)$$

Another effect which comes into the picture at high concentration is energy migration between donors (see 10.2.2.4). Because of the rapid non-radiative relaxation between levels above $^4F_{3/2}$ (see 10.2.1), we think the following energy path for diffusion to be probable

$$Nd_A^{3+}(^4F_{3/2}) + Nd_B^{3+}(^4I_{9/2}) \rightarrow Nd_A^{3+}(^4I_{9/2}) + Nd_B^{3+}(^4F_{3/2}) \qquad (10.33)$$

Since transition probabilities involved in the above relations for resonant transfer are not expected to vary considerably from matrix to matrix (85), the role played by the overlap integral should be the most important in comparing strong and small quenching materials. This result comes directly from Equation 10.9 of Section 10.2.2.2.

Table 10.6 presents some yet unpublished results of the work currently in progress at our laboratory showing conclusively that the measured overlap integral between $^4F_{3/2} \rightarrow ^4I_{15/2}$ and $^4I_{9/2} \rightarrow ^4I_{15/2}$ is systematically smaller for small quenching materials. From this it can be inferred that small quenching or strong quenching materials behave in a way that can be predicted from known energy transfer processes without resorting to new ones. Some recent experimental work confirms this point of view for NdLaUp (86); it is found that τ varies with concentration at room temperature, according to

$$\tau = \tau_0/(1+kC_{Nd}^2) \quad (10.34)$$

Table 10.6 Spectral overlap integrals between $^4F_{3/2} \rightarrow {}^4I_{15/2}$ and $^4I_{9/2} \rightarrow {}^4I_{15/2}$ transitions for strong and small quenching materials

Materials	Quenching rate: $^\tau C_{Nd} < 10^{19}$ cm^{-3} / $^\tau C_{Nd} > 10^{21}$ cm^{-3}	Overlap integral ($\times 10^{-4}$) Kayser^{-1}
NdNa (WO$_4$)$_2$	22	6.5
YAG:Nd	> 22	4
NdP$_5$ O$_{14}$ (NdUP)	2.7	0.65
Na$_2$Nd$_2$Pb$_6$(PO$_4$)Cl$_2$ (CLAP)	1.8	1.45

k being a constant, which is the same equation as Equations 10.17 and 10.19 of Section 10.2.2.4 for the 'limited diffusion' case. The same form was given for YAG:Nd (87), though for NdLaUp the cross-relaxation is mostly due to multiphonon-assisted transfer (86) which is in accordance with our finding of a small overlap. However, in the case of NdLaUp and for other small quenching self-activated materials, the lifetime can also be of the form (84, 88)

$$\tau = \tau_0/(1+KC_{Nd}) \quad (10.35)$$

K being a constant, which is the same equation as Equations 10.17 and 10.20 of Section 10.2.2.4 for the fast 'diffusion case'. The discrepancy for NdLaUp can be linked with the following result.

For K$_3$Nd(PO$_4$)$_2$ (KNP), at concentration $< 3.2 \times 10^{20}$ cm^{-3}, a law in $(1+KC_{Nd}^2)^{-1}$ is found whereas for $C_{Nd} > 3.2 \times 10^{20}$ cm^{-3} the law in $(1+KC_{Nd})^{-1}$ predominates (89).

This may be ascribed to a passage from limited to fast diffusion with increasing concentration. Also, as can be seen in 10.2.2.4, the distinction between the two regimes for fixed concentration is related to the strength of the cross-relaxation energy transfer with respect to the diffusion one: a strong overlap would give a 'limited diffusion', whereas a small overlap would give a 'fast diffusion'. But for a particular crystal growth, if the relative probabilities of diffusion and cross-relaxation are for some reason modified, we think one could expect a change in the expected diffusion behaviour.

10.4.2 Systematic research for new Nd^{3+} self-activated laser materials

Although I do not believe the above self-quenching explanation to be 'so unsatisfactory in term of established models' for energy transfer as was suggested recently (90), it is clear that a method of predicting low-quenching self-activated materials is not yet available. However, at CNET we have developed a simple powder method to find materials with such properties by a systematic search. The usual method employed is the following.

Once a self-activated material has been synthesised in a polycrystalline powder form, lifetime measurements are performed for the self-activated, full Nd^{3+} concentration, followed by measurements for diluted concentration. Dilution is obtained by an 'optically neutral' ion such as La^{3+}, Y^{3+}, Gd^{3+}, Lu^{3+}, the choice being in order that there is no change of structure while replacing Nd^{3+} by the 'neutral' ion.

But this result alone is insufficient, because at low concentration one does not know whether the measured τ is near τ_0, the radiative lifetime, or not. An efficiency measurement is then necessary and such measurements are difficult.

A global simpler method consists in a relative fluorescence intensity measurement at around 1.06 μm on a powder (optically thick sample).

The gain obtained in a laser material with population inversion ΔN is

$$g = \sigma_{\text{laser}} \Delta N l \qquad (10.36)$$

where σ_{laser} is the laser cross section, and l is the length of material.

For a given pumping intensity

$$\Delta N \propto \sigma_{\text{abs}} \eta C_{\text{Nd}} \qquad (10.37)$$

For a given excitation intensity one has also for the fluorescence intensity

$$I_{\text{f}} \propto \sigma_{\text{abs}} C_{\text{Nd}} \qquad (10.38)$$

where σ_{abs} is the absorption cross section at excitation energy, and η is the quantum yield of $^4F_{3/2}$.

On the other hand, we have shown theoretically that for the laser transition at 1.06 μm:

$\int \sigma_{\text{laser}}(\nu) d\nu$ is nearly a constant K independent of the material
$(K \simeq 3 \times 10^{-18}$ cm) (85),

and:

$$\tau_{laser} = K/\Delta\nu, \Delta\nu \text{ being an effective emission line width.}$$

From this and from Equations 10.36 and 10.37 we have

$$(g/l)_{max} \propto \frac{K}{\Delta\nu} I_{f\,max} \tag{10.39}$$

$\Delta\nu$ can be taken to be (91)

$$\Delta\nu = \frac{\int I_f(\nu)d\nu}{I_{f\,max}}$$

and

$$(g/l) \propto I_{f\,max}^2 \Big/ \int I_f(\nu)\,d\nu \tag{10.40}$$

This shows the importance of $I_{f\,max}$ in selecting such a material. Measurement of $I_{f\,max}$ are obtained on samples of selected grain diameter ($\simeq 100$ μm) and are compared to $I_{f\,max}(NdUp) = 100$.

Such study has led us to propose two new minilaser materials: $Na_2Nd_2Pb_6(PO_4)_6Cl_2$ (CLAP) (92) and $NdTa_7O_{19}$ (93). The first synthesized crystals of CLAP (94) made it possible to obtain laser threshold of about 4.5 mW at 0.5 μm (Argon laser pump) (95). Figure 10.16 presents $I_{f\,max}$ (C_{Nd})

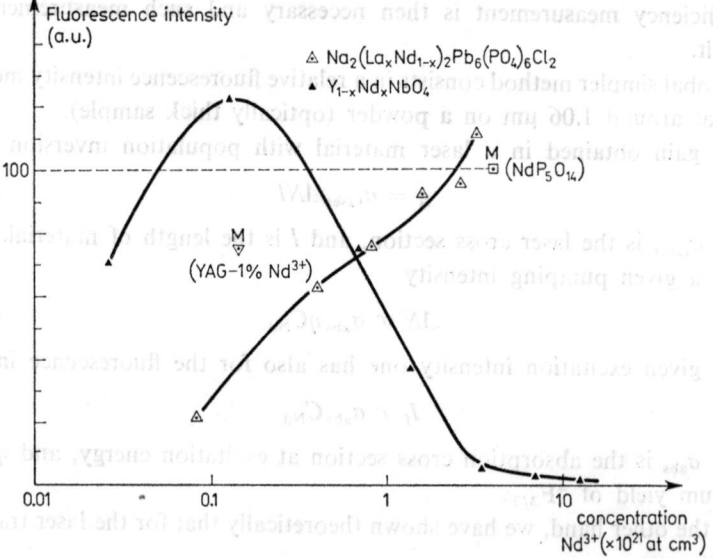

Figure 10.16 Typical behaviour of emission intensity at 1.06 μm versus Nd^{3+} concentration for a strong ($YNdNbO_4$) and a small self-quenching (CLAP) material; reference points are given for NdUP and YAG

Table 10.7 Relative figure of merit for different low self-quenching Nd^{3+} self-activated materials in powder form

Materials	C_{Nd} (relative to NdUP) ($\oslash$ grain $\simeq 100$ μm)	C_{Nd} (cm^{-3})	$\Delta\nu_{eff}$ (cm^{-1})	Figure of merit (relative to NdUP)	Quenching rate τ low C_{Nd} ($< 10^{19}$) τ high C_{Nd} ($> 10^{21}$)
NdP$_5$O$_{14}$ (NdUP)	100	3.9×10^{21}	127	1	2.7
Na$_2$Nd$_2$Pb$_6$(PO$_4$)Cl$_2$ (CLAP)	120	3.4×10^{21}	134	1.1	1.8
NdTa$_7$O$_{19}$	50	3.1×10^{21}	110	0.6	2.2
Na$_5$Nd(WO$_4$)$_4$	230	2.6×10^{21}	125	2.4	2.4
NdAl$_3$(BO$_3$)$_4$ (NAB)	70	5.4×10^{21}	72	1.3	2.5
KNdP$_4$O$_{12}$ (KNP)	190	4.1×10^{21}	131	1.8	2.8
LiNdP$_4$O$_{12}$ (LNP)	190	4.4×10^{21}	85	2.9	2.4
YAG: 1% Nd*	80	1.4×10^{20}	68	1.5	> 20

* YAG: Nd is given for comparison

for strong and small quenching self-activated materials showing the distinctive behaviour of the small quenching one for which the optimum is obtained for $C_{Nd} > 10^{21}$ cm^{-3}.

The different self-activated materials are compared in Table 10.7. Also, for small concentration materials, there can be discrepancies, because absorption can take place over several grains of the powder, the sample being then no longer an optically thick one.

It is interesting to compare the factor of merit proposed by Singh et al. (84) with our relative figure of merit on powder as given by $I_{f\,max}/\Delta \nu$: From Reference 84 one has:

$$M = \alpha \sigma_{laser} \tau$$

where: $\alpha = \sigma_{abs} C_{Nd}$. That is,

$$M = \sigma_{abs} \eta C_{Nd} \tau_0 \sigma_{laser} \propto (g/l)_{max}$$

from Equations 10.36 and 10.37, or

$$M \propto I_{f\,max}/\Delta \nu$$

only if one assumes τ_0 to be a constant independent of the type of material.

10.4.3 Results of diode side pumped Nd^{3+} self-activated materials

All small-quenching materials obtained to date in single crystal form have shown CW room temperature laser effect with Argon or dye laser pumping with threshold of the order of $\lesssim 5$ mW of longitudinally absorbed power.

Besides the first experiments where diode-pumped long YAG rods were used (96–101), there have been some experiments with small ($\lesssim 10$ nm in length) YAG rods or fibre end-pumping (102–104). It is difficult, however, to improve the threshold and efficiency of YAG lasers for side pumping because of a poor pump light coefficient, which may be insufficient for round trip, the non-resonant losses being higher than 0.4% (84). This shows the usefulness of self-activated materials for side-pumped mini-lasers, which is the more compact pumping configuration. Another advantage is that a LED array can be used and the total absorbed pump power increases with the number of LEDs. To the best of my knowledge only some preliminary results are available, and cooling has been necessary either for both the pumping diode and the crystal (77) or at least for the thermal tuning of the diode alone (76, 105). The results are summarized in Table 10.8. Before closing this section, an important question can be asked: what are the advantages of diode pumped rare-earth lasers over direct semiconductor sources for fibre optics telecommunication?

Table 10.8 Diode side-pumped Nd^{3+} self-activated lasers

Self-activated material		0.81 μ pumping diode	Threshold		Output	Ref.
Matrix	Temperature	Type	Electrical	at 0.81 μ		
NdP_5O_{14} $C_{Nd} = 4 \times 10^{21}$ cm^{-3} 565×78 μm	300 K	$Al_xGa_{1-x}As$; doubleheterostr. 296×100 μm optical imaging	Pulsed 300 μs $\simeq 160$ mW	Single laser diode $\simeq 220°K$	7.2 mW 3.4 mW	(76) (105)
$NdAl_3(BO_3)_4$ $C_{Nd} = 5.45 \times 10^{21}$ $L = 173$ μm	300 K	-idem-	Pulsed 300 μs $\simeq 370$ mW		16.5 mW $\simeq 1.3$ mW	(76) (110)
LiNd P_4O_{12} $4.85 \times 4 \times 19$ mm $C_{Nd} = 4.4 \times 10^{21}$	238 K	Array of 5 domed incoherent diodes $Al_xGa_{1-x}As$; 238 K optical imaging	Pulsed 450 μs $\simeq 660$ mW			(77)
$Nd_{0.25}La_{0.75}P_5O_{14}$	260 K	Array of 18 diodes $Al_xGa_{1-x}As$; 300 K optical imaging	Pulsed 600 μs		450 mW	(107)
$10 \text{ mm} \times 3 \times 2$ mm	223 K		CW		175 mW	
Room temperature extrapolations						
NdP_5O_{14} Length: 760 μm	300 K	$Al_xGa_{1-x}As$ (close coupling) (8×LEDs), junction area: 2×50 μm $Al_xGa_{1-x}As$ (30×lasers) Double hetero-structure, junction area: 3×13 μm	3–6 W		16 mW 1.7 mW 7.5 mW 21 mW	(76) (76)
$LiNdP_4O_{12}$	300 K	$Al_xGa_{1-x}As$ (LEDs)			32 W/cm^2	(77)

There are several of them (76, 106):

(a) narrower bandwith ($\simeq 1$ Å as compared with 20–50 Å for semiconductor lasers and $\simeq 300$ Å for incoherent diodes);
(b) mode control; single mode is readily achievable and this will be needed for high-speed transmission in single mode fibres.
(c) the transverse pumping scheme permits an easy choice of output power just by modifying the pumping diode array for a given crystal configuration.

A major drawback is that direct modulation is not available and the cost of the implementation of an efficient modulator may be the criterion of the use of Nd^{3+} self-activated laser (70, 106).

10.5 CONCLUSION

I have described the processes involved in some high concentration rare-earth materials which can be useful adjunctions to semi-conductors, emitting diodes, or lasers. Applications involve either APTE materials to produce light sources and displays or Nd^{3+} self-activated materials to obtain compact lasers in the useful transmission range of optical fibres. Since Nd^{3+} can also emit at 1.32 µm such a material could also be adapted (108) to current progress in fibre optics technology (109). Materials have been somewhat optimized and the theory, at least for APTE materials, indicates a new direction: more covalent host with $\hbar\omega_m \simeq 360$ cm^{-1}. In the case of laser materials our figure of merit indicates that $Na_5Nd(WO_4)_4$ and LNP should be the best.

But application of known theory in order to have general predictions is still under way. Work at room temperature is essentially a question of wavelength matching the pumping diodes. More generally, for both applications the critical point in improving the overall efficiency is in the coupling with high brilliance pumping sources. Close coupling without imaging devices is necessary for cost and compactness, certainly in the case of APTE emitters, but also for lower threshold side pumped self-activated lasers. In this case epitaxial growing of the rare-earth material could lead to quasi-monolithic technology (106).

REFERENCES

1. F. E. Auzel, *Proc. I.E.E.E*, **61**, 758 (1973).
2. H. G. Danielmeyer, in *Solid State Physics XVI*, Pergamon Press, Vieweg (1975), p. 253.
3. K. H. Hellwege, *Ann. D. Phys.*, **40**, 529 (1942).

4. G. E. Barasch and G. H. Dieke, *J. Chem. Phys.*, **43**, 988 (1965).
5. F. Varsanyi, in *Quantum Electronics*, edited by P. Grivet and N. Bloembergen, p. 787. Dunod (1964).
6. L. A. Riseberg and H. W. Moos, *Phys. Rev.*, **174**, 429 (1968).
7. F. E. Auzel, *Phys. Rev.*, B **13**, 2809 (1976).
8. S. Murata, C. Iwanaga, T. Toda, and H. Kokubun, *Chem. Phys. Lett.*, **15**, 152 (1972).
9. N. F. Mott, E. A. Davis, and R. A. Street, *Phil. Mag.*, **32**, 961 (1975).
10. M. J. Weber, *Phys. Rev.*, **171**, 283 (1968).
11. M. J. Weber, *Phys. Rev.*, **157**, 262 (1967).
12. F. E. Auzel, Thesis, Paris (1968). Also in: *Ann. Telecom.*, **24**, 199 (1969).
13. M. J. Weber, *Phys. Rev.*, **138**, 54 (1973).
14. J. M. Flaherty and B. Di Bartolo, *J. of Lum.*, **8**, 51 (1973).
15. R. Reisfeld, L. Boehm, Y. Eckstein, and N. Lieblich, *J. of Lum.*, **10**, 93 (1975).
16. T. Förster, *Ann. Phys. (Germany)*, **2**, 55 (1948).
17. T. Förster, *Radiat. Res. (Supplement)*, **2**, 326 (1960).
18. D. Curie, in *Luminescence Cristalline*, Dunod, Paris, 1960.
19. D. L. Dexter, *J. Chem. Phys.*, **21**, 836 (1953).
20. J. D. Axe and P. F. Weller, *J. Chem. Phys.*, **40**, 3066 (1964).
21. R. Orbach, in *Optical Properties of Ions in Crystals*, p. 445, Intersience, New York, 1967.
22. F. Auzel, *C.R. Acad. Sci. (Paris)*, **263**, 819 (1966).
23. L. G. Van Uitert and L. F. Johnson, *J. Chem. Phys.*, **44**, 3514 (1966).
24. T. Miyakawa and D. L. Dexter, *Phys. Rev.*, B **1**, 2961 (1970).
25. N. Yamada, S. Shionoya, and T. Kushida, *J. Phys. Soc. Japan*, **32**, 1577 (1972).
26. L. A. Riseberg and H. W. Moos, *Phys. Rev. Lett.*, **19**, 1423 (1967).
27. L. A. Riseberg and H. W. Moos, *IEEE J. Quantum Electron.*, QE-4, 609 (1968).
28. M. Inokuti and F. Hirayama, *J. Chem. Phys.*, **43**, 1978 (1965).
29. E. Nakazawa and S. Shionoya, *J. Chem. Phys.*, **47**, 3211 (1967).
30. L. G. Van Uitert, *J. of Lum.*, **4**, 1 (1971).
31. M. Yokota and O. Tanimoto, *J. Phys. Soc. Japan*, **22**, 779 (1967).
32. M. J. Weber, *Phys. Rev.*, B **4**, 2932 (1971).
33. N. Krasutsky and H. W. Moos, *Phys. Rev.*, B **8**, 1010 (1973).
34. M. V. Artamonova, C. M. Briskina, A. L. Burshtein, L. D. Zusman, and A. G. Skleznen, *Sov. Phys. JETP*, **35**, 457 (1972).
35. One has: $D = (8/4\pi) Css/R^4$ (33); Css being the sensitizer-sensitizer transfer constant. For limited diffusion we take R to be the average distance between sensitizers: $R = (3/4\pi)^{1/3} Ns^{-1/3}$, therefore: $D\varrho = 2C^{1/4} Css^{3/4} Ns$. For fast diffusion we take R to be the minimum distance between sensitizers as permitted by the lattice: $R = R_{min}$ therefore $D\varrho = (8/4\pi) Css R_{max}^{-4}$, a constant.
36. E. Okamoto, M. Sekita, and H. Masui, *Phys. Rev.*, B11, 5103 (1975).
37. J. F. Pouradier and F. Auzel in *Proceedings of the 13th Rare Earth Research Conference, Olgebay Park W. VA, USA october 16–20, 1977*, to be published.
38. R. A. Hewes and F. F. Sarver, *Phys. Rev.*, **182**, 427 (1969).
39. T. Miyakawa and D. L. Dexter, *Phys. Rev.*, B1, 70 (1970).
40. W. J. C. Grant, *Phys. Rev.*, B4, 648 (1971).
41. J. F. Pouradier and F. Auzel, *Journal de Physique* (1978), to be published.

42. T. Kushida, *J. Phys. Soc. Japan*, **34**, 1318 (1973).
43. F. Auzel, *C.R. Acad. Sci. (Paris)*, **263**, 819 (1966).
44. F. Auzel and D. Pecile, *J. of Lum.*, **11**, 321 (1976).
45. V. V. Ovsyankin and P. P. Feofilov, *Sov. Phys. JETP Lett.*, **4**, 317 (1966).
46. E. Nakazawa and S. Shionoya, *Phys. Rev. Lett.*, **25**, 1710 (1970).
47. Y. Mita, *J. Appl. Phys.*, **43**, 1772 (1972).
48. F. Auzel and D. Pecile, *J. of Lum.*, **8**, 32 (1973).
49. D. Pecile, *Thèse*, Paris (1976) unpublished.
50. F. Auzel, *C.R. Acad. Sci. (Paris)*, **262**, 1016 (1966).
51. F. Auzel and O. Deutschbein, *Z. Nat.*, **24 a**, 1562 (1969).
52. M. R. Brown and W. A. Shand, *J. Phys. C (Solid State Physics)*, **2**, 1908 (1969).
53. F. Auzel and D. Pecile, in *Proceeding of the 12th Rare Earth Research conference*, Vail, Col. USA (1976).
54. F. Auzel, D. Pecile and D. Morin, *J. Electro. Chem. Soc.*, **122**, 101 (1975).
55. F. Auzel, D. Pecile, C. R. Acad. Sci. (Paris), **B277**, 155 (1973).
56. S. V. Galginaitis and G. E. Fenner, in *Proc. 2nd Int. Conf. Gallium Arsenide*, Dallas, Tex., USA (1968).
57. L. F. Johnson, H. J. Guggenheim, T. C. Rich, and F. W. Ostermayer, *J. Appl. Phys.*, **43**, 1125 (1972).
58. A. M. Barnett and F. K. Henmann, *Electronics P.*, **89**, May 11th (1970).
59. F. Auzel and D. Pecile in *Final Report DGRST N° 72.7.0516*, 1973, unpublished.
60. J. P. Wittke, I. Ladany and P. N. Yocom, *J. Appl. Phys.*, **43**, 595 (1972).
61. F. Auzel, P. Leclerc and J. C. Reymond, *Brevet Français* **PV 71-44031**.
62. J. E. Geusic and H. E. Scovil, *US Patent* **3593-055** (1971).
63. J. C. Reymond and F. Auzel, in *Proc. Colloque Int. Dispositifs et Affichage alpha-numérique*, Paris, 9–10 Avril, 1973.
64. D. Leroy, E. Andre, and D. Diguet in *Contrat CNET N° 759, B 48900790, 9245*, 1977, unpublished.
65. F. Auzel, *Brevet Français*, **1532**, 609 (1967).
66. L. G. Van Uitert, H. J. Levinstein, and W. H. Grodkiewicz, *Mat. Res. Bull.*, **4**, 381 (1969).
67. T. Kushida and M. Tamatani, *J. Japan Soc. Appl. Phys.*, **39**, 241 (1970).
68. T. Kano, A. Suzuki, S. Minagawa, T. Saitoh in *Proc. Colloque Int. Dispositifs et Affichages alpha-numérique*, Paris, April 9–10th, 1973.
69. D. Pecile and F. Auzel, in *Rapport CNET RP/PEC/RPM/OAM/326*, Février 1978, unpublished.
70. J. P. Noblanc, *Appl. Phys.*, **13**, 211 (1977).
71. K. Otsuka and T. Yamada, *Appl. Phys. Letters*, **26**, 311 (1975).
72. Z. I. Alferov, V. M. Andreev, D. Z. Garbuzon, N. Y. Davidyuk, V. R. Larionov, V. M. Marakhonov, E. K. Smirnova, and G. N. Shelonova, *Sov. Phys. Tech. Phys.*, **20**, 234 (1975).
73. H. P. Weber, *Optical and Quantum Electronics*, **7**, 431 (1975).
74. H. G. Danielmeyer and H. P. Weber, *IEEE J. Quant. Electron.*, **QE8**, 805 (1972).
75. A. A. Kaminskii, in *Lasernie Kristally*, Nauka, Moscow, 1975.
76. S. R. Chinn, J. W. Pierce, and H. Heckscher, *Appl. Optics*, **15**, 1444 (1976).
77. M. Saruwatari and T. Kimura, *IEEE J. of Quant. Electron.*, **QE-12**, 584 (1976).

78. H. G. Danielmeyer, *J. of Lum.*, **12/13**, 715 (1976).
79. F. Auzel, *IEEE J. of Quant. Electron.*, **12**, 779 (1976).
80. G. E. Peterson and P. M. Bridenbaugh, *J. Optic. Soc. Am.*, **54**, 644 (1964).
81. C. K. Asawa and M. Robinson, *Phys. Rev.*, **141**, 251 (1966).
82. H. G. Danielmeyer, in *Advances in Lasers*, Vol. IV, Decker, New York, 1975.
83. J. Chrysochoos, *J. Chem. Phys.*, **61**, 4596 (1974).
84. S. Singh, D. C. Miller, J. R. Potopowicz, and L. K. Shick, *J. Appl. Phys.*, **46**, 1191. (1975).
85. F. Auzel and J. C. Michel, *C.R. Acad. Sci (Paris)*, **B 279**, 187 (1974).
86. W. Strek, C. Szafranski, E. Lukowiak, Z. Mazurak, and B. Jeżowska-Trzebiatowska, *Phys. Stat. Sol.*, **41**, 547 (1977).
87. H. G. Danielmeyer and M. Blatte, *Appl. Phys.*, **1**, 269 (1973).
88. S. R. Chinn, H. Y. P. Hong, and J. W. Pierce, *Laser Focus*, P. (64 May 1976).
89. H. Y. P. Hong and S. R. Chinn, *Mat. Res. Bull.*, **11**, 421 (1976).
90. A. Lempicki, *Optics Com.*, **23**, 376 (1977).
91. O. K. Deutschbein, C. C. Pautrat, and I. M. Svirchesky, *Rev. Phys. Appl.*, **2**, 29 (1967).
92. J. C. Michel, D. Morin, and F. Auzel, *C.R. Acad. Sci (Paris)*, **B281**, 445 (1975).
93. J. C. Michel, D. Morin, J. Primot, and F. Auzel, *C.R. Acad. Sci. (Paris)*, **B 284**, 555 (1977).
94. B. Joukoff, M. Fadly, J. Ostorero, and H. Makram, *J. of Cryst. Growth.*, **43**, 81 (1978).
95. J. P. Budin, J. C. Michel, and F. Auzel, *J. Appl. Phys.*, to be published.
96. R. B. Allen and S. J. Scalise, *Appl. Phys. Lett.*, **14**, 188 (1969).
97. F. W. Ostermayer, R. B. Allen, and E. G. Dierschke, *Appl. Phys. Lett.*, **19**, 289 (1971).
98. N. P. Barnes, *J. Appl. Phys.*, **44**, 230 (1973).
99. G. I. Farmer and Y. C. Kiang, *J. Appl. Phys.*, **45**, 1356 (1974).
100. L. J. Rosenkrantz, *J. Appl. Phys.*, **43**, 4603 (1972).
101. Z. I. Alferov, V. M. Andreev, D. Z. Garbuzov, N. Y. Davidyuk, V. R. Larionov, P. P. Pashinin, A. M. Prokhorov, V. D. Rumyantsev, V. M. Tuchkevich, and M. M. Khaleev, *Sov. Phys. Tech. Phys.*, **20**, 231 (1975).
102. C. A. Burrus and J. Stone, *Appl. Phys. Lett.*, **26**, 318 (1975).
103. R. B. Chesler and D. A. Draegert, *Appl. Phys. Lett.*, **23**, 235 (1973).
104. D. A. Draegert, *IEEE J. Quant. Electron.*, **QE-9**, 1146 (1973).
105. S. R. Chinn, J. W. Pierce, and H. Heckscher, *IEEE J. Quant. Electron.*, **QE-11**, 747 (1975).
106. G. Winzer, L. Vite, K. Krühler, R. Plättner, P. Möckel, and H. Pink, *Siemens Forsch. u. Entwickl. Ber.*, **5**, 287 (1976).
107. J. P. Budin, private communication (1978).
108. M. Saruwatari, K. Otsuka, S. Miyazawa, T. Yamada, T. Kimura, *IEEE J. of Quant. Electron.*, **QE-13**, 836 (1977).
109. T. Kimura and K. Daikoku, *Opt. Quantum Electron.*, **9**, 33 (1977).
110. S. Chinn, H. Y. P. Hong, and J. W. Pierce, *IEEE J. of Quant. Electron.*, **QE-12**, 189 (1976).

Chapter 11
Reversible Optical Memory on MNOS-structures

Yu. M. Popov

11.1 INTRODUCTION

The first investigations of holographic writing by means of lasers gave rise to hopes as to the possibility of constructing high-capacity holographic memory devices which would be free from the defects present in modern memory systems. It was hoped that new memory devices would be characterized by high reliability of storage, the possibility of data exchange by pages, fast and easy access to any part of the information stored.

By now holographic memory systems have been rather successfully investigated, and some progress has been made. By using high resolution photomaterials (of more than 2000 lines/mm), it became possible to obtain more than 10^7 bits of the constant holographic memory recorded on one photoplate (1). It has been found that quite a number of limitations resulting from imperfections in the optical systems, photodetector matrix, and the laser beam deflectors do not permit one (2) to increase the capacity of holographic memory on one photoplate (more than 10^8 bits). In contrast to permanent holographic memory systems, the design of reversible holographic memory systems continues to pose considerable problems.

This is first of all due to the absence of a suitable reversible medium with high light sensitivity, and the resolution of no less than 1000 lines per mm. The latter requirement excludes from consideration quite a number of substances known to be highly-sensitive and otherwise suitable for memory-multi-layer semiconductor structures, ferroelectrics with the photoconducting layer and so on. For these reasons, although the use of the binary bit data storage might cause a certain decrease in resolution, it should permit the construction of an essentially new reversible optical memory. It therefore seems very promising to use semiconductor structures, because they will simultaneously have the functions of photodetector matrices. As in the case of holographic memory, it is advisable to perform paginal exchange of information. But in this case the page would be formed not by a data matrix but by a system of passive splitters of the laser beam. The high sensitivity of semiconductor structures makes it possible to split the laser beam of 1 W power into 16 thousand beams, i.e. one page will consist of 16 thousand bits. The page is changed by deflecting the laser beam in front of the splitters. When the beam is deflected in mutually perpendicular directions into 256×256 positions, the total memory capacity will be 10^9 bits.

11.2 PHYSICAL PRINCIPLES OF DATA RECORDING IN MNOS-STRUCTURES

The reversible medium represents a multi-layer structure homogeneous in plane comprised of Si_3N_4 (dielectric, which contains traps), SiO_2 (a thin layer of dielectric, allowing the tunnelling of current carriers under certain conditions), and Si (the semiconductor, in which the light beam produces electron-hole pairs). Outer sides of the structure are coated with metallic contacts. On the Si_3N_4 side the coating is transparent to the light beam (Au) (Figure 11.1). The carriers produced in the silicon by the light under the

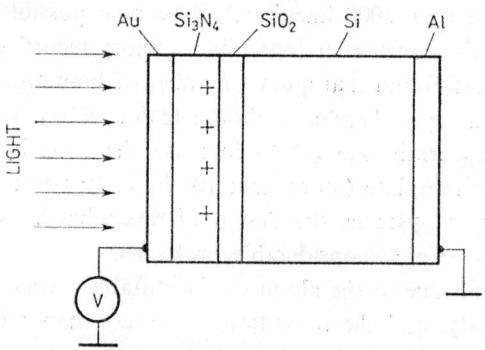

Figure 11.1 Scheme of MNOS-structure

simultaneous action of the electric field, tunnel through SiO_2, and produce a potential relief of the charged traps in Si_3N_4, which gives a memory effect. Switching of the structure is explained by the overcharging of traps, whose energy levels correspond to the forbidden band of silicon. When the voltage is applied to the metallic electrode (relative to the silicon substrate), there occurs a displacement of trap energy levels. These turn out to be higher than the bottom of the conduction band, or lower than the upper level of the valence band (the polarity of voltage being opposite). As a result, the traps in the dielectric are being overcharged due to the tunnelling of charges through the thin SiO_2 layer (Figure 11.2).

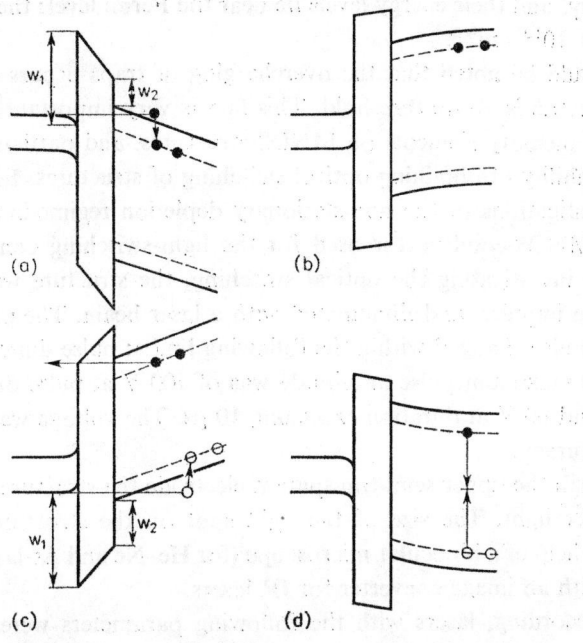

Figure 11.2 MNOS-structure zone diagrams, illustrating the change of the trap charge value in dielectric at opposite polarities of switching pulses

As a rule, the voltage applied to the electrodes of a structure falls almost entirely on a dielectric layer (for example, MNOS-transistors). Hence, the overcharging of dielectric traps takes place whenever a voltage above threshold is applied to the electrodes of the structure. By using the MNOS-structure, one can establish non-equilibrium conditions when the dielectric-semiconductor boundary does not have the majority carriers, as is the case with MNOS-transistors. This means that although the voltage, which is above threshold, is applied to the structure, the traps in the dielectric do not, nevertheless, overcharge. Such a situation can be observed when the voltage impulse causes the formation of a non-equilibrium layer in the boundary region of the semiconductor-dielectric, where a significant part of the applied voltage is concentrated. This part of the voltage in the boundary layer depleted by majority carriers can be changed by illuminating the structure with the light absorbed in this layer. This phenomenon is taken advantage of in optical data recording.

By studying the kinetics of trap overcharging it was found that the traps responsible for the overcharging of a structure are located at the SiO_2–Si_3N_4 boundary, and their energy levels lie near the Fermi level; the density of traps is about 10^{13} cm^{-2}.

It should be noted that the overcharging of traps occurs if the voltage in the dielectric is above threshold. This fact is very important for the development of memory elements on MNOS-structures, and particularly, in view of the possibility of obtaining optical switching of structures. Below we describe the investigations of the non-stationary depletion regime in a semiconductor of the MNOS-condensator used for the light-switching control.

When investigating the optical switching, the structure was supplied with a voltage impulse, and illuminated with a laser beam. The parameters of the electric pulse changed within the following limits: pulse duration, from 0.1 μs to 2 ms; maximum pulse amplitude was of 100 V at pulse duration less than 10 μs, and 60 V at duration exceeding 10 μs. The voltage was measured with 10% accuracy.

Through the upper semi-transparent electrode the structure was illuminated with laser light. The size of the light spot on the structure was measured with the help of a binocular microscope (for He–Ne and Ar-lasers), or a microscope with an image converter for IR lasers.

For recording, lasers with the following parameters were used:

	Wavelength	Pulse duration	Power
GaAs (heterostructure)	0.9 μm	100 ns	1–4 W
GaAs (78 K)	0.87 μm	0.5–100 μs	0.2 W
He–Ne	0.63 μm	—	30 mW
Ar	0.51 μm	—	0.1–0.2 W

Gas lasers were used in combination with an electrooptical modulator having the band from 0 to 5 MHz.

The emission power was controlled by a photodiode calibrated for two wavelengths (0.63 μm and 0.9 μm).

The density of the light flux on the sample could be smoothly varied by using visual and IR light polarizers (polaroids, Glan–Thomson prisms).

An important part of the experiments was observation of the dynamics of the voltage pulse redistribution during illumination of the structure with light. As has already been noted, in order to switch the structure under the action of the voltage pulse of a given duration, it is necessary for the voltage on the dielectric to exceed threshold. When investigating the optical switching of the structure, it is important to know the initial voltage distribution between dielectric and semiconductor, the influence of various parameters on the initial voltage distribution (concentration of majority carriers in the semiconductor, thickness of the dielectric, surface charge at the semiconductor-dielectric boundary, and so on).

To observe the dynamics of voltage redistribution in the structure the following technique was developed.

Preliminary measurements of the capacity of the condenser formed by a dielectric layer of the structure were taken. The measuring scheme included a test capacity C_t connected in series with the structure (Figure 11.3). The

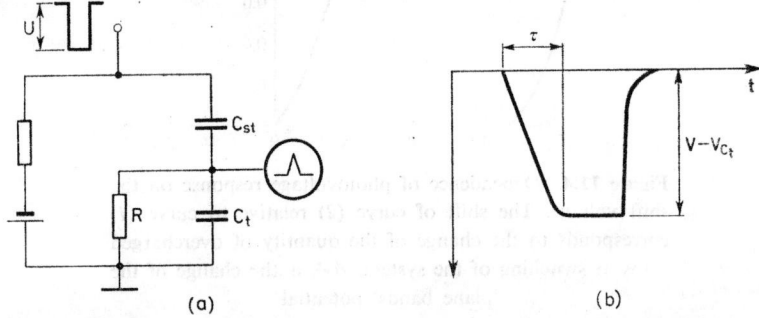

Figure 11.3 Scheme for measuring the dynamics of the voltage redistribution during illumination of the structure (a); a diagram of the voltage change on the dielectric during illumination of the structure by He–Ne laser (b)

value of the capacity was chosen in such a way that the greater part of the pulse voltage applied fell on the structure. To maintain the same regime under constant current in the system, the capacity C_t was connected in parallel with the resistor R. It should be remembered that the capacity formed by the dielectric C_d and the test capacity were charged by the same current since they were connected in series. The change of voltage on the dielectric can be written as

$$\Delta V_d = 1/C_d \int i(t)\,dt$$

where i is the charge current. Accordingly, the change of voltage on the test capacity is: $V_{C_t} = 1/C_t \int i(t)\,dt$, so

$$\Delta V_d = \Delta V_{C_t} \frac{C_t}{C_d}$$

The time dependence of voltage on the test capacity was recorded using an oscillograph. Figure 11.3(b) shows an oscillogram of the changing voltage on the dielectric resulting from laser light illumination.

Changes in the trap charge were recorded by the 'plane bands' potential method. The plane bands potential is determined by measuring the photovoltage response. To improve the measurement accuracy, the dependence of photovoltage response on the voltage shift was determined before and after trap charging (Figure 11.4). The change of the 'plane bands' potential cor-

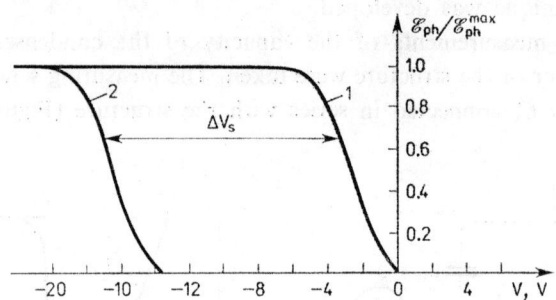

Figure 11.4 Dependence of photovoltage response on the shift voltage. The shift of curve (2) relative to curve (1) corresponds to the change of the quantity of overcharged traps at switching of the system. ΔV_s is the change of the 'plane bands' potential

responds to the shift of curve (2) relative to curve (1), and characterizes the change of the trap charge during data recording.

The structure was supplied with a pulsed voltage of 50 V amplitude. As a result of illumination, the voltage on the dielectric increased linearly with time, and reached the value of the voltage applied to the structure.

The time of space charge screening during illumination by He–Ne and GaAs laser light was measured, the measurements being taken at different light flux intensities which permitted a more accurate determination of the screening time. By using GaAs lasers a relaxation time was obtained that was twice as long as that obtainable with a He–Ne laser at the same number of quanta absorbed in Si. The diffusion length determined from these data was equal to 10 μm. The measurement error was 30%.

Thus, to produce a surface charge which would cause voltage redistribution in the structure, one has to use light whose absorption depth in Si does not exceed 10 μm. The velocity of the surface charge change can be considered constant and proportional to the density of the light flux.

Suppose that the wavelength chosen is such that the condition $\alpha L_d \gg 1$ is fulfilled (α is the absorption coefficient). In this case the charge σ of minority carriers at the Si–SiO$_2$ surface during illumination is connected with the light pulse energy $\sigma = bqkE/h\nu$, where E is the density of the light pulse energy, $h\nu$ the quantum energy, k the quantum yield, and b a coefficient which takes account of the reflection of incident radiation and of absorption in the metallic electrode.

An experimental study was made of the space charge relaxation, where the structure was illuminated with a He–Ne laser beam (3, 4, 5). The purpose of the experiment was to define the light pulse energy required for space charge relaxation and determination of the dependence of relaxation time on the light flux power. Experimental and theoretical results were also compared.

Figure 11.5 shows the dependences characterizing the penetration depth of He–Ne and GaAs laser light in silicon. Figure 11.6 illustrates the

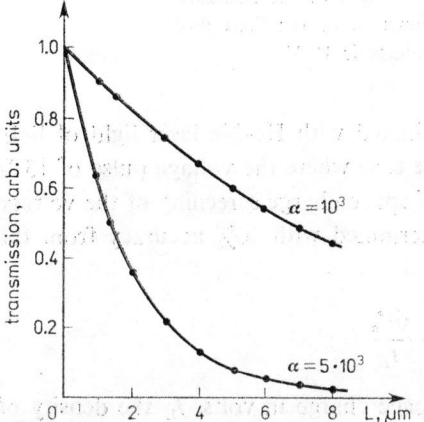

Figure 11.5 GaAs ($\alpha = 10^3$) and He–Ne ($\alpha = 5 \times 10^3$) laser light transmittance as the function of plate thickness

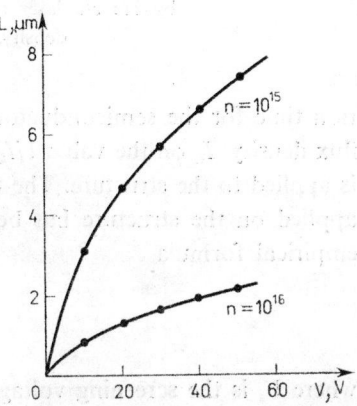

Figure 11.6 Dependence of depth of the non-equilibrium depletion region in semiconductor on the voltage of the metallic electrode of the structure

dependence of the depth of the non-equilibrium depletion region in semiconductors on the pulsed voltage applied to the structure. The initial state of the structure corresponds to zero potential of the 'plane bands'.

Taking the penetration depth of light to be α^{-1}, we find that the light of the He–Ne laser is fully absorbed in a 4 μm-layer, and that of the GaAs laser—in a 20 μm-layer. Note that for the He–Ne laser, the velocity of charge variation, i.e. the velocity of voltage growth in the dielectric, remains constant until the relaxation of the non-equilibrium space charge in the semiconductor is completed. Figure 11.7 shows the dependence of the space charge relaxa-

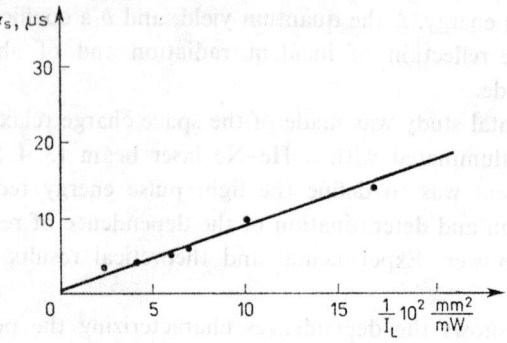

Figure 11.7 Space charge relaxation time in the semiconductor during illumination of the structure by He–Ne laser as the function of the light flux density. The voltage is 15 V

tion time for the semiconductor illuminated with He–Ne laser light of light flux density I_L on the value $1/I_L$ in the case where the voltage pulse of 15 V is applied to the structure. The time of space charge screening of the voltage applied on the structure has been determined with 10% accuracy from the empirical formula

$$\tau_{scr} = \frac{6V_s}{I_L}$$

where V_s is the screening voltage of space charge in volts, I_L the density of the light flux in mW/mm², and τ_{scr} is the screening time in μs.

In order to increase the voltage on the dielectric by 30 V (this value approximately corresponds to threshold), the required energy of the light pulse should be $I_L \tau_{scr} = 1.8 \times 10^{-7}$ Jmm⁻², in accordance with the experimental data. The estimated value is

$$E = \frac{h\nu\sigma}{bqk}$$

Taking the transmittance of the metallic electrode $b = 0.5$, the quantum yield equal to 1, and $\sigma = VC_d$, $C_d \sim 10^{-9}$ F·mm⁻², we get $E = 1.2 \times 10^{-7}$ J/mm⁻², which agrees well with the experimental data.

A linear dependence between τ_{scr} and $1/I_L$ (Figure 11.7) maintained until the pulse duration of the voltage applied to the structure reaches 2 ms. Thus, one can operate under weak light flux and long relaxation time. For a fixed voltage applied to the structure the light pulse energy required for screening of the electric field in the semiconductor should remain constant, regardless of the light flux density.

Note that when the light flux density reaches 0.18 Wmm^{-2}, the voltage on the dielectric can be switched within 1 μs.

11.3 METHODS OF INFORMATION READING

Optical reading of information stored in the form of a potential relief in MNOS-structure can be performed taking advantage of the influence of charged traps in the dielectric on the value of changes in the curvature of the semiconductor bands near the surface ('plane bands' potential). This potential of the semiconductor can be determined by measuring the photovoltage response (PE) appearing on the structure contacts under the illumination of the structure by the light, which is absorbed in the surface region of the semiconductor (6).

In a general case, the value of the surface potential can be derived from a common solution for Poisson equations describing dielectric and semiconductor regions.

In particular, the problem can be significantly simplified if the charge of traps is assumed to be localized at the SiO_2–Si_3N_4 boundary, and if it can be determined by changing the 'plane bands' potential. By the 'plane bands' potential we understand such a potential on the metallic electrode, which produces no changes in the curvature of semiconductor bands. This case corresponds to a zero response of the photovoltage. It is clear that a change in the 'plane bands' potential value ΔV_s is connected with a change in charge density of traps $\Delta\sigma$, and capacity of the layer Si_3N_4 calculated on cm^2 C_d, as $\Delta V_s = \Delta\sigma/C_N$. It is important to consider two regimes of the signal readout.

(1) The regime of measuring the photovoltage response (Figure 11.8(a)), and (2) the regime of photocurrent recording (Figure 11.8(b)).

In the first case the signal amplitude is proportional to the changes in the value of the 'plane bands' potential during illumination, and in the second, to the value of the recharging current through the structure. A method of measuring the photovoltage response was first proposed in (6). The studies that followed were concerned with the photovoltage response ($\mathscr{E}_{ph}$) depending on the light flux intensity (7), and temperature (8). This method has been

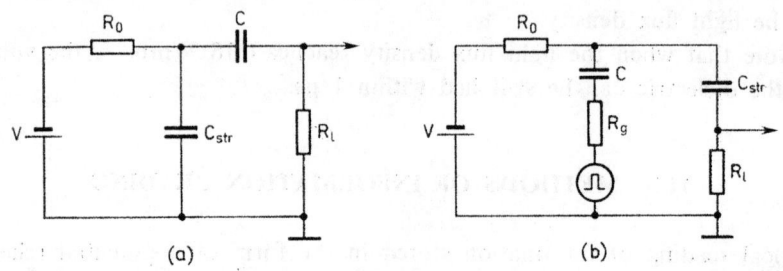

Figure 11.8 Scheme for measuring: (a) Photovoltage response; C_{str} capacity of the structure; R_l resistance; C separative capacity; $C \gg C_{str}$; $(R_l \| R)C_{str} \gg t$; t light pulse duration. (b) Photocurrent signal; R_g internal resistance of voltage pulse generator

successfully applied to studies of traps in Si–SiO$_2$ systems (9). Here we shall consider the possibility of using it for data readout in MNOS-structure.

A scheme for measuring the photovoltage response is shown in Figure 11.8(a). The photovoltage response is measured on the resistance R_l connected in parallel to the structure. The outer source of the bias voltage is connected with the structure through the resistance R_0. The signal $\mathscr{E}_{ph}$ can be observed without the outer bias on the control electrode if the dielectric has a sufficiently large trap charge.

The structure is illuminated with light pulses of duration t. The time constant of the capacity charging is chosen in such a way that the condition $(R_l\|R)C_{str} \gg t$ is fulfilled (assuming that $C \gg C_{str}$). The assumptions made show that the charge on a metallic electrode under the action of the light pulse does not change in any significant way: $Q_M = $ const.

When the structure is illuminated with intensive light, the semiconductor bands are fully straightened, i.e. the photovoltage response is equal to the initial potential difference.

The photovoltage signals $\mathscr{E}_{ph}$ were measured when the structure was illuminated with He–Ne and GaAs laser light pulses.

The dependence of the photovoltage amplitude $\mathscr{E}_{ph}$ on energy density is shown in Figure 11.9. The saturation of the signal $\mathscr{E}_{ph}$ can be explained by straightening of semiconductor bands during illumination. For the structures under study the maximum value does not exceed 300 mV and the donor concentration is about 10^{15} cm^{-3}. The linear part of the curves in Figure 11.9 shows the response to the He–Ne laser light, which is six times greater than

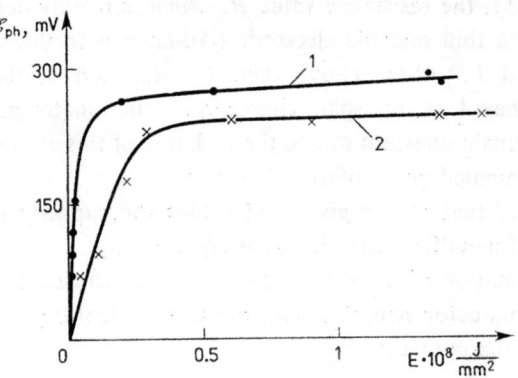

Figure 11.9 Dependence of photovoltage amplitude on energy density; *1*, readout by He–Ne laser; *2*, readout by GaAs laser

for GaAs. The decrease in response, in the case of GaAs laser, is caused by two factors:

(1) due to illumination with $\lambda = 0.9$ μm the absorption in a 200 Å-thick metallic electrode is 3 times greater than at illumination with $\lambda = 0.63$ μm;
(2) the depth of GaAs laser light penetration is greater, and hence a smaller number of carriers can reach the region of the space charge. Using the photo-electric readout, it is advisable to operate in the linear region of the dependence, shown in Figure 11.9. The dependences enable us to determine the energy of the light pulse required for reading. Thus, the signal $\mathscr{E}_{ph} = 100$ mW is observed when the energy density of the He–Ne laser is about 2.5×10^{-10} J mm^{-2}.

When a MNOS-structure is used in the optical memory, then it is advisable to place $\sim 10^4$ bits of information in an area with one metallic electrode. In this case the region corresponding to one bit of information occupies a small part of the whole metallic electrode area. When one reads out only one bit of information, the neighbouring regions will affect the photovoltage response. The parallel-connected unilluminated parts of the structure (with an area 10^4 times greater) will recharge the illuminated region with the time constant

RC_b, where R is the effective resistance determined by the upper metallic electrode resistance, and the volume resistance of the silicon substrate; C_b is the capacity of the structure region where one bit is recorded. The value of $C_b \sim 10^{-11} - 10^{-13}$ f, the resistance value R, which is mainly determined by the conductivity of a thin metallic electrode (Au-layer with thickness ~ 200 Å), does not exceed 100 ohm. Thus, when the duration of the readout light pulse is more than 1 ns, the space charge in the illuminated part of the structure does not remain constant due to the recharge of this region by the neighbouring unilluminated parts of the structure.

Let us assume that at the given outer bias the capacity of the structure with the area of metallic electrode S_0, is equal to C_0.

Let the illumination of the structure with area S_1 cause a complete straightening of semiconductor bands. Then, due to illumination, we get the value of response on the contacts

$$\mathscr{E}_{ph} \approx V_s \frac{S_1}{S_0} \frac{C_d}{C_0}$$

where C_d is the capacity of the dielectric layer with the area S_0, $S_1/S_0 \ll 1$, and $C_d \gg C_0$ (depletion of the semiconductor surface). The ratio C_d/C_0 is constant at the given parameters of the semiconductor, and the fixed constant bias voltage on the structure. The result obtained shows that the value of the photovoltage response decreases proportionally to the S_1/S_0 ratio. The decrease in the photovoltage response due to the influence of the neighbouring regions of the structure will be termed the photovoltage response bypassing effect.

In order to decrease this effect, it is necessary to use samples with a possibly higher concentration of majority carriers.

The rise times of the photovoltage response up to the maximum value is determined by the density of the light flux power, and by the time of separation of the light generated carriers in the space charge region, which does not exceed 10^{-10} s. It follows from Figure 10.9 that the maximum signal $\mathscr{E}_{ph}$ is observed when the pulse energy density is about $10^{-10} - 10^{-8}$ J mm^{-2}. Thus, the laser of 1 W power can provide data readout from an area of 1 mm^2, at pulse duration $t = 10^{-9} - 10^{-10}$ s.

The response rise time in our experiments was limited by the frequency of the output amplifier (1–10 MHz).

The photovoltage response decay time is determined either by the time it takes to recharge the structure with a source of fixed bias, or by the lifetime of non-equilibrium minority carriers near the dielectric-semiconductor boundary.

The maximum cycle frequency during readout is determined by the larger one of those times.

It follows from the above speculations that information readout using the $\mathscr{E}_{ph}$ signal measurements method involves major restrictions on the maximum cycle frequency. The electron-hole pairs produced are divided in the field of space charge. Minority carriers are found to be in a potential well whose depth equals the value of the 'plane bands' potential (~ 0.3 V). In this case the recombination of non-equilibrium carriers is suppressed by the potential barrier dividing the electrons and holes.

An experimental study has been undertaken to determine the maximum readout frequency depending on the 'plane bands' potential. The structure was illuminated with two light pulses of the same power and duration, but shifted in time. The energy of the light pulse was sufficient for a complete straightening of semiconductor bands. If the delay time between light pulses is shorter or equal to the lifetime of non-equilibrium carriers at Si–SiO$_2$ interface, then the response from the second light pulse should decrease. Figure 11.10 shows the ratio of photovoltage responses from the first and

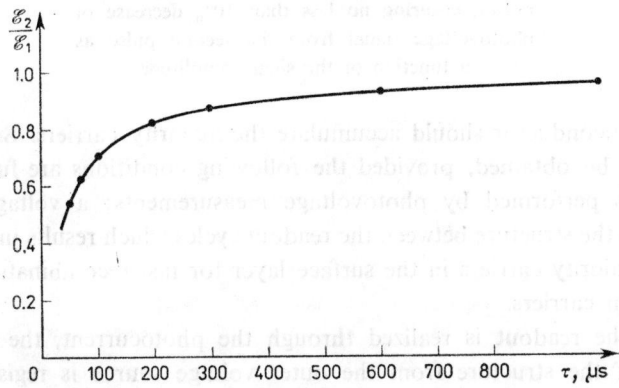

Figure 11.10 Ratio of photovoltage response signals of two time-shifted light pulses as a function of time delay between these pulses

second light pulses depending on the delay time between them. A decrease in the time interval between the two light pulses leads to a decrease in photovoltage response from the second pulse, because in this case the stationary 'plane bands' potential cannot be settled for a time between the light pulses. Figure 11.11 shows the value of time delay τ providing a decrease of the second signal by no more than 10%, as a function of equilibrium 'plane bands' potential. The maximum readout cycle frequency determined by measuring the photovoltage response, is proportional to $1/\tau$, and increases linearly with a decrease in the value of the stationary 'plane bands' potential. If that potential is 10 mV, then the maximum cycle frequency of data readout will not exceed 20 kHz. To heighten the cycle frequency, the surface region

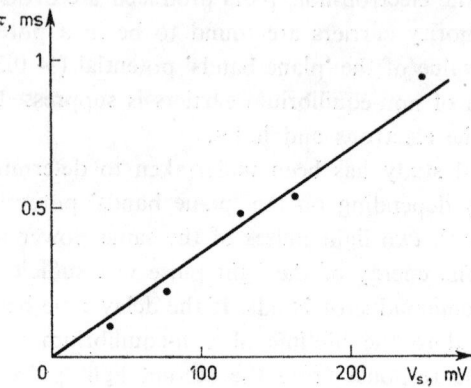

Figure 11.11 Time delay between two light pulses, ensuring no less than 10% decrease of photovoltage signal from the second pulse as a function of the signal amplitude

of the semiconductor should accumulate the majority carriers. Such a situation can be obtained, provided the following conditions are fulfilled: the readout is performed by photovoltage measurements; a voltage pulse is applied to the structure between the readout cycles which results in accumulation of majority carriers in the surface layer for fast recombination of non-equilibrium carriers.

When the readout is realized through the photocurrent, the recharging current of the structure from the outer voltage source is registered. The measurement scheme is shown in Figure 11.8(b). In the initial state the semiconductor surface is in accumulation, which provides for a fast recombination of the non-equilibrium minority carriers produced by light. By readout, the structure is fed with voltage pulses, which lead to the formation of a depletion region in the semiconductor. The constant charging time $R_1 C_{str}$ is chosen to be much less than the voltage pulse duration and the duration of the light pulse. The light pulse is applied with a delay sufficient to complete the relaxation processes caused by the field pulse. The light-generated carriers are separated in the field of the space charge (the separation time is 10^{-10} $< R_1 C_{str}$.

The experiments were performed in order to determine the relation between the photocurrent and the light flux.

The structure with n-type substrate and concentration of majority carriers $n_0 = 10^{15}$ cm^{-3} was supplied with a negative voltage pulse of 10 V amplitude. At the same time, the structure was illuminated with He–Ne or GaAs laser light.

The dependence of the photocurrent on the light flux is shown in Figure 11.12. Readout by the method of photocurrent measurement shows that the

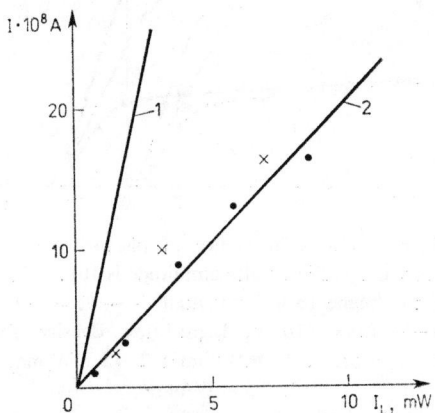

Figure 11.12 Dependence of photocurrent on light flux. The structure is illuminated by: 1, He–Ne laser; 2, GaAs laser

response to the light of He–Ne laser (~ 0.1 A/W) is five times longer than the response to GaAs laser light.

Figure 11.13 shows the dependence of photocurrent on the bias voltage for two states of the structure with different values of 'plane bands' potential. The changing trap charge results in a shift of the curve along the voltage axis by a value, which is equal to the change of the 'plane bands' potential. The dependence depicted in Figure 11.13 shows that the position of the 'plane bands' potential changed by 4 V, is followed by an alteration of the photovoltage response in comparison with the initial value in the ratio higher than 1 : 10.

The speed of photocurrent measurement readout is determined by the feasibility of obtaining short pulses of electric field and light. The use of lasers makes it possible to obtain short light pulses with sufficient power. For

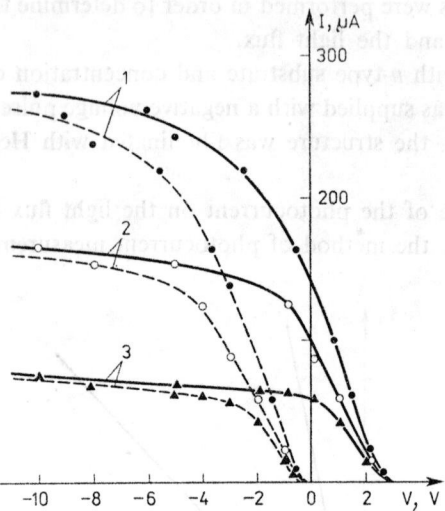

Figure 11.13 Dependence of photocurrent on bias voltage. Field pulse amplitude is $10V$. Values of the 'plane bands' potential: ——— $V_s = -7.0$, ——— $V_s = -10$ V. Light flux density for GaAs laser: 1, 26 mW/mm²; 2, 15 mW/mm²; 3, 7.5 mW/mm²

example, GaAs heterostructures provide pulse duration of 10^{-7} s and light powers of 1–10 W; shorter pulse durations can also be obtained.

Electric voltage can be applied to the structure within the time $t \sim 2RC_s$, where R is the total resistance load of the generator and contacts, and C_s is the capacity of the structure. At $R = 10^2$ ohm, and the metallic electrode area ~ 1 mm², the value of t is about $2 \cdot 10^{-7}$ s. The time constant of recharging the structure by the voltage source is much higher. In fact, it is impossible to get $R < 10^2$ ohm because of the resistance of the semitransparent electrode and current-carrying contacts. Thus, the maximum readout frequency can be a few MHz.

Using this method of readout, we find that the voltage of load resistance R_1 (Figure 11.8(b)) depends on the ratio of the illuminated part of the structure to the total area of the electrode. Actually, the mean value of the photocurrent during illumination of a part of the structure can be represented as $i = (\Delta V_s/t) C_d^*$ where ΔV_s is the change of the 'plane bands' potential in the illuminated part of the structure; C_d^* is the dielectric capacity in this part, and t is the light pulse duration. When using the photocurrent measurement readout method, the inequality $RC_d \ll t$ should be fulfilled, where R is the resistance of the external circuit, and C_d is the capacity of the dielectric with the area of the metallic electrode.

Taking $RC_d \approx 0.1t$, and assuming that the load resistance is much greater than the resistance of the generator ($R_l \gg R_g$, Figure 11.8(b)), we get $V_{R_l} \approx 0.1 \Delta V_s C_d^* / C_d$, i.e. the voltage load is proportional to the ratio of the illuminated part of the area to the area of the whole structure. But in contrast to reading by the photovoltage measurement method, the signal of the photocurrent is independent of the charge in the neighbouring regions of the structure. In fact, the current in the external circuit is determined by the change in the metallic electrode charge dQ_M/dt. The value of the current can be represented as $dQ_M/dt \approx d\sigma/dt$ where $d\sigma$ is the change of the surface charge density at the Si–SiO$_2$ interface.

If the voltage of the structure is kept constant and there is no diffusion of the minority carriers along the Si–SiO$_2$ surface, then the density of the surface charge varies only in the illuminated part of the structure.

Space charge of non-illuminated regions remains constant (although it is different for different trap charges), and $i \approx d\sigma/dt$, where σ is the surface charge density for the illuminated parts of the structure.

Thus, when the information is read out by the method of photocurrent measurement, one can provide cycle frequency of about 1 MHz.

Let us now try to find out which of these methods is better for information readout. When reading by means of $\mathscr{E}_{ph}$ measurements, we get $RC_d \gg t$, where t is the light pulse duration, and RC_d is the constant of the structure recharging. When measuring the photocurrent, we get $RC_d \ll t$, so that if all other things are equal, the cycle frequency at measuring the photocurrent will always be higher (about 10^2 times). Let us show what energy of the light pulse is needed for readout by any of these methods, and in what way it is connected with the power of the electric signal.

It has been found that the maximum signal $\mathscr{E}_{ph}$ equal to 0.3 V can be obtained when the light pulse energy density equals 10^{-9} Jmm^{-2} (see Figure 11.9).

Let us estimate the value of the maximum signal on the load resistance during readout by photocurrent measurement. Let the amplitude of the readout pulse V_r be equal to 10 V ($V_r < V_t$, where V_t is the threshold voltage causing recharging of traps). Suppose that the value of the light flux provides a complete relaxation of the space charge in the semiconductor during the time of the light pulse. As noted above, in structures with $n_0 \sim 10^{16}$ cm^{-3} the greater part of the voltage is applied to the space charge of the semiconductor. And one can assume that the initial surface potential is approximately equal to V_r. The mean value of the photocurrent per pulse is

$$\bar{i}_{ph} = V_r C_d / t,$$

where C_d is the dielectric capacity.

Taking the following parameter of the structure circuit, $R_1 C_d = 0.1t$, we obtain $\bar{i}_{ph} \approx 0.1 V_r / R_1$, and $V_{R_1} \sim 0.1 V_r$. This means that the value of the maximum signal for the load resistance is about 1 V. Experimentally defined energy density of the light pulse required for the space charge relaxation is $E = \sigma \cdot V \times 10^{-9}$ Jmm^{-2} = 6×10^{-8} Jmm^{-2}; it is 60 times greater than the energy density corresponding to the maximum signal $\mathscr{E}_{ph}$. Let us compare the power of the electric signal for the load resistance if readout by both methods utilizes light pulses of equal energy. Electric pulse power can be written as $P = V^2 / R_1$. When measuring the photovoltage $R_1 C_d \gg t$, we take $R_1 C_d = 10t$. The ratio of signal powers is

$$\frac{P_{\mathscr{E}}}{P_i} = \frac{V_{\mathscr{E}}^2 R_{1i}}{V_i^2 R_{1\mathscr{E}}} \sim 3.6$$

The energy of light pulses being equal, the power of the electric signal on R_1 in the case of photovoltage measurements is 3.6 times greater than in the case of photocurrent measurements. However, by increasing the energy of the light pulse the power of the signal in the case of photocurrent can be increased 3600 times ($P_{\mathscr{E}} / P_i \sim 10^{-3}$). This circumstance seems to be very important because the photoelectric signal in optical memory will decrease in proportion to the ratio of the illuminated part to the area of the metallic electrode. On the other hand, if the same volume of information in each cycle is recorded and read out, then the requirements with respect to the laser pulse energy will be determined mainly by the first process and the lower light pulse energy will be insufficient for readout by the photovoltage response method.

Thus, one can conclude that readout by the photocurrent method seems to be more promising, because: (i) a higher cycle frequency is provided, (ii) maximum power of electric signal delivered on R_1 may be three orders higher than in the case of photovoltage response, (iii) the value of the signal is independent of the charge of traps in the neighbouring regions of the structure.

11.4 INVESTIGATION OF SPATIAL RESOLUTION FOR INFORMATION RECORDING AND READOUT

Space resolution of the medium determines such basic parameters as maximum density of data storage (and, as a consequence, the volume of the whole memory), and the energy required for the recording of one bit of information. In optical memory on MNOS-structure it is necessary to have the maximum number of information bits recorded on the area with a common metallic electrode in order to minimize the number of contacts in the memory device.

When using MNOS-structures as the information media, one can expect more difficulties to be involved in high-density data recording than in data readout.

Actually, data recording requires much higher voltage than readout, and hence, the edge effects are more important. The concentration of minority carriers at the Si–SiO$_2$ boundary is considerably higher. Accumulation of the surface charge at the Si–SiO$_2$ boundary results in the increase of surface potential of the illuminated parts of the structure. As a result, one can observe the formation of a potential well for minority carriers in the neighbouring non-illuminated regions. This can lead to a spreading out of the light-produced surface charge along the Si–SiO$_2$ boundary. The situation is similar to the processes occurring in charge-coupled devices (CCD). To estimate the resolving ability one can use theoretical and experimental results obtained when studying charge-coupled devices. In the calculation of charge transfer in CCD (10), two mechanisms are discussed, which result in the charge spreading out over the Si–SiO$_2$ boundary: thermal diffusion of carriers, and drift of carriers in the electric field due to the gradient of charge concentration along the Si–SiO$_2$ boundary.

At the same time it is well known that in devices based on ZnS, ZnO, and Bi$_{12}$SiO$_{20}$, carriers are trapped at the surface. However, the number of such centres in the Si–SiO$_2$ system should be very small because the procedure of material preparation always aims at a reduction of surface density and an increase in surface conductivity.

We have conducted experiments on defining the resolution of optical recording and photoelectric information readout (11). Since the recorded resolution is determined by photoelectric readout, it was necessary to determine, initially, the resolution of the readout.

A spectral potential relief was produced on the structure. The relief was scanned by a 10 μm spot of the laser beam. The readout was performed by photovoltage and photocurrent measurements. To produce a potential relief,

296 Yu. M. Popov

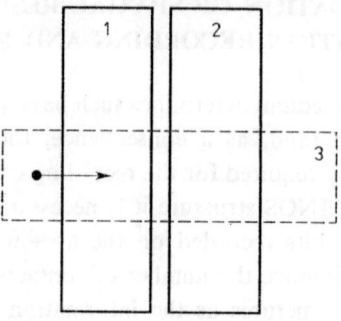

Figure 11.14 Electrode configuration in the experiments on defining space resolution of optical information recording and readout on MNOS-structure. The arrow shows the direction of scanning of the structure

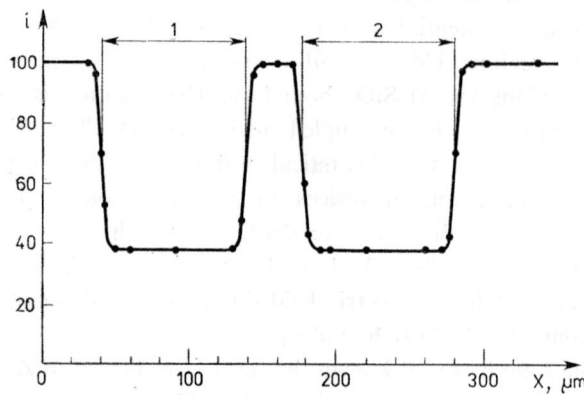

Figure 11.15 Signal variation for scanning along the surface of the metallic electrode. The traps in regions 1 and 2 are overcharged, $V_s = -6$ V

two metallic electrodes were deposited on the structure. The electrodes were 100 μm wide and they were separated by 20–40 μm from each other (electrodes 1 and 2, Figure 11.14). By applying a constant voltage, the charged states under the electrodes were changed. The potential shift for plane bands was 6 V relative to the initial value. When the dielectric is 1000 Å thick, the traps are overcharged under metallic electrodes only. Following that the stripes were electrically connected with the help of electrode 3. Figure 11.15

shows the curves describing the signal variation for scanning along the surface of transverse electrode 3. The resolution was actually determined by the diameter of the scanning beam, and the difference was in the absolute value of the signal only.

The contrast determined by the formula

$$\frac{i_{max} - i_{min}}{i_{max} + i_{min}}$$

was equal to 0.45. The rather small contrast is due to the parasitic exposure of the electrode during the experiment. Experimental results show that the resolution at readout is not worse than 10 μm. Similar results were obtained in the case of readout by the photovoltage measurement method. On the basis of the results of this experiment one can conclude that the photoelectric readout method allows one to record the variation in the trap charge at the distance of 10 μm. The recording of space resolution in optical data recording experiments was made under the configuration of electrodes in a structure corresponding to the case of Figure 11.14.

The energy needed for data storage in one stripe was determined. When the intensity of He–Ne laser was 3.8 mW, voltage pulse amplitude, 40 V, and pulse duration, 10 μs, the potential of plane bands changed by 4 V. The structure was discharged down to the initial value of the plane band potential. Following that, the stripes were electrically connected by means of evaporation of the transverse electrode. An attempt was made to charge stripe I by illuminating it with light of 3.2 mW intensity, and by applying a 40 V-voltage pulse of 10 μs duration. Recording of the photovoltage response showed no change in trap charge in region I. When the light intensity reached 10 mW (voltage pulses of 40 V and 10 μs), it was possible to observe charged traps over the whole area of the metallic electrodes. The potential of plane bands changed by about 4 V. A threefold increase in storage threshold intensity corresponds to a three-times increase of the electrode area in comparison with the area of the first stripe. Since the traps are charged as a result of voltage increase in the illuminated parts of the dielectric, which is due to the storage of light-produced minority carriers in Si at the Si–SiO$_2$ boundary, the absence of space resolution means that minority carriers are distributed under the whole surface of metallic electrode during the 10 μs of data recording. Thus, in order to obtain space resolution, one should provide conditions preventing carrier diffusion.

The minority carriers located at the semiconductor-dielectric boundary are in a potential well. The depth of the potential well is determined by the surface potential V_s of the semiconductor. By changing the value V_s along the semiconductor-dielectric surface, one can produce potential barriers preventing the minority carriers from diffusion. The value of the surface potential is determined by

$$V_s = V - V_d = V - d\sqrt{\frac{2\varepsilon_s q n_0 V}{\varepsilon_d^2 \varepsilon_0}}$$

As can be seen from the formula, under fixed voltage V, the depth of the potential well can be changed by changing the dielectric thickness d.

Thus, by varying the dielectric thickness one can produce potential barriers for the minority carriers. In order to avoid the disappearance of potential barriers at illumination, the depth of the potential relief should be higher than the value of changed voltage in the semiconductor for the regions exposed to illumination.

An experiment with optical data storage on the structure with potential barriers has been performed. Seven parallel SiO stripes 10 μm wide and 1 μm thick were deposited on the nitride surface (Figure 11.16). The nitride thick-

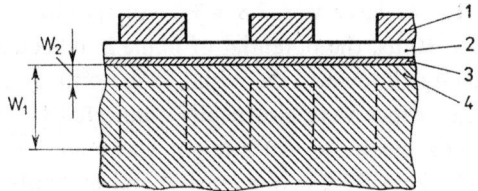

Figure 11.16 Scheme of MNOS-structure with additional dielectric layers: 1, SiO; 2, Si_3N_4; 3, SiO_2; 4, Si (W is the thickness of the space charge layer in semiconductor)

ness was about 900 Å. A narrow metallic electrode was deposited from above perpendicular to the stripes. When the structure was supplied with a voltage pulse whose polarity causes depletion of the semiconductor surface, then six barrier-separated potential wells were formed for the minority carriers under the electrode. Figure 11.17(a) shows a picture of photocurrent signals obtained on the structure contacts in the case of He–Ne laser beam scanning along the metallic electrode, when the 'plane bands' potentials for all six regions were about the same. Then the structure was supplied with a voltage pulse 50 μs long and 50 V amplitude. Regions 1, 3, 5 and 7 were illuminated. Figure 11.17(b) shows the signals obtained in scanning of the structure after recording.

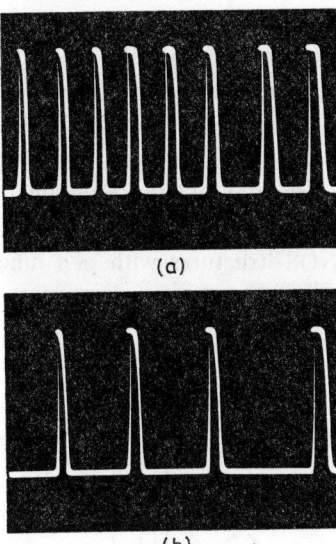

Figure 11.17 Photos of photocurrent signals obtained in scanning of the structure before (a) and after (b) the recording. Structure parameters: $SiO_2 = 18$ Å, $Si_3N_4 = 900$ Å, SiO 1 μm

Regions 1, 3, 5 and 7 showed a considerable change of the signal (more than 10 times). In other regions the signal did not change. The experiments performed at different recording times (from 1 μm to 1 mm) produced identical results. Thus, it turns out to be possible to obtain space resolution by depositing additional dielectric layers with the formation of potential barriers, which prevent minority carriers from spreading out. It is not difficult to deposit dielectric stripes, and electrodes with the size of ~ 10 μm, by making use of the photolithographic method.

11.5 OPTICALLY CONTROLLED INFORMATION ERASURE

In optical memory on MNOS-structure it is desirable to record and erase information by parts. As shown above, data recording can be performed in the illuminated regions of the metallic electrode only, which permits storage of 10^3–10^4 bits under the area of one electrode. To erase information it is necessary to apply to the structure a voltage pulse of opposite polarity in comparison with the recording. In this case the majority carriers from the semiconductor volume are concentrated at the dielectric-semiconductor interface. They fully screen the electric field in the semiconductor. Almost all of the voltage applied falls on the dielectric layer of the structure, and one cannot control switching of the structure by light. Experiments on the control switching of the structure under both polarities of the switching voltage pulses were performed on MNOS-structures with p–n junction (Figure 11.18).

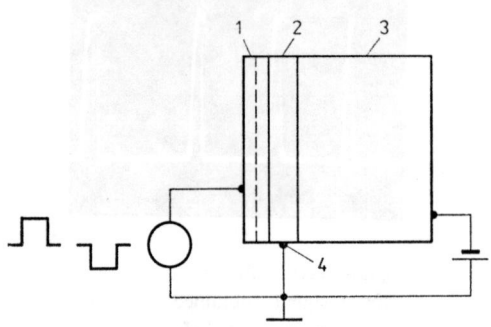

Figure 11.18 Preparation of MNOS-structure with optical data recording and erasure; 1, SiO_2 and Si_3N_4 dielectric layers; 2, epitaxial Si layer with n-type conductivity; silicon substrate of p-type

A thin ($\sim$ 6 μm) epitaxial Si layer of opposite conductivity was grown on a typical semiconductor substrate.

Normally, the substrates had a p-type conductivity while the epitaxial layer had an n-type conductivity. The p–n junction substrate was deposited with SiO_2 and Si_3N_4 layers.

Consider the operation of such a structure, when it is supplied with voltage pulses of different polarity. Since numerical calculations will be needed, let us assume the following parameters of the silicon substrate: a 200 μm thick p-layer, $p_0 = 10^{15}$ cm^{-3}, n-layer, thickness $d_1 = 6$ μm, $n_0 = 5 \times 10^{15}$ cm^{-3}.

In the case of a negative voltage with 40–50 V amplitude applied to the control electrode the thickness of the depletion region in silicon will be 5 μm (Figure 11.6). If the p–n junction is supplied with the shut-off voltage, the structure will operate in almost the same way as in the cases considered above. Under the action of the positive voltage on the control electrode, by concentrating at the Si–SiO_2 boundary, the majority carriers of the silicon *n*-type layer will lead only to a partial screening of the field in the semiconductor. Let us now estimate the density of the majority carriers at the dielectric-semiconductor boundary: $n_0 d_1 = 5 \times 10^{15} \times 6 \times 10^{-4}$ cm^{-2} = 3×10^{12} cm^{-2}.

Remembering that the dielectric capacity is $C_d = 10^{-7}$ fcm^{-2}, such density of the charge is sufficient for the screening of 10 V only.

The voltage supplied to the n-region should be barrier-like in order to prevent the injection of the majority carriers from the contact to the semiconductor. When the contact operates in the shut-off direction, the maximum density of current injected through the contact can be determined from the following simulations. Equilibrium concentration of the electrons near the contact with the potential barrier W_{ms}, is determined by the formula:

$$n_s = n_0 \exp\left(-\frac{qW_{ms}}{kT}\right)$$

where W_{ms} is the difference of work functions in the metal-semiconductor. The maximum value of the current flowing from the metal to the semiconductor is: $j/q = \frac{1}{4} V_t n_s \sim 10^{12}$ cm^{-2}s^{-1}, where V_t is the heat velocity, and W_{ms} equals 0.65 eV, which corresponds to the difference of work functions for the case of gold deposition on an n-type semiconductor (Si) with the concentration of majority carriers $n_0 = 5 \times 10^{15}$ cm^{-3}.

If the duration of the switching voltage pulse is 10^{-4} s, then the number of electrons entering the semiconductor from the contact is about 10^8 cm^{-2}, which results in the change of voltage applied to the dielectric, $\Delta V_d \sim 2 \times 10^{-4}$ V.

The current of minority carriers through p-n junction is negligibly small as well. One can estimate the maximum value of current from $j/q = n_p V_T$, where n_p is the concentration of electrons in the p-layer. At room temperature

$$n_p = \frac{n_i^2}{p_0} = \frac{1.6 \times 10^{20}}{p_0}$$

and for the concentration of majority carriers in p-layer, $P_0 = 10^{15}$, we get $j/q = 1.6 \times 10^{12}$ cm^{-2}s^{-1}. This means that the current of the minority carriers through the p-n junction does not exceed the calculated current flowing from the shut-off contact to silicon.

Thus, the epitaxial Si-layer structure with a p-n junction can be optically controlled both under the positive and negative polarities of the voltage applied.

11.6 SCHEME OF THE OPTOELECTRONIC MEMORY DEVICE ON MNOS-STRUCTURES

We consider here a scheme of an optoelectronic memory device based on the bit data storage. The operation of such a memory is schematically shown in Figure 11.19. Information recording and readout is paginal. Memory cells

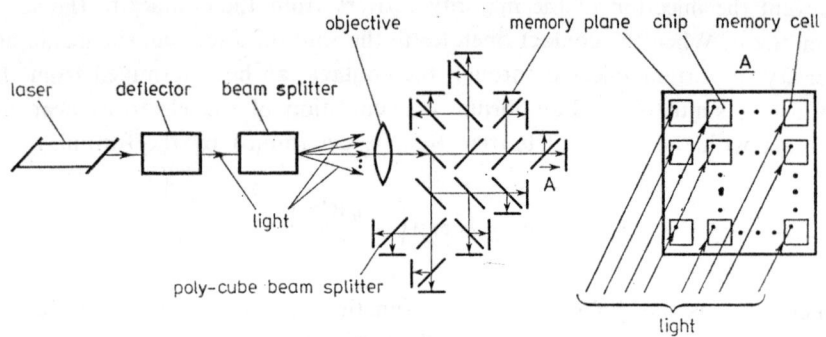

Figure 11.19 Scheme of an optoelectronic memory device

are addressed with the use of light. To do this a laser beam deflected in accordance with the page address is divided by the splitter into beams of equal intensity spaced at equal angles. They are then focused by an objective into similar cells of memory chips (the number of chips in one plane is equal to the number of divided beams). The separation of beam to different planes is performed by light-splitters of an additional splitter (image multiplier) which has a split coefficient of 1:1 and ensures equality of beam energies. Each memory chip is addressed by one beam shifting to another cell of the same chip when the deflector is switched. Thus, the number of memory cells of each chip is equal to the number of position of the deflector and determines the volume of data pages stored in optical memory. The total number of chips is equal to the number of bits in one page.

To record a bit of information the chip is simultaneously exposed to the action of light and electric voltage which provides the recording of 'one'. For data readout one uses the photocurrent of each chip. Its value, as mentioned above, depends on which bit of information ('0' or '1') is read out by the light beam.

In such a memory the insufficiently high resolution of the medium seems to be compensated by a multiplication of each bit. In addition, the reflecting surfaces of the light splitters as if multiply the objective apertures, i.e. a memory with synthesized aperture is obtained, which results in an increase in the maximum data capacity. The value of the memory capacity is determined as

$$N = N_d N_c N_p,$$

where N_d is the number of memory cells of each chip (the total number of positions produced by the deflector), N_c the number of chips in each memory plate (the number of split beams), and N_p the number of memory plates.

Calculation results for real optical systems and laser powers with regard to the material response can be presented. For example, the total capacity of the optical memory device is 1.1×10^9 (16 400 bits in a page), provided that the objective is 80 mm in diameter, laser power 1 W, the number of deflector positions 256×256, beam splitting 8×8, and 8 light multiplicator splitters available in the way of any beam from the objective to the corresponding plate.

REFERENCES

1. A. L. Mikaelyan, V. I. Bobrinev, *Radiotechnika i elektronika*, **19**, 898 (1974).
2. A. A. Akaev, S. A. Maiorov, *Kogerentnye opticheskie vychislitelnye mashiny*, Mashinostroienie, Leningrad, 1977.
3. A. B. Kravchenko, A. F. Plotnikov, V. H. Seleznev, V. E. Shubin, *Kratkie soobshcheniya po fizike, FIAN*, **10**, 7 (1973).
4. A. B. Kravchenko, A. F. Plotnikov, Yu. M. Popov, V. N. Seleznev, V. E. Shubin, *FTP*, **8**, 810 (1974).
5. A. F. Plotnikov, V. N. Seleznov, V. E. Shubin, C. P. Ferchev, *Kvantovaya elektronika*, **1**, 1885 (1974).
6. E. O. Johnson, *Phys. Rev.*, 3, 133 (1958).
7. H. Yamagishi, *J. Phys. Soc. Jap.*, **25**, 766 (1968).
8. Y. W. Lam, *J. Appl. Phys.*, **4**, 1970 (1971).
9. A. F. Plotnikov, V. S. Vavilov, *FTP*, 7, 878 (1973).
10. J. E. Cornes, W. E. Kosonocky, E. G. Rambery, IEEE Trans., **ED-19**, 798 (1972).
11. A. B. Kravchenko, A. F. Plotnikov, V. N. Seleznev, D. N. Tokarchuk, V. E. Shubin, *Kvantovaya elektronika*, **1**, 10 (1974).

Chapter 12

Quantum Efficiency and Radiative Lifetimes in GaAs and AlGaAs

D. Z. Garbuzov

12.1 INTRODUCTION

This chapter deals with the results of investigations of the radiative properties of GaAs and $Al_xGa_{1-x}As$ direct-gap materials which represent the active region of double AlGaAs-heterostructures. Double heterostructures with narrow-gap active region of width d ($< L_D$) enclosed between two wider-gap emitters are not only the base for the fabrication of high-efficiency coherent and spontaneous injection radiative sources, but also an ideal object to study the luminescence properties of direct energy-gap semiconductors. When routine investigations of photoluminescence of direct-gap materials and electroluminescence of p–n-homojunctions obtained on their basis are carried out in order to determine intrinsic recombination parameters by observed external characteristics, it is necessary to evaluate first the spatial profile of excess-minority carriers which in turn depends upon such factors as surface recombination velocity, mobility and total lifetime of the minority carriers. Computations become especially complex if the material investigated has η_i of approximately 100% and processes of self-absorption and photon recycling produce a substantial effect on the luminescence properties observed. As was shown by our investigations reported at the first School (1), this is the situation which exists over a wide range of temperatures and doping levels for GaAs and $Al_xGa_{1-x}As$ ($x < 0.3$) grown by epitaxy methods. When the internal efficiency (η_i) of radiative transitions is high, the sequential computation of the relationship between the intrinsic luminescence properties and the observed ones can be effected only in the case of double heterostructures. The investigation is simpler here because the interface recombination velocity is negligible and the spatial profile of the minority carriers in the active region of AlGaAs heterostructures can be assumed to be uniform provided that $d < L_D$. The results of computation of the external efficiency $\eta_e = f(\eta_i)$ and the ratio of the observed effective lifetime to the net lifetime

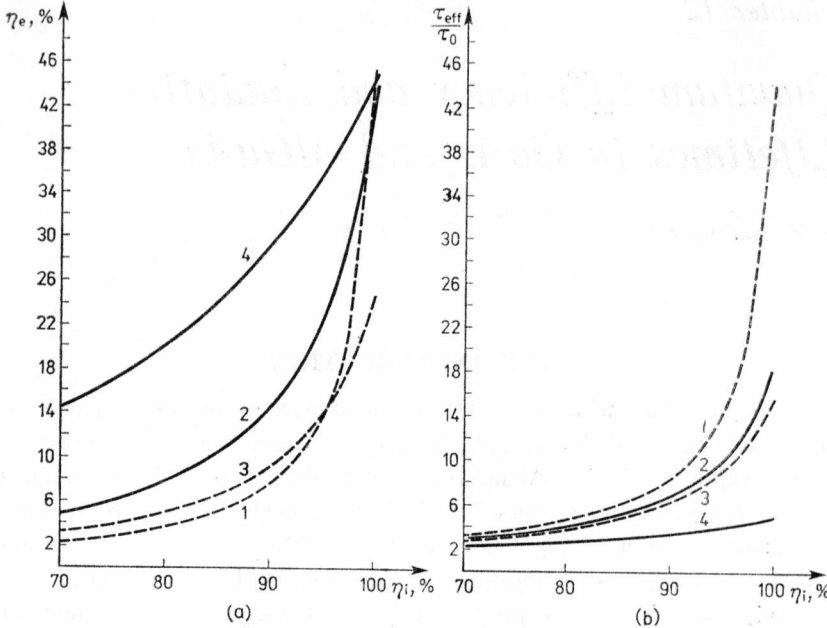

Figure 12.1 (a), (b) Theoretical dependences $\eta_e = f(\eta_i)$ and $\tau_{eff}/\tau_0 = f(\eta_i)$ estimated for double heterostructures on the assumption that the active region edge absorption data coincide with Sturge's results (5). The structures have no substrates and their active region widths are about 3 μm. Curves 1 and 3 refer to structures with mirror surfaces at 294 K and 90 K, respectively. Curves 2 and 4 are calculated for structures with scattering surfaces at 249 and 90 K

versus the internal quantum efficiency $\tau_{eff}/\tau_0 = f(\eta_i)$ for the double AlGaAs-heterostructures are given in (2–4) as well as in a report read at the *Summer School on optoelectronics and integrated optics* held in the Czechoslovak Socialist Republic (1976). Data on the spectral dependence of the absorption coefficient of the active region materials $k = f(h\nu)$ that are necessary for such evaluations may be found in the literature or determined by experiment. When evaluating the characteristics $\eta_e = f(\eta_i)$ and $\tau_{eff}/\tau_0 = f(\eta_i)$ used here, data on absorption edge for GaAs of different conductivity types and doping levels for temperatures of 300 K and 77 K were obtained from (5–8). The results of evaluations illustrating the effect of temperature variations on the course of characteristics $\eta_e = f(\eta_i)$ and $\tau_{eff}/\tau_0 = f(\eta_i)$ for heterostructures with different treatment of external surface are shown in Figures 12.1(a), (b). Without going into any detailed analysis of the question, it must be noted

Quantum Efficiency and Radiative Lifetimes in GaAs and AlGaAs

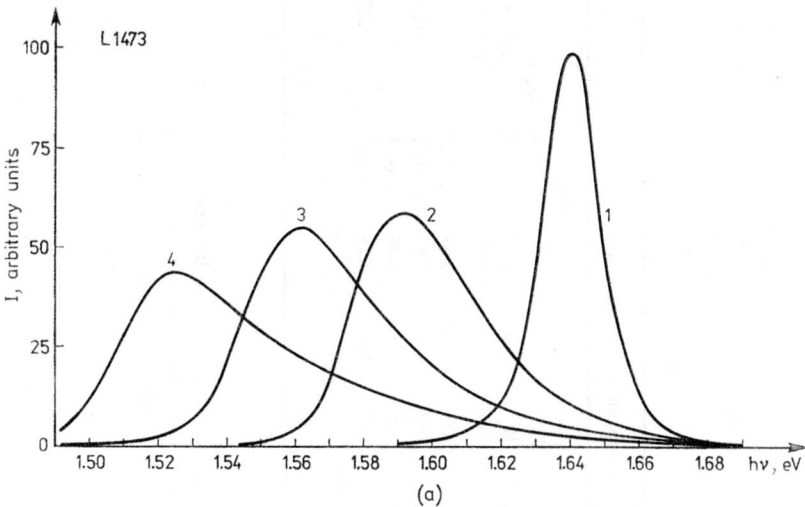

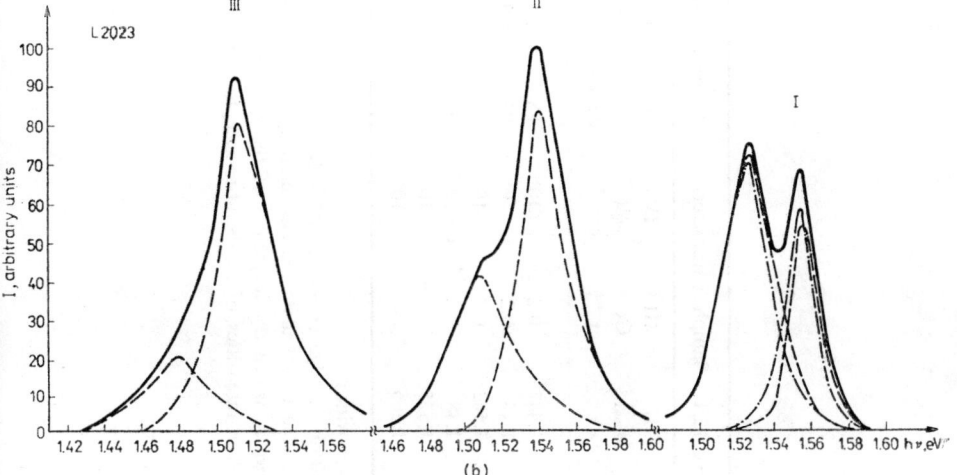

Figure 12.2 (a), (b) Luminescence spectra for double heterostructure (L 1473) with undoped active region and for L 2023 with germanium doped active region. ($N_A = 6 \times 10^{17}$ cm^{-3}) at different temperatures. The spectrum areas correspond to the radiation intensity at different temperatures. The results of two ways of approximation permitting determination of the intensity of each peak are shown in Figure 12.1 (b) by dashed lines

Figure 12.2 (a) 1—78 K, 2—206 K, 3—300 K, 4—373 K
Figure 12.2 (b) I—78 K, II—130 K, III—200 K

Table 12.1 Sample parameters and the results of temperature dependence investigation

No	Wafer	I d μm	II $h\nu_{max}$ eV	III Ge, at %	IV N_A cm^{-3}	V η_i (300 K)	VI K	VII τ'_0 (300 K)	VIII T_0 K	\multicolumn{4}{c}{IX I_{BA}/I_{BB}}
										80 K	120 K	200 K	300 K
1	L1453	1.8	1.442	0.2	6×10^{17}	88	3.7	6.0	100	1.88	0.69	0.27	
2	L2072	2.0	1.45	0.2	6×10^{17}	90	4.0	7.0	100	1.85	0.67	0.37	0.29
3	L2048	1.4	1.503	0.2	6×10^{17}	92	3.6	6.0	85	1.9	0.48	0.38	
4	L2023	0.9	1.48	0.2	6×10^{17}	93			105	1.9	0.74	0.37	0.27
5	L2053	1.1	1.525	—	10^{17}	80	2.5	20					
6	L1473	1.3	1.566	—	10^{17}	93	4.0	26					

I d—active layer width
II ($h\nu$)—peak position of edge emission at 300 K
VI K—coefficient used to calculate the net minority carriers lifetime from the observed effective lifetimes $k = \tau_{eff}/\tau_0$
VIII T_0—temperature at which the intensities of the band-to-band and the band-to-acceptor transitions are equal
IX I_{BA}/I_{BB}—ratio between intensities of the band-to-acceptor and the band-to-band transitions at different temperatures

Table 12.2 Parameters used and calculated temperature dependences for the efficiency and radiative lifetimes (τ_e^t calculated radiative lifetime at 300 K)

No	Conductivity Type	N_A cm^{-3}	N_D cm^{-3}	τ^t (300 K)	T_0 K	I_{BA}/I_{BB} 80 K	120 K	200 K	300 K
1	p	6×10^{17}	0	6.1	100	1.36	0.84	0.54	0.39
2	p	6×10^{17}	2×10^{17}	9.2	150	4.9	1.5	0.69	0.43
3	p	6×10^{17}	0	6.1		0.72	0.54	0.39	0.30
4	p	6×10^{17}	2×10^{17}	9.1	105	1.68	0.82	0.47	0.33
5	p	6×10^{17}	10^{17}	7.0	105	1.8	0.85	0.52	0.36
6		—	10^{17}	29					
7		5×10^{16}	10^{17}	56					

only that the change in the course of the characteristics is due to the variation in the rate of the luminescence self-absorption in the active region with temperature. In general, an increase in the active region doping level causes the same changes in the considered characteristics as in the case of temperature decrease, but the absolute value of these variations is considerably smaller and does not exceed 10–20%.

12.2 SAMPLES AND MEASUREMENT PROCEDURES

The heterostructures studied were grown by the LPE-method at 800°C. The active region of structures, used to study the relationship between luminescence properties and the doping level, represented solid solutions with AlAs content of approximately $10\pm3\%$. The active region width of the structures studied was about 1 µm. Table 12.1 shows the parameters of the active region of structures used for temperature investigations. The AlAs-content of both emitters was approximately the same for all structures and amounted to 50–60%. Because of this emitter composition, the GaAs-substrate could be selectively etched and the red and yellow radiation that excited the luminescence was absorbed directly in the active structure region.

The photoluminescence properties of both undoped structures (more precisely, structures not intentionally doped) and structures having the Ge- or Te-doped active region were studied. In order to provide uniform doping and to avoid formation of p–n junctions, Ge- or Te-doping was effected not only for the active structure region but also for the adjoining emitters. The Ge content in the active region and the free-hole concentration were evaluated by the use of calibration curves shown in (9). Data from References 10, 11 were used to determine the electron concentration in the active region of Te-doped structures. Undoped and slightly doped n-type GaAs layers were grown on semi-insulating substrates and measurements of carrier concentration by the Van der Pauw method were made to take into consideration the effect of uncontrollable residual donor impurities.

The grown structure was separated into slices. The photoluminescence spectrum of the active region was studied by standard procedures on one of the slices that had a GaAs-substrate. A He–Ne-laser ($I \simeq 10 \text{ Wcm}^{-2}$) was used as a source of excitation radiation and a Si-cell of known spectral photoresponse as a photodetector. Spectrum recording was effected in the temperature range of 80–400 K. Figures 12.2(a) and 12.2(b) show radiation spectra at several temperature values for a structure with undoped active region and for a structure with Ge-doped active region ($N_A = 6 \times 10^{17}$ cm^{-3}), respectively. Spectrum areas are in agreement with the luminescence intensity at each temperature value. The radiation spectrum of the structure with undoped n-type region consists of a single band-to-band radiation peak. In addition to the band-to-band radiation peak, the p-type material spectra have a second longer-wavelength peak caused by conduction band-to-acceptor transitions. It should be emphasized that no other peaks, except for the active region edge emission, were observed in luminescence spectra of the structures investigated over the above temperature range. For this reason the integral investigations of the intensity and kinetics of luminescence could be made without using a spectrometer.

In order to make absolute measurements of the external quantum efficiency, the GaAs-substrate was removed from one of the structure parts. The sample under study was placed directly on the surface of a calibrated Si-cell of a large area. The external quantum efficiency was determined as the ratio of photocell current values in the presence of the structure and at direct illumination of the photocell surface with the He–Ne-laser excitation radiation. This procedure of η_e-determination has been considered in detail in (3). Aside from the absolute measurements of the external luminescence efficiency, measurements of relative values of η_e could be made for the same sample over a wide range of excitation levels (10–5 × 10^4 Wcm^{-2}) and temperatures (78–400 K) in another way. In this case a Kr-laser made by Spectra Physics Company was used as an excitation radiation source focused on the structure surface, the spot being about 30 μm in diameter. The small excitation area and a decrease in the average power by a factor of 100 due to an additional chopper enabled the measurements to be carried out at high excitation levels without a considerable heating of the material studied. Comparison between results of absolute and relative measurements of η_e made it possible to determine first the dependence of the external luminescence efficiency upon the temperature and excitation level. Following that the internal quantum efficiency of the radiative recombination in the material of heterostructure active region was determined by the use of calibration curves of $\eta_e = f(\eta_i)$. The η_i values determined in such a manner were then used to define the minority carrier net lifetimes (τ_0) by characteristics of $\tau_{eff}/\tau_0 = f(\eta_i)$ and by total lifetimes for radiative-type transitions: $\tau_0' = \tau_0/\eta_i$.

The effective lifetimes of minority carriers were determined in the active-structure region (τ_{eff}) on the basis of measurements of the luminescence pulse decay time. A photomultiplier with a 50-ohm load was used as a photodetector. The output signal of the latter was applied to a stroboscopic oscillograph and then recorded by a two-axis recorder. A N_2-laser provided with an additional unit with Rodamin σ–G dye ($\lambda = 0.573$ μm) was used as an excitation radiation source for pulse measurements. The excitation radiation pulse half-width amounted to no more than 10 ns with the use of the same equipment that had been applied for recording the luminescence pulses.

The density of excitation radiation was varied by changing the degree of the laser beam focusing and by using neutral attenuators. In every case of τ_{eff} determination, the luminescence pulses of the investigated heterostructure were recorded at 5 to 10 excitation levels over the range from the minimum possible value for these recording conditions to the greatest achievable value. The determination of τ_{eff}-value was made on the basis of only those recordings for which the luminescence pulse time-decay was independent of the excitation level and had a nearly exponential shape. For structures with undoped or slightly doped active region and high excitation densities, the luminescence pulse time-decay differed from the exponential shape and an initial portion of steeper slope could be distinguished. The appearance of this portion was due to a change to bimolecular recombination at high excitation levels ($\Delta n = \Delta p > n_0$). The analysis of time-decay shape at high excitation levels permits Δn and Δp to be evaluated. When τ_{eff} is known, one can determine the excitation levels which are realizable in the heterostructure active region at a given intensity of pulse excitation. Knowledge of the excitation level, at which the pulse measurements have been made, is necessary to use the characteristics $\tau_{\text{eff}}/\tau_0 = f(\eta_i)$ if η_i depends on the excitation level.

12.3 THEORETICAL CALCULATION OF LIFETIMES AND PROBABILITIES OF RADIATIVE TRANSITIONS AND THEIR TEMPERATURE RELATIONSHIPS

In the following, the results of experimental determination of lifetime for radiative transitions at various temperatures and doping levels of the active region material will be compared with data of theoretical calculations in terms of simple quantum mechanical theory. When computing the dependences of lifetimes and transition probabilities upon donor and acceptor concentrations, we applied general formulas valid for both the non-degenerate and degenerate semiconductors. Computer calculations were carried out using programs prepared by V. B. Khalfin. We will not list here all the cumbersome formulas where the quantities of interest are represented through integrals that include the Fermi–Dirac distributions. For the non-degenerate semiconductor, the corresponding expressions simplify to general formulas. Their development can be found, for example, in monographs 12, 13. These formulae were used during the first part of the study where temperature dependences of lifetimes were investigated in a slightly doped material of the n-type and p-type conduction. Now let us briefly discuss the model, the assumptions made and the results of quantum mechanical calculations of lifetimes and probabilities of radiative transitions of various types. We considered band-to-band transitions and transitions involving majority shallow impurities (donors in n-type material and acceptors in p-type material). The considerable doping levels ($n, p \geqslant 5 \times 10^{16}$ cm^{-3}) and the relatively high temperatures (50–400 K) justify the fact that radiative transitions of other types and effects due to the Coulomb interaction of recombining carriers have been neglected.* The conduction band and light-hole and heavy-hole bands were regarded as parabolic ones and wave-vector selection rules for recombination transitions were treated as fulfilled. Under these assumptions Equations 12.1 through 12.3 similar to the equations given in monograph 13 may be applied for calculating the lifetimes of radiative transitions of interest to us:

$$\tau_{BB} = \frac{1}{B_{BB}(n+p)} = \frac{32C^2(\pi kT)^{3/2}m_0^{1/2}m_e m_r^{3/2}}{C_0(n+p)n_r E_g^2 h^3} \tag{12.1}$$

$$\tau_{BA} = \frac{1}{B_{BA}N_A^0} = \frac{\sqrt{2}C^2 m_r^{3/2} m_0^{1/2}}{8C_0 N_A^0 \pi n_r m_e^{1/2}(E_g - E_A)^2 a_A^{*3} G(\alpha)} \tag{12.2}$$

$$\tau_{DB} = \frac{1}{B_{DB}N_D^0} = \frac{\sqrt{2}C^2 m_r^{3/2} m_0^{1/2}}{8C_0 N_D^0 \pi n_r m_e^{1/2}(E_g - E_D)^2 a_D^{*3} G'(\alpha)} \tag{12.3}$$

Equation 12.1 refers to band-to-band transitions, and Equations 12.2 and 12.3 refer to band-to-acceptor and donor-to-band transitions; τ_{BB}, τ_{BA} and τ_{DB} are used in the sense of average probabilities for corresponding transitions. Most of the quantities appearing in Equations 12.1 through 12.3 are given in Table 12.3, where values of the respective quantities for GaAs are also listed. Let us give certain additional explanations as to the application of Equations 12.1 through 12.3.

1. When these formulae were used in the computations, experimental values relating to a slightly-doped material $E(Te) = 5.2$ meV; $E(Ge) = 40$ meV were taken as 'optical' activation energies of shallow impurities E_D and E_A

---

* The effect of the Coulomb interaction at $n, p < 5 \cdot 10^{16}$ cm^{-3} will be considered at the end of the fourth section of this chapter.

Table 12.3 Symbols used in Equations 12.1 through 12.5 and their values used for calculations

E_g	Energy gap	1.42–1.55 meV
n_r	Refractive index	3.6
E_A	Activation energy for acceptors	40 meV
E_D	Activation energy for donors	5.2 meV
a_D^*	Donor Bohr radius	90 Å
a_A^*	Acceptor Bohr radius	13 Å
m_e	Electron effective mass	$0.07\, m_0$
m_{lh}	Light-hole effective mass	$0.09\, m_0$
m_{hh}	Heavy-hole effective mass	$0.5\, m_0$
$m_r^{3/2}$	$m_r^{3/2} = \left(\dfrac{m_{hh} m_e}{m_{hh}+m_e}\right)^{3/2} + \left(\dfrac{m_{lh} m_e}{m_{lh}+m_e}\right)^{3/2}$	$0.023\, m_0^{3/2}$
$m_v^{3/2}$	$m_v^{3/2} = m_{lh}^{3/2} + m_{hh}^{3/2}$	$0.38\, m_0^{3/2}$
N_c	$N_c = 2(2\pi m_e kT/h^2)^{3/2}$	$N_c \simeq 8.8 \times 10^{13}(T^{3/2})$
N_v	$N_v = 2(2\pi m_v kT/h^2)^{3/2}$	$N_v \simeq 1.8 \times 10^{15}(T^{3/2})$
H_{cv}	$\lvert H_{cv}\rvert^2 = \dfrac{m_0^2 E_g}{2m_e}$	
C_0	$C_0 = \dfrac{32\pi^2 e^2 (2m_r)^{3/2}}{3ch^3 v m_e'^{1/2} m_0^{3/2}} \lvert H_{cv}\rvert^2$	$C_0 = 3.4 \times 10^{15}\ \text{cm}^{-1}\text{j}^{-1/2}$

appearing explicitly in Equations 12.2 and 12.3. The Bohr radius of impurity states was found from the expression $a^* = e^2/2\varepsilon E_{\text{imp}}$. The same values of E_D and E_A were used in computations of Dumke integrals designated as $G(\alpha)$ and $G'(\alpha)$ (13).

2. The constant in Equations 12.1–12.3 is proportional to the square of the optical matrix element

$$\lvert H_{cv}\rvert^2_{\vec{k}=0} = \dfrac{m_0^2}{2m_e} E_g$$

computed with the aid of the Kane–Dumke–Stern approximation (13). When using this approximation it is assumed that

$$\dfrac{1}{m_{hh}} + \dfrac{1}{m_e} \approx \dfrac{1}{m_e} \quad \text{and} \quad \dfrac{1}{m_{lh}} + \dfrac{1}{m_e} \simeq \dfrac{2}{m_e}$$

The same assumption was employed to calculate $m_r^{3/2}$ in Equations 12.1–12.3.

3. The neutral acceptor concentration was computed utilizing the formula

$$N_A^0 = N_A - N_D - p = N_A + \frac{1}{2}\left(\frac{N_v}{4}e^{-E_A^*/kT} - N_D\right) -$$

$$- \sqrt{\frac{1}{4}\left(\frac{N_v}{4}e^{-E_A^*/kT} + N_D\right)^2 + \frac{N_v}{4}(N_A - N_D)e^{-E_A^*/kT}} \quad (12.4)$$

Here, p is the free-hole concentration, N_D compensating donor concentration, N_A total acceptor concentration, N_v density of states for the valence band, calculated for effective masses m_{hh} and m_{lh} given in Table 12.3, and E_A^* thermal activation energy of Ge-acceptor centre. As opposed to the 'optical' activation energy E_A, the thermal activation energy was assumed to be dependent

Table 12.4 Activation energy of the shallow acceptor centre in Ge-doped GaAs (9, 14)

N_D cm^{-3}	2×10^{16}	3×10^{16}	1.5×10^{17}	1.9×10^{17}	5×10^{17}	2×10^{18}	3×10^{19}
E_A^* meV	35	32	31	28	23	19	13

on acceptor concentration. Table 12.4 lists the E_A^* values for different acceptor concentrations as taken from References 9, 14.

4. The neutral donor concentration was computed from the Equation 12.5 as

$$N_D^0 = N_D - n - N_A = N_D + \frac{1}{2}\left(\frac{N_c}{2} - N_A\right) -$$

$$- \sqrt{\frac{1}{4}\left(\frac{N_c}{2} + N_A\right)^2 + \frac{1}{2}N_c(N_D - N_A)} \quad (12.5)$$

where N_D^0 is the total donor number, N_c concentration of electrons involved in band-to-band transitions ($N_c < N_D$), and N_A compensating acceptor concentration. As no temperature dependence of free electrons was observed at $n > 5 \times 10^{16}$ cm^{-3}, it is assumed that E_D is equal to zero for donor centres. In addition to this model, assuming that a portion of electrons is located on donors and is involved in impurity radiative transitions despite the fact that $E_D = 0$, the case of complete neglect of donor states ($N_c = N_D$) was considered for the n-type material as well.

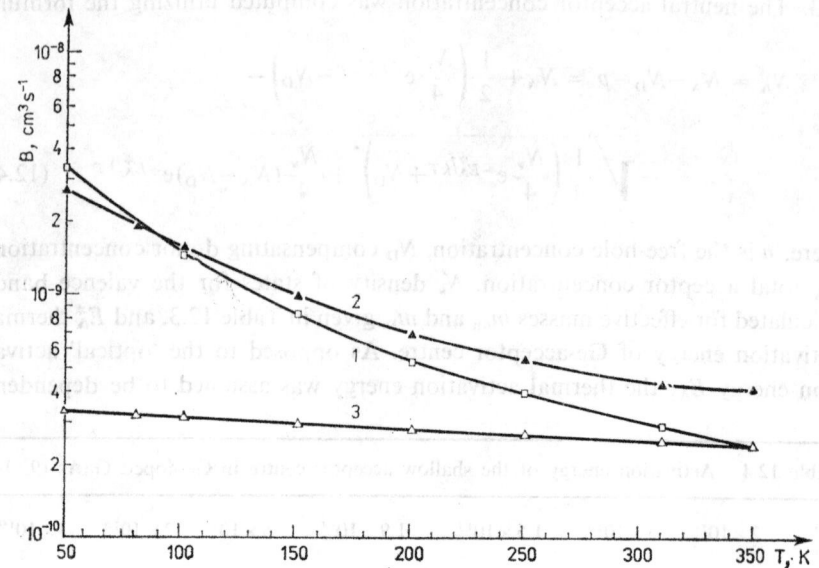

Figure 12.3 Calculated temperature dependences of the probabilities for different types of radiative transitions: 1, band-to-band transitions; 2, donor-to-valence band transitions; 3, conduction band-to-acceptor transitions. The parameters used are given in Table 12.3

Let us now consider the results of the calculations and begin from the temperature and concentration dependences of transition probabilities. Figure 12.3 shows the results of calculations of transition probabilities B_{BB}, B_{BA}, and B_{DB} by Equations 12.1–12.3. The temperature dependence of probabilities for band-to-band radiative transitions (curve 1, Equation 12.1) varies as $T^{-3/2}$. The temperature dependences of impurity transition probabilities are related to temperature dependences of Dumke integrals. The physical cause of the decrease in band-to-band transition probability with increasing temperature is related to a broadening of carrier distribution in quasi-momentum space and, consequently, to a reduction of the probability that each pair of states involved in transitions will be filled. The momentum spread of the acceptor state is comparable with the width of free-hole momentum spectrum at 300 K ($E_A \simeq kT$). The latter spectrum is considerably more spread than the free-electron spectrum. That is the reason why further narrowing of electron momentum distribution with a decrease in the temperature from 400 K down to 50 K does not give rise to practically any increase in the conduction band-to-acceptor transition probability. As a consequence of the foregoing, the conduction band-to-acceptor transition probability is comparable to the band-to-band transition probability in the high-temperature range, where $E_A \simeq kT$, and the value of B_{BA} is less than the value of B_{BB} by one order of magnitude at 50 K (curves 1 and 3, Figure 12.3). By contrast, the donor-to-valence band transition probability is comparable to the band-to-band transition probability at 50–80 K, where $E_D \simeq kT$. The values of B_{DB} decrease with increasing temperature, because of the spread of the free-hole momentum spectrum, nearly at the same rate as the band-to-band transition probability.

Due to the differences in the effective masses of electrons and holes, the concentration dependences of band-to-band transition probability in a material with n-type and p-type conduction are different as well. In the case of the n-type material, relative transition probability remains nearly unchanged, as n grows and degeneracy occurs, in spite of the corresponding broadening of the electron momentum spectrum. Only when the Fermi electron level reaches the value F_n, for which

$$F_n > \frac{m_{hh} + m_e}{m_e} kT \simeq 8kT$$

does band-to-band transition probability begin to fall with increasing electron concentration. The hole momentum spectrum is broader than the electron momentum spectrum. Because of this, in contrast to the n-type material, an increase in hole concentration leads to a decrease in the radiative band-to-band transition probability before degeneracy occurs (on the assumption that impurity transition probabilities are independent of the doping level of the material).

Let us now consider the concentration and temperature dependences of lifetimes for radiative transitions. Aside from lifetimes for band-to-band and impurity transition, the total lifetimes for electronic radiative transitions (τ_e^t) in the p-type and those of hole radiative transitions in the n-type material will also be considered. By definition, the total lifetimes are

$$\tau_e^t = \frac{\tau_{BB} \cdot \tau_{BA}}{\tau_{BB} + \tau_{BA}}, \quad \tau_h^t = \frac{\tau_{BB} \cdot \tau_{DB}}{\tau_{BB} + \tau_{DB}} \quad (12.6)$$

Deviations from linearity in relationship $\tau_{BB} = f(n)$ and $\tau_{BB} = f(p)$ (curve 2, Figure 12.4(a); curve 3, Figure 12.5(a)) may be associated only with variations in the radiative transition probability discussed above.*

---

* The same is true of the total hole lifetimes in the case of the n-type semiconductor model in which donor states are neglected.

In the range of low doping levels ($n, p < 10^{17}$ cm^{-3}) the lifetimes for impurity transitions exceed considerably the lifetimes for band-to-band transitions, as shallow impurities are almost completely ionized at 300 K (Figures 12.4(a), 12.5(a)). However, as the free-carrier concentration grows, the lifetimes for impurity transitions must become shorter at a greater rate than those for the band-to-band radiative transitions (curve 1, Figures 12.5(a), 12.4(a)). This is due to the fact that the neutral impurity concentration is proportional (if compensation is neglected) to the square of the free-carrier concentration (see Equations 12.4 and 12.5). Computations made in terms of the simple theory used predict that the lifetimes for impurity radiative transitions are to become less than the lifetimes for band-to-band transitions at $n, p > 5 \times 10^{17}$ cm^{-3} and are to determine the total minority carrier radiative lifetimes for both the n-type and p-type materials.

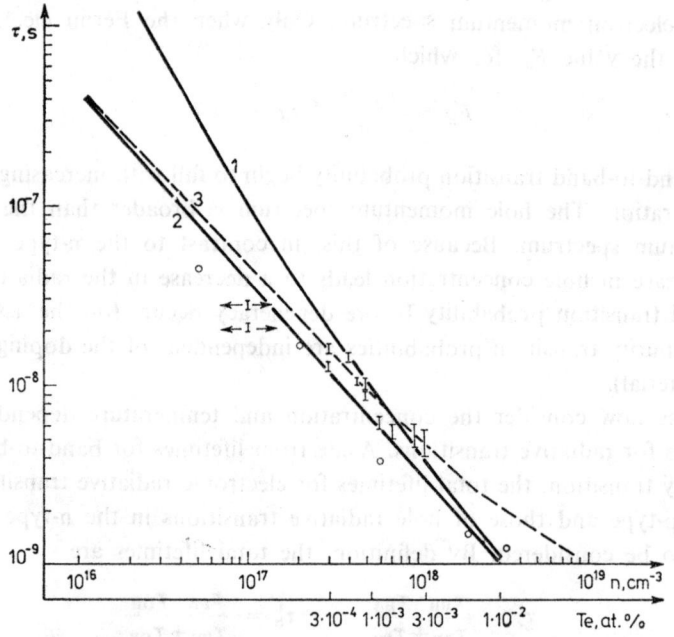

Figure 12.4 (a) Experimental and theoretical data for the radiative lifetime dependences on the concentration of the electrons in the active region of the double heterostructures: 1, lifetimes for donor-to-valence band transitions; 2, total hole radiative lifetimes; 3, band-to-band lifetimes in the model neglecting the donor states. The data calculated by the Shockley–van Roosbroeck method are marked by circles

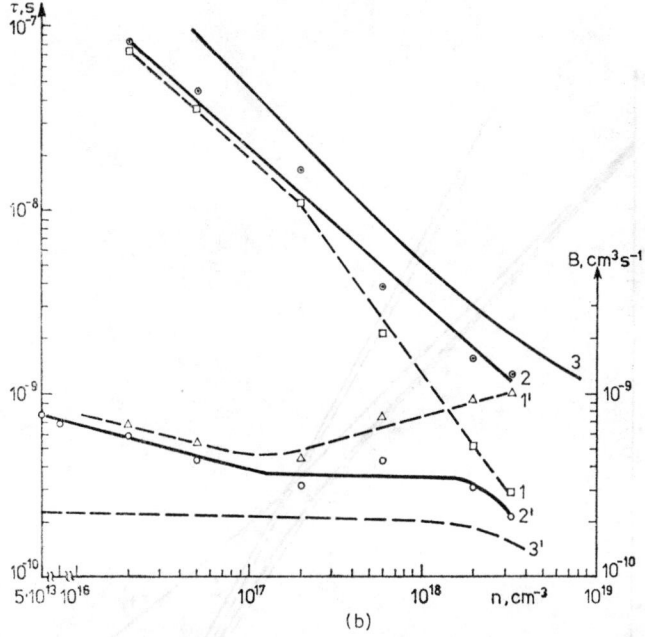

Figure 12.4 (b) Curves 1 and 1' represent lifetimes and probabilities for radiative transitions in n-GaAs calculated by the Shockley–van Roosbroeck method without the correction for the energy gap shrinkage due to doping. Curves 2 and 2' are results of similar calculations with the correction for the energy gap shrinkage. Curves 3 and 3'— calculated lifetimes and probabilities for radiative transitions in the n-type material model neglecting the donor levels

Figures 12.6 and 12.7 show theoretical temperature dependences of lifetimes for radiative transitions for the n-type material ($N_D \simeq 10^{17}$ cm^{-3}) and the p-type material ($N_A = 6 \times 10^{17}$ cm^{-3}).

Table 12.2 (lines 5, 6, and 7) lists material parameters used for these calculations. Curves 4 through 6 in Figure 12.6 are computed for the case of shallow acceptors providing a 50 per cent compensation ($N_A = 0.5 N_D$). Curves 1 through 3 refer to an uncompensated n-type material. Curve 7 was computed on the assumption that donors are completely ionized and are not involved in radiative transitions. The course of curve 7 indicates a temperature dependence of probability for the band-to-band transitions $B_{BB} \sim T^{-3/2}$. In a model that takes into account the part played by donors, the lifetimes for band-to-band transitions (curves 2 and 5, Figure 12.6) change at a slightly lower rate and the lifetimes for donor-to-valence band transitions (curves 1 to 4)

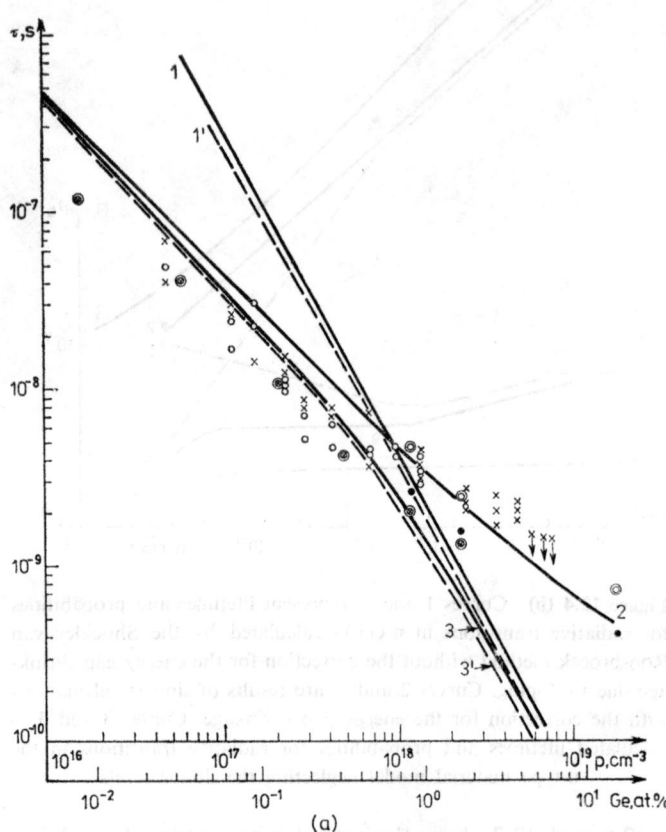

Figure 12.5 (a) Calculated and experimental results for the radiative lifetimes in the active region of the double-heterostructures as a function of germanium doping level. 1, calculated lifetimes for conduction band-to-acceptor transitions in the absence of compensation (τ_{BA}); 1′, the results of the calculations of similar values under the assumption of 20% material compensation ($N_D/N_A = 0.2$); 2, calculated values of band-to-band lifetimes (τ_{BB}); 3 and 3′, total electron radiative lifetimes in the absence of compensation and with the assumption that $N_D/N_A = 0.2$. Circles represent the values of the electron lifetimes determined from the experiments on structures with substrates, crosses—the results of the experiments on structures without substrates, double circles with crosses—the data of the calculation by the Shockley–van Roosbroeck method. The large light and dark circles in the range of $10^{18} < p < 10^{19}$ cm^{-3} are the results of calculations in (18)

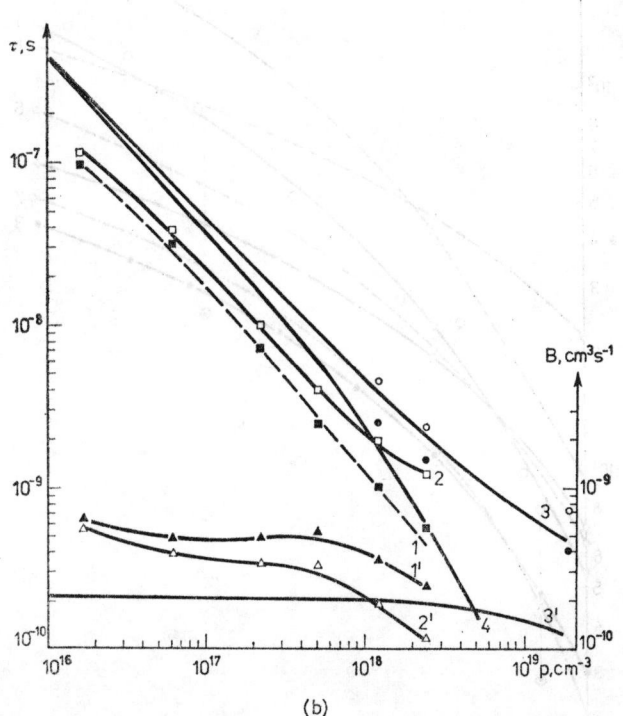

Figure 12.5 (b) Curves 1 and 1'—lifetimes and probabilities for radiative transitions in p-GaAs calculated by the Shockley–van Roosbroeck method without the correction for the energy band shrinkage due to doping. Curves 2 and 2'—the same values calculated with the correction for the energy gap shrinkage. Curves 3 and 3'—the results of quantum-mechanical calculations of lifetimes and probabilities for the band-to-band radiative transitions. 4—the result of the quantum-mechanical calculation of the total electron radiative lifetime in p-type material. Three pairs of light and dark circles in the concentration range $10^{18} < p < 10^{19}$ cm^{-3} represent the results of calculation of the electron lifetimes in Reference 18

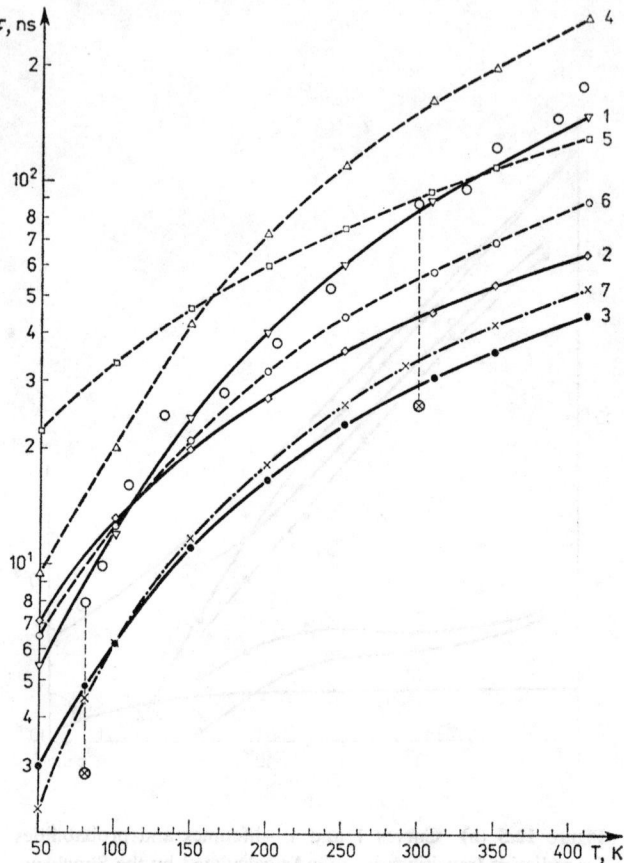

Figure 12.6 Experimental and theoretical data on temperature dependence of radiative lifetime for undoped n-type material ($N_D \simeq 10^{17}$ cm^{-3}). The circles are experimental data for hole lifetime in the active region of structure L1473. Curves 1–3 represent calculated results for uncompensated n-type material ($n = N_D = 10^{17}$ cm^{-3}). Curves 4–6 are computed for the case of 50% compensation ($N_A = 5 \times 10^{16}$ cm^{-3}, $N_D = 10^{17}$ cm^{-3}). Curves 1, 4—lifetime for donors-to-valence band transition; 2, 5—lifetime for band-to-band transition; 3, 6—total hole radiative lifetime (τ_h^t); 7—calculated results for total radiative lifetime for a model in which transitions involving donors are neglected. The crosses in circles represent values of the total hole lifetime determined from experimental data at 300 K and 77 K

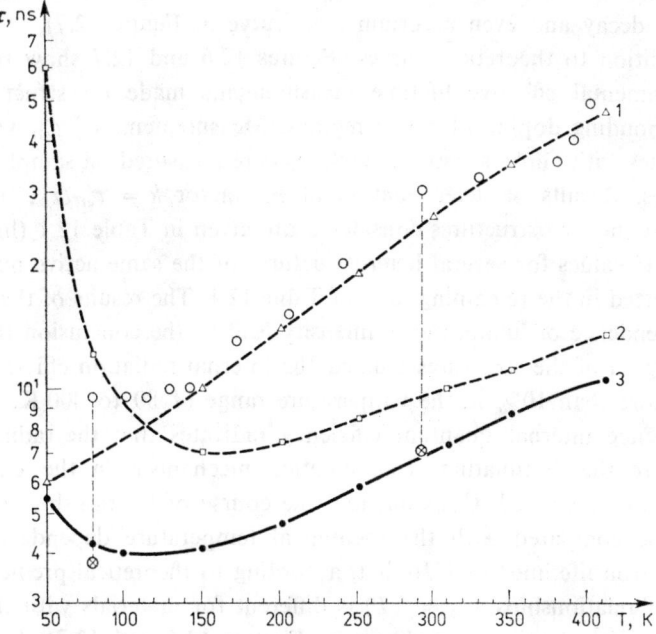

Figure 12.7 Experimental and theoretical data on the temperature dependence of radiative lifetime for a p-type material ($N_A = 6 \times 10^{17}$ cm^{-3}). The circles are the results measured for effective electron lifetime in the active region of the structure L 2072. The crosses in the circles represent calculated values of τ'_0 from experimental results. Curves 1, 2—calculated values of the lifetime for conduction band-to-acceptor and band-to-band transitions. Curve 3—calculation for total radiative lifetime (τ^t_0)

change at a higher rate than it would follow from the law of $T^{3/2}$ due to electron redistribution between the conduction band and the donor level.

In the case of the p-type material (Figure 12.7) the lifetimes for band-to-band radiative transitions become shorter with a decrease in the temperature from 400 K down to 150 K at a lower rate than according to the law of $T^{3/2}$. The theory predicts a sharp rise of τ^t_e due to localization of the holes on acceptor centres (curve 2) in the low temperature range ($T < 150$ K). As we have already noted, according to a simple quantum mechanical theory at low temperatures the conduction band-to-acceptor transition probabilities are lower than the band-to-band transition probabilities by one order of magnitude. That is why at low temperatures we must expect that the temperature dependence of total lifetimes for electron radiative transitions will exhibit a slower decay and even a certain rise (curve 3, Figure 12.7).

In addition to theoretical curves, Figures 12.6 and 12.7 show the results of experimental effective lifetime measurements made on structures with a corresponding doping of active regions. Measurements of τ_{eff} were made on samples with GaAs substrate, while η_i were measured on samples without substrates. Results of determination of η_i, factor $k = \tau_{\text{eff}}/\tau_0$, and τ_0 at 300 K for the two structures considered are given in Table 12.1 (lines 2 and 6); similar values for several heterostructures of the same active region doping are listed in the remaining lines of Table 12.1. The results of the temperature dependence of luminescence intensity lead to the conclusion that in the active region of the structures studied the internal radiation efficiency varies by no more than 10% in the temperature range of 80 to 400 K. The high luminescence internal quantum efficiency indicates that the radiative transitions are the dominating recombination mechanism in the entire temperature range studied. Consequently, the course of relationship $\tau_{\text{eff}} = f(T)$ should be compared with the theoretical temperature dependence of the total electron lifetime* (τ_e^t). In fact, according to theoretical predictions, the course of relationships $\tau_{\text{eff}} = f(T)$ is different for materials with the n-type and p-type conductions (as shown in Figures 12.6 and 12.7). In order to determine the net lifetimes for radiative transitions on the basis of experimental data and to compare them with results of theoretical computations it would be necessary to know the characteristics $\tau_{\text{eff}}/\tau_0 = f(\eta_i)$ for all temperatures at which measurements of τ_{eff} were made. As was mentioned in the *Introduction*, the characteristics $\tau_{\text{eff}}/\tau_0 = f(\eta_i)$ were computed at 300 K and 77 K only, and τ_0' can be defined only for these temperatures. The values of τ_0' at 300 K and 77 K, shown in Figures 12.6 and 12.7 are in sufficiently good agreement with the computed values of the total lifetimes for radiative transitions. Figures 12.6 and 12.7 show that the factor $k = \tau_{\text{eff}}/\tau_0$ decreases as the temperature drops due to self-absorption reduction. Because of this, the experimental values of τ_{eff} fall with a decrease in temperature at a slightly higher rate than it is theoretically predicted for total radiative transitions lifetimes.

The temperature dependence of the lifetime ratio for impurity and band-to-band transitions should lead to a change in the relative intensity of band-to-band and acceptor radiation peaks. Appropriate experimental and theoretical

---

* Agreement between experimental data shown in Figures 12.6 and 12.7 and theoretical curves referring to other transition types is quite accidental.

Quantum Efficiency and Radiative Lifetimes in GaAs and AlGaAs

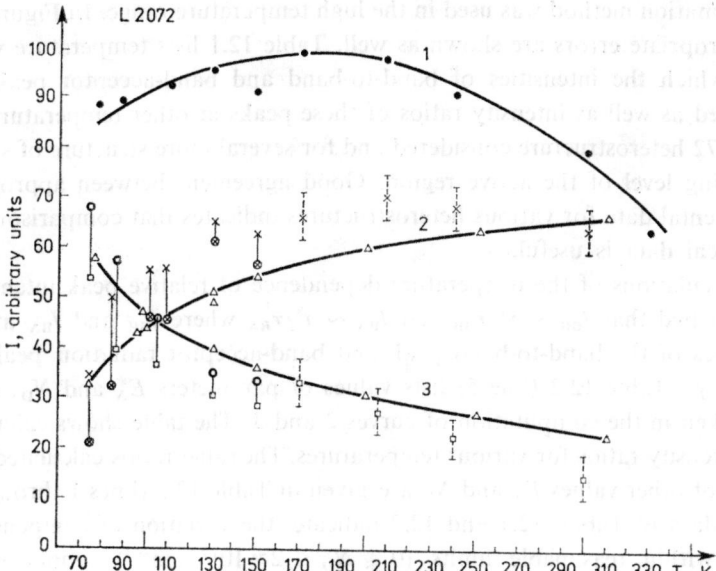

Figure 12.8 Experimental and theoretical data for the temperature dependence of luminescence intensity in p-type materials ($N_A = 6 \times 10^{17}$ cm^{-3}). The experimental results for the total intensity of the edge emission are shown by curve 1. The intensities for the band-to-band transitions are indicated by crosses and by crosses in circles. The squares (and squares in circles) correspond to the intensity of the band-to-acceptor transitions. The interlinked pairs of experimental points correspond to two ways of approximation used to distinguish the intensity of each of the edge emission components. Approximation errors for the high temperature region are shown by a dashed line. Curves 2 and 3—the calculated result for the temperature dependences of the intensity of each of the luminescence components obtained on the assumption that $E_A = 23$ meV and $N_D = 10^{17}$ cm^{-3}

data are compared in Figure 12.8. The experimental values were obtained by processing the spectra shown in Figure 12.1(b). Approximations of two types, shown in Figure 12.2(b) (1), were used to distinguish the peaks and to re-establish the true shape of the band-to-band and band-acceptor peaks. The band-to-band peak shape was used as the initial one for the first approximation method, and the impurity peak shape was used as the initial one for the second approximation method. In Figure 12.8 the results of computation of integral intensity for each peak as obtained with the two approximation methods are marked with dots and interlinked. Only the first approximation method was used in the high temperature range. In Figure 12.8 the appropriate errors are shown as well. Table 12.1 lists temperature values T_0 at which the intensities of band-to-band and band-acceptor peaks are compared as well as intensity ratios of these peaks at other temperatures for the L2072 heterostructure considered and for several more structure of similar Ge-doping level of the active region. Good agreement between appropriate experimental data for various heterostructures indicates that comparison with theoretical data is useful.

In calculations of the temperature dependence of relative peak intensity it was assumed that $I_{BB} \sim \tau^t/\tau_{BB}$ and $I_{BA} \sim \tau^t/\tau_{BA}$ where I_{BB} and I_{BA} are the intensities of the band-to-band peak and band-acceptor radiation peak, respectively. Table 12.2 (line 5) lists values of parameters E_A^* and N_D, which were taken in the computation of curves 2 and 3. The table shows calculated peak intensity ratios for various temperatures. The same ratios calculated with the use of other values E_A^* and N_D are given in Table 12.2 (lines 1 through 4). As the data of Tables 12.1 and 12.2 indicate, the variation of compensation degree within reasonable limits ($0 \leqslant N_D \leqslant 2 \times 10^{17}$ cm^{-3}) cannot ensure agreement between calculated and experimental data for values E_A^* other than the selected value 23 meV. Values $E_A^* = 23$ meV and $N_D = 10^{17}$ cm^{-3} providing the best fit of theoretical and experimental results (curves 2 and 3, Figure 12.8 are in good agreement with values E_A^* given in Table 12.4, as well as with results of concentration measurement of uncontrollable donor impurities in undoped n-type layers).

In conclusion of this part of the chapter it should be noted that theoretical calculations based on assumed momentum conservation at radiative transitions between parabolic bands are in good agreement with the results of the study of the temperature dependence of the luminescence intensity and kinetics in materials with doping level in the range $10^{17} < n, p < 6 \times 10^{17}$ cm^{-3}. It is obvious that the radiative transition model which does not take into account the wave-vector selection rules and in which the band-to-band transition probability is in general independent of the temperature, could not explain the observed experimental relationships.

12.4 DEPENDENCES OF RADIATIVE TRANSITION LIFETIMES ON THE ACTIVE REGION DOPING LEVEL

Results of determination of lifetimes for radiative transitions in the active region for double heterostructures with various donor (Te) and acceptor (Ge) concentrations are shown by crosses and circles in Figures 12.4(a) and 12.5(a).

The data were determined by measuring τ_{eff} for samples with or without substrates, then by determining total lifetimes τ_0 by using calibration curves $\tau_{\text{eff}}/\tau_0 = f(\eta_i)$, and by subsequent computation of lifetimes for radiative transitions from the relationship $\tau'_0 = \tau_0/\eta_i$. In the high doping level region ($n, p > 10^{18}$ cm^{-3}) the measured values τ_{eff} and the values of τ'_0 computed from them may be overestimated owing to equipment time contribution (arrows in Figure 12.5(a)). At low doping concentrations ($n, p < 10^{17}$ cm^{-3}) the main measurement error is caused by inaccuracy in determination of free-carrier concentration due to the effect of uncontrollable residual impurities.

In addition to the experimental data, Figures 12.4(b) and 12.5(b) show theoretical doping concentration dependences of lifetimes for impurity radiative transitions (curve 1 in Figures 12.4(a) and 12.5(a)) and total lifetimes for radiative transitions τ^t (curves 3 in Figures 12.5(a), 12.5(b), and 12.4(b), curves 2 and 3 in Figure 12.4(a)). In the case of n-type material, curve 2 in Figure 12.4(a) is related to the model of a semiconductor with a discrete donor concentration, while curves 3 in the same figure and in Figure 12.4(b) refer to the computation of τ_h^t for the case where donor levels are neglected.

In spite of the assumptions made in the theoretical computations and a complex procedure used for determination of τ'_0 from the experimental values of τ_{eff}, the data of Figure 12.4(a) demonstrate a good agreement of experimental and theoretical results obtained for radiative lifetimes. However, to be absolutely sure that this agreement is not accidental and that the respective lifetime values are true, we thought it necessary to compute radiative lifetimes by another method; the Shockley–van Roosbroeck method was chosen for this purpose. The following expressions can be derived for hole radiative lifetime in the n-type material (τ_h^{sh}) and for electron radiative lifetime in the p-type material (τ_e^{sh});

$$\tau_h^{\text{sh}} = \frac{N_v}{R_0} \exp\left(\frac{-E_g - F_n}{kT}\right); \quad \tau_e^{\text{sh}} = \frac{N_c}{R_0} \exp\left(\frac{-E_g - F_p}{kT}\right) \quad (12.7)$$

where F_n and F_p are Fermi levels for the majority carriers in the material under investigation.

Radiative transition rate R_0 is given by the well-known expression (15)

$$R_0 = \int_0^\infty \frac{8\pi^2}{c^2} n_r^2 \nu^2 k(h\nu) \exp\left(-\frac{h\nu}{kT}\right) d\nu \quad (12.8)$$

The lifetimes τ_h^{sh} and τ_e^{sh} which are computed from Equation 12.7 are integral quantities taking into account the contribution of both the band-to-band transitions and the shallow impurity transitions. If we assume that at 300 K the band-to-band and impurity transition probabilities are approximately the same (Figure 12.2), the respective values of B_n^{sh} and B_p^{sh} for the n-type and p-type materials can be defined as follows:

$$B_n^{sh} = \frac{\tau_h^{sh}}{N_D}; \quad B_p^{sh} = \frac{\tau_e^{sh}}{N_A} \qquad (12.9)$$

For non-degenerate semiconductors with a non-high doping level where band-to-band transitions are predominant, Equation 12.9 is transformed into a simple expression which defines the band-to-band transition probability:

$$B_{BB} = \frac{R_0}{(n_i)^2} \qquad (12.10)$$

(n_i is the intrinsic carrier concentration in the material under investigation).

To compute the lifetimes and transition probabilities for a material with a different doping level, using Equations 12.7 and 12.9, one should generally know the state densities in the bands, the position of Fermi levels for the majority carriers, the energy gap and the spectral dependence of the absorption coefficient.

Figures 12.4(b) and 12.5(b) show the results of computer calculations for τ^{sh} and B^{sh} for a number of n-type and p-type GaAs-samples with the 300 K spectral dependence of the absorption coefficient given in (5, 6, 7). Curves taken from these studies were photographically enlarged, tabulated and introduced into a computer. The calculation program did not consider the variation of state density in the bands due to doping; F_n and F_p were calculated using the usual method with the values of effective masses listed in Table 12.3. The calculation results were most strongly affected by the selection of the value for E_g.

The results of similar calculations performed by Varshni (15) for a high-resistive GaAs-sample whose spectral dependence of absorption coefficient was studied by Sturge (5) are given in Table 12.5. In calculations of τ^{sh} and B^{sh} we used recent values for the GaAs-energy gap and the hole effective mass: $E_g = 1.424$ eV, $m_{hh} = 0.5\, m_0$ (see Table 12.3). Nevertheless, due to a mutually compensating effect of variations in the values of E_g and m_{hh} the results of our probability calculations for the sample studied by Sturge closely agree with Varshni's data (Table 12.5). The respective values $B^{sh} \simeq 7.5 \times \times 10^{-10}$ cm^{-3}s^{-1} as well as the results of our transition probability calculations for other slightly doped samples sharply differ from the data obtained by Casey (16) for the n-type GaAs ($N_D = 10^{16}$ cm^{-3}) which was the material for the active region of the double heterostructure (Table 12.5). This difference

Table 12.5 Parameters used for calculations of the radiative transition probabilities by the Shockley–van Roosbroeck method and their results for undoped n-GaAs

Reference	E_g, eV	$\dfrac{m_e}{m_0}$	$\dfrac{m_{hh}}{m_0}$	$\dfrac{m_{lh}}{m_0}$	$\dfrac{m_v}{m_0}$	n_i, cm^{-3}	R_0, cm^{-3} s^{-1}	B, cm^3 s^{-1}
Sturge–Varshni (15)	1.435 (294 K)	0.07	0.68	0.12	0.71	1.29×10^6	1.2×10^3	7.2×10^{-10}
Sell–Casey (16)	1.424 (297 K)	0.066			0.5	1.8×10^6	4.5×10^3	1.4×10^{-9}
Our estimations on Sturge's edge absorption data (5)	1.425 (294 K)	0.07	0.5	0.09	0.5	1.235×10^6	1.16×10^3	7.67×10^{-10}

is due probably to the absorption edge distortion caused by mechanical stresses in the investigated material in the latter case.

In the first version of the calculations for τ^{sh} and B^{sh} whose results are represented by curves 1 and 1' in Figures 12.4(b) and 12.5(b) the energy gap of GaAs was assumed to be independent of the doping level. The accelerated fall of τ_h^{sh} and the increase in B_h^{sh} with enhanced donor doping in the region of $n > 10^{17}$ cm^{-3} (Figure 12.4(b)) which are difficult to explain in terms of physical causes indicate that it is necessary to take into consideration the effect of doping on the energy gap. Three effects were taken into account when the energy gap shrinkage in the n-type GaAs was calculated: (1) electron exchange interaction (μ_{ee}); (2) lowering of electron Fermi level due to conduction band tailing (μ_{ei}); (3) shift of Fermi quasi-level (ΔF_p) for minority holes due to the valence band tailing. The equations for μ_{ee} and μ_{ei} used

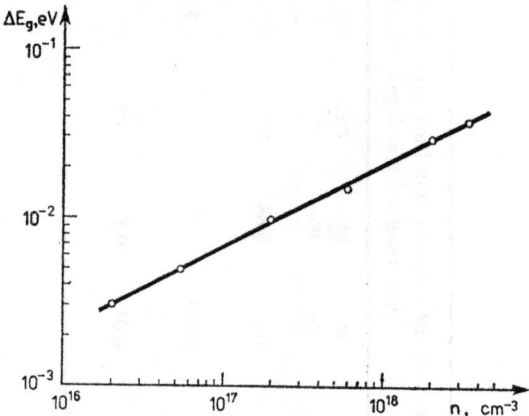

Figure 12.9 Calculated energy gap shrinkage as a function of donor concentration in n-GaAs

in these calculations may be found in (17). The shift of hole Fermi quasi-level was calculated from formula:

$$\Delta F_p = \gamma^2/4kT \tag{12.11}$$

where γ is the potential produced by the root-mean-square fluctuation of charged donor concentration in a volume with the typical size of about the screening radius r_s. Quantities γ and r_s depend upon the donor concentration; the expression used to calculate them is given in the above mentioned review (17). Figure 12.9 shows the values of $\Delta E_g = \mu_{ei} + \mu_{ee} + \Delta F_p$ calculated for different donor concentrations. In the case of the p-type conduction we used the energy gap shrinkage data from (18) (Fig. 9). Figures 12.4(b) and 12.5(b) show the values of lifetimes and transition probabilities calculated by the Shockley–van–Roosbroeck method, using correction coefficients $e^{\Delta E_g/kT}$, for the p-type and n-type GaAs samples with various doping level (curves 2 and 2').

Let us compare the values of radiative lifetimes as obtained by various methods. Keeping in mind the accuracy of theoretical calculations and experimental measurements, it is probably true to say that for the free-carrier concentration region of $(0.5–1) \times 10^{17}$ to $(0.5–1) \times 10^{18}$ cm^{-3} the experimental values of τ'_0 and lifetimes determined by the S-V-R method satisfactorily agree with the results of quantum mechanical calculation of total radiative lifetimes (curves 3 in Figures 12.4(b) and 12.5(b)). The respective quantities τ have the same value (in the range 5×10^{-8} to 5×10^{-9} s) for both the p-type material and the n-type material and, in general, appear to be defined by the band-to-band radiative transitions. This conclusion is confirmed by the results of quantum mechanical calculations which show that at $N_D, N_A < 5 \times 10^{17}$ cm^{-3} the lifetimes for transitions involving shallow impurities are longer than the lifetimes for band-to-band transitions (Figures 12.4(a) and 12.5(a)); the band-to-band radiative transition probability is $(3–4) \times 10^{-10}$ cm^{3}s^{-1}.

In the case of the p-type samples, experimental measurement data permit the conclusion that, at 300 K, band-to-band radiative transitions remain the predominant recombination mechanism in the region of $p > 5 \times 10^{17}$ cm^{-3} as well. In this respect there is a discrepancy between the experimental data and the simple theory predicting a higher transition rate for transitions involving acceptors as compared to band-to-band transition rate in the corresponding concentration range. This discrepancy might be caused by the fact that the overlap of wave functions of individual acceptor states which is important at $p > 5 \times 10^{17}$ cm^{-3} is not taken into account in the simple theory at all. The experimental data illustrated in Figures 12.5(a), (b) seem to suggest that the calculated values of probabilities B_{BA} are over-estimated for the region of $p > 5 \times 10^{17}$ cm^{-3}.

The results of radiative lifetime calculation by the S-V-R method give slightly lower values than the values obtained on the basis of luminescence measurements for the considered concentration range of $p \geqslant 10^{18}$ cm^{-3}. Nevertheless, in this concentration range the values of τ_e^{sh} are also larger than the theoretical lifetimes τ_e^t. The results of the calculations of τ_e^{sh} made by Casey and Stern (18) using the S-V-R method for radiative lifetimes in the p-type GaAs are given in Figure 12.5(b) by dark circles along with our own data for τ_e^{sh}.* Double light circles are used in Figure 12.5(a) to show the results of theoretical calculations for τ made by the same authors for the model of a strongly doped p-type material taking into account the band tailing. The complete agreement of respective calculated points with our experimental data and the results of calculation of τ_{BB} for band-to-band transitions in the simplest model may not be accidental at all; it may indicate that, regardless of the band tailing, in a strongly doped material at 300 K the radiative transition rate is defined by the recombination of electrons and holes whose wave functions do not differ considerably from the wave functions of carriers in undistorted bands. Casey and Stern (18) did not perform calculations for the strongly doped n-type material. Our experimental data (Figures 12.4(a), (b)) are also insufficient to permit definite conclusions on whether donor states are involved in radiative transitions in the n-type material; it may, however, be noted that the experimental values of τ_0' agree somewhat better with the values of τ_h^i calculated for the model neglecting donor states (curve 3 in Figures 12.4(a), (b)).

In the low range of majority carrier concentration $n, p < (0.5-1) \times 10^{17}$ cm^{-3}, the lifetimes for radiative transitions calculated by the S-V-R method increase at a lower rate than would follow from the relationship $\tau \sim n^{-1}$. Accordingly, the band-to-band radiative transition probability increases and reaches a value of 7.7×10^{-10} cm^{3}s^{-1} (Figure 12.4(b)) for the high-resistive sample studied by Sturge ($n \simeq 10^{10}$ cm^{-3}). Let us show that the observed increase of transition probability in samples with low free carrier concentrations (curve 2 in Figures 12.4 (b) and 12.5(b)) may be explained by an enhancement of the role of the Coulomb carrier interaction in the recombination or absorption event. It is obvious that since at room temperature $E_{ex} \simeq 0.1kT$, the integral intensity of radiative transitions can be mainly effected only by continuum exciton states† or, in other words, by the Coulomb interaction of free carriers. In this case, according to Elliott's calculations (19), the spectral

---

* The same experimental curves $k(hv)$ as in our calculations were used to compute transition lifetimes by the Shockley–van Roosbroeck method for samples with $p = 1.2 \times 10^{18}$ cm^{-3} and $p = 2.3 \times 10^{18}$ cm^{-3} in (18). However, in contrast to us, the authors of (18) used the theoretical value $n_0 p_0$ obtained on the basis of the band tailing model adopted by them for the strongly-doped p-type material.

† The maximum contribution which the region $k(hv)$ corresponding to continuum exciton states makes to the radiation spectrum is estimated to be no more than 20% of the total edge peak intensity.

dependence of absorption coefficient is given by the following expression

$$k(hv) = \frac{2\pi A(E_{ex})^{1/2}}{1-\exp[-2\pi\sqrt{E_{ex}/(hv-E_g)}]} \quad (12.12)$$

where E_{ex} is the energy of the ground exciton state and A is a constant which determines the value of the absorption coefficient when no Coulomb interaction takes place. At $hv - E_g > 50 E_{ex}$ the Coulomb interaction is unimportant, and Equation 12.12 is transformed into the conventional root dependency:

$$k = A(hv - E_g)^{1/2}. \quad (12.13)$$

According to Sturge, Equation 12.12 describes well the spectral dependency of absorption coefficient in the high-resistive GaAs at $hv > E_g$.

Taking from the literature that in GaAs at 300 K E_{ex} is 3.8 meV, the lifetimes for radiative transitions for the theoretical absorption spectra described by Equations 12.12 and 12.13 were calculated by the Shockley–van Roosbroeck method, and it was found that, when the Coulomb interaction is taken into account, τ^{sh} is lower by a factor of 3.6 than in the case where the Coulomb interaction is not present. The calculated variation of τ^{sh} agrees satisfactorily with our data on the transition probability increase when the screening-carrier concentration is reduced (Figure 12.4(b)).

The fact that the assumption of screening of the Coulomb interaction for recombining carriers at $n, p > 10^{17}$ cm^{-3} is correct is additionally evidenced by the plots of Figures 12.10(a), (b). The absorption spectrum of pure sample from (6) was taken as an initial one for plotting absorption spectra 2 and 2' since the exciton effects are most evident for this pure sample spectrum. Following that the experimental values of $k(hv)$ were multiplied by the theoretical factor S indicating the absorption coefficient reduction when the Coulomb interaction is completely screened:

$$S = \frac{(hv-E_g)^{1/2} 2\pi (E_{ex})^{1/2}}{1-\exp[-2\pi\sqrt{E_{ex}/(hv-E_g)}]} \quad (12.14)$$

Curve 3 in Figure 12.10(a) is an absorption spectrum measured experimentally for a sample with $n = 2 \times 10^{17}$ cm^{-3}. The radiation spectra plotted on the basis of curves 3 and 2' with the use of the Shockley–van Roosbroeck relationship are illustrated in Figure 12.10(b). Their areas differ by less than 10%. Thus, the result obtained shows that at $n = 2 \times 10^{17}$ cm^{-3} in the energy range $E_g + 2kT$ which is typical for recombining carriers their Coulomb interaction is almost completely screened. This result agrees well with the theoretical estimates and with the known experimental data for germanium

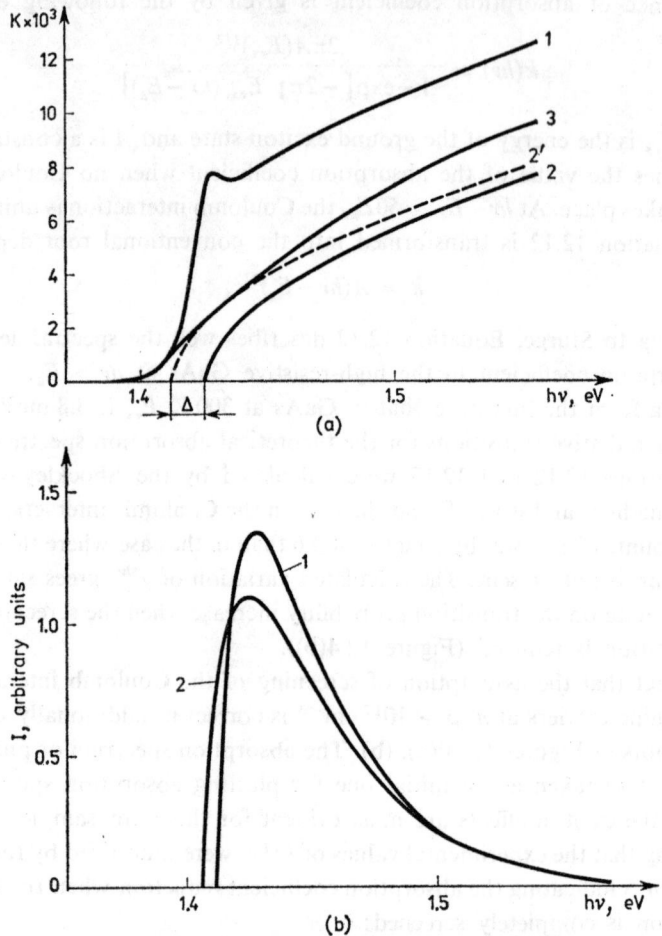

Figure 12.10 Absorption coefficients and the luminescence spectra calculated by the Shockley–van Roosbroeck method

(a) 1, spectral dependence of the absorption coefficient for the purest n-GaAs samples ($n = 5 \cdot 10^{13}$ cm^{-3}) from (6);

2, spectral dependence of the absorption coefficient obtained from curve 1 multiplied by S (see Equation 12.13);

2', spectrum 2 shifted along the energy axis;

3, absorption edge of n-GaAs with $n = 2 \times 10^{17}$ cm^{-3} (6).

(b) 1, Luminescence spectrum calculated with the use of the absorption coefficient represented by curve 2';

2, calculated luminescence spectrum for n-GaAs with $n = 2 \times 10^{17}$ cm^{-3}.

(20). The absorption edge caused by direct transitions in germanium approaches the 'root' shape at $n, p = 10^{17}$ cm^{-3} (20).

Thus, to sum up the results concerning the relationship between the lifetimes for recombination transitions at 300 K in a direct-band semiconductor and the doping level, the following may be stated.

(1) The current level of technology allows structures to be made for which radiative transitions are a predominant recombination mechanism over a wide range of active region doping levels.

(2) As carrier concentration increases from 10^{16} to 10^{17} cm^{-3}, the recombination of electron-hole pairs with the Coulomb interaction changes to the recombination of completely free carriers. In this concentration range the radiative transition probability is not constant, but decreases with the growth of carrier concentration.

(3) In the concentration range of $(0.5–1) \times 10^{17}$ to $(0.5–1) \times 10^{18}$ cm^{-3}, radiative transitions in both the n-type and the p-type material are caused mostly by band-to-band transitions whose probability is $(3–4) \times 10^{-10}$ cm^{3}s^{-1} for GaAs.

(4) The experimental data for doping levels of about 10^{18} cm^{-3} obtained only for the p-type material indicate a reduction of probabilities of quasi-band-to-band transitions at high doping levels and a respective slowing of the fall of τ with an increase of p. This fact may be explained both in terms of the simple theory taking into account the occurrence of degeneracy in the valence band and by more complex calculations considering the band structure distortion due to the fluctuation of concentration of charged impurities (18).

12.5 EFFECT OF DOPING ON INTERNAL RADIATIVE TRANSITION EFFICIENCY

The second section of this chapter dealt with measurement procedures which made it possible to determine the internal quantum efficiency of radiative recombination in the active region of double heterostructures in a wide range of excitation levels ($I \leqslant 5 \times 10^4$ Wcm^{-2}). As was shown by respective studies of structures with Ge-doped active region ($5 \times 10^{16} < p < 10^{19}$ cm^{-3}), the values of η_i are practically independent of the excitation level at $I > 10^2$ Wcm^{-2}. The values of η_i found at excitation level $I = 1.5 \times 10^2$ Wcm^{-2} for structures with different hole concentration in the active region are rep-

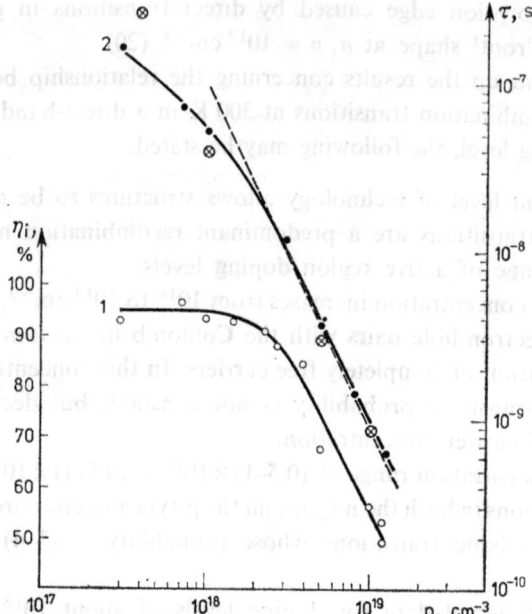

Figure 12.11 Internal quantum efficiency (curve 1) and non-radiative lifetimes (curve 2) in p-materials. Non-radiative lifetimes calculated from Equation 12.15. The Auger lifetimes for transitions involving neutral acceptor are calculated from Equation 12.19 and marked with crosses

resented by curve 1 in Figure 12.11. In the case of structures with the active region strongly doped with tellurium ($n > 10^{18}$ cm^{-3}), a slight increase of η_i takes place even at the maximum excitation level $I \simeq 5 \times 10^4$ Wcm^{-2}. Nevertheless, we shall assume that curve 1 in Figure 12.12 is referred to $I = 3.5 \times 10^4$ Wcm^{-2} and that it represents variation in the internal radiative transition efficiency as a function of the material donor concentration level. Curves 1 in Figures 12.11 and 12.12 show that the values of η_i are almost independent on the material doping level and amount to 92–95% at low doping levels ($n, p < 10^{18}$ cm^{-3}). As donor or acceptor concentration rises, the internal quantum efficiency of radiative recombination decreases. In the case of the n-type material, the fall of η_i begins at lower doping levels and proceeds at a higher rate than in the case of the p-type material. Lifetimes for non-radiative transitions can be computed for the n-type material at a given excitation level with the aid of experimental and theoretical values

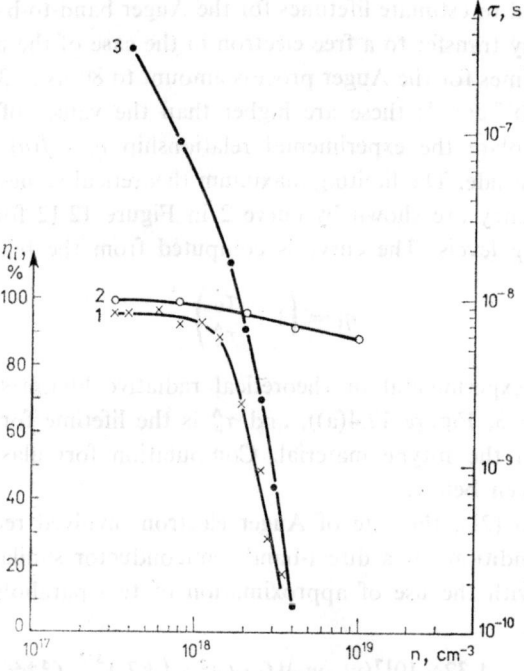

Figure 12.12 Internal quantum efficiencies and non-radiative lifetimes in *n*-materials. Curve 1 represents experimentally obtained η_i; curve 2 maximum values of the internal quantum efficiencies in n-materials, calculated from Equation 12.16; it was assumed that the non-radiative recombination is caused by the band-to-band Auger recombination (Equations 12.17 and 12.18), Curve 3 represents non-radiative lifetimes in *n*-materials, calculated from Equation 12.15 on the basis of experimental data

of lifetimes for radiative transitions in the n-type material (curve 3, Figure 12.4(a)) and of the values η_i (curve 1, Figure 12.12) from the relationship:

$$\tau_{nr} = \frac{\eta_i}{1-\eta_i}\tau_0' \tag{12.15}$$

Results of the computation are represented by curve 3 in Figure 12.12.

It is of interest to compare the lifetimes for non-radiative recombination found in such a way with the computed lifetimes τ_n^A for the Auger recombination in the n-type material. The results obtained by Landsberg and Beattie

(21) can be used to estimate lifetimes for the Auger band-to-band recombination with energy transfer to a free electron in the case of the n-type material. Estimated lifetimes for the Auger process amount to 80 ns at 300 K for GaAs with $n = 2 \times 10^{18}$ cm^{-3}; these are higher than the values of τ_{nr} which are required to explain the experimental relationship $\eta_i = f(n)$ by nearly two orders of magnitude. The limiting maximum theoretical values of the internal quantum efficiency are shown by curve 2 in Figure 12.12 for n-GaAs with different doping levels. The curve is computed from the relationship

$$\eta_i = \left(1 + \frac{\tau_0'}{\tau_n^A}\right)^{-1} \tag{12.16}$$

where τ_0' are experimental or theoretical radiative lifetimes in the n-type material (curve 3, Figure 12.4(a)), and τ_n^A is the lifetime for the Auger recombination in the n-type material. Computation formulas for the latter lifetime are given below.

According to (21), the rate of Auger electron involved recombination at equilibrium condition for a direct-band semiconductor similar to GaAs can be expressed with the use of approximation of two parabolic zones in the following way:

$$G_{ee} = \frac{1.32 \times 10^{17}(m_e/m_0)|f_1 \cdot f_2|^2}{\varepsilon^2(1+\mu)^{1/2}(1+2\mu)} n\left(\frac{kT}{E_g}\right)^{3/2} e^{\left(\frac{1+2\mu}{1+\mu}\right)\frac{E_g}{kT}} \tag{12.17}$$

where G_{ee} is the number of recombination transitions per cm^3 and s, $\mu = m_e/m_v = 0.14$ for conventional values of effective masses in GaAs ($m_e = 0.07 m_0$ and $m_v = 0.5 m_0$); $|f_1 \cdot f_2|^2$ is the overlapping integral assumed to be equal to 0.25.

Using Equation 12.17 and substituting numerical values of the parameters, one can obtain in the case of n-type GaAs the following formula for lifetimes of the Auger recombination involving a free electron

$$\tau_n^A = \frac{n_i^2}{G_{ee}n} = \frac{0.13}{n^2} T^{3/2} \exp\left(\frac{1980}{T}\right) 10^{24} \tag{12.18}$$

If the temperature is substituted in Kelvin degrees, (12.18) gives lifetimes for the Auger recombination in seconds. Note that the Landsberg–Beattie model corrections taking into account the degeneracy of electron gas and the departure of the Γ_1-minimum of conduction band from the parabolic shape in GaAs can give only an increase in lifetimes for the Auger recombination. As to the Auger processes involving shallow impurity, they appear not to play any important role in the n-type material since all donors are ionized.

Thus, one can conclude that the fall of the internal quantum efficiency of radiative recombination in the n-type GaAs with Te-doping increase is caused not by fundamental processes of the Auger recombination type, but by recombination processes involving deep levels. Complexes V_{Ga}–Te whose concentration in the n-type material can considerably exceed the shallow donor concentration at $n > 10^{18}$ cm^{-3} may become centres of non-radiative recombination. In the n-type GaAs the concentration ratio between tellurium atoms included in the complexes and those playing the role of shallow donors depends upon the conditions of material growth. The assumption of the predominant role of the recombination involving deep centres in a strongly doped n-type material is also supported by the increase of internal quantum efficiency of radiative recombination observed in such a material at high excitation levels. Further investigations are required to establish the particular mechanism of carrier energy dissipation in the recombination event involving deep centres in the n-type material.

Let us consider the dependence of η_i on the acceptor doping level of the p-type material (Figure 12.11). Curve 2 of Figure 12.11 represents the dependence of lifetimes for non-radiative transitions on hole concentration; this curve is plotted in the same way as curve 3 of Figure 12.12 for the n-type material. It is evident from the comparison of Figures 12.11 and 12.12 that lifetimes for non-radiative processes which can explain the experimental relationship $\eta_i = f(p)$ are larger in the p-type material than in the n-type material. Changes in values of τ_{nr} in the range of $p > 2 \times 10^{18}$ cm^{-3} can be approximated by the relationship $\tau_{nr} \sim p^{-2}$ (dotted straight line in Figure 12.11).

Let us try to compare lifetimes τ_{nr} with calculated data for the Auger process involving holes in the p-type material in the same way as for the n-type material. Consider first the data on the Auger 'common' band-to-band recombination in which the energy of the free electron-hole pair is transferred to a second hole. Several studies (22, 23, 24) have been devoted to calculations of the Auger band-to-band recombination in the p-type GaAs. Reference 22 neglects the complex structure of the valence band and uses the simplest Landsberg–Beattie approach leading to a formula for the Auger recombination rate, identical to Equation 12.17 as to its form for the calculation of τ_p^A. However, the quantity μ' is now a reciprocal of the quantity which has been used in the description of the Auger electron process. The physical sense of this circumstance lies in the fact that due to the high effective mass of holes the requirement of momentum conservation lowers sharply the probability for the Auger process and respective lifetimes should be longer in the p-type material than in the n-type material by a factor of about $e^{E_g/kT}$ within the scope of this model. Note that the author of paper 22 made an error in designations of quantity μ, and the calculated values of μ' obtained in his work are lower than the true ones by a factor $e^{E_g/kT}$.

The effect of the light and split valence bands, which was taken into account in References 23, 24, leads, as it must, to a decrease in values of τ_p^A as compared with the results of calculation using Equation 12.17. Nevertheless, the values of τ_p^A predicted by the calculations are sufficiently high and the Auger band-to-band process appears to be able to compete with the radiative recombination processes only at $p > 10^{20}$ cm^{-3} (23). In the range of concentration $p < 10^{19}$ cm^{-3} investigated by us, the theoretical values of τ_p^A are longer than the values of τ_{nr} given in Figure 12.12 by three to four orders of magnitude. For example, according to one of the recent studies (24), τ_p^A is approximately 2×10^{-4} s at $p = 2 \times 10^{18}$ cm^{-3} and $T = 300$ K. It is difficult to expect that further refinements of the theoretical calculations for the p-type material (of the kind made recently for InAs (25)) will eliminate the difference in values of τ_p^A and τ_{nr}.

In addition to the Auger band-to-band recombination in the p-type material as distinct from the n-type material, one may consider the 'Auger process involving impurity' in which an electron from the conduction band recombines with a hole on a neutral acceptor with energy transfer to a free hole in the valence band. The probability for such a process in the direct-band material is calculated in (26). The formula for calculating the lifetimes for the Auger transitions involving acceptor centre with a Coulomb-like potential is given in the study mentioned above (22):

$$\tau_p^{AA} = \frac{2.9 \times 10^{16} \varepsilon^{3/2} m_v kT(E_A)^{3/2}}{N_A^0 p^{1/2} m_0 (E_g - E_A)^{1/2}} \left[1 + 0.9 \frac{m_v}{m_0} \frac{E_g - E_A}{E_A} \right]^4 \quad (12.19)$$

In (12.19) the designations are changed in accordance with those used here: E_A is the acceptor ionization energy, and N_A^0 the number of neutral acceptors, and so on. If kT, E_A and $E_g - E_A$ in (12.19) are expressed in electron-volts, then τ_p^{AA} will be in seconds.

The calculations of τ_p^{AA} with the use of Equation 12.19 were made with two types of assumptions as to the acceptor ionization energy. In the first case it was assumed that quantity E_A included explicitly into Equation 12.19 does not depend upon the acceptor concentration and is 40 meV as in the calculations of radiative transitions. The values of N_A^0 and p were calculated with the aid of Equation 12.4 taking into account the reduction of acceptor activation energy with the increase of N_A. The values of τ_p^{AA} in the range of

$p > 10^{18}$ cm^{-3} which were obtained in this case appeared to be lower than the values of τ_{nr} by one order of magnitude; they are not shown in Figure 12.11.

In the second case, it was assumed that quantities E_A included explicitly into (12.19) also vary with N_A (Table 12.4). The results of calculations of τ_p^{AA} for several hole concentrations are shown by crosses in circles in Figure 12.11. The agreement of the 'experimental' values of τ_{nr} and the calculated values of τ_p^{AA} is almost too good to be true. It is probable that more refined calculations of τ_p^{AA} taking into account, for example, the complex nature of wave functions for acceptor in $A^{III}B^V$, would lead to a worse agreement between the calculated and experimental data for non-radiative recombination lifetimes.

The investigation results of the temperature dependence of internal quantum efficiency of radiative recombination in the p-type material, which we will not consider in detail here, cannot also be explained by the effect of the Auger band-to-band process and do not contradict the assumption that the non-radiative mechanism in the p-type material is caused by the Auger transitions involving acceptors. The calculation using Equation 12.17 shows that in the case of the Auger band-to-band process for a material with $N_A = 2 \times 10^{18}$ cm^{-3} a drop of values of τ_{nr} should be expected with the temperature increase from 300 to 400 K by more than two orders of magnitude. The temperature investigations of luminescence intensity (Figure 12.8, curve 1) and lifetimes for transitions show that lifetimes for non-radiative transitions in the p-type material vary with the temperature increase at a considerably lower rate. Note that the temperature dependences of the Auger transitions involving impurities (see (12.19)) are much weaker than for the Auger band-to-band processes. Refinement of the experimental data and theoretical calculations for the temperature dependence of the Auger process involving impurities can apparently lead to a satisfactory agreement between theory and experiment.

12.6 CONCLUSION

The principal results of the investigations considered in this chapter are as follows:

(1) Measurements are made of efficiency and kinetics of luminescence in the range of 80 to 400 K for double AlGaAs-heterostructures with various levels of active region doping by donors and acceptors.

(2) Values of the internal efficiency and lifetimes for radiative transitions in GaAs and AlGaAs are calculated on the basis of these measurements, taking into account the self-absorption and photon recycling effects.
(3) The lifetime values obtained and the character of their temperature dependence for doping levels $n, p < 10^{18}$ cm^{-3} agree well with the results of lifetime calculations taking into account band-to-band transitions and transitions involving shallow acceptors. This result indicates that in theoretical calculations one can apply a model which takes account of radiative transitions with quasi-momentum conservation between parabolic bands or between a band and an acceptor centre with Coulomb-like potential.
(4) The comparison of experimental and theoretical data points to the necessity of taking into account Coulomb interactions of recombining carriers at $n, p < 5 \times 10^{16}$ cm^{-3} ($T = 300$ K).
(5) The determination of lifetimes and efficiency of radiative transitions in a strongly doped material ($n, p > 10^{18}$ cm^{-3}) makes it possible to calculate lifetimes of non-radiative recombination processes and to make assumptions as to their nature. Thus, for the p-type material the most probable mechanism of non-radiative transitions is the Auger recombination involving holes on neutral acceptors.

REFERENCES

1. D. Z. Garbuzov, in 'Semiconductor Sources of Electromagnetic Radiation', *Proceedings of the First International School on Semiconductor Optoelectronics, Cetniewo, Poland 1975*, p. 91, Polish Scientific Publishers, Warszawa, 1976.
2. V. B. Halfin, D. Z. Garbuzov, N. Yu. Daviduk, *Fiz. Tekhn. Poluprov.*, **10**, 1490 (1976).
3. Zh. I. Alferov, V. G. Agafonov, D. Z. Garbuzov, N. Yu. Daviduk, V. B. Halfin, *Fiz. Tekhn. Poluprov.*, **10**, 1497 (1976).
4. D. Z. Garbuzov, A. H. Ermakova, V. D. Rumyantsev, M. K. Trukan, V. B. Halfin, *Fiz. Tekhn. Poluprov.*, **11**, 717 (1977).
5. M. D. Sturge, *Phys. Rev.*, **127**, 768 (1962).
6. H. C. Casey, D. D. Sell, K. W. Wecht, *J. Appl. Phys.*, **46**, 250 (1975).
7. C. I. Hwang, *J. Appl. Phys.*, **40**, 3731 (1969).
8. D. E. Hill, *Phys. Rev.*, **133**, 866 (1964).
9. C. I. Hwang, J. C. Dyment, *J. Appl. Phys.*, **44**, 3240 (1973).
10. H. Kressel, F. Z. Hawrylo, M. S. Abrahams, and C. S. Buiocchi, *J. Appl. Phys.*, **39**, 5139 (1968).
11. P. D. Greene, *Sol. St. Commun.*, **9**, 1299 (1971).
12. H. B. Bebb, E. W. Williams, in *Semiconductors and Semimetals*, **8**, 181, Academic Press, 1972.
13. T. S. Moss, G. J. Burrell, B. Ellis, in *Semiconductor Opto-Electronics*, Butterworth, 1973.

14. F. E. Rosztoczy, F. Ermanis, I. Hayashi, B. Schwarz, *J. Appl. Phys.*, **41**, 264 (1970).
15. Y. P. Varshni, *Phys. Stat. Sol.*, **19**, 459, 20.9 (1967).
16. D. D. Sell, H. C. Casey, *J. Appl. Phys.*, **45**, 200 (1974).
17. A. L. Efros, *Usp. Fiz. Nauk*, **111**, 451 (1973).
18. H. C. Casey, F. Stern, *J. Appl. Phys.*, **47**, 631 (1976).
19. R. J. Elliott, *Phys. Rev.*, **108**, 1384 (1957).
20. A. A. Rogachev, in *Materials of IV Winter School on Semiconductor Physics, F.T.I. imeni A. F. Ioffe AN SSSR*, Leningrad, 1972.
21. A. R. Beattie, P. T. Landsberg, *Proc. Roy. Soc., Ser. A*, **249**, 16 (1959).
22. L. R. Weisberg, *J. Appl. Phys.*, **39**, 6096 (1968).
23. K. H. Zschauer, *Sol. St. Commun.*, **7**, 1709 (1969).
24. M. Takeshima, *J. Appl. Phys.*, **43**, 4114 (1972).
25. M. P. Mihajlova, A. A. Rogachev, I. N. Essievich, *Fiz. Tekhn. Poluprov.*, **10**, 1460 (1976).
26. L. Bess, *Phys. Rev.*, **10**, 1460 (1976).

Chapter 13
Electroluminescent Image Transformers and Amplifiers

S. V. Svechnikov

13.1 INTRODUCTION

Semiconductor optoelectronics opens up possibilities for light emission intensification and transformation at the level of elementary circuits (1, 2). Thus, the idea of creating optoelectronic image intensifiers and transformers (OET), i.e. solid state analogues of electron-optical transformers (EOT) arose. In comparison with the latter, OETs have the following advantages: (i) small dimensions and the absence of the vacuum; (ii) low supply voltage; (iii) interference stability, i.e. preservation of the efficiency after strong illumination at the input; (iv) uniform resolution over the working field of the device; (v) wide spectral range of working, i.e. from the infrared and visible up to X-ray radiation.

The interest in OETs is caused by the possibility of creating, on this basis, a new generation of optical image transformers: noctovision devices (image intensifiers–transformers); 'spectacles' for weak-eyed people (image intensifiers); high resolution X-ray screens (image transformers).

Each of the above applications poses its own special requirements with regard to the set of characteristics and parameters of OETs (intensification and transformation factors, threshold sensitivity, resolution, time of response, dynamic range, etc.), which may be based on different principles of transformation and amplification of radiation. To deal with them one can resort to the phenomena of luminescence (3–6), electroluminescence (7, 8), electrooptics (9), acousto- and magneto-optics (10–12).

If the base radiation source is not provided in the OET, then the use of the electroluminescence combined with the photoconductivity is the only way to obtain the light intensification factor

$$K_B = \frac{B_{out} - B_b}{L_{in}} \quad (13.1)$$

available at present, which exceeds significantly the unity (13).

In (13.1) B_{out}, L_{in}, B_b are the output, input and background brightness, respectively. This means that the intensifier structure corresponds to the elementary optron with an electro-optical coupling (Figure 13.1) (2). The

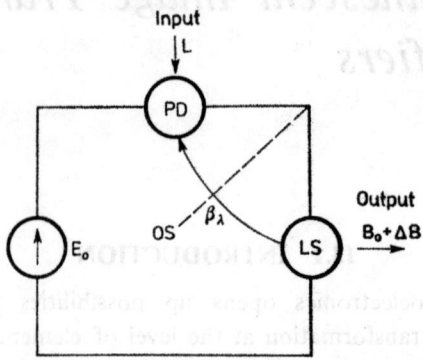

Figure 13.1 The optron with electro-optical coupling in the intensifier-transformer radiation regime; PD, photodetector; LS, light source; β_λ, optical feedback; OS, optical screen

monitored light source LS, i.e. the output element of the intensifier, is connected in series with the photodetector PD which plays the part of the optical input of the intensifier. Optical feedback β_λ is possible between LS and PD. This feedback will be positive if the spectral characteristics of LS and PD are correlated (monochromatic amplification). Its degree may be regulated by the electrically monitored transparent (ferroelectrical, electrochromic, liquid crystal) materials. If the spectral characteristic of the photoconductivity does not overlap with the spectral characteristic of the light source, then the transformation regime takes place (heterochromic amplification). Within the heterochromic amplification the optical feedback is absent, which excludes the necessity of using an optical screen making the device structure simpler.

The LS and PD in the diagram of Figure 13.1 form a compound potentiometer whose arms have to be strictly matched as to their electrical parameters: voltage, current, dissipated power. Since the quantum efficiencies of the internal photoeffect and the electroluminescence do not exceed unity, the condition $K_B \gg 1$ corresponds to the superlinear electro-optical characteristic of at least one of elements of the optronic pair.

Electroluminescent phosphor, for which

$$B \sim \exp\left(-\frac{b}{\sqrt{V}}\right) \quad \text{or} \quad B \sim V^{\gamma_e} \tag{13.2}$$

where b, γ_e are constants, V is the excitation voltage characterized by a highly non-linear brightness-voltage dependence. For powdered luminescent phosphors one has $\gamma_e = 4, ..., 6$, for film luminescent phosphors $\gamma_e = 10, ..., 19$. The latter are of the greatest interest as light sources for high resolution image intensifiers.

Light emitting diodes (LEDs) possess linear brightness-current characteristics ($B \sim i_{LS}$) in the range of current densities corresponding to the optical efficiency. In this case for the condition $K_B \geqslant 1$ to be fulfilled the intermediate amplification of the photocurrent should be realised with the gain

$$K_a \geqslant \frac{K_B}{\eta_p \cdot \eta_s} \tag{13.3}$$

where η_p, η_s are the efficiencies of the internal photoeffect and the electroluminescence, respectively, the order of magnitude being η_p, $\eta_s \sim 10^{-2}, ..., 10^{-3}$. Thus, it is necessary to have $K_a \geqslant 10^6$ to obtain $K_B = 10^4$. The use of an intermediate amplifier of electrical signals (integrated one) in light intensifiers, even with such a high gain, does not meet serious obstacles. In an image amplifier, where light is the main functional element and is integrated in it to a degree determined by the resolution required (25–40 lines/mm), intermediate amplification can only be built in. The efficiency of the spontaneous light source $K_B \geqslant 1$. Therefore, built-in amplification in an image intensifier circuit can take place only in the photodetector (photoresistor, phototransistor, avalanche photodiode, etc.). This photodetector has to allow integration with the light source. The pair chosen determines the maximum value of K_B which can be obtained. At small signal level in the linear approximation (the amplification regime) (14)

$$K_B = \frac{\Delta B_{out}}{L_{in}} = \frac{S_p \cdot S_{LS} \cdot \beta_\lambda Y_{11}}{Y_{11} \mp Y_{21} + Y_{22}} \tag{13.4}$$

where

$$\beta_\lambda = \frac{\Delta B_{out}}{B_0 + \Delta B_{out}}$$

S_p, S_{LS} are the integral sensitivity of the PD and the slope of the power-current characteristic of the light source, respectively. In this case

$$i_p = S_p L_{in}, \quad i_{LS} = \frac{Y_{11}}{Y_{11} \mp Y_{21} + Y_{22}} i_p$$
$$\Delta B_{out} = S_{LS} \beta_\lambda i_{LS} \tag{13.5}$$

$Y_{ij} = \partial i_i / \partial V_j$ are the elements of the conductivity matrix of the quadrupole (Figure 13.1):

$$Y_{11} = \frac{\partial i_{LS}}{\partial V_{LS}}; \quad Y_{22} = \frac{\partial i_p}{\partial V_p}; \quad Y_{21} = \frac{\partial i_p}{\partial V_{LS}} \qquad (13.6)$$

Expression 13.4 characterizes the most general case of the optical signal transformation in the OET when feedback is present. In this case the regenerative amplification regime takes place when

$$(Y_{11} - Y_{21} + Y_{22}) \to 0 \qquad (13.7)$$

The deep positive feedback in the OET essentially increases the amplification factor K_B, but narrows the dynamic range and increases the time of response. Because of this attempts to use the positive optical feedback in image intensifiers were not successful (15–16).

In image intensifiers with high resolution and contrast feedback is absent as a rule. If it does take place, its degree is insignificant. To ensure this a non-transparent optical screen (OS) is placed between the PD and LS. The conditions $Y_{21} = 0$, $\beta_\lambda = 1$ correspond to the absence of optical feedback. In this case according to (13.4)

$$K_B = K_{B0} = \frac{S_p S_{LS}}{1 + Y_{22}/Y_{11}} \qquad (13.8)$$

If the current-voltage characteristic of the PD is linear, as e.g. in the case of the photoresistor, then

$$\frac{Y_{22}}{Y_{11}} \sim \frac{Y_0}{\gamma Y_0} \sim \gamma^{-1} \ll 1$$

where Y_0 is the light source static conductivity in the vicinity of the working point, and γ is the non-linearity factor of its current-voltage characteristic. In the limit $\gamma \to qV/nkT$, as in the case of light emitting diodes. This makes it possible to take for the evaluation of the maximum amplification factor

$$K_{B\,max} = S_p S_{LS} \qquad (13.9)$$

For the PD with built-in amplification $S_p \approx 10^1, \ldots, 10^3$ A/W (17). In modern light emitting diodes with green, yellow and red emission in the range of optimal efficiencies ($\eta \approx (1, \ldots, 5) \times 10^{-2}$), the slope $S_{LS} \sim 10^{-2}$ W/A. This defines $K_{B\,max} \leqslant 1, \ldots, 10$, according to (13.9).

The effectiveness of electroluminescent phosphors is of the same order as that of light emitting diodes (18). Under these conditions one can take in the first approximation

$$S_{LS(e)} \approx \frac{S_{LS(d)} \cdot j_{(d)}}{j_e} \sim (10^2, \ldots, 10^3) S_{LS(d)}. \quad (13.10)$$

Because of this we can expect to obtain the amplification factor of the order of magnitude of 10^2–10^4 in the PD-EL system (13). The smaller values refer to powder electroluminescent phosphors, the larger ones refer to sublimated phosphors.

The photodetector selected should have the maximum value of S_p, all other conditions being equal.

In the visible and near infrared spectral regions (0.5, ..., 1 µm) the photoresistors based on $A^{II}B^{VI}$ materials and Si (17) possess the highest sensitivity at the small signal level ($\leq 10^{-4}$ lx, 10^{-10} W), namely $10^2, \ldots, 10^3$ A/W. The same order of sensitivity holds for the avalanche photodiodes (19). However, the manufacture of multi-element photodetectors with a high degree of integration on the basis of such materials is not at all simple. Of all photodetectors the photoresistors are characterized by the greatest dynamic range of the input optical signal and by the highest degree of the possible microminiaturization.

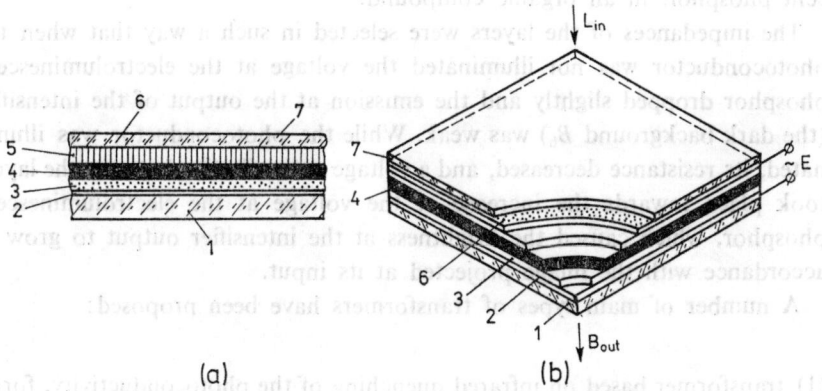

Figure 13.2 A schematic of the optoelectronic image intensifier-transformer: (a) structure; (b) construction; 1, 7—glass bases; 2, 6—electrodes; 3—(LS) electroluminescent phosphor; 4—optical screen; 5—distributed photoresistor (PD)

It follows from the above that the most promising optronic pair for the solid state image intensifier is a thin film photoresistor-electroluminescent phosphor sandwich with the LS and PD of the intensifier being the distributed structures divided by the optical screen (Figure 13.2).

In the case of OET, the weak light relief projected on the photolayer of the photodetector is transformed in the strong light on the electroluminescent layer of the light source, with the given degree of resolution, contrast and spectral composition.

In this chapter we will be concerned with this type of image intensifier.

For the transformers working at high input signal level another possible PD-LS pair is a diode–diode one, which permits a high degree of microminiaturization and integration in the transformer.

13.2 THE STATE OF THE ART AND MAIN DEVELOPMENT TRENDS

The first report concerning electroluminescent image intensifiers goes back to 1953 (20). The first intensifier designs (7, 8, 21) corresponded to the canonical variant shown in Figure 13.2 and were based on high-ohmic powder electroluminescent and photoconductive layers. The latter are suspensions of the powder of the active substance, i.e. of the photoconductor or electroluminescent phosphor, in an organic compound.

The impedances of the layers were selected in such a way that when the photoconductor was not illuminated the voltage at the electroluminescent phosphor dropped slightly and the emission at the output of the intensifier (the dark background B_b) was weak. While the photoconductor was illuminated, its resistance decreased, and a voltage redistribution between the layers took place towards the increase of the voltage at the electroluminescent phosphor, which caused the brightness at the intensifier output to grow in accordance with the image projected at its input.

A number of main types of transformers have been proposed:

(1) transformer based on infrared quenching of the photoconductivity, forming a negative image (22);
(2) intensifiers with a positive optical feedback and pulse supply, preventing bistable regime (23–24);
(3) intensifiers where photoelectroluminescent and thermochromic materials are used as the LS (25–26);
(4) intensifiers where the supply is carried out by the voltage which is a complex of several harmonical voltages of the different frequencies that linearize the transmission characteristic $B_{out} = f(L_{in})$ (27);

(5) intensifiers where the electrode in the form of a grid with small cells inserted in the photoconductive layer is used instead of the transparent electrode in order to reduce the influence of the PD capacitance and to increase K_B (28).

The use of powder electroluminescent phosphor and photodetector did not make it possible to obtain high K_B and resolution in the OET designs under discussion for the following reasons:

(a) The high resistance of powder electroluminescent phosphors which demand, for effective running, photoconductive layers with still higher dark impedance values. With the high photosensitivity and high resistance materials used (CdS, CdSe, CdTe) this requirement could only be met at the expense of a significant increase of the photolayer thickness $h_p \gg \alpha^{-1}$, where α was the absorption coefficient of the emission. In this case the significant bulk resistance, reducing S_p, is found to be connected in series with the photosensitive part of the PD.

(b) The significant part of the reactive component of the impedance of the PD. In powder photoconductive layers the relation of the active and capacitive dark conductivities is such that at the working frequencies the capacitive current makes up a considerable part of the total current of the intensifier. This leads to an increase in the dark background and thereby to K_B reduction.

(c) The large size of the grains in the structure, which limits the resolution of the intensifier by the units of lines per mm less than what we have when the thickness of the layer is large.

It has been shown in References 29, 30 that the $K_B > 1$ regime requires the fulfilment of the condition

$$C_e = (20, ..., 30)C_p, \qquad (13.11)$$

where C_e, C_p are the driving capacities of the electroluminescent layer and of the photolayer, respectively.

Since in the plane intensifier design the areas of both layers are equal and the dielectric constants for the $A^{II}B^{VI}$ materials differ insignificantly from each other, the role of the tuning parameter of the layers is played by their thickness. In accordance with (13.11)

$$h_e \approx (0.03, ..., 0.05)h_p \qquad (13.12)$$

The thicknesses of the powder layers are restricted by the grain size. For ZnS electroluminescent phosphors, $h_{e_{min}} \approx 15 \div 20$ μm. The light absorption

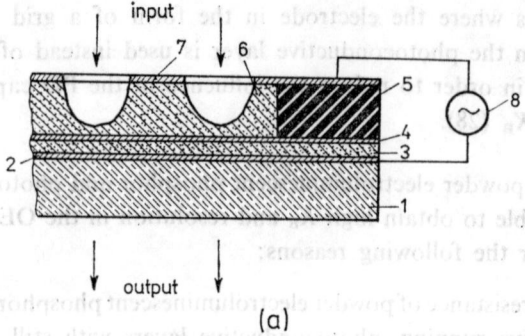

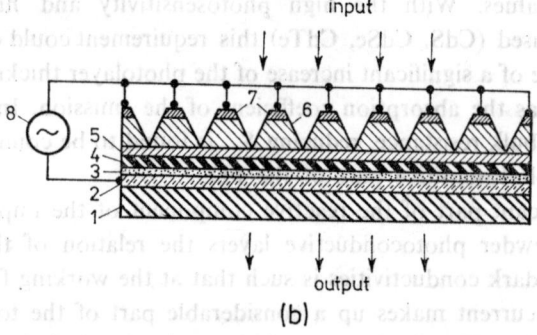

Figure 13.3 OET design based on powder elements: (a) with discrete elements performed as hollows in the photoconductive layer (1—transparent base; 2—transparent electrode; 3—electroluminescent phosphor; 4—optical screen; 5—isolating support; 6—photoconductor; 7—electrode; 8—supply source); (b) with 'comb-like' photoconductor (1—transparent base; 2—transparent electrode; 3—electroluminescent layer; 4—optical screen; 5—current dissipating layer; 6—photoconductive layer; 7—conducting strips; 8—supply source)

coefficient for the photoconductors based on the compounds between Cd and S, Se, Te in the region of their maximum photoconductivity is 10^3 cm^{-1} in order of magnitude. From condition (11) this gives $h_p \sim (400...450)$ μm while the photoactive thickness equals $1/\alpha \sim 20, ..., 30$ μm. Thus, the bulk resistance becomes the dominant one in the PD resistance. The high photolayer thickness reduces its longitudinal resistance, leads to the spread of the charge in it, limits the intensifier resolution and its contrast factor.

In principle, the low resolution still remains a defect of the OET based on powder elements. To overcome the low intensification a number of solutions have been proposed, the idea being to extend radiation along the entire photoconductive layer thickness and to reduce the influence of its capacitance.

One such solution is the substitution of the distributed PD amplifier by a multi-element one (7, 8, 13, 31, 32). Figure 13.3(a) demonstrates a typical design of such an OET, where a number of hollows in the photoconductive layer act as discrete photodetectors. In such a case the capacitance of the photoconductive layer is considerably reduced and, as a result, the dark impedance increases and sensitivity is improved at the expense of the photosensitivity of the slopping walls of the hollows.

This idea was further developed with the 'comb-like' photodetector as shown in Figure 13.3(b) (2, 7, 8, 13, 32–37). The improvement of OET characteristics is achieved here in essentially the same way as in the amplifiers with discrete elements. In the design shown in Figure 13.3(b) the new element is the current dissipating layer placed between the photoconductive and electroluminescent layers. Its purpose is to smooth the striped structure of the output image.

Together with the main design and the usual supply scheme of the OET a number of variants of the intensifier supply and of the performance of its photodetector whose function is to increase the S_p and to improve the conditions of the matching of the PD and LS impedance are known. Usually, the modifications involve an additional d.c. supply of the PD (8, 38–42), replacement of the photoresitor with a photocapacitor, making use of the resonance characteristics of the OET circuit, division of the supply circuits of the PD and LS, and the use of bridge compensational supply schemes (43, 44, 45–50).

The division of the supply circuits makes it possible to optimize the PD and LS characteristics of the S_p and S_{LS}, using the direct voltage for the PD and the alternating voltage of a given frequency for the electroluminescent phosphors.

The most successful solution of this kind is the use of the dividing ferroelectric capacitors (cf. Figure 13.4). An intensifier of this type is a double-circuit electric scheme. The first circuit is formed by the divider R-PD with ferroelectric capacitors (FEC) connected to it. It is supplied by the direct voltage source. The second circuit is made up of electroluminescent capacitors (ELC) and ferroelectric capacitors connected in series. It is supplied from the secondary coil of the transformer T_1. At the position of the switches K, as shown in Figure 13.4, the resistance $R_p \gg R$, and the potential of the point O is about zero. The FEC capacitance has the maximum value and ELC shine brightly. The potential of the point O increases with illumination

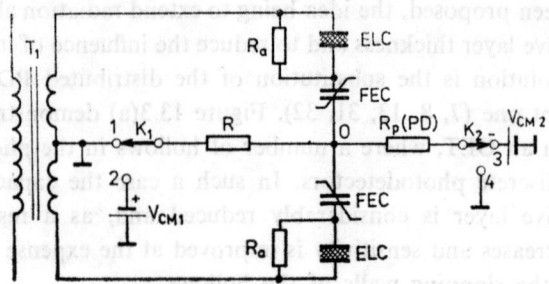

Figure 13.4 Equivalent intensifier scheme with dividing ferroelectric capacitors

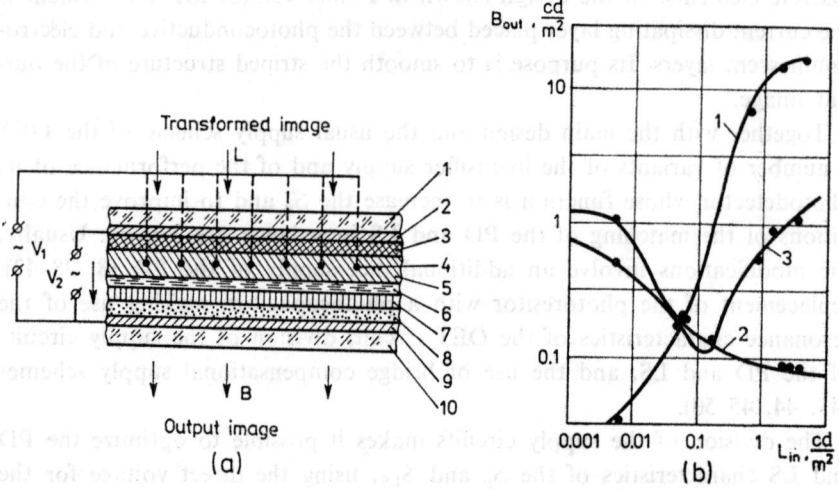

Figure 13.5 Transformer with a monitoring grid; (a) schematic of the design, 1—glass; 2—transparent electrode; 3—dielectric film; 4—coils of the grid; 5—photodetector layer; 6—non-transparent layer; 7—reflecting layer; 8—electroluminescent layer; 9—transparent electrode; 10—glass; (b) its characteristics

of the PD, the FEC capacitance decreases and the emission intensity of the ELC decreases as well. The negative image of the intensifier is formed in this way. To achieve a positive image, the switches K are to be in positions 2 and 4. An analysis of the work of the scheme shows that coordination of the PD and OET impedances is not necessary in the double circuit scheme. Further improvement of the bridge schemes of the OET with simultaneous simplification of their topology under the moderate demands towards the PD and LS leads to the construction of intensifiers with a monitoring grid (Figure 13.5(a) (8, 13, 47–51)). In an intensifier with a grid the coils of the grid electrode 4 are mounted in the photoconductive layer.

Voltage V_1 is applied to the coils of the grid electrode whose phase is opposite to that of voltage V_2 applied to the transparent electrode 2. In this case the intensifier regime is defined by the magnitude and phase of the monitoring voltage V_1. The latter may be chosen in such a way that the voltage at the electroluminescent phosphor is equal to zero when the input signal is absent. The back-ground emission ($B_b = 0$) disappears and the image contrast increases there. The transmissive characteristic of the intensifier in this regime (curve 1, Figure 13.5(b)) is typical of light intensifiers with coordination of the impedances. However, the amplification factor of such an intensifier is not high, because the working voltage of the luminescent phosphor decreases by the same magnitude which compensates the background voltage. Its maximum value does not exceed 10–15. For the same reason the sensitivity threshold of the intensifier increases, which defines the comparatively high level of the input signals (more than 10^{-2} cd/m^2).

Thus, the characteristics of the intensifier with a grid better correspond to the regime of the image transformer, which is where it is mainly used.

If voltage V is chosen to compensate for the effect of voltage V_2 under conditions of maximum illumination of the photoconductor, the reduction of the illumination on the input will be accompanied by an increase of the emission brightness of the electroluminescent phosphor (curve 2). In this working regime the transformer transmits the image projected at the photodetector in the negative form. The intermediate regime (curve 3) is also possible when the character of the image transmission (negative or positive) depends on the irradiation intensity of the PD.

Table 13.1 represents some of the main variants of the OET with their simplified equivalent schemes, which illustrate the main development trend in this area. However, in spite of the variety of technical solutions and the essential improvement of the characteristics of powder electroluminescent phosphors in the last 10 years, the OET based on them could not solve the problem of designing optoelectronic image intensifiers with high amplification factor, high sensitivity and high resolution, which would be comparable with electron-optical transformers.

The highest values of the parameters achieved are close to the following figures: $K_B = 2200$ in the case of the maximum resolution $N = 0.8$ lines/mm (38); $K_B = 500$ in the case of $N = 5$ lines/mm (52, 53); the threshold in

Table 13.1 OET scheme variants

Intensifier–transformer type	Simplified equivalent scheme	Notes
1 Double-electrode type with monitored photoconductivity	L, PD, EL, B_{out}, V	In widespread use, usually based on thin films
2 Triple electrode type with monitoring grid	PD, EL, C_s, i_1, i_2, V_1, V_2	In widespread use; based on powder materials
3 Double-electrode type with monitoring photocapacity	C_p, C_L, L, B_{out}, V	Not in widespread use because of the low quality factor and sensitivity of the photocapacitor
4 Double-electrode type with resonant circuit and monitoring photocapacity	C_p, R_p, L, V~	Not in widespread use because of the difficulty of obtaining the inductance component and the low quality factor of the circuit
5 Double-circuit type with dividing ferroelectric and d.c. bias on the photodetector	R, C_p, V=PD, V~	Advantageous for thin film construction but technologically complicated
6 Double-frequency type with background compensation by means of current generator	$f_1 \gg f_0$, R_p, C_p, C_e, i_1, i_2, V_0, f_0, V_1, f_1	It was used in the film variant of the construction. It is not widely used because of the complicated supply

the IR-transformer regime (0.7–1.2 µm) is 2×10^{-10} W/cm² at $N = 1$, ...
..., 4 lines/mm (54), the threshold is 0.1 R_{min} in the regime of the X-ray image transformer (54).

Powder layers of CdS, CdSe, CdTe and their solid solutions are used as PDs. Attempts have been made to use layers of this kind on the base of CdHgTe as a low-ohmic part of the photoconductor (55). The radiation source is formed on the base of ZnS activated with Mn, Cu, Cl, Er and other impurities. The voltage supply of powder electroluminescent phosphors is 150, ..., 400 V with hightened frequency (400 Hz). This creates some complications in the exploitation of such OET. The main expectations for the improvement of OET characteristics were connected with the use of thin film structures instead of the powder ones, notably sublimated phosphors and thin film photoresistors. The thin film technology makes it possible to change the thicknesses of the structure films simultaneously without producing any major changes of their characteristics, to obtain high non-linearity coefficients of the brightness-voltage characteristics of electroluminescent phosphors (10–19), to obtain a high ratio of the light-to-dark photoconductor resistance in the range of low excitation levels (10^{-4} lx), and to obtain high resolution which is not limited by the grain size. When $h_e = 0.3$–1 µm, the photolayer thickness is not yet limited by condition (13.12) but by the breakdown strength of the photodetector and by the resolution of the intensifier as a whole. However, there remain a number of unresolved difficulties with the thin film variant, and the question of producing efficient OETs is still open.

13.3 AN ANALYSIS OF OET CHARACTERISTICS AND PARAMETERS

13.3.1 The transmissive characteristic and the transformation factor

It follows from the above description that the basic OET design is a three-layer system consisting of electroluminescent phosphor, optical screen and photoresistor (see Figure 13.6(a)). The equivalent scheme consisting of an RC-circuit made of three sections (Figure 13.6(b)) corresponds to it.

The parameters and characteristics of OETs are defined by the parameters of each of the layers and the parameters of the equivalent circuit as a whole as well as by the working regime of the intensifier, by the level of the input illumination and by the value of the voltage applied. Figure 13.7 is a scheme illustrating this interconnection.

The light intensifier regime is limited by the value of the transformation (amplification) factor (2, 56)

$$\eta_B = \frac{\pi(B_{out}-B_b)}{L} \qquad (13.13)$$

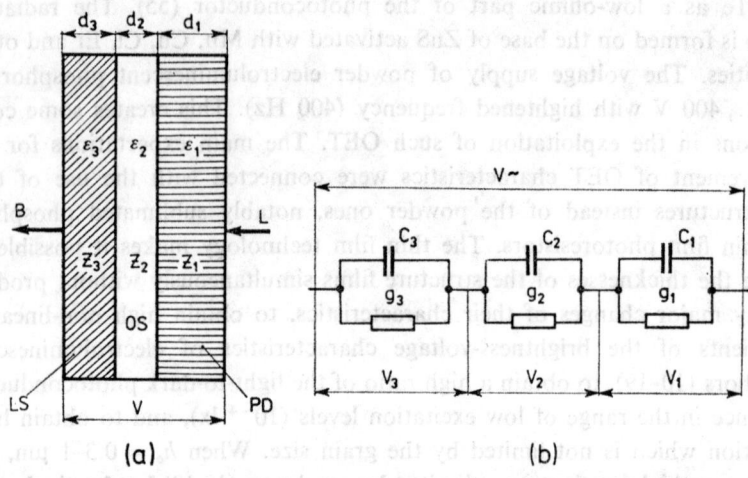

Figure 13.6 Simplified structure (a) and the equivalent scheme (b) of layered OET design

The usefulness of the structure for image transmission is limited by its resolution N [lines/mm] and by the contrast factor

$$\gamma = \frac{B_{out}-B_b}{B_{out}} \qquad (13.14)$$

To determine both factors ((13.13) and (13.14)) it is necessary to have the transmissive characteristic of the OET, given by the dependence $B = f(L)$. Taking then

$$Y_1 = Y_d + AL^\alpha$$
$$B_{out} = bV_3^{\gamma_c} \qquad (13.15)$$
$$V_3 = \frac{Z_3}{\sum\limits_1^3 Z_i} V$$

where Y_1, Y_d are the whole and dark conductivities of the photoresistor,

Electroluminescent Image Transformers and Amplifiers

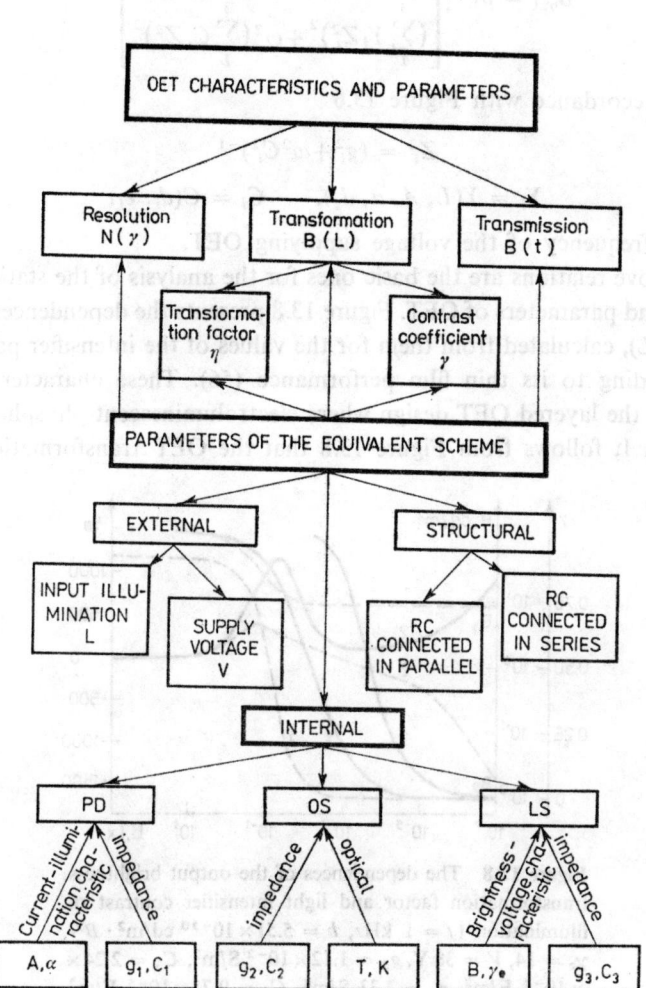

Figure 13.7 Scheme illustrating the interconnection between OET characteristics and parameters

respectively, A, α, γ_e, b are constants, we get for the transmissive characteristic:

$$B_{out} = bV^{\gamma_e}\left[\frac{Z_3^2}{(\sum_1^3 Y_i Z_i^2)^2 + \omega^2(\sum_1^3 C_i Z_i^2)^2}\right]^{\gamma_e/2} \quad (13.16)$$

Here in accordance with Figure 13.6

$$Z_i^2 = (g_i^2 + \omega^2 C_i^2)^{-1}$$

$$Y_i = Y(L, A, \alpha, d_1), \quad C_i = C(d_i, \varepsilon_i)$$

ω is the frequency of the voltage supplying OET.

The above relations are the basic ones for the analysis of the static characteristics and parameters of OET. Figure 13.8 presents the dependences $B_{out}(L)$, $\eta_B(L)$, $\gamma(L)$, calculated from them for the values of the intensifier parameters corresponding to its thin film performance (56). These characteristics are typical of the layered OET design where electroluminescent phosphor is used as an LS. It follows from Figure 13.8 that the OET transformation factor

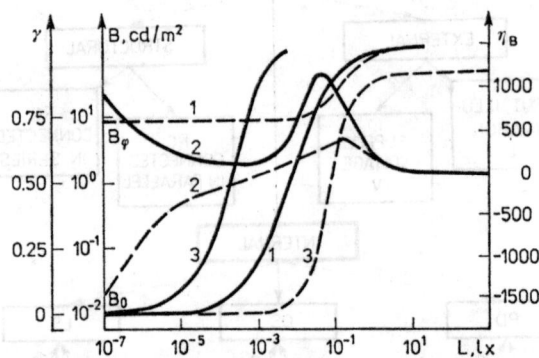

Figure 13.8 The dependences of the output brightness, transformation factor and light intensifier contrast on illumination ($f = 1$ kHz, $b = 5.51 \times 10^{-20}$ cd/m² · B^{γ_e}, $\gamma_e = 14$, $V = 38$ V, $g_3 = 1.12 \times 10^{-2}$ S/m², $C_3 = 2.24 \times 10^{-4}$ F/m², $g_2 = 3.33$ S/m², $C_2 = 9.73 \times 10^{-5}$ F/m², $\alpha = 0.7$, $Y_d = 0.92$ S/m², $A = 30$ S/m² · lx$^\alpha$): 1—B_{out}; 2—η_B; 3—γ (solid—$C_1 = 9.6 \times 10^{-7}$ F/m², dashed—10^{-3} F/m²

$\eta_B(L)$ reaches its maximum value when the transmissive characteristic is nonlinear. This narrows the intensifier dynamic range considerably according to the input intensity L near the extreme values of η_B and K_B.

Expressions 13.15–13.19 make it possible to analyse the influence of the parameters of the structural elements of OET and its working regime on its characteristics: (a) the parameters of the current-illumination characteristic (α, A) and the photodetector impedance ($g_1 c_1$); (b) the parameters of the brightness-voltage characteristic (b, γ_e) and the light source impedance; (c) the optical screen impedance (g_2, c_2); (d) the degree of coordination of the structure impedances:

$$\theta = \frac{Z_{1d}}{Z_2 + Z_3} = [(Y_d^2 + \omega^2 C_1^2)(g_S^2 + \omega^2 \chi_S^2)]^{-1/2} \qquad (13.17)$$

where

$$g_S = \sum_{i=2}^{3} \frac{g_i}{g_i^2 + \omega^2 C_i^2}, \qquad \chi_S = \sum_{i=2}^{3} \frac{C_i}{g_i^2 + \omega^2 C_i^2}$$

(e) the character of the photodetector and light source impedances.

The results of such an analysis are illustrated by the diagrams of Figure 13.9 (56). It follows from the data collected there that to obtain high values of the transformation (amplification) factor preserving a good contrast in the small illumination range ($L < 10^{-2}$ lx) it is necessary:

1. To use a photoresistor with small α ($\alpha \leqslant 1$) and high integral sensitivity ($A \geqslant 10^{-1}$ S/m²lux$^\alpha$). However, neither an extreme decrease of α nor

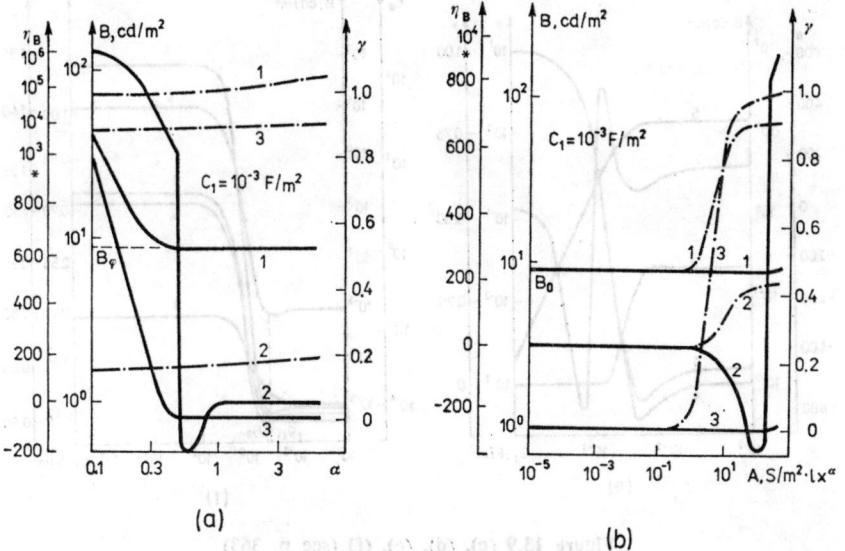

Figure 13.9 (a), (b), (see p. 363)

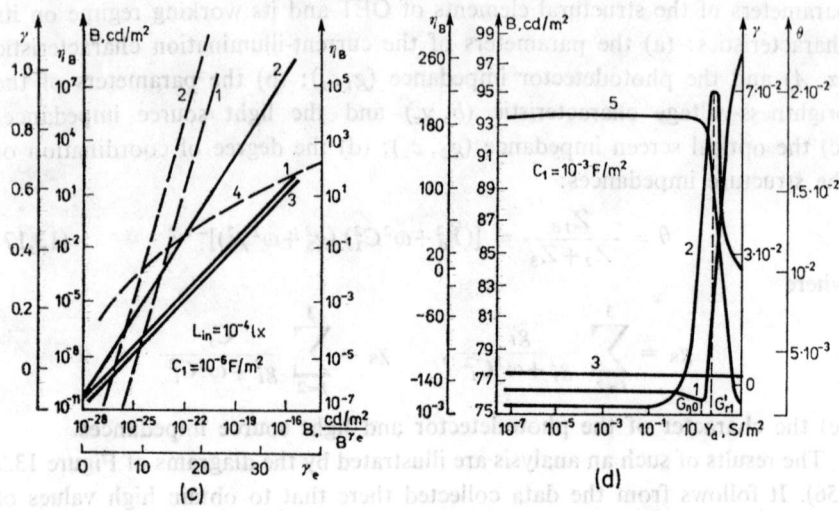

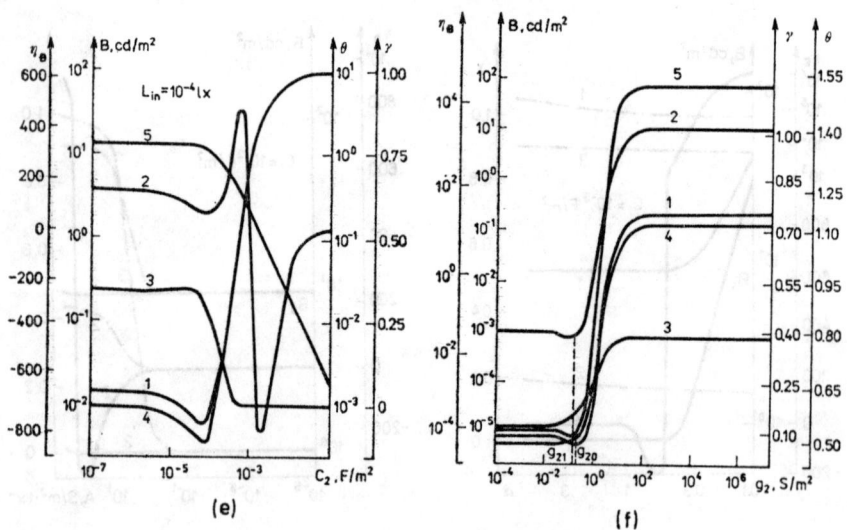

Figure 13.9 (c), (d), (e), (f) (see p. 363)

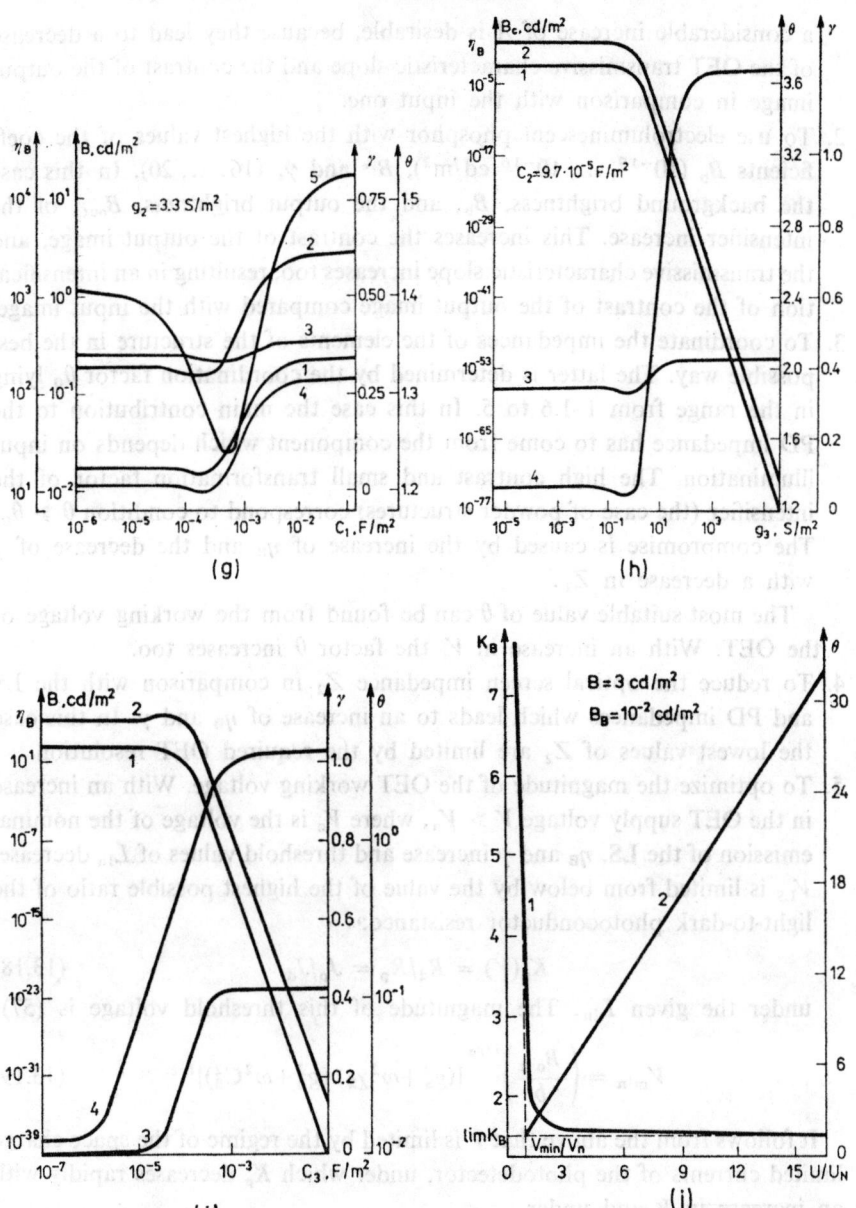

Figure 13.9 Comparative OET characteristics in the function of the parameter elements of its: (a), (b) 1—B_{out}; 2—η_B; 3—γ; $L_{in} = 10^{-4}$ lx (solid curves), $L_{in} = 1.5$ lx (dashed curves); (c) 1—B_{out}, 2—η_B; 3—γ (dashed curves); 1—B_{out}, 2—η_B; 3—B_b (solid lines); (d), (e), (f), (g) 1—B_{out}; 2—η_B; 3—γ; 4—B_b; 5—θ; (h), (i) 1—B_{out}; 2—η_B; 3—γ; 4—θ; (j) 1—K_B; 2—θ

a considerable increase of A is desirable, because they lead to a decrease of the OET transmissive characteristic slope and the contrast of the output image in comparison with the input one.

2. To use electroluminescent phosphor with the highest values of the coefficients B_b (10^{-19}, ..., 10^{-16} cd/m^2), B^{γ_e} and γ_e (16, ..., 20). In this case the background brightness, B_b, and the output brightness, B_{out}, of the intensifier increase. This increases the contrast of the output image, and the transmissive characteristic slope increases too, resulting in an intensification of the contrast of the output image compared with the input image.

3. To coordinate the impedances of the elements of the structure in the best possible way. The latter is determined by the coordination factor θ_0 lying in the range from 1–1.6 to 5. In this case the main contribution to the PD impedance has to come from the component which depends on input illumination. The high contrast and small transformation factor of the intensifier (the case of powder structures) correspond to condition $\theta \gg \theta_0$. The compromise is caused by the increase of η_B and the decrease of γ with a decrease in Z_1.

The most suitable value of θ can be found from the working voltage of the OET. With an increase in V, the factor θ increases too.

4. To reduce the optical screen impedance Z_2 in comparison with the LS and PD impedances which leads to an increase of η_B and γ. In this case the lowest values of Z_2 are limited by the required OET resolution.

5. To optimize the magnitude of the OET working voltage. With an increase in the OET supply voltage $V > V_n$, where V_n is the voltage of the nominal emission of the LS, η_B and γ increase and threshold values of L_{in} decrease. V_{LS} is limited from below by the value of the highest possible ratio of the light-to-dark photoconductor resistance:

$$K_p(V) = R_d/R_p = J_p/J_d \qquad (13.18)$$

under the given L_{in}. The magnitude of this threshold voltage is (57):

$$V_{min} = \left(\frac{B_{out}}{b}\right)^{1/\gamma_e} [(g_S^2 + \omega^2 \chi_S^2)(g_3^2 + \omega^2 C_3^2)]^{1/2} \qquad (13.19)$$

It follows from the above that V is limited by the regime of the space charge limited currents of the photodetector, under which K_p decreases rapidly with an increase in V and under

$$V = V_c = \frac{2\pi |\varrho_c| d_1^2}{\varepsilon_1}, \qquad K_B = 1 \qquad (13.20)$$

and also by the breakdown strength of the photoconductor in the dark.

In the given OET design a photocapacitor may be used instead of the photoresistor (1). To obtain the same OET output parameters (B_{out}, B_p) in both cases the following condition has to be satisfied (56)

$$\frac{R_d}{R_p} = \frac{\chi_d}{\chi_1} \left[\frac{1 + \frac{2}{\chi_d}\left(\frac{1}{\omega C_2} + \chi_3\right)}{1 + \frac{2}{\chi_1}\left(\frac{1}{\omega C_2} + \chi_3\right)} \right]^{1/2} \qquad (13.21)$$

This is correct if we have a pure reactivity electroluminescent phosphor ($g_3 = 0$, $Z_3 = X_3 = 1/\omega C_3$). It is also assumed that for the photoresistor $Z_1 = R_1$, $C_1 = 0$; X_d and X are the impedances of the photocapacitors in the dark and in the light.

It follows from (13.21) that under the conditions adopted

$$R_d/R_p < \frac{\chi_d}{\chi_1}$$

and the photoresistor has advantages over the photocapacitor in that it provides a high transformation factor at a given threshold or a lower threshold at a given transformation factor. Examining the situation where the LS is represented by the active resistance ($X = 0$, $Z_3 = R_3$), one can show that in this case photocapacity is a more effective photodetector for the OET. The condition $X_3 = 0$ corresponds to electroluminescent phosphors working at d.c. supply (58), which excludes the use of the capacitor in series with it. As was shown earlier, the use of the light emitting diodes in image intensifiers has so far been ineffective.

13.3.2 Resolution and frequency-contrast characteristic

The ability of the light intensifier to transmit a high quality image is determined by its resolution and frequency-constant characteristic. The image projected on the PD will lose the contrast on the intensifier output because of the spread of the charge from the illumination regions to the dark ones.

Resolution can be evaluated when the image of the absolutely contrasted grid with the rectangular law of the distribution of the light intensity between the lines of equal sizes (optical tuning chart) is projected on the PD. In this case infinite extent of the PD and the grid is assumed as are the electrical isotropy of the layers of the intensifier and the linearity of their current-voltage characteristics. The equivalent scheme of the intensifier design shown in Figure 13.6(a) can be represented in this case in terms of a net of distributed conductivities, as shown in Figure 13.10; index '1' corresponds to the

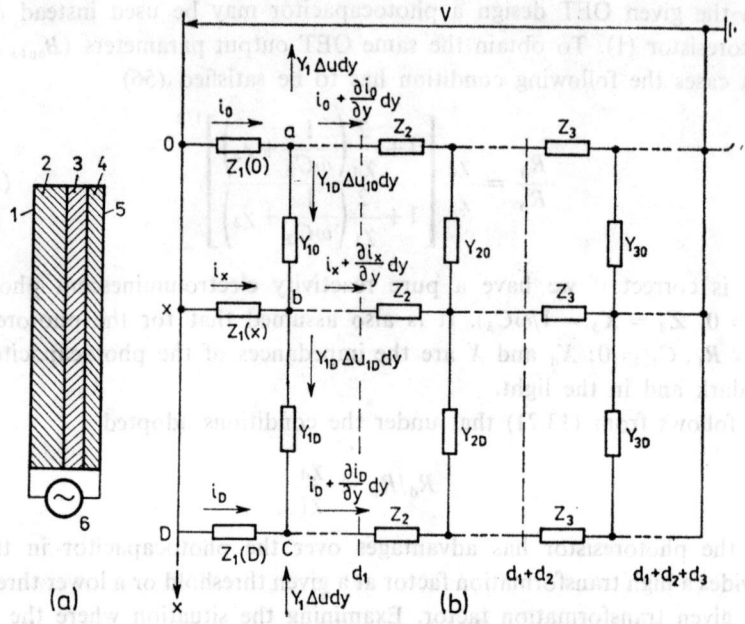

Figure 13.10 Equivalent scheme of the OET element in the regime of image transformation of the absolutely contrasted grid

distributed conductivity of the photolayer with the thickness d_1, and indexes '2' and '3' correspond to the optical screen and electroluminescent phosphor, whose thicknesses are d_2 and d_3, respectively. The X axis is perpendicular to the lines; coordinate $X = 0$ corresponds to the centre of the illuminated line, coordinate $X = D/2$ corresponds to the boundary between the illuminated and non-illuminated lines, and coordinate $X = D$ corresponds to the centre of the non-illuminated line, whose width is D. In this case the width of the line and the resolution of the intensifier are related as:

$$N = 1/2D \qquad (13.22)$$

Such a grid of black and white lines represents a symmetrical and periodical picture, and it is therefore sufficient to restrict the solution to the region

$$0 \leqslant X \leqslant D \qquad (13.23)$$

In this region we determine the specific conductivities of the photoconductor $\sigma_1(0)$, $\sigma_1(X)$, $\sigma_1(D)$ with $\sigma_2 = $ const, $\sigma_3 = $ const, corresponding to the impedances per unit length

$$Z_i = [\sigma_i + j\omega\varepsilon_i]^{-1} \qquad (13.24)$$

where $i = 1, 2, 3$ and whole conductivities per unit length

$$Y_{10} = \left[\int_0^x \frac{dx}{Z_1(x)}\right]^{-1}, \quad Y_{i-10} = (\chi Z_{i-1})^{-1}$$

$$Y_{1D} = \left[\int_0^D \frac{dx}{Z_1(x)}\right]^{-1}, \quad Y_{i-1D} = [(D-\chi)Z_{i-1}]^{-1} \quad (13.25)$$

$$Y_1 = \left[\int_0^D \frac{dx}{Z_1(x)}\right]^{-1}, \quad Y_{i-1} = (DZ_{i-1})^{-1}$$

To determine the contrast characteristics

$$\gamma(L, N) = \frac{B(0) - B(D)}{B(0)} \quad (13.26)$$

it is necessary to know the values of the brightnesses of the intensifier output at the points corresponding to the coordinates $X = 0$, $X = D$.

Since

$$B(\chi) \sim V_3^{\gamma e}(\chi), \quad V_3(\chi) = Z_3(\chi) i(\chi)$$

then to find dependence 13.26 it is necessary to know the values of the currents in the electroluminescent phosphor layer, which are determined by the light relief on the photoconductor. The solution is to be found relative to the currents in the illuminated i_0 and the dark i_D regions for all three layers of the intensifier.

In this case the initial equations are the equations of the currents in nodes a and c, and the Kirchhoff equations for the elementary circuit $OacD$ (Figure 13.10)

$$\sum_a i = 0, \quad \sum_c i = 0, \quad \sum_{OacD} \Delta V = 0 \quad (13.27)$$

which give the following differential equations

$$\frac{d^3 i_0}{dy^3} - 2Y_1[Z_1(0) + Z_1(D)] \frac{di_0}{dy} = 0$$

$$\frac{d^3 i_A}{dy^3} + 2Y_1[Z_1(0) + Z_1(D)] \frac{di_A}{dy} = 0 \quad (13.28)$$

Solution of the equations similar to 13.27 and 13.28 for all three layers of the structure gives the following expressions for their currents (59)

$$i_D = A + A_0(e^{ay} - e^{-ay})$$
$$i_d = A_1 - A_0(e^{ay} - e^{-ay}) \quad 0 \leqslant y \leqslant d_1 \quad (13.29)$$
$$\text{(PD)}$$
$$a^2 = 2Y_1[Z_1(0) + Z_1(D)]$$

$$i_0 = A + A_2 e^{a_1 y} + A_3 e^{-a_1 y}$$
$$i_D = A_1 - A_2 e^{a_1 y} - A_3 e^{-a_1 y} \quad d_1 \leqslant y \leqslant d_1 + d_2 \quad (13.30)$$
$$\text{(OS)}$$

$$i_0 = A + A_4 \left[e^{a_1 y} + e^{2a_1 (\sum_1^3 d_i + y/2)} \right]$$
$$i_D = A - A_4 \left[e^{a_1 y} + e^{2a_1 (\sum_1^3 d_i + y/2)} \right], \quad d_1 + d_2 \leqslant y \leqslant \sum_i^3 d \quad (13.31)$$
$$\text{(LS)}$$
$$a_1^2 = 4/D$$

The constants $A, \ldots, A_4$ are determined from the boundary conditions

$$i_j|_{y=d_1} = i_j|_{y=d_1+d_2}, \quad j = 0, D; \quad \sum_1^3 Z_i i_j = \sum_1^3 \Delta U_i = V$$

$$\Delta U_i = -\frac{1}{2Y_2} \cdot \frac{di_{02}}{dy}\bigg|_{y=d_1} \quad (13.32)$$
$$\Delta V_3 = -\frac{1}{2Y_2} \cdot \frac{di_{02}}{dy}\bigg|_{y=d_1+d_2}$$

The above formulae make it possible to calculate $\gamma(L)$ and $\gamma(N)$ in accordance with (13.26). Both characteristics are represented in Figure 13.11 for the case of OET film elements (59). It follows from the curves that the high resolution ($N = (1, \ldots, 5) \cdot 10^2$) corresponds to the high contrast ($\gamma = 1, \ldots, 0.8$). Thus, the reserve of N is sufficiently high in the thin film OET. This parameter is not critical for the intensifier.

Dependences (24) make it possible to analyse the influence of the OS and LS parameters on the $\gamma(L)$ and $\gamma(N)$. It has been shown in (59) that the changes of the relative dielectric constants of the OS and EL from 1.8 to 320 and also of σ_2 in the range of 10^{-7}–10^{-3} S/m, and of σ_3 in the range of 10^{-10}–10^{-5} S/m do not have any significant effect on the contrast and resolution values. This means that one can use OS and EL which do not meet the requirements expected of dielectric layers, their specific conductivities can be lowered which simplifies the choice of the materials, especially for the OS and the conditions of the correlation of the impedances. For the perception of the image distribution of the brightness $B(x)$ on the output of the intensifier in the function of the realized resolution is also of great importance. The

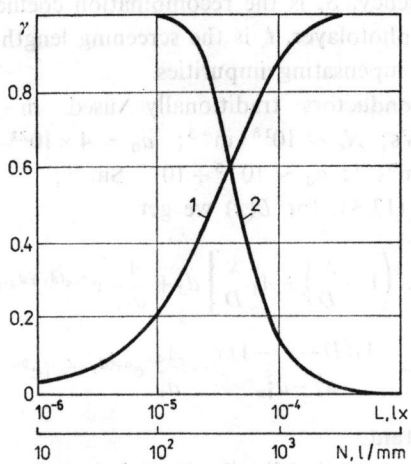

Figure 13.11 Dependence of the contrast on input illumination and the amplitude frequency-contrast characteristic of the intensifier ($f = 320$ Hz; $\gamma_e = 14.8$, $\sigma_0 = 10^{-6}$ S/m, $\varepsilon = 1.9$, $d_1 = 10^{-4}$ m, $\sigma_2 = 10^{-4}$ S/m; $\varepsilon_2 = 11$, $d_2 = 10^{-6}$ m, $\sigma_3 = 10^{-8}$ S/m, $\varepsilon = 8.1$ $d_3 = 3.2 \times 10^{-7}$ m): $1-\gamma(L)$, $N = 38$ lines/mm; $2-\gamma(N)$, $L = 9 \times 10^{-4}$ lx

magnitude of $B(x)$ is determined by the value of $\sigma(x)$, where x is a coordinate, taken at will.

It has been shown in (60) that for one-level non-compensated monopolar semiconductor (the case which approximates the high-resistivity photoconductors of $A^{II}B^{VI}$) under low levels of illumination L, when $K_p \leqslant 1.1$, we have

$$\sigma_1(x) = \begin{cases} \sigma_d \left[K_p - \dfrac{(K_p-1)\mathrm{e}^{-D/2l_s}}{1+\mathrm{e}^{-D/2l_s}} \cosh \dfrac{x}{l_s} \right], & 0 \leqslant x \leqslant D/2 \\ \sigma_d \left[1 + \dfrac{(K_p-1)\mathrm{e}^{-D/2l_s}}{1+\mathrm{e}^{-D/2l_s}} \cosh \dfrac{x-D}{l_s} \right], & D/2 \leqslant x \leqslant D \end{cases} \quad (13.33)$$

where

$$K_p = \left[1 + \frac{a_0 L}{S_r n_d^2}\right]^{1/2}, \quad l_s = \left[\frac{\varepsilon_1 kT}{q^2(2n_d + N_c)}\right]$$

$n_d = \sigma_d/q\mu$ is the ratio of the change of the photoconductor resistance, a_0 is its quantum efficiency, S_r is the recombination coefficient, σ_d is the dark conductivity of the photolayer, l_s is the screening length, and N_c is the concentration of the compensating impurities.

For the photoconductors traditionally used in the OET $A^{II}B^{VI}$ $\mu = 10^5 - 10^6$ m²/Vs; $N_c \sim 10^{18}$ m^{-3}; $a_0 = 4 \times 10^{21} \div 10^{22}$ m^{-3}s^{-1}lx^{-1}; $S_r \sim 10^{-4} \div 10^{-10}$ m³s^{-1}; $\sigma_d \sim 10^{-5} \div 10^{-7}$ Sm^{-1};

Using expression (13.33) for $B(x)$ we get

$$B(x) = b \left| Z_3 \left\{ \left[A\left(1 - \frac{x}{D}\right) + A_1 \frac{x}{D} \right] d_3 + \frac{A_5}{a_{1x}} e^{a_{1x}(d_1+d_2)}(e^{2a_{1x}d_3} - 1) + \right. \right.$$
$$\left. \left. + \frac{1/(D-x) - 1/x}{a_1^2 - a_{1x}^2} \cdot \frac{A_4}{a_1} e^{a_1(d_1+d_2)}(e^{2a_1 d_3} - 1) \right\} \right|^{\gamma_c} \quad (13.34)$$

where A_5 is a constant;

Figure 13.12 represents the distribution of the brightnesses on the OET output calculated according to (34)

$$\Delta B/\Delta B_{st} = f(x/D)$$

where

$$\Delta B = B(x) - B_b, \quad \Delta B_{st} = B_{st} - B_b$$

B_{st} is the brightness on the output of the intensifier in the case of uniform illumination on the input.

The resolution serves as the parameters of the curves.

As can be seen from the curves, the clearness of the transmitted image deteriorates with an increase in N and a decrease in L. However, it weakly depends on the excitation level up to resolutions 25–40 lines/mm in the region of small $L \leqslant 1$ lx.

The experimental values of γ and N are noticeably lower than the theoretical ones. The reason is the roughness of the analysis where no account is taken of the non-linear OET characteristics, the inhomogeneity of its layers, deflection of its impedances from the optimal matching conditions, the degree of approximation of the model in determining $\sigma_1(X)$, etc. The qualitative coincidence is quite sufficient and makes it possible to predict and compare the results.

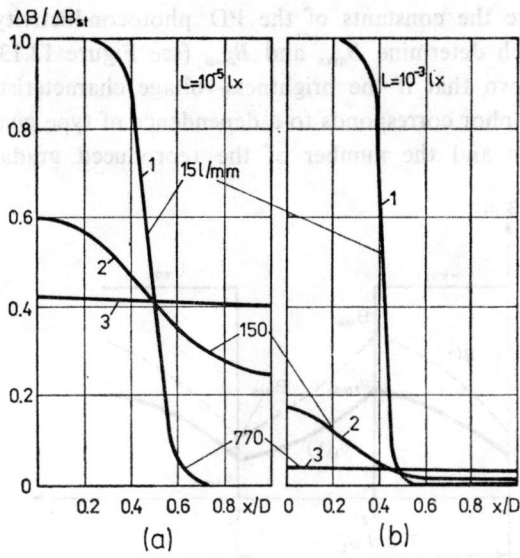

Figure 13.12 Brightness distribution on the image intensifier output ($f = 320$ Hz $\gamma_e = 14.8$, $\sigma_0 = 10^{-6}$ S/m; $\varepsilon_1 = 1.9$, $d_1 = 10^{-4}$ m, $\sigma_2 = 10^{-4}$ S/m, $\varepsilon_2 = 11$, $d_2 = 10^{-6}$ m, $\sigma_3 = 10^{-8}$ S/m $\varepsilon_3 = 8.1$, $d_3 = 3.2 \times 10^{-7}$ m)

The regime under discussion is static. The picture changes essentially if the moving image has to be intensified. In this case the contrast, the number of the transmitted gradiations, and the mean output brightness will depend on OET inertia, τ_{OET}, and on the speed of the input image shift V.

The problem is discussed in the approximation of the model (58):

$$Z_3 = 1/\omega C_3, \qquad Z_1 = R_1, \qquad Z_2 = 0, \qquad \tau_{OET} = \tau_{PD}$$

The transmitted moving image is represented by the sequence of the rectangular light pulses of the arbitrary on-off time ratio. The frequency ω of the supply voltage V is significantly higher than ω_L, the frequency of the exiting light pulses.

The magnitude of the photolayer modulated conductivity $\sigma_1(t)$ determines $B(t)$. Its maximum and minimum values are:

$$\sigma_{min} = \sigma_d \left[1 + (K_p - 1) \frac{(1 - e^{-t_2/\tau_{de}}) e^{-t_1/\tau_g}}{1 - e^{-t_2/\tau_{de}} e^{-t_1/\tau_g}} \right]$$

$$\sigma_{max} = \sigma_d \left[1 + (K_p - 1) \frac{1 - e^{-t_2/\tau_{de}}}{1 - e^{-t_2/\tau_{de}} e^{-t_1/\tau_g}} \right] \qquad (13.35)$$

where τ_g, τ_{de} are the constants of the PD photoconductivity growth and decay time, which determine B_{max} and B_{min} (see Figure 13.13).

It can be shown that if the brightness-voltage characteristic of electroluminescent phosphor corresponds to a dependence of type given by formula 13.2, the contrast and the number of the reproduced gradations will be

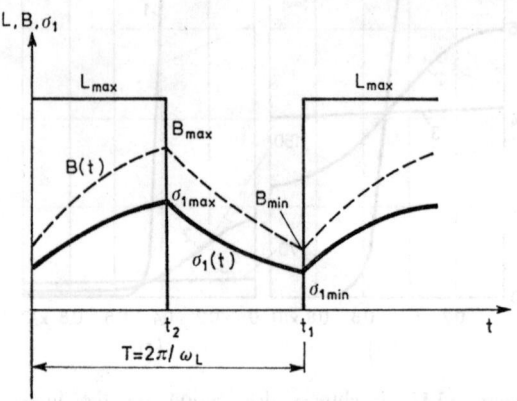

Figure 13.13 Change of the photoconductivity and input brightness with time under OET pulse excitation

determined by the range of the voltage change on the light source $V_{3max}/V_{3min} = f(\sigma_{max}/\sigma_{min})$

$$\gamma^* = \frac{B_{max}}{B_{min}} = \left(\frac{V_{3max}}{V_{3min}}\right)^{\gamma_e}, \quad \xi = 46 \lg \frac{V_{3max}}{V_{3min}} \quad (13.36)$$

For the model under discussion

$$V_3 = \left[1 + \frac{j\omega C_3}{\sigma_1(t)}\right]^{-1} V \quad (13.37)$$

Joint solution of (13.35) and (13.37) makes it possible to find

$$(V_{3max}/V_{3min})^2 = (x_1 N_t K_p, \sigma_d/\omega C_3) \quad (13.38)$$

where $x = 1 + t_2 \tau_{de}/t_1 \tau_g$ is the analogue of the on-off time ratio, and $N_t = t_1/\tau_{de}$ is the analogue of the temporal resolution.

Figure 13.14 illustrates dependence 13.38 (57). It follows from an analysis of the curves that the function $V_{3max}/V_{3min} = f_1(K_p)$ has an extremum which occurs at $K_p > 2$. The higher the temporal resolution required, the stronger is this inequality. Thus, the contrast of the output image is the highest at a certain level of the alternating input signal. The contrast value increases

Electroluminescent Image Transformers and Amplifiers

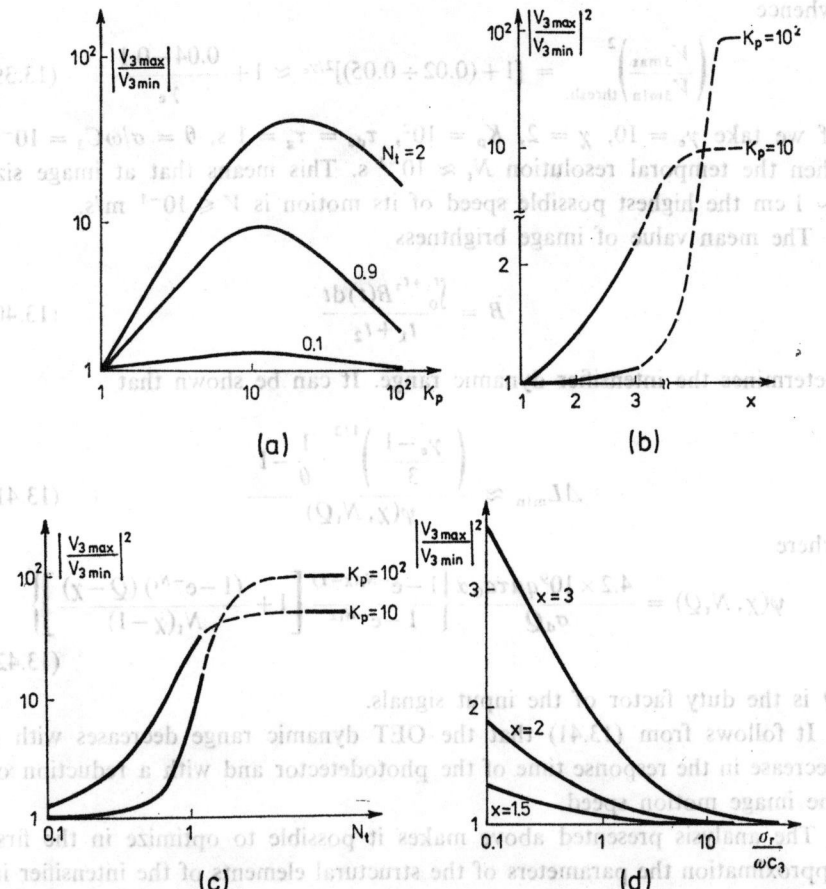

Figure 13.14 Typical shape of the dependence $(V_{3max}/V_{3min})^2 = f_1(f, N_t, \theta, K_B)$ for layered OET designs; (a), (b) $\theta = \sigma_d/\omega C_3 = 10^{-1}$; $\chi = 3$; (c) $\theta = 10^{-1}$, $N_t = 0.3$; (d) $N_t = 0.5$; $K_B = 10$

with an increase in the on-off time ratio of the signal and of the temporal resolution, reaching its static values as a limit. The range of change of V_{3max}/V_{3min} is limited from below by the threshold value of γ^*. In accordance with the Weber–Fechner law the relative brightness increment perceived by the human eye

$$\frac{\Delta B}{B_b} = \frac{B_{out} - B_b}{B_b} = 0.02 \div 0.05$$

whence

$$\left(\frac{V_{3\text{max}}}{V_{3\text{min}}}\right)^2_{\text{thresh.}} = [1+(0.02 \div 0.05)]^{2/\gamma_e} \approx 1+\frac{0.04-0.1}{\gamma_e} \quad (13.39)$$

If we take $\gamma_e = 10$, $\chi = 2$, $K_p = 10^2$, $\tau_{de} = \tau_g = 1$ s, $\theta = \sigma/\omega C_3 = 10^{-1}$, then the temporal resolution $N_t \approx 10^{-1}$ s. This means that at image size ~ 1 cm the highest possible speed of its motion is $V \leqslant 10^{-1}$ m/s.

The mean value of image brightness

$$\bar{B} = \frac{\int_0^{t_1+t_2} B(t)\,dt}{t_1+t_2} \quad (13.40)$$

determines the intensifier dynamic range. It can be shown that

$$\Delta L_{\text{min}} \approx \frac{\left(\frac{\gamma_e-1}{3}\right)^{1/2} \cdot \frac{1}{\theta} - 1}{\psi(\chi, N_t Q)} \quad (13.41)$$

where

$$\psi(\chi, N_t Q) = \frac{4.2 \times 10^9 q \mu \tau a_0 \alpha}{\sigma_d Q} \left\{ \frac{1-e^{-N_t(\chi-1)}}{1-e^{-N_t\chi}} \left[1 + \frac{(1-e^{-N_t})(Q-\chi)}{N_t(\chi-1)}\right] \right\} \quad (13.42)$$

Q is the duty factor of the input signals.

It follows from (13.41) that the OET dynamic range decreases with a decrease in the response time of the photodetector and with a reduction of the image motion speed.

The analysis presented above makes it possible to optimize in the first approximation the parameters of the structural elements of the intensifier in the function of its purpose and working regime.

13.4 THE INTENSIFIER MODEL AND ITS LIMITATIONS

It follows from what has been said above that the most advantageous OET design is the thin film variant of the layered one of all known photosensitive and electroluminescent materials, the $A^{II}B^{VI}$ materials are to all intents and purposes the only ones to correspond best to the OET photodetectors and light sources with respect to their parameters (except time response). As for photoresistors, the best are the compositions of the Cd with the S, Se, Te and their solid solutions, and for electroluminescence sources it is ZnS doped with Mn.

Table 13.2 Some parameters of the ELS based on ZnS:Mn films

No.	Parameter	Type I SnO$_2$–ZnS·Mn– SnO–Al dSi$_0$ ≪ dZnS	Type II SiO$_2$–ZnS·Mn– SiO–Al dSi ≫ dZnS	Type III SnO$_2$–Y$_2$O$_3$– ZnS.–Mn–Y$_2$O$_3$–Al dY$_2$O$_3$ 0.4dZnS	Type IV
1.	Active layer thickness, μm	~ 0.3	~ 0.6	~ 0.6	
2.	Working voltage, V	25–30 eff. f = 2000 Hz	70÷80 under $B_2 = 10^3$ cd/m^2 f = 10 Hz	90÷120 under $B_2 = 10^3$ cd/m f = 10^3 Hz	12÷20
3.	Maximum brightness cd/m^2	30–60	1000–1500	3000–5000	10÷20
4.	The non-linearity coefficient of the brightness-voltage characteristic	14÷16	7÷8 $B = 10^2$–10^3 cd/m^2	20 $B = 10^2$–10^3 cd/m^2	6÷8
5.	The margin of electrical strength	1.3 B = 30 cd/m^2	1.1 $B = 10^3$	1.5 $B = 10^3$	
6.	Emission yield, lm/W	0.3–0.5	0.5–1.0	0.5–1.0	10-15
7.	General current on the working frequency, mA/cm^2	—	45÷55 $B = 10^3$ cd/m^2	60÷65 $B = 10^3$	10-15
8.	Active current		12÷15	10÷4	

The choice of the parameters of the structural elements is determined by the light source. Table 13.2 presents comparative characteristics of electroluminescence sources of yellow-orange emission based on ZnS:Mn films (58, 62). The choice of source is determined by the requirements of all OET parameters. Table 13.3 presents mean OET data when it is used as an inten-

Table 13.3 Parameters of OET when used as the intensifier of an optical prosthetic appliance

N	Parameter	Magnitude
1.	Working area (mm^2)	$\geqslant 400$
2.	Resolution (lines/nm)	$\leqslant 25$
3.	Threshold sensitivity (lx)	1
4.	The transformation factor at the threshold signal	250
5.	The output image contrast in the large–scale details at the threshold signal	80:1
6.	The contrast transformation factor	$\geqslant 1$
7.	Time of response at the threshold signal (seconds)	0.1+0.3

sifier of an optic prosthetic device. The regime of the threshold intensification corresponds to a decrease of L_{in} (to 10^{-2}–10^{-5} lx) and to an increase of η_B and N. The regime of the transformer corresponds to an increase of B_{out} and γ. In the case of intensifiers of the optic prosthetic appliance films of the second type may be used (Table 13.2). For the threshold intensifiers the maximum output brightness can be limited by 20–30 cd/m^2. The highest possible non-linearity factor of the brightness-voltage characteristic is required ($14 \div 16$) over the entire dynamic range, the maximum working voltages of the electroluminescent LS in the dark should not exceed the breakdown voltages of the PD, optimized with respect to its thickness, and the space charge limited currents should not upset the correlation between the PD and LS. For the photoresistor thicknesses $15 \div 25$ μm (CdS, CdSe, $\varrho \sim 10^7$ ohm cm) the breakdown voltages (taking into account the possible imperfection of the photosensitive film) do not exceed 40–60 V. The film-based ELS of the first type correspond best to this regime.

The ELS of the third type can be used in principle in transformers (for instance of the X-ray image) as well as in threshold intensifiers with comparatively small dynamic range. A complication is involved in their use in OET, arising from the high working voltages ($100 \div 180$ V). This leads to a considerable growth of the PD thickness, decrease of its sensitivity, deterioration of the matching conditions, and finally, to a small η_B.

However, in the literature available to us we could find no data on the use of films of this type in the image intensifier designs, which makes it impossible to make a correct evaluation of the perspectives for progress in OETs.

A promising possibility is the use of ELSs with a d.c. supply, owing to their advantageous parameters (especially the resistance value), and the working regime (direct current). They are inferior to the low-voltage ELS with an a.c. supply with respect to the coefficient γ_e which limits η_B with all other conditions being equal. In this case it is desirable to exclude the optical screen from the OET design. The optical screen here introduces into the OET scheme the ballast resistance of the same order of magnitude as the resistances of the PD and LS. As a result of this η_B decreases as it does under bad impedance matching.

In the case under discussion there remains the problem of the resistance matching of the OET components which is related to the complicated technology required to execute the design.

Table 13.4 Mean parameters of film photoresistor of OET

Dark resistance (Ω cm)	$10^4 \div 10^5$
Resistance in the light (Ω cm) ($L = 300$ lx, $V_1 = 10$ V)	$10^1 \div 10^3$
Light-to-dark conductivity ratio at the threshold signal ($10^{-2} \div 10^{-4}$ lx)	$1.05 \div 1.1$
Time of response at the threshold signal	< 10 s
Breakdown voltage in the dark	$\geqslant 50$ V

Table 13.4 presents film photoresistor parameters according to the matching condition in the OET with the low-voltage ELS of the d.c. supply. The choice of the material (CdS, CdSe, CdTe, etc.) is determined by the spectral range of the OET sensitivity, its response time as the threshold signals, and also the photolayer thickness.

For the range of wavelengths from 520 to 950 nm the best results are obtained with the solid solutions CdS_xSe_{1-x} and $CdSe_xTe_{1-x}$. The shortest response time at signal $L_{in} = 10^{-2}$ lx equal to 2s was obtained with CdTe and $Cd_xHg_{1-x}Te$ photoresistors. However, in the latter case one cannot ensure $\eta_B > 5$ because of the poor matching of PD and LS.

The most essential disadvantage of the high ohmic photosensitive $A^{II}B^{VI}$ materials is their significant inertia (Figure 13.15), which increases rapidly as the excitation level decreases:

$$\tau \sim L^{-\alpha}, \qquad (13.43)$$

where $\alpha \leqslant 1$. However, no alternative uncooled photoconductors for the film or powder OET have been found yet.

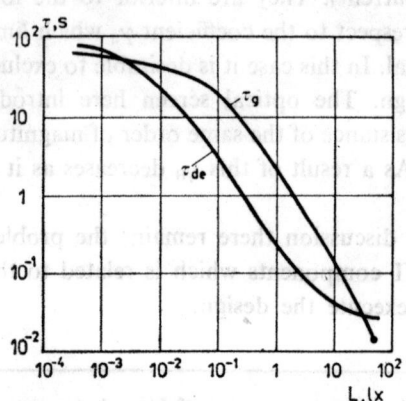

Figure 13.15 Typical dependences of growth and decay time constants on the level of illumination for photoresistors based on CdS, CdSe

Correlation (13.43) is fundamental: for the known photoconductors which can work in pair with ELS, it practically upsets the hopes of obtaining OET design of the type discussed with thresholds $L_{in} < 1$ lx under the response time of 0.1–0.3 s.

The choice of LS and PD determines the choice of the optical screen material. Table 13.5 presents comparative characteristics of the materials most commonly used.

In addition to the proper electrical and optical characteristics of the optical screen material it must be technologically compatible with the PD and LS materials and must adhere well to them over the entire range of technological temperatures.

Figure 13.16 shows an example of the characteristics of the transmission T of OS made of non-photosensitive CdTe 150÷200 nm thick compared with the spectral characteristics of ZnS:Mn–LS and CdSe:Cu, Cl-PD. As

Table 13.5 Some parameters of OET optical screen materials

Parameters	OS materials		
	CdTe	Si	Ge
Resistivity (Ω cm)	3×10^7	1.4×10^5	1.3×10^3
Absorption coefficient in the wavelength range of 0.55–0.65 µm, cm^{-1}	$(1.05 \div 8.23)10^4$	$(4.01 \div 1.46)10^4$	$(9 \div 8.2)10^4$
Relative dielectric constant	10.9	11.8	16

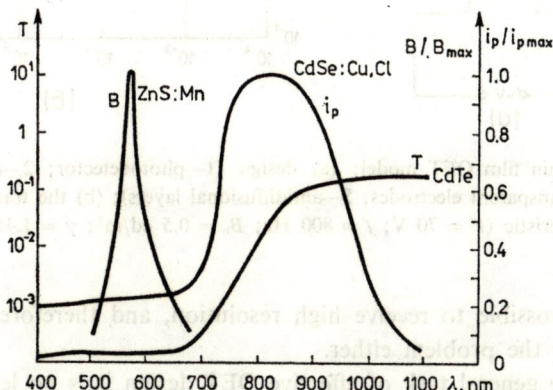

Figure 13.16 Comparative electro-optical characteristics of OET light source, photodetector and optical screen

can be seen, in the emission region of the electroluminescent phosphor one can neglect the positive optical feedback through OS up to $\eta_B = 10^4 \div 10^5$.

The peculiarities of the thin film $A^{II}B^{IV}$ materials under discussion make it technologically difficult to manufacture OETs based on them. First of all, a two-stage process is necessary; other problems include differences in the activation regime of the electroluminescent phosphor and photoconductor, high coefficients of the mutual diffusion, preparation of a number of solid ZnS–CdS(Se) solutions. As a result we are forced to introduce antidiffusional layers between the working elements of OETs. These do not solve the technological problem in any radical way, but they lower the diffusion coefficients by a few orders of magnitude. The use of different non-isotropic glues does

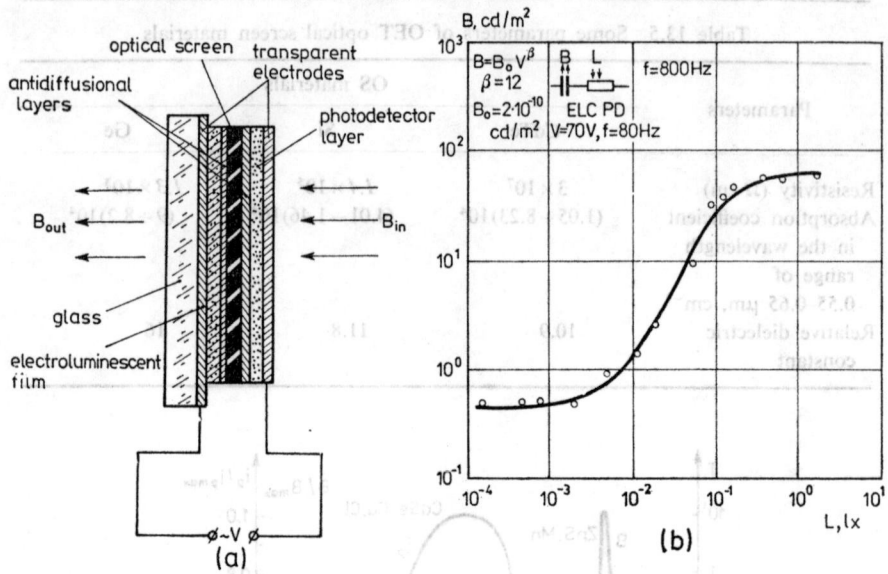

Figure 13.17 Thin film OET model: (a) design (1—photodetector; 2—optical screen; 3—source; 4—transparent electrodes; 5—antidiffusional layers); (b) the transmissive characteristic ($V = 70$ V; $f = 800$ Hz; $B_b = 0.5$ cd/m²; $\gamma = 1.45$)

not make it possible to receive high resolution, and therefore it does not radically solve the problem either.

Thus, in the general task of effective OET design it is no less important to overcome the technological barrier than it is to overcome the inertia barrier.

Figure 13.17(a) shows the transverse structure of a thin film OET design where the concepts given above were taken into account, and Figure 13.17(b) presents its transmissive characteristic. The highest transformational factor attained was $\eta_B \leqslant 200$ under $L_{in} = 10^{-2}$ lx, which is two orders of magnitude less than that obtained on a light intensifier model composed of the same but technologically non-coordinated elements. The threshold sensitivity obtained on this model is also lower (by three orders of magnitude). However, the inertia in this case reaches some units—tens of seconds. In the thin film structure a uniform resolution of 40 lines/mm is achieved over the entire intensifier area. Figure 13.18 presents a picture of the intensifier screen working in the transformational regime $L_{in\ \lambda=0.9\mu m} \rightarrow B_{out\ \lambda=0.58\mu m}$. As can be seen, both the resolution and contrast of thin film OETs are good.

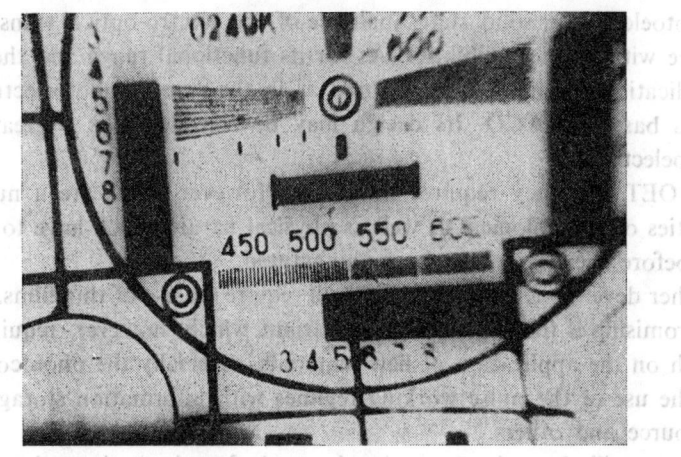

Figure 13.18 Photographs of the image on the intensifier output

13.5 CONCLUSION

The optoelectronic solid state analogue of the electro-optical transformer is a device with great possibilities. As for its functional range and the spheres of application the outlook is as good as in the case of photoelectric transformers based on CCD. Its design may be considered as a great success of optoelectronics.

The OET efficiency requires no proof. However, there are a number of difficulties of technological as well as physical nature which have to be overcome before the device can be manufactured.

Further developments in this area will require the use of thin films. Particularly promising is the triple-electrode variant which, however, requires more research on the application of new materials, especially the photoconductive ones, the use of the pulse working regimes with information storage on the light source and others.

Progress will depend on continual search for alternative solutions; one such solution may involve the bionic principle of visual information processing (63).

REFERENCES

1. S. V. Svechnikov, *Photoelectric One-Port Devices*, Publ. House Technica, Kiev, 1965.
2. S. V. Svechnikov, *Optoelectronics Elements*, Sov. radio Publ. House, Moscow, 1971.
3. J. G. Rabatin, R. A. Sieger, 'X-ray image convertors utilizing lantanum and gadolinium oxyhalide luminescent materials activated with terbium', *Us-Patent, cl.* **250-71**, (G01t 1/20, H01f3150), N 3617743, 2.11.71.
4. N. R. Nail, 'Thermally coupled image amplifier', *US-Patent, cl.* **250-213**, (H01f, G01 t), N 3453436, 1.07.69.
5. J. Wojciechowski, 'Elektroluminescencyjno-photoluminescencyjny przetwornik promieniowania', *Przegl. elektron.*, **9**, 1, 21–33 (1968).
6. S. V. Galginaitis, 'The use of the up-converter phosphors in semiconductor light emitting devices and displays', *Met. Trans.*, **2.3**, 757 (1971).
7. G. Henish, *Electroluminescence*, Mir Publ. House, Moscow, 1964.
8. V. P. Derkatch, V. M. Korsunsky, *Electroluminescent Devices*, Naukova Dumka Publ. House, Kiev, 1968.
9. A. Z. Dun, 'Photoelectro-optical Image Transformers (on the materials of native and foreign literature 1965–74)', *Reviews on the Electronic Technic*, series 4, *N* **4**, (350), Central Scientific Research Institute 'Electronica' Publ. House, Moscow, 1975.
10. G. A. Massey, 'An optical heterodyne ultrasonic image converter', *Proc. IEEE*, **56**, 12, 2157 (1968).
11. 'Acousto-Optical Imaging. Developed by TRW for Nondestructive Testing', Mater. Eval., **28**, 9, 57A (1970).

12. T. Motidzuki, 'Device for transformation of the created by the thermal emission temperature distribution into the visible image with colour gradation', *Patent Jap.*, cl. **99**(5) KO, N 22107, 23.12.66.
13. N. I. Krasnikov, S. V. Svechnikov, 'Solid-state image intensifiers and transformers' *Poluprov. Tekh. Microel.*, 17–27 (1972).
14. V. S. Kretulis, P. F. Oleksenko, V. M. Sorokin, 'Schemal functions of the optoelectronic quadrupoles', *Poluprov. Tekh. Microel.*, **16**, 89–95 (1974).
15. E. I. Adirovitch, A. G. Vishnevetsky, P. M. Karageorgy–Alkalaev, G. I. Neimark 'Optron as a superregenerative light intensifier', *Isv. Akad. Nauk Uzb. SSR, Ser. fiz.* N **2**, 39–46 (1972).
16. I. Ya. Lyamitchev, I. N. Orlov, N. I. Taborko, 'On the contrast transmission in the electroluminescent image intensifier', *Fiz. Tekhnol. Vopr. Kibern.*, **4**, 56–64 (1968).
17. S. V. Svechnikov, 'Photodetectors in optoelectronics', *Proceedings of the Summer School, on Optoelectronics and Integrated Optics*, Edited by V. Prosser, Part III (Supplement), Universiteta Karlova, Praha, 1977, pp. 139–196.
18. S. V. Svechnikov, 'The device basis of the optoelectronics and perspectives of their development. Semiconductor sources of electromagnetic radiation', *Proc. of the International Autumn School on Semiconductor Optoelectronics*, Cetniewo, 1975, Edited by M. A. Herman, pp. 543–600. Polish Sci. Publishers, Warszawa, 1976.
19. T. Ya. Gorbatch, S. V. Svechnikov, 'Main tendencies of development of avalanche photodiodes as the high-speed photodetectors', *Poluprov. Tekh. Microel.*, **11**, 3–20 (1973),
20. W. C. White, 'X-Ray intensification and method', *US-Patent*, cl. **250-71**, N 2650310. 25.VII.1953.
21. R. K. Orthuber, L. R. Ullery, 'A solid-state image intensifier', *J. Opt. Soc. Amer.*, **44**, 4, 297–299 (1954).
22. A. Lempicki, 'Image converter', *US-Patent*, cl. **250-213**, N 2890350, 9.VI.1959.
23. B. Kazan, 'Half-tone image production', *US-Patent*, cl. **250-213**, N 2896087, 21.VII.1959.
24. T. R. Nisbet, 'Radiation amplifier', *US-Patent*, cl. **250-213**, N 3019345, 30.I.1962.
25. F. H. Nicoll, 'Radiant energy translating device', *US-Patent*, cl. **250-213**, N 2908824, 13.10.1959.
26. R. D. Kell, 'Display device', *US-Patent*, cl. **250-213**, N 3064134, 13.XI.1962.
27. G. Diemer, H. A. Klasens, 'Device comprising a photoconductive part and an electroluminescent part', *US-Patent*, cl. **250-213**, N 2922892, 26.I.1960.
28. J. M. N. Hanlet, 'Electroluminescent devices', *US-Patent*, cl. **250-213**, N. 3035177, 15.5.1962.
29. M. V. Fok, *Theory of the Electroluminescent Image Transformers*, Sov. radio Publ. House, Moscow, 1961.
30. I. N. Orlov, N. I. Taborko, M. V. Fok, 'About the influence of the photoconductor capacity and of the active losses in the electroluminescent layer on the image transformer parameters', *Fiz. Tekhnol. Vopr. Kibern.*, **1**, 27–43 (1968).
31. G. Diemer, H. A. Klasens, van J. G. Santen, 'A solid-state image intensifier', *Phys. Res. Rep.*, **10**, 6, 401–424 (1955).
32. F. Pshibyl, 'New in the development of the image the electronic methods of receiving of the non-vacuum systems of the image reproducing', *Haboprondy Obzor*, **23**, 6, 33–40 (1962).

33. B. Kazan, F. H. Nicoll, 'Solid-state light amplifier', *J. Opt. Soc. Amer.*, **47**, 10, 887–894 (1957).
34. G. F. J. Garlik, 'Solid-state image amplifiers', *J. Sci. Instr.*, **34**, 12, 473 (1957).
35. T. B. Tomlinson, 'Principles of the light amplifier and allied devices', *J. Brit. IRE* **17**, 3, 141 (1957).
36. B. Paszkowski, A. Swit, J. Wojciechowski, H. Wierzba, 'Dwuwarstwowy półprzewodnikowy przetwornik obrazu', *Przegląd Elektroniki*, **4**, 8, 459–462 (1963).
37. B. Paszkowski, A. Swit, J. Wojciechowski, H. Wierzba, 'Two-layer solid-state image converter', *Bull. Acad. Polon. Sci., Sér. Sci. Techn.*, **11**, 5, 27–30 (1963).
38. Ya. Uekhara, 'Solid-state light intensifier', *Tosiba Review*, **15**, 6, 619–622 (1960).
39. 'Solid light intensifier', *Fr. Patent, cl.* **H011**, N 1257765, 27.02.61.
40. J. G. Van Santen, 'Solid-state image intensifier', *US-Patent, cl.* **250-213**, N 381402, 12.03.63.
41. B. Kazan, 'Circuit for energizing light amplifier devices', *US-Patent, cl.* **250-132**, N 2839690, 17.06.58.
42. E. E. Loebner, *Recent Developments in Solid-State Image Transducers. Solid-State Physics in Electronics and Telecommunications*, 4. *Magnetic and Optical Properties*, **2**, Acad. Press., London–New York, 1960, pp. 762–775.
43. I. Uchide, K. Satake, 'Improvements in or relating to photoelectric devices', *Brit. Patent*, *cl.* 37, N 904779, 29.08.62.
44. I. Uchide, K. Satake, 'Photoelectronic devices', *Jap. Patent, cl.* **100D0**, N 6410, 20.05.63.
45. R. K. Orthuber, 'Solid-state radiation amplifier', *US-Patent, cl.* **250-213**, N 3054900, 18.09.62.
46. 'Electroluminescent device and circuits thereof', *US-Patent, cl.* **250-213**, N 3043961, 10.01.62.
47. I. Ya. Lyamitchev, I. N. Orlov, G. G. Pershin, N. I. Taborko, 'Experimental study of the possibility of the creation of the multielement electroluminescent devices with ferroelectric materials usage, *Izv. Akad. Nauk SSSR, Ser. fiz.*, **25**, 4, 492–500 (1961).
48. P. N. Wolf, E. A. Sack, 'Solid-state image—Producing screens', *US-Patent, cl.* **250-213**, N 2988646, 13.6.61.
49. F. Pzhibil, 'Electroluminescent image intensifier', *Chech.-Patent, cl.* 21g, 29/40, N 102986, 15.03.62.
50. J. S. Winslow, B. Kazan, 'Solid-state image intensifiers with ferroelectric amplification', *Proc. IEEE*, **52**, 11, 1381–1382 (1964).
51. I. Keidzi, 'Solid-state image brightness intensifier', *Jap.-Patent, cl.* **99(5)K0**, **(H05b)**, N 45715, 17.11.72.
52. T. Kohashi, Mijaii, 'EL-PC image intensifier used control grid', *Electronics*, **36**, 38, 30–34 (1963).
53. T. Ohasi, T. Nokamura, 'Semiconductor electron-optical transformer-pholicon', *Dency Kogjo*, **13**, 6, 437–440 (1964).
54. T. Kohashi, 'Solid-state image transformers', *Nat. Techn. Rept.*, **17**, 6, 633–646 (1971).
55. T. Ohasi *et al.*, 'Solid-state device for the image transformation', *Jap.-Patent, cl.* **99(5)k10**, (G02*f*), N 48-33358, 13.10.73.
56. N. I. Krasnikov, 'The study of the solid-state light intensifier static characteristic', *Poluprov. Tekhn. Microel.*, **14**, 20–37 (1974).

57. Yu. G. Pismenniy, S. V. Svechnikov, V. P. Stepanchuk, 'The study of the solid-state image intensifier parameters', *Poluprov. Tekhn. Microél.*, **26**, 3–8 (1977).
58. N. A. Vlasenko, 'Electroluminescent films as the light sources for the optoelectronic aims', *Poluprov. Tekn. Microel.*, **13**, 93–101 (1973).
59. N. I. Krasnikov, 'Solid-state image intensifier resolution', *Poluprov. Tekn. Microel.*, **8**, 81–90 (1972).
60. A. N. Zyuganov, N. I. Krasnikov, S. V. Svechnikov, 'Distribution of the carriers concentration at the boundary between light and darkness in the non-compensated quasi-monopolar semiconductors', Ukr. fiz. Zh., **14**, 6, 1036–1039 (1969).
61. S. M. Rivkin, *Photoelectrical Phenomena in the Semiconductors*, Fizmatgiz Publ. House, Moscow, 1963.
62. T. Inogushi and S. Mito, Phosphor Films, Topics in Applied Physics, "Electroluminescence'", edited by J. I. Pankove, Springer-Verlag, 1977, v. 17, pp. 197–210.
63. S. V. Svechnikov, A. M. Shkvar, 'Neuronik structures models in the devices of the optical information processing', *Znanie Ukr. SSR Publ. House*, 1977.

Chapter 14

Photodetectors Based on Heterostructures

V. I. Korol'kov

14.1 INTRODUCTION

In most cases, optical radiation is the carrier of information about the physical phenomenon or process under study. Radiation detection is the initial link in the information processing chain. Considering the process of radiation interaction with the substance from a practical point of view, one finds radiation and measurement to be of most interest. Various semiconductor photoelectric devices have long and successfully been used for this purpose. Requirements with respect to radiation detector parameters and characteristics are growing continuously in connection with their increasing sphere of application. At present, they have become of special importance for the creation of effective coherent and spontaneous radiation sources what has resulted in the development of optical information processing and transmission systems. Further progress in optical communication systems calls for the creation of photodetectors with a high sensitivity in the appropriate spectral interval, with the necessary band width, and with minimum noise. In a number of cases, one needs photodetectors which are able not only to convert a light signal into an electric one, but also to convert a light signal into one of higher intensity, and to convert light of one spectral composition into another. However, conversion of infra-red radiation into visible appears to be of most interest. It is obvious that these multifunctional detectors are impossible on the basis of p-n homojunctions. Application of heterostructures makes it possible not only to improve the parameters and characteristics of radiation detectors, but also to create entirely new types of detectors. It is not accidental that the development of electronics, and of optoelectronics in particular has, in the last decade, been connected primarily with the application of heterojunctions.

Creation and study of GaAs–AlAs heterostructures, the properties of which appeared to be very close to those of ideal ones (1), have to a great extent promoted the more and more expanding sphere of their application in semiconductor electronics. These heterojunctions have created a possibility of realizing numerous interesting suggestions to be applied in various devices, they have stimulated the development of absolutely new devices, and have indicated the ways of a search for new ideal heterostructures.

At present heterojunctions are used in the GaAs–AlAs system in the production of low-threshold lasers (2, 3) and high-efficiency LEDs (4). The development of various heterophotodetectors has not yet been treated in detail, and the aim of this chapter was to fill this gap, at least to some extent.

It should be noted that most of the photodetectors considered in this chapter are more complicated in structure than those based on single heterojunctions. However, their operation is based on a peculiarity of the minority carriers separation in single heterostructures, which are to be considered below.

14.2 PHOTOELECTRIC PROPERTIES OF HETEROSTRUCTURES

The photoelectric properties of heterostructures based on various materials have been studied in numerous papers. The study of these properties gives a lot of information about the character of the non-equilibrium carriers separation, the interface effect, and the energy-band diagram. As a rule, interpretations of the results proceed from Anderson's qualitative ideas developed in his early works on heterojunctions (5, 6). According to Anderson, spectral response has a plato, i.e. a region of constant spectral sensitivity between the absorption edges of a narrow-gap and a wide-gap heterostructure material.

However, it has been found in a number of studies that the behaviour of spectral response differs from the predicted one (in the anomalous photoresponse sign, in the change of the photoresponse sign, and in the fact that sensitivity is not observed in the whole range of incident light energies within the 'window'). The analysis of the experimental data shows that these anomalies may be due to both the surface states and the peculiarities of the energy-band diagram, and that in most cases it is rather difficult to say which one it is.

Therefore, our purpose is to analyse and point out the distinctive peculiarities of photoelectric phenomena typical of heterostructures only. They will be considered using as an example ideal GaAs–AlAs heterostructures with well-known energy-band diagram parameters and a well-developed fabrication technique.

14.2.1 A band model and photoelectric properties

Let us consider an interaction of light with the contact of two semiconductors with different forbidden gaps, which results in the appearance of non-equilibrium carriers near the interface. The appearance of the photocurrent (photo e.m.f.) on such a contact can evidently be connected only with the separation of positive and negative mobile charges in space, i.e. with minority carriers passing through the interface. Therefore, photocurrent in heterojunctions is determined by the form and thickness of the potential barrier, and by the state of the interface. Hence, in addition to creating promising perspectives for the development of effective photodetectors with a controlled spectral sensitivity region the study of photoelectric properties also provides a method of investigating heterojunction energy-band diagrams and the interface effect.

14.2.1.1 Abrupt heterostructures

(a) *An ideal model with a zero discontinuity in one of the bands*

In this case, the absence of a discontinuity in the p–n heterojunction conduction band or in the n–p heterojunction valence band ensures complete separation of non-equilibrium carriers created both in the wide-gap and in the narrow-gap material of the pair (Figure 14.1(a)).

Heterojunctions of the GaAs–AlGaAs system are described by the energy-band diagram with $\Delta E_v = 0$. The photocurrent occurring at illuminating a heterojunction through a wide-gap window with monochromatic light of energy $E_{g_1} < h\nu < E_{g_2}$ is composed of a photocurrent due to carrier generation in the depletion layer region and of a photocurrent due to light absorption in the volume of the narrow-gap material. The generation rate of non-equilibrium carriers (neglecting the absorption in the wide-gap region) is

$$g = a_1 k_1 \Phi \exp(-k_1 x), \tag{14.1}$$

where a_1 is quantum efficiency of carrier excitation, Φ is the incident light flux, and k_1 is the absorption coefficient in the narrow-gap material. The photocurrent due to generation in the depletion layer region, taking into account the change in the absorption coefficient of the depletion layer region (i.e. the Franz–Keldysh effect), will be

$$I^w = \int_0^w q a_1 k_1 \Phi \exp(-k_1 x) dx = q a_1 \Phi \left[1 - \exp\left(-\int_0^w k_1(x) dx \right) \right] \tag{14.2}$$

Beyond the space-charge layer the distribution of non-equilibrium carriers is described by an ordinary diffusion equation

$$\frac{\partial p}{\partial x} = g - \frac{p}{\tau_p} + D_p \frac{\partial^2 p}{\partial x^2} = 0 \tag{14.3}$$

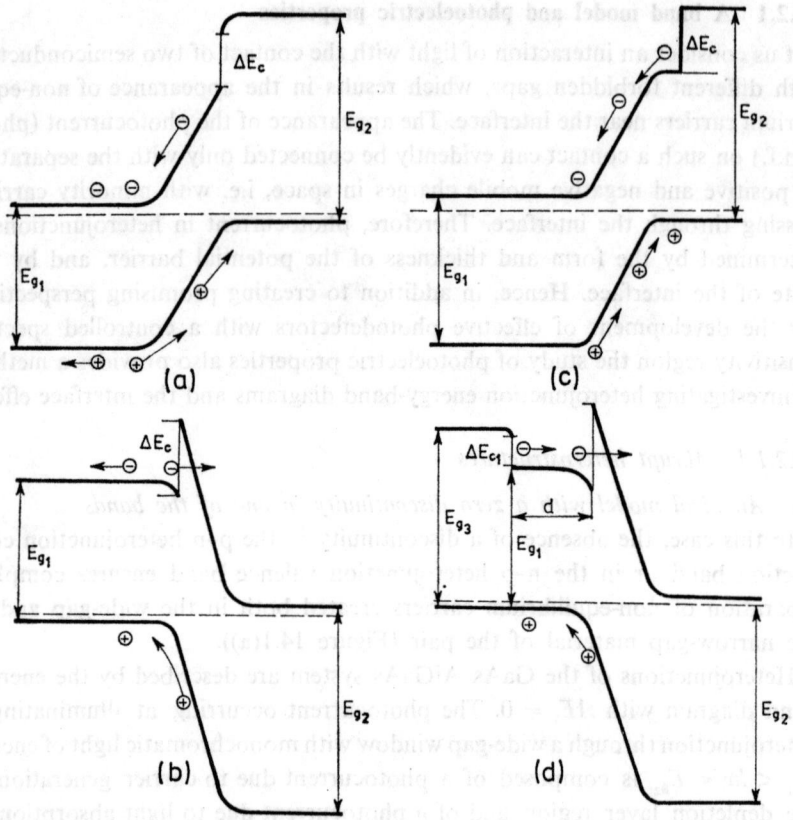

Figure 14.1 Energy-band diagram of various types of heterophotodetectors; (a) anisotype heterojunction with $\Delta E_v = 0$; (b) anisotype heterojunction with ΔE_c and $\Delta E_v \neq 0$; (c) structure with a p–n homojunction; (d) double heterostructure

with the boundary conditions $p = 0|_{x=w}$ and $p = 0|_{x=\infty}$, and the photocurrent is described by

$$I_p = -qD_p \frac{\partial p}{\partial x}\bigg|_{x=w} = \frac{qa_1 k_1 L_p}{1+k_1 L_p} \Phi \exp\left[-\int_0^w k_1(x)\,dx\right] \quad (14.4)$$

The summary photocurrent, therefore, is

$$I_{ph} = I^w + I_p = qa_1 \Phi \left(1 - \frac{\exp[-\int_0^w k_1(x)\,dx]}{1+k_1 L_p}\right) \quad (14.5)$$

The above expression differs from that for the photocurrent in p–n homojunctions in the absence of terms corresponding to recombination losses in p-regions and to the velocity of surface recombination on the illuminated surface.

Near the long-wave sensitivity boundary $k_1 L_p \ll 1$ and $\int_0^w k_1(x)dx < 1$. The photocurrent then is

$$I_{ph}(V) \sim \left(\int_0^w k_1(x)dx + k_1 L_p\right)$$

and the change of photocurrent with voltage will be determined primarily by the change of the absorption coefficient in the space-charge layer region, which is favourable for studying the Franz–Keldysh effect (7).

It is observed from (14.5) that with $N_d \ll N_a$ the expression for photocurrent will be similar to that in p–i–n photodiodes (8):

$$I = qa_1 \Phi\left(1 - \frac{1}{1+k_1 L_p} e^{-k_1 w}\right) \tag{14.6}$$

Hence, to obtain the maximum photocurrent it is necessary that $k_1 w \gg 1$ and $k_1 L_p \gg 1$, i.e. that lightly doped materials with large minority carriers diffusion lengths be used as narrow-gap materials.

The use of a lightly doped narrow-gap material increases the depletion layer thickness and, therefore, the transit time of minority carriers. The latter, with certain doping levels, may become comparable to the life-time of minority carriers which is most sensitive to all kinds of defects and imperfections. Therefore, e.g., nGaAs–pAl$_x$Ga$_{1-x}$As heterojunctions (with different Al content) on the base of a lightly doped GaAs, yield an experimental value of Q which makes it possible to estimate the effect of a heterostructure interface in the best way possible (9).

Figure 14.2 shows quantum efficiency as a function of the incident light energy for specimens with different doping levels of the narrow-gap material (a) $n = 10^{17}$ cm^{-3}; (b) $n = 10^{16}$ cm^{-3}; (c) $n = 10^{14}$ cm^{-3} and with different composition of the wide-gap material. The solid lines show values of Q calculated on the assumption that the minority carriers separation coefficient is independent of the Al$_x$Ga$_{1-x}$As composition and equal to unity. Experimental results agree with the calculated ones particularly well.

It follows from the analysis of these curves that only a lightly doped narrow-gap region ensures high collection of minority carriers in the whole of the spectral sensitivity region of a heterophotodetector, lest a certain amount of light should miss the active region of the device, that is, be absorbed at distances greater than the space-charge layer thickness and the diffusion length. That is why use is made of heterostructures (Figure 14.1(c)) with the GaAs p-region thickness of (0.5–1.0) μm (i.e. n–p(GaAs)–pAl$_x$Ga$_{1-x}$As structures)

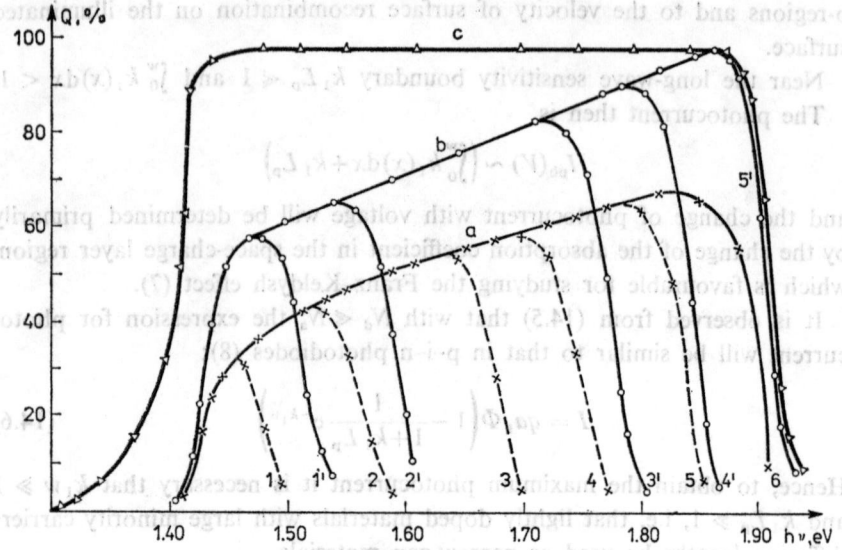

Figure 14.2 Spectral responses of nGaAs–pAl$_x$Ga$_{1-x}$As heterophotodiodes; n, cm^{-3}; a, 10^{17}; b, 10^{16}; c, 10^{14}; x: 1—0.04; 2—0.1; 3—0.17; 4—0.22; 5—0.3; 6—0.33; 1'—0.06; 2'—0.12; 3'—0.25; 4'—0.31; 5'—0.35

for conversion of solar energy. Photoelectric properties of such structures are considered in detail in (10, 11). The physical processes taking place in these structures hardly differ from those in the single heterostructures discussed above.

(b) *Carrier separation in the presence of band discontinuity*

In a general case, from the energy-band diagram of p–n or n–p heterojunctions it is observed that discontinuities in the conduction band ΔE_c or in the valence band ΔE_v prevent separation of narrow-gap minority carriers at their illumination through a wide-gap window. As has been noted before, the influence of a discontinuity in the conduction band upon the photosensitivity of heterojunctions was first observed in Ge–GaAs heterojunctions (6). However, in a number of cases heterojunctions with an energy-band diagram shown in Figure 14.1(b), have noticeable photosensitivity (12).

Let us consider the influence of a discontinuity in the conduction band ΔE_c upon the minority carriers separation coefficient in heterojunctions of the p–GaAs–nAl$_x$Ga$_{1-x}$As type (12), the spectral photosensitivity characteristic of which is often similar to those in Figure 14.2. More often than not photosensitivity depends on the value of the applied bias. Quantum efficiency increases up to certain values of the applied bias, further increase of which does not influence the value of Q.

In an abrupt heterojunction whose conduction band profile is shown in Figure 14.3, photocurrent is explained by two mechanisms—those of subbarrier and of tunnel transitions through a barrier height ΔE_c. In the case

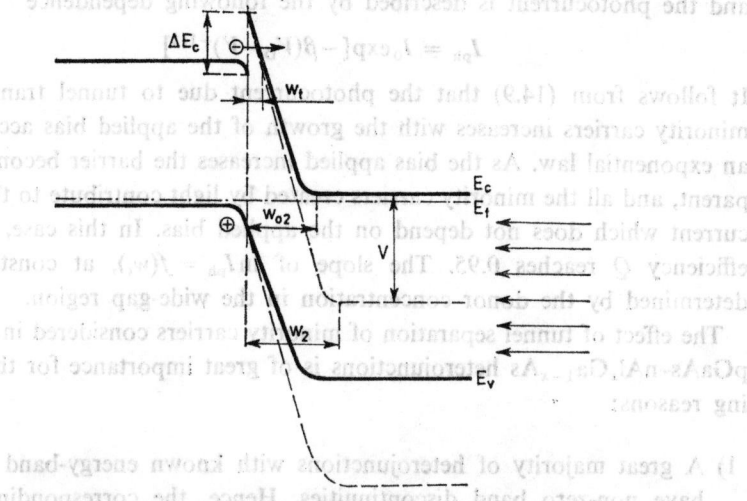

Figure 14.3 Schematic representation of the tunnel separation of carriers

under consideration, when $\Delta E_c \gg kT$ and the incident light energy is $E_g^{GaAs} < h\nu < E_g^{GaAs} + \Delta E_c$, a major role should be played by the tunnel current component.

With a bias applied, the conductivity band discontinuity ΔE_c remains constant; only the conductivity band edge is biased and also the space-charge layer thickness changes. It can be shown that if the total space-charge layer thickness increases—at a negative bias being applied—according to $w_2 \sim (V_D - V)^{-1/2}$ ($N_d \ll N_a$) the barrier thickness along the tunnelling line (at the height ΔE_c) decreases as follows:

$$w_t \simeq \frac{\Delta E_c}{V_D - V} w_2 = \Delta E_c \left(\frac{8\pi\varepsilon_2}{qN_d}\right)^{1/2} (V_D - V)^{-1/2} \qquad (14.7)$$

The flux of electrons tunnelling through barrier ΔE_c mostly depends on the barrier transparency. The barrier transparency is

$$T(V) = \exp\left[-2\int_0^{w_t} \sqrt{2m_n^*\Delta E_c}\,\frac{dx}{\hbar}\right]$$
$$= \exp\left[-\frac{16}{\hbar}\left(\frac{m_n^*\varepsilon_2}{qN_d}\right)^{1/2}\Delta E_c^{1/2}(V_D-V)^{-1/2}\right] \quad (14.8)$$

and the photocurrent is described by the following dependence

$$I_{ph} = I_0 \exp[-\beta(V_D-V)^{-1/2}] \quad (14.9)$$

It follows from (14.9) that the photocurrent due to tunnel transitions of minority carriers increases with the growth of the applied bias according to an exponential law. As the bias applied increases the barrier becomes transparent, and all the minority carriers created by light contribute to the photocurrent which does not depend on the applied bias. In this case, quantum efficiency Q reaches 0.95. The slope of $\ln I_{ph} = f(w_t)$, at constant V, is determined by the donor concentration in the wide-gap region.

The effect of tunnel separation of minority carriers considered in detail on pGaAs–nAl$_x$Ga$_{1-x}$As heterojunctions is of great importance for the following reasons:

1) A great majority of heterojunctions with known energy-band diagrams have non-zero band discontinuities. Hence, the corresponding doping level of a wide-gap material makes it easy to obtain photodetectors with a broadened spectral sensitivity region, which is of importance for the creation of solar heterophotocells.

In photodiodes the dependence of carrier separation on the applied bias makes it possible to create photodetectors with a given sensitivity regulated by the applied negative bias.

(2) Multilayer structures permit to further improve some of the photodetector characteristics, which is to be discussed using double heterostructures as an example.

14.2.1.2 Graded heterostructures

The above photoelectric properties of abrupt heterostructures show that the short-wave photosensitivity boundary of the photodetectors based on them is determined by the forbidden gap and the 'window' thickness. New

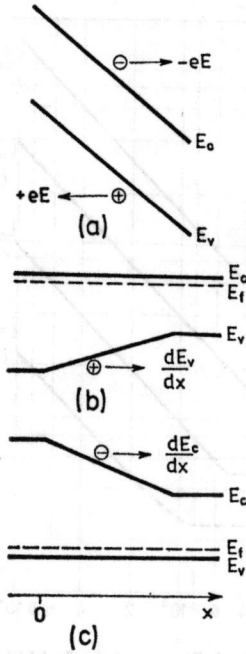

Figure 14.4 Schematic representation of the effect of the electric field on mobile carriers; (a) homogeneous material; (b), (c) graded-band gap materials

possibilities of broadening the short-wave photosensitivity boundary are found in graded heterostructures, which are particularly important for increasing the efficiency of solar energy converters.

Crystals with a graded forbidden gap width were first considered by Kroemer (13) in 1957. Kroemer has developed the Bardeen and Shockley's deformation potential theory, and has proved its applicability independently of the form and nature of the interatomic potential changes. The solution of the equation describing the behaviour of electrons in such a crystal has proved the existence of fields called 'quasi-electric'. The effect of these fields is different from that of an external electric field applied to a crystal, and that of the electric field of a p–n junction. The 'quasi-electric' field influences only one type of carriers (Figure 14.4).

Kroemer has shown that in these crystals the local density of states and distribution of electrons over energy at each point of a crystal may be described in a way similar to that applied to a homogeneous semiconductor of the same composition.

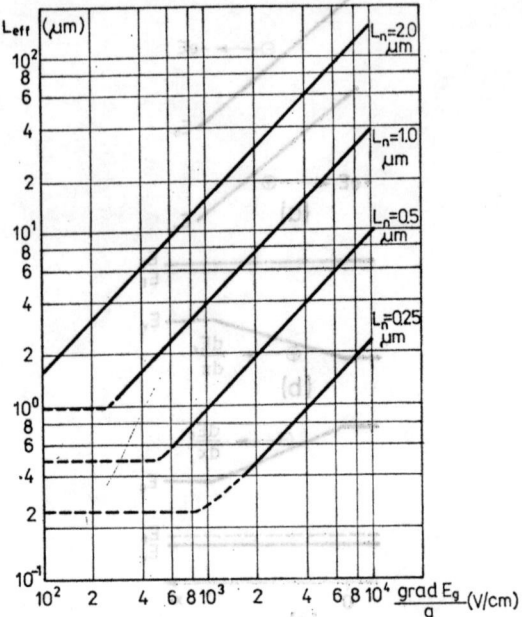

Figure 14.5 The influence of the forbidden gap gradient on the diffusion length of minority carriers

Recently, the development of methods of obtaining graded GaAs–AlAs heterostructures has attracted attention to the study of their photoelectric properties.

It should be noted that electric fields due to the impurity gradient have long been known and widely used in semiconductor electronics. However, their values are not high, and their application in semiconductor devices improves some of the parameters but deteriorates others. It is graded heterostructures only that permit control of the movement of minority carriers without changing the doping level. The forbidden gap gradient being present in a uniformly doped material, the minority carriers diffusion length increases while they move in the direction of decreasing the forbidden gap width. The effective diffusion length is

$$L_{\text{eff}} = \frac{1}{\frac{\operatorname{grad} E_g}{2kT} \left[\sqrt{1 + \left(\frac{2kT}{L_n \operatorname{grad} E_g}\right)^2} - 1 \right]} \qquad (14.10)$$

The greater the diffusion length in the initial material of constant composition, the greater this increase.

Calculated values of diffusion length in the initial material of constant composition versus E_g gradient and L_n in $Al_xGa_{1-x}As$ solid solutions are given in Figure 14.5.

Photoeffect and photocells on the basis of crystals with graded forbidden gap were first discussed in References 14, 15.

Various photocells on the basis of graded GaAs–AlAs heterostructures (Figure 14.6) have been suggested and created by now (16, 17, 18, 19) Their

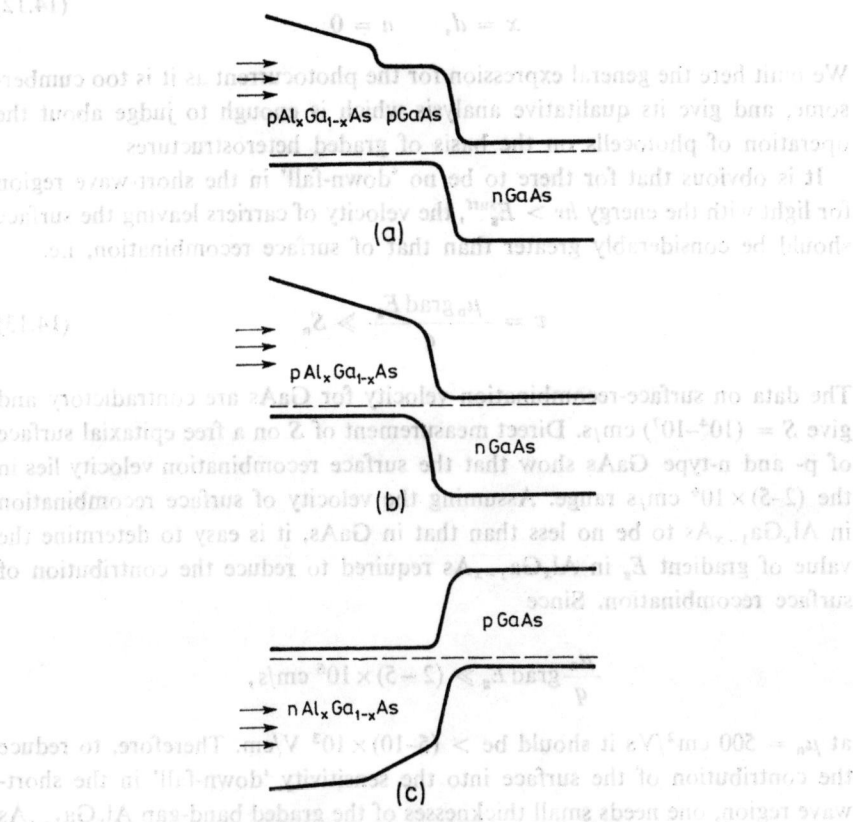

Figure 14.6 Various structures of graded band-gap heterojunction solar cells

spectral characteristics were calculated taking into account concrete properties of $Al_xGa_{1-x}As$ solid solutions with a graded forbidden gap. Thus, calculation of the photocurrent in the structure in Figure 14.6(b) (16) implies that diffusion-drift equation for electrons in the p-region ($m_n^* = $ const) can be calculated

$$D_n \frac{\partial^2 n}{\partial x^2} + \mu_n \operatorname{grad} E_g - \frac{n}{\tau_n} + g(x) = 0 \qquad (14.11)$$

with the boundary conditions

$$x = 0, \qquad D_n \frac{dn}{dx} + n\mu_n \operatorname{grad} E_g = S_n$$

$$x = d, \qquad n = 0 \qquad (14.12)$$

We omit here the general expression for the photocurrent as it is too cumbersome, and give its qualitative analysis which is enough to judge about the operation of photocells on the basis of graded heterostructures.

It is obvious that for there to be no 'down-fall' in the short-wave region for light with the energy $h\nu > E_g^{\text{surf}}$, the velocity of carriers leaving the surface should be considerably greater than that of surface recombination, i.e.

$$v = \frac{\mu_n \operatorname{grad} E_g}{q} \gg S_n \qquad (14.13)$$

The data on surface-recombination velocity for GaAs are contradictory and give $S = (10^4 - 10^7)$ cm/s. Direct measurement of S on a free epitaxial surface of p- and n-type GaAs show that the surface recombination velocity lies in the $(2-5) \times 10^6$ cm/s range. Assuming the velocity of surface recombination in $Al_xGa_{1-x}As$ to be no less than that in GaAs, it is easy to determine the value of gradient E_g in $Al_xGa_{1-x}As$ required to reduce the contribution of surface recombination. Since

$$\frac{\mu_n}{q} \operatorname{grad} E_g \gg (2-5) \times 10^6 \text{ cm/s},$$

at $\mu_n = 500$ cm^2/Vs it should be $> (5-10) \times 10^3$ V/cm. Therefore, to reduce the contribution of the surface into the sensitivity 'down-fall' in the short-wave region, one needs small thicknesses of the graded band-gap $Al_xGa_{1-x}As$ layers (0.2–10 μm). The study of graded nGaAs–pGaAs ($d = 0.5$ μm)––$pAl_xGa_{1-x}As$ heterophotocells with x increasing from 0 to $x = 0.3$–0.4 (Figure 14.7(a)) (and with different variable composition layer thicknesses), has shown that only pulling fields of 10^4 V/cm make the surface recombination contribution minimum and ensure high light sensitivity in the energy interval (1.9–3.1) eV, i.e. exceeding E_g^{surf} ($h\nu > E_{g_2}$, Figure 14.7(a)).

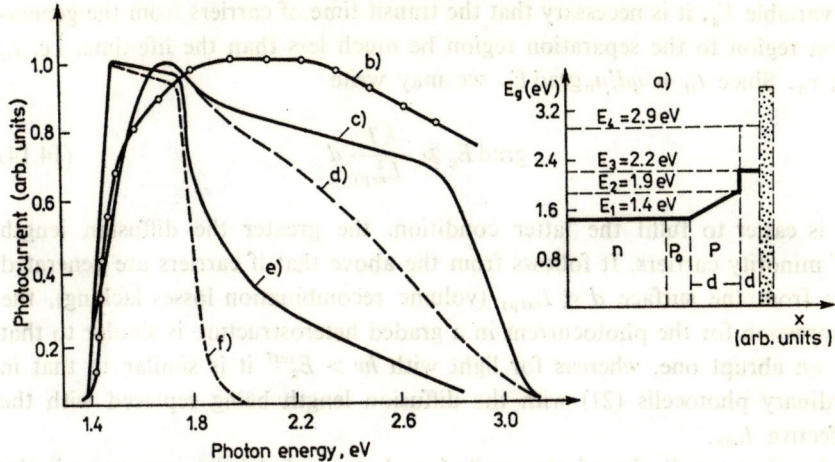

Figure 14.7 Spectral response of solar cells on the basis of GaAs-AlAs heterostructures; (a) forbidden gap-vs.-distance diagram; (b) $I/q\,\mathrm{grad}\,E_g \sim 10^4$ V/cm; (c) $I/q\,\mathrm{grad}\,E_g \sim 10^3$ V/cm and $Al_{0.8-0.9}Ga_{0.2-0.1}As$ layer; (d) $I/q\,\mathrm{grad}\,E_g \sim 10^3$ V/cm without $Al_{0.8-0.9}Ga_{0.2-0.1}As$ layer; (e), (f) abrupt heterostructures nGaAs-pGaAs-pAl$_{0.25}$Ga$_{0.75}$As ($d = 2$ μm) without and with the $Al_{0.8-0.9}Ga_{0.2-0.1}As$ layer

Photoresponse spectra of photocells on the base of graded heterostructures with different d-region thicknesses, as well as those of 'ordinary' heterophotocells on the base of abrupt heterostructures, are shown in Figure 14.7. It can be seen that although photocells with a thin $Al_xGa_{1-x}As$ layer possess a high sensitivity in the shortwave region, they have an essential fault, i.e. a high surface layer resistance, which makes it difficult to use them in solar batteries, especially in concentrated light conditions. Structures with a variable composition region thickness of (4–5) μm and with a thin protective $Al_{0.8-0.9}Ga_{0.2-0.1}As$ layer, which is practically transparent to light with $hv < (1.9-3.0)$ eV (20), are more promising for solar batteries. In this case carriers created by a short-wave light with $hv = (1.9-3.0)$ eV are separated from the photocell surface by a potential barrier $\Delta E = E_3 - E_2 \gg kT$, which reduces the surface recombination contribution. The AlAs layer being etched off the sensitivity in the short-wave region (curve B) decreases due to surface recombination.

The second condition necessary for obtaining high efficiency is that volume recombination losses should be small. This condition is, naturally, a less rigid one. It is obvious that to reduce recombination losses in a layer with a variable E_g, it is necessary that the transit time of carriers from the generation region to the separation region be much less than the life-time, i.e. $t_{tr} \ll \tau_n$. Since $t_{tr} = qd/\mu_n \mathrm{grad}\, E_g$, we may write

$$\mathrm{grad}\, E_g \gg \frac{kT}{L_{n(p)}^2} d \qquad (14.14)$$

It is easier to fulfil the latter condition, the greater the diffusion length of minority carriers. It follows from the above that if carriers are generated far from the surface $d \ll L_{n(p)}$ (volume recombination losses lacking), the expression for the photocurrent in a graded heterostructure is similar to that in an abrupt one, whereas for light with $h\nu > E_g^{\mathrm{surf}}$ it is similar to that in ordinary photocells (21) with the diffusion length being replaced with the effective L_{eff}.

Another peculiarity of photocells based on graded heterostructures is the possibility of increasing the open-circuit voltage (V_{oc}) at high levels of illumination due to the volume photo e.m.f. of the variable forbidden gap crystals added. This accounts for the interest taken in photocells based on graded heterostructures.

When illuminating a graded p-type heterostructure with a light flux parallel to the forbidden gap gradient, in the region of low excitation levels there occurs photo e.m.f. (22) which considerably exceeds the Dember e.m.f. in a homogeneous material of the same concentration and is

$$V = b\frac{\Delta E_g}{qp_0}N' \qquad (14.15)$$

where $b = \mu_n/\mu_p$, and N' is the non-equilibrium carrier concentration on the illuminated surface with account taken of carrier removal in the field direction, and surface recombination.

At high illumination levels the volume photo e.m.f. increases, and at $\Delta n \gg p_0$ it reaches (14)

$$V_v = \int_0^d \left(\frac{b}{b+1} \cdot \frac{1}{q} \cdot \frac{\mathrm{d}E_g}{\mathrm{d}x} + \frac{kT}{q} \cdot \frac{b-1}{b+1} \cdot \frac{\mathrm{d}\ln \Delta n}{\mathrm{d}x} \right) \mathrm{d}x \qquad (14.16)$$

Here, d is the variable composition region thickness, and the last term represents the Dember e.m.f. due to ambipolarity of the diffusion process. It is evident that with $b \gg 1$ and neglecting the Dember e.m.f., this voltage approaches the difference in the forbidden gaps. In Reference 23, it is shown that at high-level injection the voltage-current characteristic of a graded heterostructure becomes not ohmic but more abrupt, which opens prospects for creating high-efficiency photoconverters.

Experimental observation of the volume photo e.m.f. in a graded heterostructure was first described in (24) where two types of structures with different variable composition region thicknesses were studied: (i) with $Al_xGa_{1-x}As$ thickness d of 5 μm and concentration $p_0 \sim 5 \times 10^{17}$ cm^{-3}, and (ii) with $d = 2$ μm, $p_0 = 2 \times 10^{18}$ cm^{-3}. In both cases the forbidden gap on the illuminated surface was $E_g^{surf} = 1.9$ eV and fell down to 1.4 eV.

In structures with a 'thick' variable composition layer, at low-level illumination ($P \sim (10.3$–$0.5)$ W/mm^2), the photo e.m.f. did not exceed 20–30 mV, which is in good agreement with the calculation of expression 14.15. Linear dependence was observed within this illumination range. At high illumination levels photocurrent depends linearly on illumination $I_{ph} \sim P$, and the photo e.m.f. grows linearly at first, and then tends to saturation. The maximum values obtained were $V_{oc} = 0.3$ V for 'thin' specimens and $V_{oc} = 0.18$ V for the 'thick' ones.

The loading characteristic is non-linear for the case of a high excitation level. The above experiments open new prospects for further increase of the efficiency of solar converters provided that concentrated solar radiation is used.

In conclusion, it should be emphasized that heterostructures make it possible not only to govern the carrier fluxes (i.e. one-sided injection, super injection, electron confinement) (25) but also

(1) to govern light fluxes (input and output of radiation without absorption through a wide-gap semiconductor);
(2) to introduce radiation directly into the space-charge region of a narrow-gap material, which leads to generation of minority carriers and separation of regions combined and, therefore, increases the high speed response of photodetectors. This property also permits to study the interaction of light with the strong electric field region;
(3) to create pulling and breaking electric fields and, therefore, to govern the diffusion length of minority carriers without changing the doping level.

The above possibilities are rather general and apply to heterostructures described by the ideal Anderson model; they are of great importance for all semiconductor heterojunction devices since they permit to improve their parameters and to develop new principles of their construction.

14.3 PHOTODETECTORS AND CONVERTERS BASED ON HETEROSTRUCTURES

14.3.1 The main parameters

Converters of light energy into electric energy are described by a number of characteristics the most important of which are the spectral sensitivity region, monochromatic sensitivity S_λ, high speed response, the sensitivity threshold or the inverse value of detectivity. As has been noted before, application of heterostructures in photodetectors makes it possible to improve considerably most of the above characteristics which are to be considered on a heterophotodiode.

14.3.1.1 Spectral characteristics

Spectral sensitivity distribution of a photodetector is determined as short-circuit photocurrent versus incident light wavelength. Since the interaction of radiation with the substance has a selective character, spectral characteristics of photodetectors on the base of p-n junctions are usually represented as continuous curves with one maximum. The long-wave sensitivity boundary is determined by the forbidden gap of the detector material, whereas the short-wave boundary—by numerous factors the most important of which are the depth of a p-n junction, absorption coefficient, diffusion length of minority carriers, and surface recombination velocity. The long-wave sensitivity in heterophotodiodes is also determined by the forbidden gap of the narrow-gap material of the pair. However, for light with $h\nu > E_g$ high values of the collection coefficient close to unity are observed within the 'window' limits. It follows from (14.5) that it has practically no maximum as regards k_1. This is of great practical importance for materials with small values of the minority diffusion length and a high surface recombination velocity.

Heterostructure components form a continuous series of solid solutions; control of the heterojunction component composition implies a possibility of varying the spectral sensitivity region and, therefore, of obtaining photodetectors in a given narrow region, i.e. selective photodetectors. A selective photodetector was first described in (26). Its sensitivity width was $\Delta h\nu$ = 22 mV and $Q = 0.67$. This type of selective photodiodes on GaAs–AlAs heterojunctions was obtained in the 1.4–1.95 eV range.

The constant sensitivity region of photodetectors on the base of abrupt GaAs–Al$_x$Ga$_{1-x}$As heterostructures covers the range of 1.4–2.0 eV, which shows that their spectral characteristic is in a good agreement with the most widely used modern semiconductor light sources. Photodetectors on the base of GaSb–AlGaSb (27) and InP–$Ga_xIn_{1-x}As_{1-y}P_y$ (28) and GaAs–$GaAs_{1-x}Sb_x$ (29) heterojunctions are of interest for another important spectrum region $\lambda = (1.06–1.3)$ μm.

14.3.1.2 Speed response

High speed response is one of the main requirements with regard to photodetectors in optoelectronics. With corresponding doping levels of both heterostructure parts, radiation within the 'window' is absorbed in the depletion layer of the narrow-gap material, which means that the electron-hole pair generation and the separation regions coincide. Therefore, high speed response of heterophotodiodes is determined only by the transit-time of carriers in the space-charge region ($t_{tr} \sim w/\mu E$) and the barrier capacity.

14.3.1.3 Sensitivity threshold

The least detected radiation power at which the photoresponse is equal to noise is the sensitivity threshold P_{thr}. To obtain a low sensitivity threshold it is necessary to reduce the dark current of the photodiode. In most detectors an increase in the signal/noise ratio is generally achieved by decreasing their square. Application of heterojunctions in a number of cases permits a considerable reduction of dark currents. That it is possible to considerably reduce the inverse current in heterostructures based on materials with large diffusion lengths and a high minority carriers concentration compared to that in a p–n homojunction of a narrow-band material of a pair has been demonstrated in Reference 30.

Data on threshold sensitivity of heterophotodetectors are almost completely lacking in the literature. In Reference 31 heterophotodiode detectivity for different material pairs is calculated on the assumption that the ideal model is true (but without taking into account the interface effect). These calculations are purely qualitative and yield D^* for heterophotodiodes rather higher than that for a p–n homojunction in a narrow-band material. Experimental study of the detectivity of nGaAs–pAl$_x$Ga$_{1-x}$As heterophotodetectors (32) has shown that they may be used as threshold detectors ($D^* \sim 3 \times 10^{13}$ cm W^{-1}s$^{-1}|^2$). Besides, conditions are found under which detectivity is close to the maximum possible value.

14.3.2 Heterophotodiodes

At present GaAs–AlAs heterostructures are widely used for creating high-efficiency solar energy converters (10, 11, 19, 33). A single heterostructure used as an effective fast-operated photodiode has received much less attention.

This is due to the fact that creation of heterophotodiodes with a wide bandwidth and a high sensitivity in a given spectral interval requires the use of a lightly-doped narrow-band material, i.e. GaAs. Recently, n⁺GaAs–n°GaAs–pAl$_x$Ga$_{1-x}$As structures (9) have been grown and studied which is associated with progress in LPE technology. The latter chapter analyses the choice of an optimum photodiode structure for obtaining high $Q = 1$ within the spectral sensitivity region $h\nu = (1.4$–$1.9)$eV and high speed response of $\sim (10^{-9}$–$10^{-10})$s.

As mentioned above, high speed response is determined by two processes only: the space-charge region transit-time of minority carriers and the time of circuit RC-relaxation, the two processes being interrelated. The upper frequency limit of $\sim 10^{-10}$ s, with the quantum efficiency $Q = 1$ conserved, is determined only by the life-time of the minority carriers in the space-charge region. Close to zero voltage the inverse dark currents did not exceed $\sim 10^{-10}$ A/cm², no special surface protection having been applied. Break-down voltage was $V > 100$ V.

The possibility of separating minority carriers with a discontinuity in the conductivity band of pGaAs–nAl$_x$Ga$_{1-x}$As heterojunctions opens new prospects for further improvement of heterophotodiode parameters in the long-wave spectral characteristics region due to more complete collection of light generated carriers in the narrow-band region (35). Thus, in a double heterostructure Al$_{x_1}$Ga$_{1-x_1}$As–pGaAs (Al$_{x_2}$Ga$_{1-x_2}$As)–nAl$_{x_3}$Ga$_{1-x_3}$As ($x_2 < x_1, x_3$), an energy-band diagram of which is shown in Figure 14.1(d) (the well-known heterolaser structure), with certain ratios between active region thickness and the electron diffusion length, all of the light-generated carriers are separated by the p–n junction electric field. This effect is explained by the fact that electrons are reflected by a barrier in the conductivity band of a p–p heterojunction. This is true only if the barrier in the conductivity band at a pGaAs–nAl$_x$Ga$_{1-x}$As anisotype junction interface has a 100% transparency.

Assuming the recombination rate at a p–p heterojunction interface to be equal to that in the volume and $T(V) = 1$, the photocurrent may be written as

$$I_{ph} = qa_1\Phi\left[1 - e^{-k_1 w_0} + \frac{k_1 L_n}{k_1 L_n^2 - 1}\left(k_1 L_n - \tanh\frac{d}{L_n}\right.\right.$$

$$\left.\left. - k_1 L_n\left(\operatorname{sech}\frac{d}{L_n}\right)e^{-k_1 d}\right)e^{-k_1 w_0}\right]T(V) \qquad (14.17)$$

where d is the distance between the space-charge region and the p–p-junction.

It can be seen from (14.17) that the photocurrent depends on the ratio between k_1, L_n and d. Estimations show that the photocurrent for the light absorbed behind the space-charge layer has the highest value when the potential pit width is $d \sim L_n$. Literature (36) gives descriptions of GaAs crystals of the p-type conductivity with L_n up to 20 μm, which makes it possible to broaden the long-wave limit to 1.35 eV.

Thus, the structure considered above makes it possible

(i) to increase (to some extent) the sensitivity in the long-wave region of spectral sensitivity;
(ii) to reduce dark currents in materials with a large diffusion length of the minority carriers. This is achieved by both the choice of the doping levels, and the structure geometry. The active layer thickness d should be $d \ll L_{n(p)}$ and should not exceed the light absorption depth ($d \sim 1/k$), which is particularly important in the case of narrow-gap materials;
(iii) depending on the polarity of the applied bias, to be used in the regime of both a detector and an effective light emitting diode;
(iv) to perform effective light modulation based on the Franz–Keldysh effect (37, 38).

14.3.3 Photodetectors with internal gain

These include avalanche photodiodes, phototransistors, and photodetectors with S-shaped voltage-current characteristics (avalanche phototransistors, photothyristors, S-diodes). Photodetectors with internal gain are increasingly more widely used in modern electronics which is associated with the development of optical communication systems and optical information processing methods. Compared with photodiodes, available photodetectors of this type developed on the base of Ge and Si suffer from grave defects. They have a worse temperature stability, greater dark current and noise levels, and lower maximum frequency. These defects apply to all detectors with internal gain. As will be shown below, the above defects may be avoided to a great extent if one uses wide-gap A_3B_5 materials with a direct band structure. Nevertheless, the advantages of these materials had for a long time remained unrealized which was related to difficulties in obtaining homogeneous epitaxial layers with sufficient diffusion lengths of minority carriers and of required doping level and thickness.

Thus, for instance, the gain coefficient of a phototransistor (39) is $k \simeq 2(L_{n(p)}/w_b)^2$, where w_b is the base region thickness and $L_{n(p)}$ the corresponding diffusion length of minority carriers in the base region. It is seen that to obtain $k > 100$ at $L_{n(p)} \sim 1$ μm the base region thickness should be $\ll 1$ μm.

In photothyristors where the voltage is controlled by creating non-equilibrium carriers in the base structure regions with the help of light, the number of electrons and holes reaching p–n junctions is determined by a corresponding transport factor, i.e.

$$\beta \simeq 1 - \frac{1}{2}\left(\frac{L}{L_{n(p)}}\right)^2$$

where L is the distance between the regions of minority carriers generation and the p–n junctions.

To increase the sensitivity it is necessary to minimize the recombination losses in the base regions, which means that requirements with regard to the thickness of at least one of the base regions are similar to those in phototransistors. Moreover, it should be taken into account that in direct semiconductors the effective depth of light penetration within the region of fundamental absorption is small, whereas the surface recombination velocity is great ($S \sim 10^6$–10^7 cm/s). Therefore, almost all of the light generated carriers recombine close to the surface, which imposes additional requirements on the thickness of the emitter or collector layer through which the light flux is introduced. Hence, the only way to overcome these difficulties is to use structures with heterojunctions.

Heterophototransistors and photothyristors with high sensitivity and high operation speed have been widely reported. Of special interest are electroluminescent photothyristors, since the presence of an S-shaped voltage-current characteristics in detectors considerably broadens the sphere of their application and opens new prospects for creating and simplifying optoelectronic schemes.

As for avalanche heterophotodiodes, their study has been started, in fact, just now, and the outlook is undoubtedly promising (40, 41).

14.3.3.1 Phototransistors

A heterophototransistor with a comparatively high gain coefficient was first reported in (42). It is a three-layer n–p–n GaAs structure with a wide-gap n–$Al_xGa_{1-x}As$ emitter through which radiation is introduced. This structure was chosen due to the fact that diffusion length of electrons in GaAs is considerably greater than that of holes. Transistors with the base region thickness from 0.8 to 6.0 μm have been studied. With the base thickness increasing, the gain coefficient falls from 300 at $w_b = 0.9$ μm to 10 at

$w_b = 4$ μm. Heterophototransistors with a similar energy-band diagram but smaller base thicknesses were reported in Reference 43. Their gain coefficient exceeded 2000, and their integrated sensitivity was about 700 A/W. The spectral sensitivity region lay in the 1.4–1.9 eV range which is typical of all GaAs–Al$_x$Ga$_{1-x}$As photodetectors. Dark currents depended on the gain coefficient and did not exceed $(10^{-7}–10^{-6})$ A/cm^2 at $V_{ec} = 2.0$ V; the current rise and drop-times at pulsed illumination were $\sim (10^{-6}–10^{-7})$ s.

One of the most important peculiarities of transistors with a wide-gap emitter pointed out by Shockley (44) is the constancy of the gain coefficient at high illumination levels, whereas in ordinary transistors a fall in the emitter efficiency at high injection levels leads to a fall in the gain coefficient. The above phototransistors had a constant gain coefficient within a wide illumination interval.

Operation of these heterophototransistors in an avalanche regime is reported in Reference 45 and is of great interest due to their high speed response and the possibility of obtaining high gain coefficients.

14.3.3.2 Electroluminescent photothyristors

A heterophotothyristor in the GaAs–AlAs system is proposed in (46). Figures 14.8(a), (b) show energy-band diagrams of such a thyristor. Radiation was introduced on the wide-gap side. It can be seen in the diagram that in such a structure the central junction shifted in the negative direction is immediately illuminated. This leads to an increase in the high speed response and in sensitivity, since recombination losses in the process of light generated minority carriers moving from the absorption region to the separation region are lacking. The on-voltage generally lay within the range of 10–20 V, the on-current did not exceed ~ 1 mA, and the off-current was about (5–10) mA, while the residual voltage was $V_{res} = 1.5$ V. The long-wave spectral sensitivity limit was determined by GaAs, whereas the short-wave limit was determined by the forbidden gap of the window and varied from 1.5 to 1.9 eV.

Photodetectors with an S-shaped voltage-current characteristic are advantageous for optrons due to the possibility of obtaining high gain coefficients both in current and in power. Therefore, the photothyristor governing characteristic, i.e. voltage versus incident light intensity, is of significance.

Evaluations of the minimum switching power

$$\Phi_{min} = \frac{h\nu}{qQ} I_{on} \{1 - [\alpha_1(V) + \alpha_2(V)]\}$$

show that at on-currents of about 10^{-6} A it reaches values of $\Phi_{min} = (10^{-8}–10^{-9})$ W.

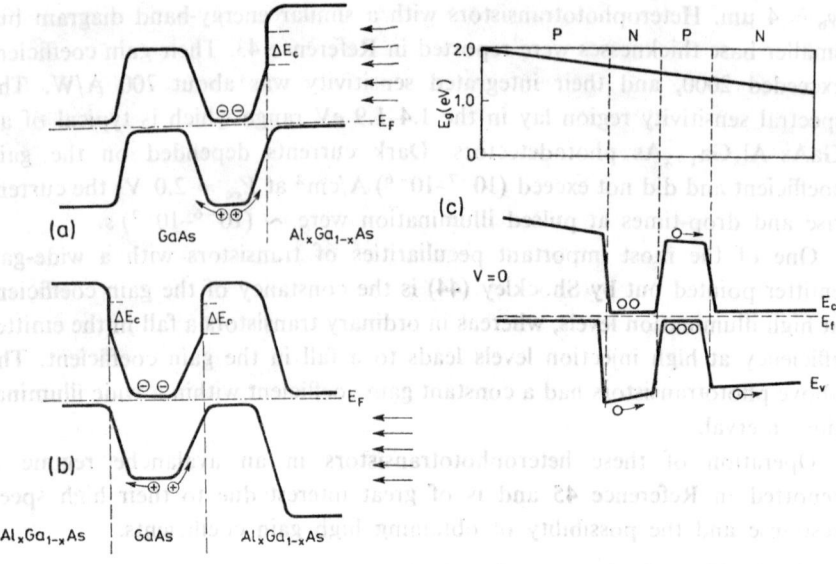

Figure 14.8 Electroluminescence photothyristors; (a), (b) energy-band diagrams on the basis of abrupt heterojunction; (c) forbidden gap-vs.-distance and energy-band diagram

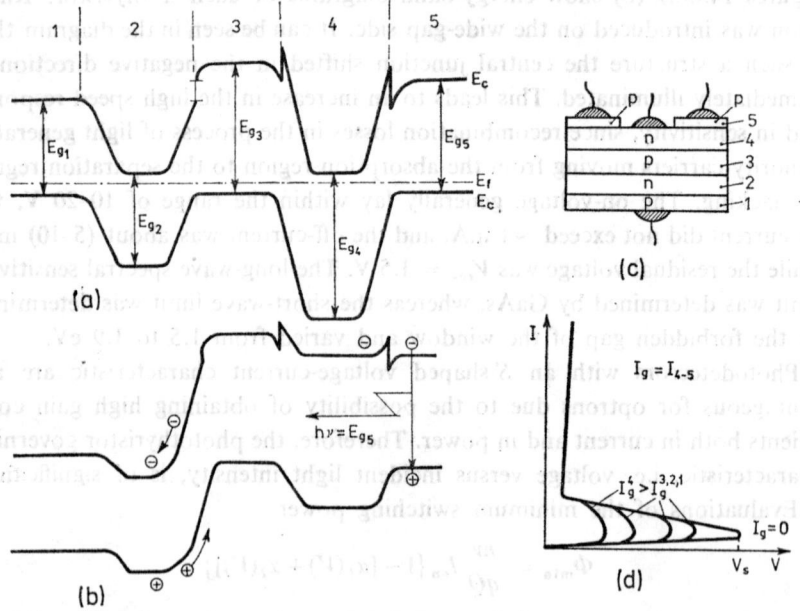

Figure 14.9 Schematic representation of a thyristor, its energy-band diagram and current-voltage characteristics

In the p–n–p–n structures under consideration, the on-time was 10–20 ns, and the off-time depended on passing current $t_{off} \sim \tau \cdot \ln j$ and was about 50 ns at $j = 150$ A/cm². A somewhat different type of electroluminescence photothyristor on the base of $Al_xGa_{1-x}As$ solid solutions, with a variable forbidden gap, doped with an amphoteric Si impurity is reported in References 47, 48. These p–n–p–n structures had the following parameters—photosensitivity within the range of 6000–7000 Å; on-voltage of 30 V, and off-time of ~ 20 ns. Figure 14.8(c) shows the behaviour of the forbidden gap change along the structure, and its energy-band diagram. When switched on the above p–n–p–n structures emit light which, taking into account their high sensitivity, allows in the simplest possible way to convert a low intensity light signal into a higher intensity one. The chosen energy-band diagram (Figure 14.8(a)) is such that, besides having a high photosensitivity, this photothyristor is also an effective light source. The use of a wide-gap emitter realizes one-sided injection and is favourable for light output without self-absorption, which shows that this structure is similar to an optimum LED structure (49).

When switched on this photothyristor emits light with the energy which, similarly as the long-wave sensitivity limit, is determined by the forbidden gap of the narrow band material. Radiation intensity grows linearly with current up to the point where the coherent radiation regime takes place. Generally, the external quantum efficiency in a flat construction is about $\eta_e \sim (0.15$–$0.25)\%$. Radiation power at $I = (0.05$–$0.1)$ A is about several milliwatt. Thus, the gain coefficient in the above structures is $k = \Phi_{em}/\Phi_{min} \sim 10^4$. This value may be increased up to $\sim 10^7$–10^8 provided that a p–n–p–n structure is used in the coherent radiation regime (50). Problems concerning creation of logical optoelectronic elements on the base of electroluminescence photothyristors are considered in Reference 51, where schemes and principles of operating different widely used logical elements are supplied. An interesting analogue of a coupling device on the basis of electroluminescence photothyristors is described in Reference 52.

In conclusion, it should be noted that high internal efficiency of radiation recombination and the possibility of governing the forbidden gap of structure layers, which ensures direct introduction of light into the space charge region of a collector junction—all these factors make it possible to consider the problem of creating a controlling electrode in thyristors (53) from an entirely new point of view. The use of an additional layer in a p–n–p–n structure (Figure 14.9) permits control of the on-voltage by means of an effective conversion of an electric signal into an optical one followed by a reversal conversion at the collector junction. The efficiency of such control is determined by the efficiency of the corresponding conversion in each stage and may be rather high if heterojunctions are applied. This method of control is of particular importance for large square thyristors in which spreading of the on-state plays a significant role.

14.3.4 Solid state infrared radiation converters

A number of interesting proposals concerning the creation of solid state infrared-to-visible converters has recently been made. This is due to the fact that converters based on heterojunctions enable one to shift the long-wave limit quite considerably compared to those based on vacuum photoelements.

The use of heterostructures for this purpose permits, in a number of cases, easy separation of the long-wave light absorption region in the narrow-gap material from that of preferential minority carriers recombination in the wide-gap part of the structure. Thus, a single isotype heterojunction with ΔE_c or $\Delta E_v \leqslant 0$ may be used as a converter in which the electric field transfers minority carriers from the narrow-gap region into the wide-gap-one.

Materials are known forming a continuous series of solid solutions on the basis of which it is possible to obtain crystals with a graded forbidden gap. It was these crystals that provided the basis for the anti–Stokes light converter (Figure 14.10(a)) proposed in Ref. 54. When an external electric field $qE > dE_c/dx$ is applied, the drift length of minority carriers increases and the light generated minority carriers from the narrow-gap region recombine into a more wide-gap region of the crystal. In such a structure the converted light wavelength varies continuously from λ_1 corresponding to the forbidden gap in the narrow-gap material, to $\lambda_{(x)}$ which is determined by the E_g of the region reached by the carriers.

An abrupt nGaAs–pGaAs–nGe heterojunction is reported in Reference 55. With positive bias, the holes created in Ge by infrared radiation diffuse in the direction of the heterojunction, and the contact field transfers them into GaAs. The long-wave sensitivity limit is $\lambda_{\text{inf}} = 1.5$ μm.

However, the efficiency of the above converters is not high since they lack an internal gain mechanism, and the amount of radiation quanta is always much smaller than that of the incident ones. Multilayer heterostructures make it possible to obtain considerable gain at infrared-to-visible conversion.

A n–p–n–p GaAs–AlAs heterostructure is used as a converter in Ref. 56. One of its base regions is prepared from a semi-insulating GaAs doped with

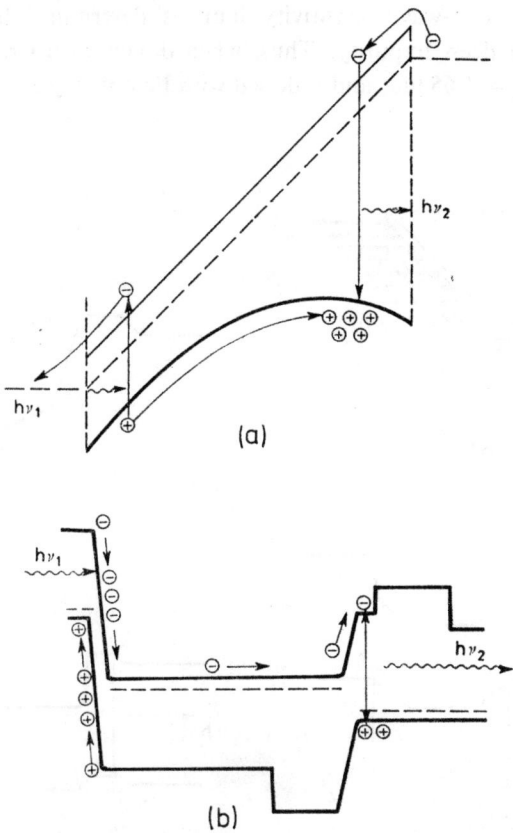

Figure 14.10 Various structures of solid-state infrared-wavelength converters based on (a) graded band gap (54); (b) abrupt heterostructures (58)

a deep lying impurity. Figure 14.11 shows a schematic picture of a solid-state converter, as well as its energy-band diagram and the operation principle. Switching conditions in such a structure are similar to those in an ordinary dinistor. Illuminated with light of energy $h\nu > E_c - E_d$, traps are being filled which leads to an increase in the minority carriers diffusion length (57). When a positive bias is applied at certain illumination levels, this structure may be switched on. In the on-state the structure emits visible light whose energy depends on $Al_xGa_{1-x}As$ composition. Radiation intensity grows linearly with current. The conversion coefficient η representing the ratio between the emitted and incident light intensities reaches values greater than 10^2. The long-wave sensitivity limit is determined by the energetic position of the deep impurity. Thus, when doped with Cr, semi-insulating GaAs yields $\lambda_{th} = 1.65$ µm, and if doped with Fe it may yield $\lambda_{th} = 3$ µm. The

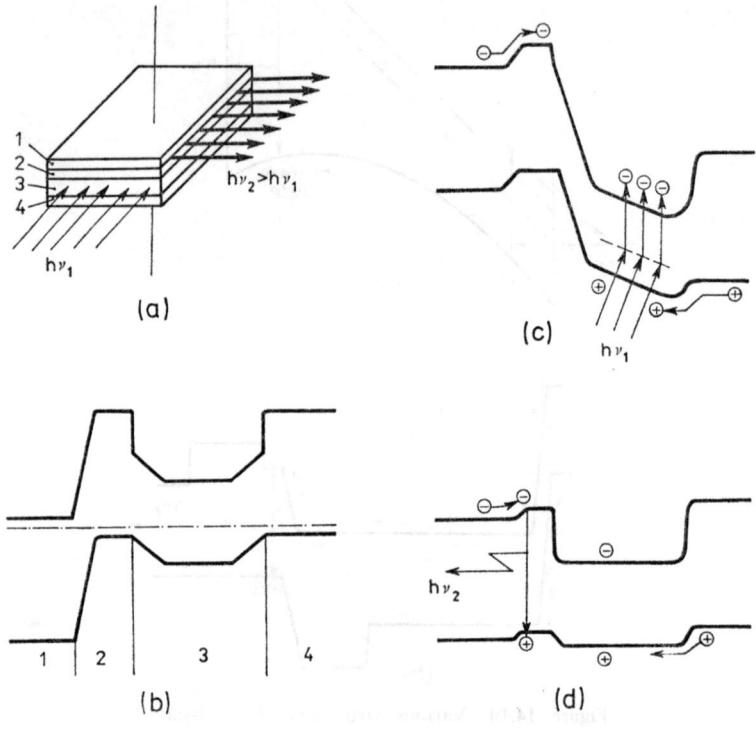

Figure 14.11 Schematic representation and operation principles of solid-state converters based on a p–n–si–p heterostructure (56); (a) schematic diagram of the structure and its energy-band diagram; (b) $V = 0$; (c) $V > 0$; (d) in the turned-on state

defect of such a converter as well as of any key device is the impossibility of obtaining characteristics with no special feeding system being used. A converter suggested in Reference 58 is free from this defect. The principle of its operation is shown in Figure 14.11(b). Here, a GaAs photodiode operating in the avalanche multiplication regime serves as an infrared radiation photodetector and as an amplifying element. The efficiency of this converter is determined by the collection and multiplication coefficients of the photodetector and by the external electroluminescence efficiency, and is about 1.5%.

14.4 CONCLUSION

A decade has nearly passed since heterojunction investigations were initiated. However, the above short review of the literature data clearly shows the progress achieved in the development of heterophotocells. These devices have better parameters than similar p–n homojunction devices, and in a number of cases they even have no analogues (as with selective photodiodes, electroluminescence photothyristors, solid state light converters). Most of the attention has been paid to photocells on the basis of GaAs–AlAs heterojunctions. The principles of their construction and the results of appropriate investigations promise to be useful for the creation of effective heterophotocells on the basis of multicomponent A_3B_5 solid solutions, as well as for the rapidly developing integrated optics.

REFERENCES

1. Zh. I. Alferov, V. M. Andreyev, V. I. Korol'kov, E. L. Portnoy, D. N. Tret'yakov, in *Physics of Electron-Hole Transitions*, Nauka, Moscow–Leningrad, 1969.
2. Zh. I. Alferov, 'Injection Heterolasers', p. 204, in *Semiconductor Devices and their Application*, issue **25**, Soviet Radio, Moscow, 1971.
3. M. B. Panish, *Proc. IEEE*, **64**, 1512 (1976).
4. D. Z. Garbuzov, 'Semiconductor Sources of Electromagnetic Radiation', *Proceedings of the First International School on Semiconductor Optoelectronics, Cetniewo, Poland 1975*, p. 91, Polish Scientific Publishers, Warszawa, 1976.
5. R. L. Anderson, *Sol. St. Electron.*, **5**, 341 (1962).
6. B. Agusta, R. L. Anderson, *J. Appl. Phys.*, **36**, 206 (1965).
7. Ya. V. Bergmann, V. I. Korol'kov, V. R. Larionov, V. G. Nikitin, *FTP*, **10**, 1933 (1976).
8. C. M. S. Ze, *Physics of Semiconductor Devices*, New York, 1969.
9. Ya. V. Bergmann, V. I. Korol'kov, H. Rakhimov, *FTP*, **11**, 1848 (1977).
10. V. M. Andreyev, T. M. Golovner, M. B. Kagan, N. S. Koroleva, T. L. Lyubashevskaya, T. A. Nuller, D. N. Tret'yakov, *FTP*, **7**, 2289 (1973).
11. V. M. Andreyev, M. B. Kagan, T. L. Lyubashevskaya, T. A. Nuller, D. N. Tret'yakov, *FTP*, **8**, 1328 (1974).
12. V. I. Korol'kov, V. G. Nikitin, D. N. Tret'yakov, *FTP*, **8**, 2355 (1974).
13. H. Kroemer, *RCA Rev.*, **18**, 332 (1957).
14. J. Take, *Rev. Mod. Phys.*, **29**, 308 (1957).
15. P. R. Emtage, *J. Appl. Phys.*, **33**, 1950 (1962).
16. M. Konagai, K. Takahashi, *J. Appl. Phys.*, **46**, 3542 (1975).
17. J. M. Hutchby, R. L. Fudurich, *J. Appl. Phys.*, **47**, 3140 (1976).
18. Zh. I. Alferov, V. M. Andreyev, M. B. Kagan, V. I. Korol'kov, T. S. Tabarov, F. M. Tadzhibayev, *Pis'ma v ZhTF*, **3**, 725 (1977).
19. J. M. Woodall, H. J. Hovel, *Appl. Phys. Lett.*, **30**, 492 (1977).
20. B. Monemar, K. Shik, G. D. Pettit, *J. Appl. Phys.*, **47**, 2604 (1976).
21. T. S. Moss, G. J. Burrell, B. Ellis, *Semiconductor Opto-Electronics*, Butteworth Co. Publishers, LTD, 1973.

22. O. V. Konstantinov, G. V. Tzarenkov, *FTP*, **10**, 720 (1976).
23. V. M. Yevdokimov, A. F. Milovanov, D. S. Strebkov, *FTP*, **11**, 2224 (1977).
24. Zh. I. Alferov, V. M. Andreyev, Yu. M. Zadiranov, V. I. Korol'kov, N. Rakhimov, T. S. Tabarov, *Pis'ma v ZhTF*, **4**, 7 (1978).
25. Zh. I. Alferov, *Proc. Int. Conf. on the Phys. and Chem. Semiconductor Heterojunctions and Layer Structures, VIII*, Budapest, 1971.
26. Zh. I. Alferov, V. M. Andreyev, O. A. Ninua, I. I. Protasov, V. G. Trofim, *FTP*, **5**, 988 (1971).
27. Zh. I. Alferov, N. S. Zimogorova, S. G. Konnikov, I. Matkova, L. D. Pramatarova, D. N. Tretyakov, *Pis'ma v ZhTF*, **1**, 641 (1975).
28. L. M. Dolginov, H. Ibrakhimov, V. Yu. Rogulin, E. G. Shevchenko, *FTP*, **10**, 1224 (1976).
29. R. C. Eden, *Proc. IEEE*, **63**, 32 (1975).
30. Zh. I. Alferov, M. Z. Zhingarev, V. I. Korolkov, N. Mursakov, L. D. Pramatarova, D. N. Tretyakov, *FTP*, **12**, 312 (1978).
31. B. L. Sharma, S. N. Mukerjee, J. K. Modi, *Infrared Physics*, **11**, 207 (1971).
32. N. B. Luk'yantchikova, B. D. Solganik, M. K. Sheinkman, Zh. I. Alferov, I. I. Protasov, V. G. Trofim, *FTP*, **6**, 2235 (1972).
33. Zh. I. Alferov, V. M. Andreyev, M. B. Kagan, I. I. Protasov, V. G. Trofim, *FTP*, **4**, 2378 (1970).
34. J. M. Woodall, H. J. Hovel, *Appl. Phys. Lett.*, **21**, 379 (1972).
35. Ya. V. Bergmann, V. I. Korolkov, V. R. Larionov, V. G. Nikitin, *FTP*, **10**, 1933 (1976).
36. M. Ettenberg, H. Kressel, S. L. Gilbert, *J. Appl. Phys.*, **44**, 827 (1973).
37. N. Bottka, L. Hutcheson, *J. Appl. Phys.*, **46**, 2645 (1975).
38. J. C. Dyment, F. P. Kapion, *J. Appl. Phys.*, **47**, 1523 (1975).
39. S. M. Ryvkin, *Photoelectronic Phenomena in Semiconductors*, Physmathgiz, Moscow, 1963.
40. J. K. Grierson, S. O. Hara, *Sol. St. Electr.*, **18**, 1003 (1975).
41. L. D'Auria, B. Cremax, *Rev. Techn. Thomson-CSF*, **6**, 919 (1974).
42. Zh. I. Alferov, F. A. Akhmedov, V. I. Korol'kov, V. G. Nikitin, *FTP*, **7**, 1159 (1973).
43. H. Beneking, P. Mischel, G. Schul, *Electr. Letters*, **12**, 395 (1976).
44. W. Shockley, *Pat. USA*, **2**, 569, 347 (1951).
45. F. A. Akhmedov, V. I. Korol'kov, L. P. Pershina, A. A. Yakovenko, *ZhTF*, 138 (1975).
46. Zh. I. Alferov, F. A. Akhmedov, V. I. Korol'kov, A. A. Yakovenko, *FTP*, **8**, 1741 (1974).
47. J. Arai, M. Sakuta, K. Sakoi, *Jap. J. Appl. Phys.*, **9**, 853 (1970).
48. S. Xano, T. Sakura, I. Inoguchi, 'Proc. 2nd Conf. on Solid-State Devices, Tokyo, 1970, *Suppl. to the Journal Jap. Soc. of Appl. Phys.*, **40**, 166 (1971).
49. Zh. I. Alferov, V. M. Andreyev, V. I. Korol'kov, E. L. Portnoy, A. A. Yakovenko, *FTP*, **3**, 930 (1969).
50. Zh. I. Alferov, V. M. Andreyev, V. I. Korol'kov, V. G. Nikitin, E. L. Portnoy, A.A. Yakovenko, *FTP*, **6**, 739 (1972).
51. V. I. Korol'kov, Yu. M. Makushenko, *Microelectronics*, **7**, 133 (1978).
52. Z. Tani, K. Murata, T. Sakura, T. Inoguchi, *Proc. of the 4th Conf. on Solid-St. Dev.*, Tokyo, 1972.

53. V. I. Korol'kov, V. G. Nikitin, N. Rakhimov, *Pis'ma v ZhTF*, **2**, 541 (1976).
54. T. Van Rayven, P. Williams, *Sol. St. Electron.*, **10**, 1159 (1967).
55. P. Kruse, *J. Appl. Phys.*, **38**, 1718 (1967).
56. Zh. I. Alferov, V. I. Korol'kov, V. G. Nikitin, D. N. Tret'yakov, *FTP*, **5**, 1503 (1971).
57. Zh. I. Alferov, V. K. Yergakov, V. I. Korol'kov, V. G. Nikitin, D. N. Tret'yakov, A. A. Yakovenko, *FTP*, **4**, 578 (1970).
58. H. Beneking, G. Schul, P. Mischel, A. Gattung, *Electron. Letters*, **10**, 347 (1974).

Part IV
Injection Lasers

Chapter 15

A Unified Approach to the Problems of Semiconductor Laser Theory

M. J. Adams

15.1 INTRODUCTION

This chapter is concerned with some of the fundamental problems of semiconductor laser theory; more specifically, let us ask the questions:

(i) *which problems remain to be solved?*
(ii) *of the remainder, which solutions are the most appropriate for given situations?*
(iii) *how much influence on laser development have the theoretical solutions had?*

A conventional starting-point would be to discuss the historical progress of the laser from the early days of a low-power, high-threshold device requiring pulsed operation and/or a low-temperature environment, to the current situation of a wide variety of laser structures adapted for CW room-temperature operation, good mode control, ease of modulation, etc. A brief history of theoretical topics should include (chronologically) the reasons for high thresholds, the temperature dependence of threshold current, the role of dielectric waveguiding, the problem of heat dissipation, the gain-current relationship, transient behaviour (noise, resonance frequencies, 'spiking', modulation, time delays, Q-switching), electromagnetic mode control, carrier transport phenomena and many other topics.

However, rather than following along the lines suggested in the preceding paragraph, I propose to take a more formal approach and discuss these topics in an ordered way, as shown in Table 15.1. The table contains three columns headed respectively 'electrons', 'photons', and 'phonons', and five rows labelled 'x-direction', 'y-direction', 'z-direction', 'time', and 'energy'. The object of this structure is to categorize each fundamental effect associated with a particular set of quasi-particles by its spatial, temporal and spectral behaviour. Thus each available space in the Table contains an entry relating

Table 15.1 Fundamental physical phenomena in semiconductor lasers

	Electrons	Photons	Phonons
x-direction (normal to the junction plane)	carrier transport normal to the junctions	waveguiding normal to the junction plane	heat transfer from active layer to heat sink
y-direction (in the junction plane, parallel to the facets)	current spreading in the junction plane (associated with stripe 'architecture')	waveguiding associated with stripe 'architecture' (real or virtual guidance)	heat diffusion associated with stripe 'architecture'
z-direction (in the junction plane, perpendicular to the facets)	longitudinal carrier diffusion	longitudinal field distribution and its influence on catastrophic degradation	longitudinal temperature variations, stimulated Brillouin emission
time	modulation, 'spiking', noise effects in carrier currents, resonance, time delays, Q-switching	modulation, 'spiking', noise effects in lasing output, resonance, time delays, Q-switching	transient heating effects, long time delays
energy	band structure effects, impurity bands	spectral distribution of emission, selection rules, allowed transitions	non-radiative recombination associated with decreased efficiency, possible effects of long-term degradation

to one or more specific physical phenomena. In the next section some of these phenomena will be discussed in more detail, whilst in Section 15.3 some of the possible interactions between these effects will be explored. Finally in Section 15.4 the conclusions are summarized in the light of the three questions posed above.

15.2 THE FUNDAMENTAL PHYSICAL PHENOMENA

In this section the principal effects indicated in Table 15.1 will be dealt with in the order suggested by each column. The basic laser structure and spatial coordinates are indicated in Figure 15.1.

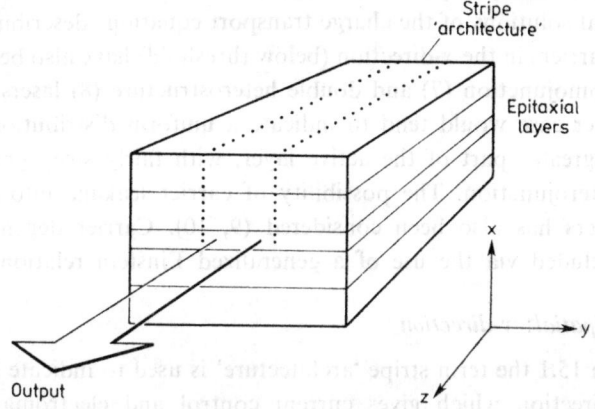

Figure 15.1 Schematic semiconductor laser structure and coordinate system

15.2.1 Electron effects

15.2.1.1 Spatial: x-direction

The simplest assumption, especially in the case of a double heterostructure laser with narrow active region, is that the carrier distributions are homogeneously distributed across the active layer. If this layer width is $2a$, then the rate of flow of carriers is simply given by $j/(2ea)$, where j is the current density and e is the electron charge. However, if a more detailed knowledge of the carrier distributions is needed, then the charge transport equations must be solved. Analytic solutions may be found only for situations of limited applicability where approximations are made to the full set of transport equations. Leaving aside the current controversy over the energy-band lineup at an abrupt heterojunction (1) and the somewhat vexed question of just how 'abrupt' such an interface may be (2), the conventional approximations to the transport equations may be listed as:

(a) electron diffusion only (3–6),
(b) electron drift only (5),
(c) electron diffusion and drift (5),
(d) drift of electrons and holes (5).

In all cases these approximations ignore Poisson's equation and assume either constant or linear variation of electric field across the active region. Whilst approximation (a) may apply to widely-spaced heterojunctions, cases (a)–(c) are suitable for heavily-doped p-type active regions, and (d) applies to double injection in a lightly-doped active layer.

Numerical solutions of the charge transport equations describing the distribution of carriers in the x-direction (below threshold) have also been published for both homojunction (7) and double heterostructure (8) lasers. The results for the latter case would tend to indicate a uniform distribution of carriers across the greater part of the active layer, with fairly steep gradients close to each heterojunction. The possibility of carrier leakage into the adjacent passive layers has also been considered (9, 10). Carrier degeneracy effects may be included via the use of a generalized Einstein relation (11).

15.2.1.2 Spatial: y-direction

In Figure 15.1 the term stripe 'architecture' is used to indicate the structure in the y-direction which gives current control and electromagnetic mode confinement. The range of possible stripe configurations is now quite large, e.g. oxide-insulated, proton-bombarded, shallow and deep diffused, buried heterostructure, TJS, mesa, rib-guide, channelled-substrate, embedded stripe, etc. rather than specify a particular structure, we are concerned here to merely enumerate the carrier transport mechanisms of relevance and indicate the cases where theoretical solutions may be found. Perhaps the simplest and most general mechanism is that of lateral diffusion of electrons under a stripe contact; in this case the carrier distribution $n(y)$ is given by (12, 13):

$$n(y) = \begin{cases} \dfrac{jL_n^2}{2eaD_n}\left[1 - \exp\left(-\dfrac{S}{L_n}\right)\cosh\left(\dfrac{y}{L_n}\right)\right] & \text{for } |y| \leq S \quad (15.1a) \\ \dfrac{jL_n^2}{2eaD_n}\sinh\left(\dfrac{S}{L_n}\right)\exp\left(-\dfrac{y}{L_n}\right) & \text{for } |y| > S \quad (15.1b) \end{cases}$$

where D_n is the diffusion constant, L_n the diffusion length, and $2S$ is the stripe width. More complicated situations will arise for specific stripe configurations, especially those involving doping control. The effects of current spreading (14–19) and stimulated emission above threshold (13, 15, 17–20) have also been analysed. The results would indicate that the effects of stimulated emission are only important for wide stripes ($2S \gtrsim 15$ μm) (15, 19) whilst for narrow stripes ($2S \simeq 10$ μm) the distribution of carriers in the y-direction is largely determined by carrier diffusion and current spreading. A refined theory of current flow has recently been published (21), together with 'pioneering measurements' of the spatial variation of junction voltage.

15.2.1.3 Spatial: z-direction

Along the cavity length of the laser structure the conventional assumption is that of a uniform distribution of carriers below threshold. At threshold and above the distribution of carriers is closely coupled to that of the photon field, although longitudinal diffusion may be expected to oppose this effect and attempt to restore a uniform carrier distribution (22).

15.2.1.4 Temporal

Below threshold the temporal evolution of the carrier populations is governed by (at most) two simple rate equations; the intrinsic time delay before lasing commences is well known (23). Above threshold the interaction of photons and carriers leads to a dynamic situation and a wide range of interesting effects, the discussion of which will be postponed until Section 15.2.2.4.

15.2.1.5 Spectral

The electron energy states associated with lasing action may be those of the parabolic bands, of isolated impurity levels (24), or more usually of impurity bands which may or may not merge with the bands (25, 26). In high-purity material a number of states associated with excitons and many-body effects at high excitation level may also be involved (27). A detailed knowledge of the densities-of-states functions of the electrons and holes involved in lasing is essential for the calculation of recombination rates, gain-current relationships, emission spectra, etc. Further discussion of these effects is given at the appropriate point in Section 15.2.2.

15.2.2 Photon effects

15.2.2.1 Spatial: x-direction

The epitaxial layers indicated on the schematic laser structure of Figure 15.1 usually provide a dielectric waveguide due to the variation of refractive index in the x-direction (28). In order to describe this waveguide in the most general

Table 15.2 Homogeneous core waveguide results for the modes (real guidance)

Waveguide	Eigenvalue equation	Confinement factor
Asymmetric 3-layer slab	$2v(1-b)^{1/2} = N\pi + \tan^{-1}\left[\dfrac{b^{1/2}}{(1-b)^{1/2}}\right] + \tan^{-1}\left[\dfrac{(b+c)^{1/2}}{(1-b)^{1/2}}\right]$	$\Gamma = \dfrac{2v + b^{1/2} + \dfrac{(b+c')^{1/2}}{(l+c')}}{2v + \dfrac{1}{b^{1/2}} + \dfrac{1}{(b+c')^{1/2}}}$
Symmetric 5-layer slab	(i) $r \leqslant (1-b) \leqslant 1$ $v(1-b)^{1/2} = \dfrac{N\pi}{2} + \tan^{-1}\left\{\left(\dfrac{1-b-r}{1-b}\right)^{1/2}\right.$ $\left.\times \tan\left[\tan^{-1}\left(\dfrac{b^{1/2}}{(1-b-r)^{1/2}}\right) - v(1-b-r)^{1/2}\left(\dfrac{d}{a}-1\right)\right]\right\}$ (ii) $0 \leqslant (1-b) \leqslant r$ $v(1-b)^{1/2} = \dfrac{N\pi}{2} + \tan^{-1}\left\{\left(\dfrac{r-1+b}{1-b}\right)^{1/2}\right.$ $\left.\times \tanh\left[\tanh^{-1}\left(\dfrac{b^{1/2}}{(r-1+b)^{1/2}}\right) + v(r-1+b)^{1/2}\left(\dfrac{d}{a}-1\right)\right]\right\}$	$\Gamma = [v(1-b)^{1/2} + \sin v(1-b)^{1/2}\cos v(1-b)^{1/2}]$ $\times \left\{v(1-b)^{1/2} - \left(\dfrac{r}{1-b-r}\right)^{1/2}\sin v(1-b)^{1/2}\cos v(1-b)^{1/2}\right.$ $\left.+ v(1-b)^{1/2}\left(\dfrac{d}{a}-1+\dfrac{1}{vb^{1/2}}\right)\right.$ $\left.\times \left[\cos^2 v(1-b)^{1/2} + \left(\dfrac{1-b}{1-b-r}\right)\sin^2 v(1-b)^{1/2}\right]\right\}^{-1}$

Notes:
(i) b is a normalized propagation constant: $b = \dfrac{(\beta/k)^2 - n_2^2}{n_1^2 - n_2^2}$, where β = longitudinal propagation constant, $k = 2\pi/\lambda$.
(ii) N is the mode number $(0, 1, 2, \ldots)$.
(iii) For the three-layer slab, c' is defined in Equation 15.3; for the five-layer slab, r is defined in (15.6), and d/a in Figure 15.3.
(iv) The result for Γ for the five-layer slab was derived only for even-order modes.

terms it is convenient to express its properties in terms of normalized variables (28, 29), the most important of which is the normalized frequency, v (29), defined as:

$$v^2 = \left(\frac{a2\pi}{\lambda}\right)^2 (n_1^2 - n_2^2) \qquad (15.2)$$

where a is the half-width of the guide (i.e. active layer), λ is the wavelength, n_1 is the refractive index of the active layer, and n_2 is the index of one adjacent layer (see inset of Figure 15.2). If n_3 is the refractive index of the passive layer on the other side of the active region, then an asymmetry factor c' may also be defined (28, 29):

$$c' = \frac{n_2^2 - n_3^2}{n_1^2 - n_2^2} \qquad (15.3)$$

The parameters v and c' defined in Equations 15.2 and 15.3 completely specify a given dielectric slab waveguide with homogeneous 'core'. Numerical solutions of the wave equation and associated boundary conditions can then yield results for the propagation constants of the various modes. For the waveguide formed in a heterostructure laser, perhaps the most useful parameter which can be found from such a calculation is the radiation confinement factor, Γ (30), defined as the ratio of power confined in the core to the total power. Table 15.2 gives results for the eigenvalue equation and confinement factor for TE modes in the homogeneous core waveguide (29, 31). Figure 15.2 shows plots of Γ versus v for various values of the asymmetry factor c' (mode TE_0); Γ varies from zero at cut-off to unity at large values of v (corresponding to complete field confinement). The net gain per unit length (G) of a guided mode may be expressed in terms of the gain g in the active core layer and the loss α associated with the passive claddings as:

$$G = g\Gamma - \alpha(1 - \Gamma) \qquad (15.4)$$

This equation may be used to compute the threshold for lasing when G is just equal to the end-loss from the cavity facets.

A refinement of the simple three-layer slab waveguide structure has been the development of multilayer lasers (LGR (32, 33), SCH (34), SCHDFB (35, 36), leaky wave (37)) with separate optical and carrier confinement. A symmetric five-layer structure is shown schematically in the inset of Figure 15.3; the dielectric variation $\varepsilon(x)$ is given by:

$$\varepsilon(x) = \begin{cases} n_1^2 & \text{for } |x| \leq a \\ n_0^2 & \text{for } a < |x| \leq d \\ n_2^2 & \text{for } d < |x| \end{cases} \qquad (15.5)$$

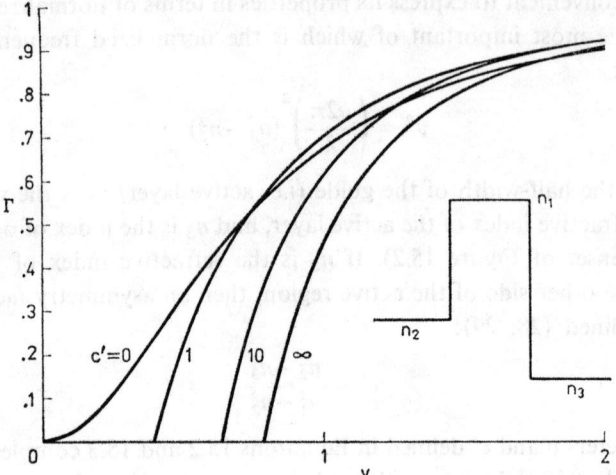

Figure 15.2 Radiation confinement factor Γ versus normalized frequency v for the lowest-order TE_0 mode of the asymmetric slab waveguide with homogeneous core; labelling parameter gives the value of the asymmetry factor c' defined in Equation 15.3

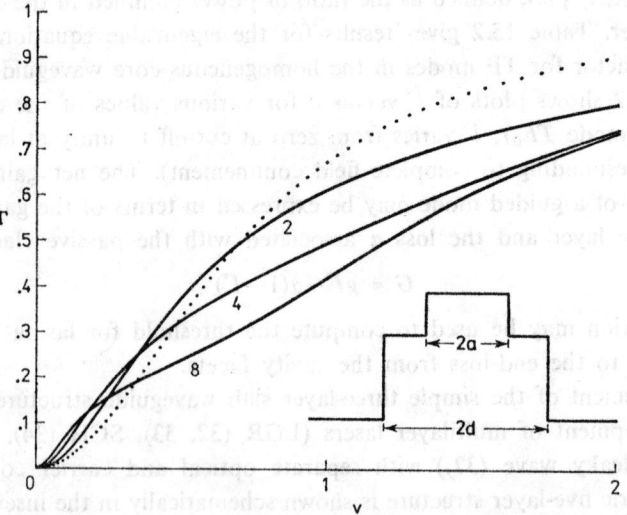

Figure 15.3 Γ versus v for the TE_0 mode of the five-layer symmetric slab waveguide (see inset) with $r = 1/3$ (defined as in Equation 15.6). Labelling parameter gives the value of the ratio d/a (see Equation 15.5). Broken line shows the variation of Γ with v for the equivalent three-layer slab $(d/a = 1)$

To analyse this structure it is convenient to retain the definition of v as in Equation 15.2, but to introduce a new measure of the asymmetry, defined as:

$$r = \frac{n_1^2 - n_0^2}{n_1^2 - n_2^2} \qquad (15.6)$$

The parameters v, r and the ratio d/a are sufficient to completely specify a symmetric five-layer waveguide structure. Results for the eigenvalue equation for TE modes and the confinement factor (for even-order modes only) are given in Table 15.2. Plots of Γ versus v for $r = 1/3$ and $d/a = 2, 4,$ and 8 are given in Figure 15.3 (solid lines). The broken line gives the variation of Γ for the equivalent three-layer slab ($d/a = 1$). For small v the values of Γ obtained with the five-layer structure are larger than those for the equivalent three-layer slab. Since at threshold the value of gain, g, is inversely dependent on Γ (cf. Equation 15.4), the result is a lower threshold for the SCH laser as compared with the equivalent DH device. Although the effect, as shown in Figure 15.3, is small, it is sufficient to produce worthwhile improvements; it may be augmented in practice by a superlinear gain-current relationship (32).

15.2.2.2 Spatial: y-direction

It was already noted in Section 15.2.1.2. that the stripe 'architecture' of Figure 15.1 would provide some degree of electromagnetic field confinement in the y-direction. In general there will be a complex dielectric waveguide, including variations of gain and loss as well as the refractive index, with a dielectric profile whose distribution in the y-direction depends on the details of the laser structure, drive current, etc. From a theoretical point of view therefore it is appropriate here to review the forms of dielectric profile for which solutions are known without the necessity of recourse to approximations or numerical techniques. For the case of real dielectric guides the symmetric profiles whose solutions are easily accessible are listed in Table 15.3; they are the step-index, linear, extended and cladded parabolic, exponential, and sech-squared waveguides. The assumption of symmetry in the y-direction seems well-founded, although it may sometimes be necessary also to consider some forms of asymmetry (38). Of the profiles listed in Table 15.3, only the step-index (39, 40), extended-parabolic (41, 42), and sech-squared (43–45) are applicable for the complex waveguide without the necessity of extensive computation.

As for the case of guidance in the x-direction (see 15.2.2.1 above) the waveguide parameter of greatest relevance to laser calculations is the confinement factor Γ. Since most of the newer laser structures are concerned

Table 15.3 Symmetric dielectric profiles with known solutions

Waveguide	Definition	Eigenvalue equation	Reference	v_{c1}
Step-index	$n^2(y) = \begin{cases} n_1^2, & \|y\| \leq a \\ n_2^2, & \|y\| > a \end{cases}$	$v(1-b)^{1/2} = \dfrac{N\pi}{2} + \tan^{-1}\left[\dfrac{b^{1/2}}{(1-b)^{1/2}}\right]$	28, 29, 31, 52, 53	1.57
Linear	$n^2(y) = \begin{cases} n_1^2\left[1 - 2\Delta\left(\dfrac{y}{a}\right)\right], & \|y\| \leq a \\ n_2^2 = n_1^2[1-2\Delta], & \|y\| > a \end{cases}$	$\dfrac{Bi'((b-1)v^{2/3})}{Ai'((b-1)v^{2/3})} = \dfrac{Bi'(bv^{2/3})}{Ai'(bv^{2/3})}$ (odd) $\dfrac{Bi'(bv^{2/3}) + b^{1/2}v^{1/3} Bi(bv^{2/3})}{Ai'(bv^{2/3}) + b^{1/2}v^{1/3} Ai(bv^{2/3})}$ (even)	46–48	2.80
Extended-parabolic	$n^2(y) = n_1^2\left[1 - 2\Delta\left(\dfrac{y}{a}\right)^2\right]$ (all y)	$b = 1 - \dfrac{(2N+1)}{v}$	13, 38, 41, 42, 53	—
Cladded-parabolic	$n^2(y) = \begin{cases} n_1^2\left[1 - 2\Delta\left(\dfrac{y}{a}\right)^2\right], & \|y\| \leq a \\ n_2^2 = n_1^2[1-2\Delta], & \|y\| > a \end{cases}$	Series solution	49	2.26
Exponential	$n^2(y) = n_1^2[1 - 2\Delta(1 - e^{-\|y\|/a})]$	$J'_{2v b^{1/2}}(2v) = 0$ (odd) $J_{2v b^{1/2}}(2v) = 0$ (even)	3, 50–53	1.20
Sech-squared	$n^2(y) = n_2^2 + 2\Delta n_1^2 \operatorname{sech}^2\left(\dfrac{y}{a}\right)$	$b = \left[\dfrac{(1+4v^2)^{1/2} - (2N+1)}{2v}\right]^2$	43–45, 52–56	1.41

Notes:
(i) b is a normalized propagation constant: $b = \dfrac{(\beta/k)^2 - n_2^2}{n_1^2 - n_2^2}$, where β = longitudinal propagation constant, $k = 2\pi/\lambda$.

(ii) Ai, Bi are Airy functions (58).

(iii) J denotes the Bessel function of the first kind (58).

Problems of Semiconductor Laser Theory

with the achievement of built-in refractive index guides in the y-direction, Γ is a meaningful parameter and Equation 15.4 for the net gain G remains valid for cases of uniform gain g and loss α in the y-direction. For the symmetric graded-index real guides listed in Table 15.3 we will review the methods of calculating Γ and give results in the case of the lowest-order TE mode for the purpose of comparison with the results of the symmetric uniform-core (step-index) guide (case $c' = 0$ of Table 15.2 and Figure 15.2). Note that the definition of v remains as in (15.2) with stripe width $2S$ replacing layer thickness $2a$.

(a) *Linear*: In this case Γ may be evaluated directly in terms of the Airy functions and their derivatives, making use of a special property of these functions (57). The results are plotted in Figure 15.4; for values of v below about 2.5, Γ is slightly below the corresponding curve for the step-index profile (broken line).

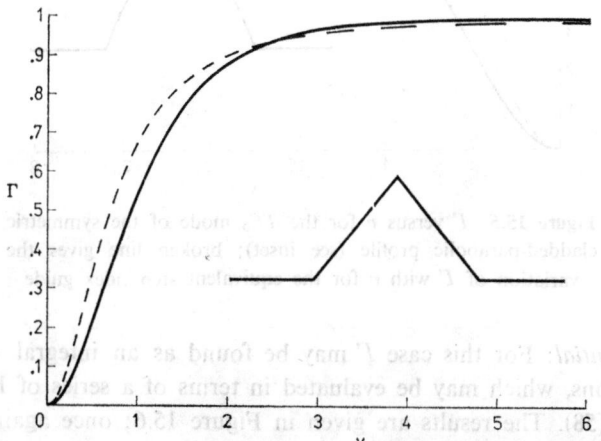

Figure 15.4 Γ versus v for the TE_0 mode of the symmetric linear profile (see inset); broken line gives the variation of Γ with v for the equivalent step-index guide

(b) *Extended-parabolic*: Computation of Γ for this profile is a straightforward integration over the Hermite–Gaussian field distributions. However, since this profile does not include the effects close to mode cut-off where the cladding layers would in practice provide a strong perturbation, no further results will be given here. Further results and a detailed critique of this profile as a laser waveguide model will be found in a recent publication (49).

(c) *Cladded-parabolic*: It is felt that this profile represents a more appropriate laser waveguide model than that of (b). Results using a series solution have been calculated here, although a number of possible other methods of analysis are available (see Reference 49 for a review). Plots of Γ versus v are given in Figure 15.5; for values of v below about 1.5, the graph lies just below that for the equivalent step-index guide (broken line).

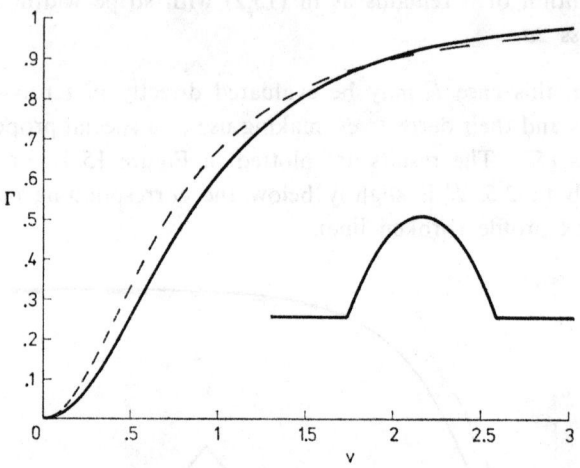

Figure 15.5 Γ versus v for the TE_0 mode of the symmetric cladded-parabolic profile (see inset); broken line gives the variation of Γ with v for the equivalent step-index guide

(d) *Exponential*: For this case Γ may be found as an integral over Bessel functions, which may be evaluated in terms of a series of Bessel functions (58). The results are given in Figure 15.6; once again the curve lies below the corresponding step-index result for a considerable range of v.

(e) *Sech-squared*: Although the field distributions take a reliability simple form for this profile, it is necessary to evaluate Γ by numerical integration. The results are shown in Figure 15.7 and again they lie just below those for the equivalent step-index curve for v less than about 3.

In view of the current interest in new laser structures for achieving good ranges of single-mode operation and 'kink'-free light-current curves, a further quantity of relevance is the value of v corresponding to cut-off of the first-order mode. The results for this quantity are given in Table 15.3 in the column

Problems of Semiconductor Laser Theory

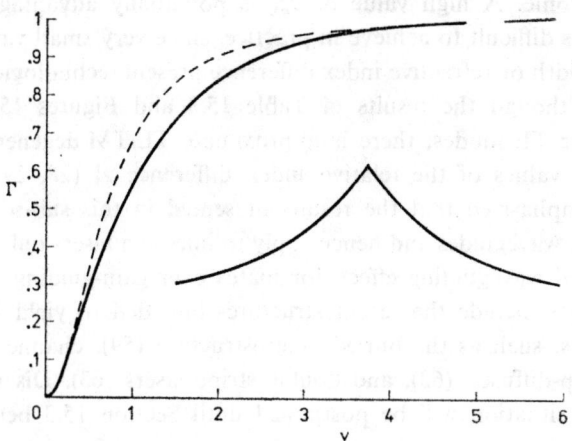

Figure 15.6 Γ versus v for the TE_0 mode of the symmetric exponential profile (see inset); broken line gives the variation of Γ with v for the equivalent step-index guide

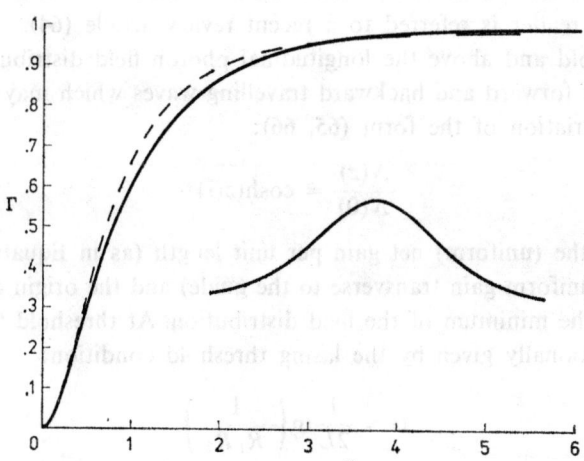

Figure 15.7 Γ versus v for the TE_0 mode of the sech-squared profile (see inset); broken line gives the variation of Γ with v for the equivalent step-index guide

headed 'v_{c1}'; from the Table the highest value of v_{c1} is seen to occur for the linear profile. A high value of v_{c1} is potentially advantageous in that a low value is difficult to achieve in practice, since very small values of either waveguide width or refractive index difference present technological problems. Note that although the results of Table 15.3 and Figures 15.4–15.7 were calculated for TE modes, there is approximate TE/TM degeneracy for sufficiently small values of the relative index difference Δ (28, 29). Finally, it should be emphasized that the results presented in this sub-section are for real dielectric waveguides and hence apply to injection lasers only in situations where the real waveguiding effect dominates over gain-guiding. Examples of such situations include the recent structures intended to yield linear output characteristics, such as the buried heterostructure (59), channelled substrate (60, 61), deep-diffused (62), and double stripe lasers (63). Discussion of the gain-guiding situation will be postponed until Section 15.3 below.

15.2.2.3 Spatial: z-direction

We will deal here with the description of the longitudinal photon field distributions in semiconductor lasers. Into this category falls also the subject of distributed-feedback (DFB) and distributed-Bragg-reflector (DBR) lasers, but these will not be discussed here: for an overview of these and related subjects the reader is referred to a recent review article (64).

At threshold and above the longitudinal photon field distribution $N(z)$ is composed of forward and backward travelling waves which may be summed to give a variation of the form (65, 66):

$$\frac{N(z)}{N(0)} = \cosh(zG) \tag{15.7}$$

where G is the (uniform) net gain per unit length (as in Equation 15.4 for the case of uniform gain transverse to the guide) and the origin of the z-axis is taken at the minimum of the field distribution. At threshold the value of G is conventionally given by the lasing threshold condition:

$$G = \frac{1}{2L}\ln\left(\frac{1}{R_1 R_2}\right) \tag{15.8}$$

where L is the cavity length and R_1, R_2 are the facet reflectivities. For a GaAs/air interface the reflectivities R_1 and R_2 are approximately 0.32, so that the maximum variation of longitudinal field distribution is given from (15.7) and (15.8) as $N(L/2)/N(0) = 1.17$. This variation is frequently neglected in calculations of laser properties; for lower values of R_1, R_2, as for example in the case of anti-reflective coatings, the longitudinal variation of photon field distribution becomes more accentuated. Under these circumstances, since the electron and photon distributions are coupled, the electron distribution and hence the gain will no longer be uniform in the z-direction and more complicated interactions may occur (67, 68). A further discussion of these phenomena will be presented at the appropriate point in Section 15.3.

There is another source of non-uniform photon distribution in the z-direction: the standing wave structure of the field along the laser cavity (22). The length L of the cavity must be equal to an integer number of half-wavelengths. Since the longitudinal gain distribution is intimately coupled to the photon distribution, this results in a periodic variation of the gain along the cavity. This effect has recently become the basis of a theory concerning the reasons for multi-longitudinal-mode operation of injection lasers (69, 70). According to this theory, the non-uniform gain distribution permits other longitudinal modes to lase, provided that the carriers do not diffuse fast enough to smooth out the non-uniformity. Hence laser structures with lightly-doped active regions where both electrons and holes are injected (see Section 15.2.1.1 above) will tend to exhibit multimode oscillations since the ambipolar diffusion constant will be dominated by the slow diffusion of the hole concentration. On the other hand, structures with electron injection into heavily-doped active regions will tend to oscillate in fewer modes, since the electron diffusion constant is relatively large. Experimental observations on heavily-doped conventional double-heterostructures (70) and on TJS lasers (71) tend to support this explanation and yield single longitudinal mode operation over large ranges of drive current.

15.2.2.4 Temporal

To discuss the transient properties of injection lasers above threshold it is necessary to include the interaction of the photon field with the electron concentration n. In keeping with the spirit of our categorization of fundamental effects in Table 15.1, for the remainder of this sub-section the photon and electron distributions will be assumed spatially uniform. With this assumption, the rate equations describing multimode laser operation become (72–74):

$$\frac{dn}{dt} = \frac{j}{2ea} - \frac{R_{sp}}{\eta} - \sum_{i=1}^{M} \frac{c}{n_1} g_i N_i \qquad (15.9)$$

$$\frac{dN}{dt} = \left(\frac{cg_i}{n_1} - \frac{1}{\tau_i}\right) N_i + \beta_i R_{sp} \qquad (15.10)$$

where M is the number of modes, N_i, g_i, τ_i and β_i are respectively the photon density, gain, cavity lifetime and spontaneous emission factor of the ith mode, c is the speed of light, R_{sp} is the total spontaneous emission rate, η is the internal quantum efficiency, and the other symbols have been defined previously. Note the distinction between the electron concentration n and the refractive index n_1 in the equations. The effects of non-uniform spatial distributions of n and N, together with the appropriately modified versions of Equations 15.9 and 15.10 will be discussed in Section 15.3.

To proceed further with the solution of the rate Equations 15.9 and 15.10 it is necessary to find explicit expressions for the spontaneous emission rate R_{sp} and the gain g_i in terms of the electron concentration n. The calculation of such expressions falls within the topics to be covered in Section 15.2.2.5. However, the conventional approximation of an electron lifetime τ_n to describe the rate R_{sp}, and a phenomenological model for g_i, e.g. as a power-law dependence on n (75) permit the general form of the solutions to (15.9) and (15.10) to be explored. The existence of at least two time constants τ_n and τ_i gives rise to characteristic resonance frequencies associated with the solutions. For the case of single-mode oscillation ($M = 1$) a small-signal analysis yields for the resonance angular frequency ω_r (75):

$$\omega_r^2 = \frac{l}{\tau_n \tau_1} \left(\frac{j}{j_{th}} - 1 \right) \qquad (15.11)$$

where l is the index in the power-law expression for g_i and j_{th} is the threshold current density. The resonance frequency manifests itself in transient 'spiking' and quantum noise effects and also plays a key role in the understanding of high-frequency modulation of injection lasers. Under conditions of large-signal modulation the resonance shifts to lower frequencies (76–78) than given by the simple expression 15.11. At the same time the modulation efficiency decreases and a strong distortion of the modulated output may occur; these effects are confirmed experimentally (79).

The fraction of spontaneous emission into a lasing mode, denoted by β_i in Equation 15.10, has an important influence on the relaxation oscillations and 'ringing' effect in laser modulation (73, 74, 79–84). Large values of β_i (10^{-3}–10^{-2}), implying a large number of oscillating modes, would tend to damp the relaxation oscillations evoked in response to a step current pulse (80–83); they would also reduce the relative modulation amplitude at the resonance frequency ω_r (74, 81). In practice, experiments on stripe-geometry double-heterostructures have resulted in measured values of β_i in the range 10^{-5}–5×10^{-4} (74, 82), although there are also reports of β_i as high as 10^{-3}–10^{-2}

(85, 86). A recent theoretical calculation of β_i (84) gave values around 10^{-5} for conventional GaAs laser parameters, and for the case of non-uniform photon distributions showed a linear dependence of β_i on the product of radiation confinement factors Γ_x, Γ_y (84, 79) defined for the x and y-directions as in Subsections 15.2.2.1 and 15.2.2.2. β_i is also found to be inversely-dependent on the volume of the active region, which may be a reason for the large β_i values, suppression of relaxation oscillations, and absence of resonance peaks in modulation characteristics of buried-heterostructure lasers (85).

The laser rate equations (15.9) and (15.10) have been used to study pattern effects in PCM of injection lasers. Pattern effects occur as a result (i) of the intrinsic time delay (23), (ii) of the values of electron population n and photon density N_i immediately before the application of a pulse (87), and (iii) the relaxation oscillations already discussed. Based on the theoretical analyses, suggestions for suppressing the pattern effects have been made; these include the use of compensation pulses (88), pre-biasing (89), pulse-shaping (90, 91), light injection (92–95), optical feedback (92), and resonant circuits (96). In all these cases the appropriate formulation of the rate equations including the spontaneous emission term in (15.10) is essential in order to permit accurate tracking of the electron and photon populations. In almost all cases, however, the assumption is made that all properties of the lasing modes are identical, so that the sum over modes in (15.9) can be replaced by the multiplicative constant M. A useful review of theoretical and experimental progress in the area of transient laser behaviour will be found in reference (79).

If Langevin noise operators are included in the rate equations 15.9 and 15.10 (97), then a detailed theory of the quantum noise in semiconductor lasers may be developed (98, 99). One important prediction of the theory is that the relative intensity noise should exhibit a maximum close to threshold and should decrease with increasing current above threshold. Experimental measurements by Paoli (100, 101) showed this stabilization in some lasers at low frequencies (< 100 MHz) but not at higher frequencies (> 2 GHz). This anomalous behaviour was tentatively attributed to long thermalization times so that the electron population could not respond quickly to the fluctuating photon density to provide quieting. However, the noise measurements of other authors (102–104) are in reasonable agreement with the theoretical predictions on this point. Further evidence of a lack of low-frequency noise stabilization above threshold has also been reported (105) and interpreted on the basis of spatially inhomogeneous gain saturation; the theoretical treatment of this problem will be discussed further in Section 15.3.

Recently, several authors (104, 106, 107) have published numerical solutions of the multimode rate equations including the Langevin noise terms, and have compared these results with the single-mode solutions. In general, the multimode case shows weaker intensity fluctuations than the single-mode situation for frequencies below resonance, the extent of the difference depending on the number of modes excited; near the resonance frequency the difference diminishes. For these calculations the properties of the modes are usually assumed identical; good qualitative agreement with experimental results has been achieved (104, 106, 107). A further development has been the extension of the theory to include the noise sources in the time-dependent equations when applied to direct modulation (108). The principal results is a shift of the noise peak to higher currents for increased modulation frequencies (shorter current pulse-widths), in general agreement with the experimentally-observed situation (108).

15.2.2.5 Spectral

In this subsection we will discuss the calculation of the spontaneous emission rate R_{sp} and the gain g as explicit functions of electron concentration n and, in the case of g, of photon energy E, for use in the rate equations 15.9 and 15.10. In the most general form, the expressions for these quantities are as follows (7, 72):

$$R_{sp} = \int_0^\infty r_{sp}(E)\,dE \qquad (15.12)$$

$$g(E) = \frac{\pi^2 c^2 \hbar^3}{n_1^2 E^2} r_{st}(E) \qquad (15.13)$$

where

$$r_{st}(E) = r_{sp}(E)\left[1 - \exp\left(\frac{E - F_i + F_j}{KT}\right)\right] \qquad (15.14)$$

$$r_{sp}(E) = \sum_{I,J} \frac{4n_1 e^2 E |M_{IJ}(E)|_{av}^2}{Vm^2\hbar^2 c^3} \varrho_{IJ}(E) f_I (1 - f_J) \delta_{E_I - E_J, E} \qquad (15.15)$$

In these expressions, $\hbar$ is Planck's constant over 2π, K is Boltzmann's constant, T is the absolute temperature, m is the electron mass, F_i and F_j are quasi-Fermi levels associated with sets of energy states denoted by i and j, f_I and f_J are the Fermi–Dirac distribution functions associated with specific states I and J ($I \in i$, $J \in j$) of energies E_I and E_J, $\varrho_{IJ}(E)$ is the density of energy states contributing to emission of a photon of energy E, $|M_{IJ}(E)|_{av}^2$ is the square of the momentum matrix element for the I–J transition suitably arranged over polarization and directions of the photon wave vector, V is the volume of the active region, and the rest of the symbols are as defined earlier.

The principal task in the evaluation of Equations 15.12–15.15 is to find suitable relations for the matrix element M_{IJ} and the density-of-states function ϱ_{IJ} for the transition of interest. The early history of calculations of this sort will be found in References 7 and 72. More recently, a rather comprehensive calculation has been presented by Casey and Stern (109) and applied to steady-state laser problems by Stern (110). This approach used the model of no k-selection rule for transitions between conduction and valence bands and their associated impurity band tails. The assumption of no k-selection rule is now fairly well-established and has found convincing experimental support (111). The density-of-states function ϱ_{IJ} thus becomes proportional to the product of density-of-states functions for each set of states i and j. The Casey–Stern model uses a density-of-states of the Kane form (25) to interpolate between the parabolic portions of the bands and the deep impurity tail states as described by the Halperin–Lax results (26). The optical matrix element M_{IJ} is calculated using empirically-determined wave functions which correctly reproduce the Bloch-like behaviour for states in the parabolic bands whilst reducing to the hydrogenic form for the localized states (24). For p-type GaAs these calculations give good agreement with experimental absorption data (109); the results are also in close accord with those from experimental studies of the spectral dependence of gain in heterostructure lasers (33, 112). A significant feature of Stern's results (110) is the confirmation that $R_{sp} = Bnp$ where B is reasonably independent of concentration at room-temperature and takes the value $\sim 2 \times 10^{-10}$ cm^{3}s^{-1} for GaAs.

Whilst the sophistication of this recent theoretical analysis (109, 110) provides excellent detail for the spectral and current dependence of the gain function in a uniform, steady-state laser model, it is also useful to have available approximate results for these quantities in order to use them in more complicated laser models allowing for non-uniformity, transience, etc. For these purposes it is, of course, possible to fit simple algebraic expressions to the numerical results for the quantities of interest. For example, the concentration dependence of gain g may be approximated by a simple power-law (75, 79, 95) as described in Section 15.2.2.4. Similarly, since the spectral dependence of g is only of interest in the region of its maximum, this may be modelled by a simple parabolic variation with energy E (20). Alternatively one may use approximations based on models of the transitions, densities-of-states, and matrix elements involved. One such approximation, based loosely on the exponential band-tail model (72, 99) yields the formula

$$\frac{g(E, n)}{g_0} = \exp\left(\frac{E-E_G}{E_0}\right)\left[\ln\left(\frac{n}{n_0}\right) - \left(\frac{E-E_G}{E_0}\right)\right] \quad (15.16)$$

where E_G is the band-gap, and E_0, n_0 and g_0 may be chosen to fit available results for $g(E, n)$, which may be theoretical (110) or experimental (111). It follows from Equation 15.16 that the gain maximum occurs at photon energy E_m, given by

$$\frac{E_m - E_G}{E_0} = \ln\left(\frac{n}{n_0}\right) - 1 \quad (15.17)$$

when the value of g is

$$\frac{g_m}{g_0} = \frac{n}{n_0}\exp(-1) \quad (15.18)$$

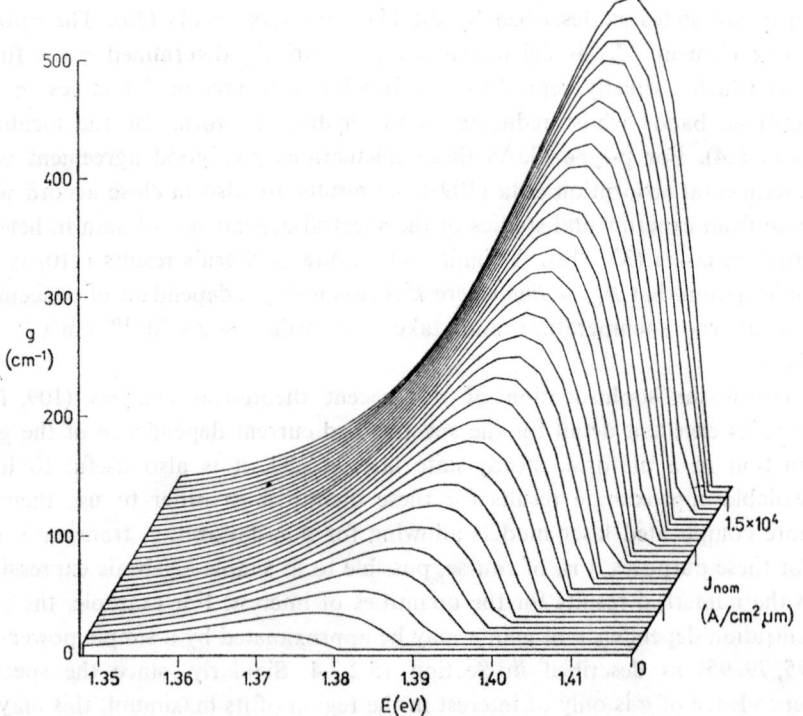

Figure 15.8 Gain g versus nominal current density J_{nom} (7, 110) and photon energy E from the simple model represented in Equation 15.16; parameters used were $E_0 = 10$ meV, $J_0 = 4 \times 10^4$ A/cm² μm, $g_0 = 2800$ cm⁻¹, $E_G = 1.424$ eV

If n is expressed in terms of Stern's nominal current density J_{nom} (7, 110) ($J_{nom} = 1.6 \times 10^{-23} Bnp = 1.6 \times 10^{-23} n/\tau_n$), then Equations 15.16–15.18 may be rewritten in terms of J_{nom} and an adjustable parameter J_0. Some results are given in Figure 15.8 in the form of a three-dimensional plot showing g as a function of E and J_{nom}. It should be emphasized that the simple model represented here cannot be expected to reproduce all the subtleties of more detailed numerical treatments (109, 110) but may be usefully applied in problems where the complications of non-uniformity and transient development are of interest. In this respect the model reasonably represents the main physical effects associated with the concentration and spectral dependencies of the gain.

15.2.3 Phonon effects

15.2.3.1 Spatial: x-direction

A primary source of heat in the semiconductor laser structure is non-radiative recombination occurring in the active region. Near threshold the power dissipated as heat at the junction is given approximately by $(jV_a 2SL)$ where j is the current density, V_a is the applied voltage, $2S$ is the stripe width, and L is the laser length. In this expression the radiative output has been ignored and all electrical input power assumed converted to heat; other heat sources such as contact resistance, distributed bulk resistance, self-absorbed spontaneous emission, etc. have been neglected. If it is assumed that heat production is limited to a plane at the centre of the active layer and then flows in the x-direction to the heat-sink, then a simple one-dimensional solution of the heat-flow equation will describe the situation (113, 114). Such a model applied to oxide-insulated stripes (113), buried heterostructures (59) and mesa-structures, where heat flow in the y-direction may be neglected. The resulting temperature distribution through the layers from the active region to the heat-sink is amenable to comparison with experimental measurements (115, 116); in addition the model may be used to calculate the CW lasing range for given laser parameters (113, 114). The temperature gradient across many epitaxial layers with different thermal expansion coefficients may lead to a strain distribution (117) which may play a role in the initial degradation behaviour of lasers (118).

15.2.3.2 Spatial: y-direction

For structures other than those listed in the previous subsection, heat flow in the y-direction cannot be ignored. In these cases the Laplace equation for two-dimensional heat flow from a planar source situated in the active layer is to be solved. The most general solutions take the form of Fourier transforms for a laser model assumed infinitely wide (119), and Fourier series for a model of finite width (120). The result is a temperature profile across the active area in the y-direction with a variation of typically a few degrees between the centre and edge of the stripe (120). Since the refractive index $n(y)$ of the active layer increases with temperature at the rate of $5 \times 10^{-4} \mathrm{K}^{-1}$ (42), such a temperature profile may give rise to a corresponding refractive index profile with a positive guidance effect in the y-direction.

15.2.3.3 Spatial: z-direction

In the z-direction the conventional assumption is that the thermal distribution is uniform; the longitudinal expansion of a laser calculated on this basis agrees qualitatively with the result of measurements by an interferometric technique (115).

Another effect which falls into this category is a possible explanation of catastrophic degradation in high-power lasers. This is the excitation of acoustic waves by the optical radiation in the laser cavity (stimulated Brillouin emission) (121); according to this theory the acoustic phonons travel along the laser beam and cause damage ('burn-off') at the facets as a consequence of the weaker mechanical properties of the surfaces. However, it should also be noted that an alternative mechanism has been postulated to account for facet damage and catastrophic degradation, viz. optical absorption at inhomogeneities near the facet resulting in thermal runaway or a 'micro-explosion' (122).

15.2.3.4 Temporal

For pulsed operation with relatively short pulses ($\simeq 100$ ns) the effects of heat diffusion away from the active region may be ignored and the heating may be assumed adiabatic. For this situation, the temperature rise ΔT of the active region in time t is given simply by (123):

$$\Delta T = \frac{jV_a SL}{\varrho Ca} t \qquad (15.19)$$

where ϱ is the density, C the specific heat, and a the half-width of the active layer. For longer pulse widths it may be necessary to include thermal diffusion and find solutions to the time-dependent equation for heat conduction (123, 124). An associated phenomenon which has been reported is the existence of longitudinal and flexural vibrations attributed to shock heating and thermal expansion of the laser (117).

Problems of Semiconductor Laser Theory

It is perhaps worth noting also in this sub-section that the combination of Equation 15.19 together with the thermal coefficient of refractive index noted in 15.2.3.2 leads to a mechanism for transient evolution of the refractive index of the active region. When this effect is considered in the context of a single-heterostructure laser where the guidance is very weak on the n-side, then it implies a time-dependent waveguide which may vary from below to above cut-off for the fundamental mode within the course of a current pulse. Such a mechanism has been invoked to account for long time delays and internal Q-switching in single heterostructure lasers (125); this latter model is to be contrasted with the other principal explanation of these phenomena, viz. that involving saturable absorption as the time-dependent 'driving force' leading to loss of optical confinement (126).

15.2.3.5 Spectral

Finally in this section we should include the fundamental features of phonon emission viewed as a non-radiative recombination route, i.e. as a mechanism causing reduced quantum efficiency η (as defined in Equation 15.9). The principal non-radiative mechanisms of this kind in laser materials are multi-phonon emission (127) and the Auger effect (128). Confining attention here to the former effect, a recent calculation (127) based on the configurational coordinate theory and applied to deep levels in GaAs yields values for the capture cross sections which may be as large as 10^{-14}–10^{-15} cm^2; experimental values also fall in this region (127).

15.3 SOME INTERACTIVE EFFECTS

In discussing the fundamental physical phenomena of Table 15.1, it has already been necessary to include some limited interaction between electrons, photons and phonons in order to give the description some validity. If we were to adopt a purely didactic approach it would now be possible to consider all interactive effects from the Table: 105 pairs, 455 triplets, etc.! However, within the limits of current or conceivable applications we are confined to rather fewer combinations. Since any limited choice must be largely determined by personal preference, I make no apology for restricting attention to the following four effects only.

15.3.1 Threshold calculations in symmetric double heterostructures

This simple example illustrates well the interactive nature characterizing the effects included in this section. The calculations of 15.2.2.1 provide the waveguide information, 15.2.2.4 the (steady-state) rate equations, and 15.2.2.5 the gain-current relationship. For fundamental mode threshold, Equations 15.4, 15.8 and 15.18 provide the constitutive relations; if we include also the expression for the flow of carriers given in 15.2.1.1, then the threshold current density j_{th} is given by (in normalized variables):

$$j_{th} = Av\left[\left(\frac{1}{\Gamma}\right)\frac{1}{2L}\ln\left(\frac{1}{R_1 R_2}\right) + \left(\frac{1}{\Gamma} - 1\right)\alpha\right] \quad (15.20)$$

where A is a constant and the other symbols are as defined previously.

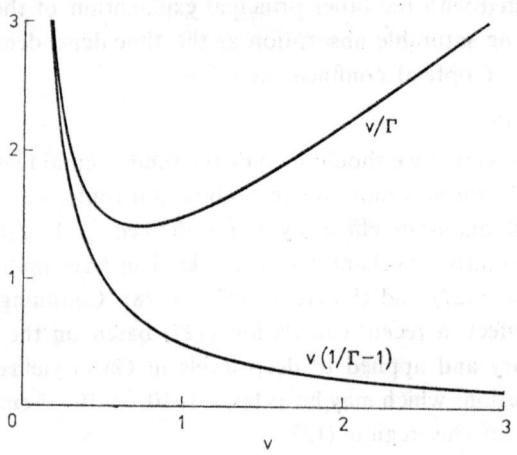

Figure 15.9 Threshold parameters v/Γ and $v(1/\Gamma - 1)$ for the TE_0 mode of the three-layer slab waveguide plotted as functions of v

From the form of Equation 15.20 it is clear that all the waveguide information is contained in the normalized factors v/Γ and $v(1/\Gamma - 1)$. Plots of these quantities versus v for the TE_0 mode of the symmetric three-layer slab waveguide ($c' = 0$ of Subsection 15.2.2.1) are given in Figure 15.9. Note that although Equation 15.20 was derived for the simple gain-current relationship of (15.18), it is a trivial extension to include the more general case $g_m \propto (n - n')$ (12, 112) where n' represents a carrier density necessary for cavity transparency. Hence Figure 15.9 may be said to contain all the necessary information for calculations of threshold current density when it is used in conjunction with an equation like 15.20. Whilst such calculations have already been presented in some detail for GaAs/GaAlAs lasers (30, 112, 129), the novelty of the present approach lies in the use of normalized variables, so that the results are easily applied to new laser materials, e.g. those intended for use in optical communication systems at longer wavelengths where the absorption and dispersion properties of optical fibres are greatly improved (130). With reference to possible optimization of such materials it is also worth noting that for the simplest case where (15.20) applies and end-loss dominates strongly over α, the optimum v-value is given by the minimum of v/Γ on Figure 15.9 as $v = 0.71$ (131). For the conventional GaAs/GaAlAs double heterostructure with $n_1 = 3.6$ and $n_2 = 3.4$ this corresponds to an optimum active layer width of about 0.17 µm; when refractive index values of the newer laser materials become available, similar predictions of optimum layer thicknesses can be made. For example, for InGaAsP lasers grown on InP substrates (132), using measured InP refractive index data (133) and estimating the InGaAsP values from the effective band-gap shift (134), operation at $\lambda \simeq 1.3$ µm would imply an optimized active layer thickness of about 0.3 µm.

15.3.2 Guidance in stripe geometry lasers

Some discussion of this topic has already been given in Subsection 15.2.2.2 together with some considerations of real built-in guides; in addition, Subsection 15.2.3.2 included one mechanism for guidance under a conventional stripe contact, viz. a thermally-induced refractive index profile. However, there are also changes of real and imaginary parts of the dielectric permittivity associated with the injection of carriers underneath a stripe contact. Increased carrier density leads to an increase of gain (cf. Equation 15.18) and to a decrease of the refractive index (134, 135); in each case, to a first approximation, the variation is linear with electron density. Bearing in mind the remarks in 15.2.2.2 concerning dielectric profiles for which analytic solutions are available, it is therefore feasible to fit the electron distribution under the stripe to a profile whose modal solutions are well known. Figure 15.10 shows two such possible profiles—the extended-parabolic and sech-squared laws—which have been fitted to the electron distribution described by Equation 15.1. The parabolic profile was fitted to the small-signal expansion of (15.1) around $y = 0$ (19), whilst the sech-squared profile was matched to Equation 15.1 at $y = \pm S$; other methods of determining the appropriate fitting process have also been discussed in the literature (13, 38, 45). In view of Figure 15.10 and the remarks on the extended-parabolic profile in 15.2.2.2, it is clear that this distribution, although widely used in the literature (13, 19, 38, 41, 42, 136) is applicable only to modes well above cut-off which are largely confined to the central region under the stripe. The sech-squared profile, on the other hand, may be applicable for modes over a wide range, including close to cut-off, and it is understood that an appropriate analysis is currently in progress (137).

Once a suitable model has been established for the dielectric distribution in the y-direction, this must be combined with the refractive index change

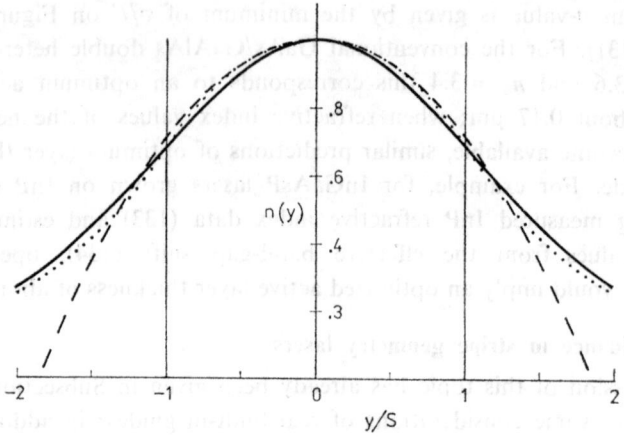

Figure 15.10 Carrier distributions under a stripe contact: *dotted line*—Equation 15.1 normalized to unity at $y = 0$, with $S/L_n = 1$; *solid line*—sech^2 profile, matched at $y = \pm S$; *dashed line*—parabolic approximation, fitted to small-signal expansion of (15.1) near $y = 0$

in the x-direction in order to find solutions to the wave equation in two dimensions. The most frequently used approximation for dealing with this problem is the effective dielectric permittivity method (138). The technique consists of replacing the actual two-dimensional dielectric distribution by an effective one-dimensional profile (19, 61, 135, 136, 138, 139), the solutions of which are usually fairly simple to obtain. For the stripe-geometry double heterostructure the guidance in the x-direction is usually much stronger than that in the y-direction where the dielectric profile is relatively slowly-varying. Hence the solution for the x-direction is usually obtained first (as discussed in Section 15.2.2.1) and then assumed to have a weak y-dependence; an effective dielectric permittivity may then be defined for the y-direction which incorporates the x-direction solutions. If the active and passive layer dielectric permittivities are denoted by $\xi_1(y)$ and $\xi_2(y)$, respectively, then the effective permittivity is given by (140)

$$\varepsilon_{\text{eff}}(y) = b\varepsilon_1(y) + (1-b)\varepsilon_2(y) \tag{15.21}$$

where b is a (weakly y-dependent) normalized propagation constant (29) found from the x-direction field solutions as

$$b = \frac{(\beta/k)^2 - \varepsilon_2(y)}{\varepsilon_1(y) - \varepsilon_2(y)} \qquad (15.22)$$

and β is the longitudinal propagation constant. Solutions of the one-dimensional wave equation with permittivity given by (15.21) may then be used to calculate the net gain G of the laser. Allowing for the y-dependence of the gain of the active layer, $g(y)$, and the loss of the passive layers, $\alpha(y)$, the analogous equation to 15.4 becomes:

$$G = \Gamma \frac{\int_{-\infty}^{\infty} g(y)|\Phi(y)|^2 dy}{\int_{-\infty}^{\infty} |\Phi(y)|^2 dy} - (1-\Gamma)\frac{\int_{-\infty}^{\infty} \alpha(y)|\Phi(y)|^2 dy}{\int_{-\infty}^{\infty} |\Phi(y)|^2 dy} \qquad (15.23)$$

where Γ is the confinement factor for the x-direction guide and $\Phi(y)$ is the y-dependent part of the (separable) field solutions.

Calculations based on the model described in this subsection have been performed by Buus (140); rather than use approximations to the electron and dielectric distributions, his model was based on the calculated profiles and employed a numerical technique for solving the wave equation.

15.3.3 Longitudinal and transverse field distributions in stripe geometry lasers

A further degree of complexity is introduced into the problem of describing guidance associated with stripe contacts when the longitudinal variation of carrier and photon populations is included. Since the photon distribution is maximum at the facets (Equation 15.7) and the gain varies inversely with the photon population for a uniform current distribution (Equation 15.9), it follows that the gain distribution is minimum at the facets. If we consider, for simplicity, the case of gain-guiding only (40–42), then it becomes clear that the strength of guidance in the y-direction must vary along the cavity length, being at its weakest at the facets (141). To quantify these ideas, note that from Equations 15.7–15.9 in steady-state and for single-mode operation far above threshold (so that spontaneous emission may be ignored), a first approximation for the gain distribution $g(z)$ is given by (141)

$$g(z) = \frac{g(0)}{\cosh\left[\frac{z}{2L}\ln\left(\frac{1}{R_1 R_2}\right)\right]} \qquad (15.24)$$

For a z-dependent dielectric distribution such as that given in (15.24), solutions to the scalar wave equation may be found by the method of Kogelnik

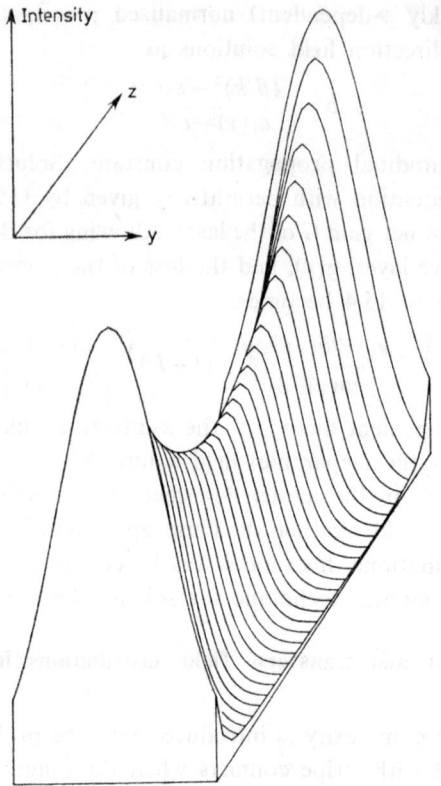

Figure 15.11 Photon field intensity distribution in the y- and z-directions calculated for a 10 μm stripe-geometry GaAs laser with facet reflectivities of 0.05 and cavity length 500 μm

(142). Assuming a parabolic variation for g in the y-direction and using the effective index approach to allow for guidance in the x-direction, it is then possible to calculate the photon field distribution as a function of all three spatial coordinates. The results differ significantly from those assuming uniformity in the z-direction, especially for low values of facet reflectivity R_1, R_2. A typical result is shown in Figure 15.11, where the photon field intensity is plotted as a function of y and z for $R_1 = R_2 = 0.05$ and $L = 500$ μm. As anticipated from the arguments given above, the field is best confined in the lateral direction near the centre of the cavity and spreads to its maximum breadth at the mirrors; the corresponding distribution of the gain is shown in Figure 15.12.

For stripe-geometry lasers, especially with anti-reflection coated facets, it follows from this theory (141) that:

(a) For the gain-guiding situation the fields are spread at the laser facets, hence giving a lower power density which may assist in inhibiting both catastrophic (121, 122) and long-term degradation (143).

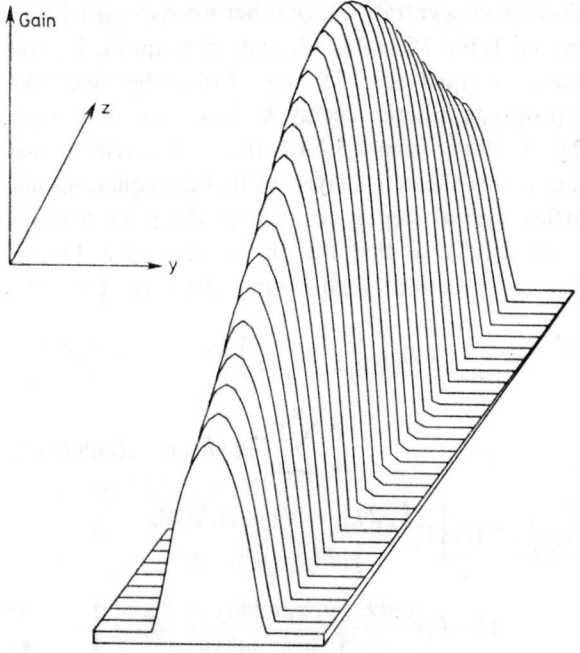

Figure 15.12 Distribution of gain corresponding to the field variation of Figure 15.11

(b) In the opposite situation of real guidance produced either by dips in the carrier concentration (spatial hole burning (15, 19)) or by a specific fabrication process, the behaviour will be reversed in that the fields will be strongly focused at the facets; similar calculations to those presented above support this. An analogous situation may exist in the x-direction in high-power single-heterostructure lasers.

(c) Experiments designed to study guidance mechanisms in a.r.-coated lasers (42) should be interpreted with caution, lest the effect of the non-uniform z-variation masks the mechanism operating in the y-direction.

(d) The interactive effect studied here could in principle be used to produce specific distributions of the laser output beam by suitable profiling of the stripe contact along the length of the laser (141).

15.3.4 Transient evolution of spatial and spectral distributions of photons and electrons

In this subsection we shall include the temporal and spectral variation of the photon and electron concentrations; in other words, non-uniformity is allowed in all five rows of Table 15.1. An adequate description is given by allowing all terms in the rate equations 15.9 and 15.10 to be functions of x, y, z, t, and, where appropriate, photon energy E. A certain amount of simplification is achieved by recalling from 15.2.1.1 that the carrier concentrations are usually constant across the active layer, so that the equations may be averaged over x. A further logical step is to average the photon rate equation also over y (144) and z (22), so that the photon density N becomes a function only of t. The rate equations then assume the form (20, 79, 86):

$$\frac{dn(y,z,t)}{dt} = \frac{j}{2ea} + D_n\left(\frac{\partial^2}{\partial y^2} + \frac{\partial^2}{\partial z^2}\right)n(y,z,t) - \frac{R_{sp}(n)}{\eta}$$

$$- \frac{c}{n_1}\sum_{i=1}^{M}\Gamma_i g_i(E_i,n)N_i(t)|\Phi(y,z)|^2 \qquad (15.25)$$

$$\frac{dN_i(t)}{dt} = \frac{c}{n_1}N_i(t)\left\{\Gamma_i\frac{\int g_i(E_i,n)|\Phi(y,z)|^2 dy dz}{\int |\Phi(y,z)|^2 dy dz}\right.$$

$$\left. -(1-\Gamma_i)\frac{\int \alpha(E_i,y,z,t)|\Phi(y,z)|^2 dy dz}{\int |\Phi(y,z)|^2 dy dz}\right\} - \frac{N_i(t)}{\tau_i}$$

$$+ \Gamma_i\frac{\pi^2 c^3 \hbar^3}{V n_1^2 E^2}\frac{\int r_{sp}(E_i,n)|\Phi(y,z)|^2 dy dz}{\int |\Phi(y,z)|^2 dy dz} \qquad (15.26)$$

where all the symbols are as defined previously. In particular, Γ_i is the confinement factor in the x-direction for mode i, E_i is the photon energy of mode i, $\Phi(y,z)$ is the normalized field amplitude, and $g_i(E_i,n)$, $r_{sp}(E_i,n)$, and $R_{sp}(n)$ are the corresponding gain, spontaneous emission rate, and total spontaneous rate (integrated over photon energy). Note that a comparison of Equations 15.10 and 15.26 may be used to obtain an expression for the spontaneous emission factor β_i (84). The second term in Equation 15.25 allows for lateral and longitudinal diffusion of electrons (20, 22, 79, 86, 145). Equation 15.26 has been derived on the assumption that the cavity lifetime

τ_i is independent of position; it usually corresponds to the end-loss which is given explicitly in Equation 15.8. In the steady-state and ignoring spontaneous emission, Equation 15.26 may be used to calculate lasing thresholds, when it becomes equivalent to the combination of Equations 15.8 and 15.23 with the z-dependence neglected. Note also that the mode index i in (15.25) and (15.26) may apply to transverse, lateral, and longitudinal mode numbers.

For single-mode operation the rate equations 15.25 and 15.26 have been solved numerically by Cross (146). In this analysis the effects of diffusion were ignored in (15.25), the loss term α in (26) was assumed constant, and the effects of spontaneous emission on the lasing mode (last term in (15.26)) were neglected; in addition longitudinal uniformity was assumed (no z-dependence). The results indicated the usual 'spiking' oscillations in response to a step current pulse, with the additional feature of a corresponding oscillation in field distribution in the y-direction; a time-dependent real waveguide in the y-direction was assumed as a consequence of the fluctuating carrier density (see Section 15.3.2). These results have recently been confirmed by experimental studies of broad oxide-stripe lasers (17–20 μm) (147).

When multimode operation is permitted in the rate equations 15.25 and 15.26, the response to a step current pulse is more complicated. Since the electron concentration overshoots its equilibrium value during the relaxation oscillations, the gain is temporarily sufficient for a number of modes to be above threshold. When the number of modes is increased, the power in the dominant mode is consequently reduced. The net effect is a spectral broadening which decreases with increasing time after the onset of the current pulse; detailed theoretical analyses have been made by Ikegami (148) and Buus *et al.* (20, 86). This broadening (of order 3 nm) may also occur in sinusoidal modulation and PCM (79, 148, 149) where it might be a severe limitation on the bandwidth of an optical communications system, as a result of material dispersion in the fibre. The effect is diminished somewhat by diffusion of carriers, which tends to favour the dominant mode (20, 22, 86, 145, 148), or by a high rate of spontaneous emission into the modes (see Section 15.2.2.4). It may be controlled by external light injection (73, 92–95) or by biasing the laser above threshold (150).

Equations similar to 15.25 and 15.26 but ignoring the z-dependence, have also been used to study the low-frequency intensity noise in CW stripe-geometry lasers (105). For two oscillating modes with different thresholds, the results indicate a lack of quieting above threshold owing to spatially-inhomogeneous gain saturation; this result is in qualitative agreement with experimental finding (see Section 15.2.2.4).

15.4 CONCLUSION

This contribution has been concerned with theoretical descriptions of the phenomena occurring in semiconductor lasers. We have seen in Section 15.2 that almost all the fundamental physical effects are now amenable to theoretical analysis and, in Section 15.3, that the prime areas of current concern are associated with interactive effects involving one or more of the possible combinations from Table 15.1. In the areas of wave-guidance, the design of new laser structures, and transient phenomena, the theoretical descriptions have contributed strongly to the mainstream of semiconductor laser development. A possible exception to this rule has been the subject of long-term degradation studies, where an understanding of some of the fundamental mechanisms is still outstanding, in spite of the enormous effort which has been invested.

It is perhaps appropriate to conclude by listing a few specific effects of current interest where further theoretical study could be profitable:

(i) *Guidance in stripe-geometry lasers.* There are at least five mechanisms influencing the guidance in the y-direction:

(a) thermally-induced variation of the refractive index (120, 151),
(b) gain-guiding due to injected carriers (40–42),
(c) index-antiguiding associated with carrier distribution (19, 38),
(d) spatial hole-burning (self-focusing) in broad stripe lasers (6, 19, 38, 135, 146),
(e) strain effects associated with the contact (38).

A comprehensive understanding of the relative roles played by these mechanisms has not yet been achieved; in particular, the reported existence of guiding action even below lasing threshold (136, 151) poses a new challenge to theoretical analysis. An associated problem would be the description of the facet reflectivities and far-field patterns associated with these mechanisms; the presence of gain-guidance and astigmatic beams implies that these quantities are functions of injected current density.

(ii) *New laser structures.* Structures with a built-in waveguide in the y-direction are now being designed (59–63, 71, 152–154) with the objectives of stable single-mode operation and 'kink'-free output characteristics. In many cases these designs utilize the effective permittivity concept (described in 15.3.2) to analyse the strength of guidance obtained in a given structure. A point that arises is therefore concerned with the accuracy of this technique; other than a comparison with numerical solutions for the step-index guide

(138), there has been no effort to estimate the accuracy of the effective permittivity method or its applicability to a given situation.

(iii) *New laser materials for long-wavelength operation.* In this respect there is now a similar situation to that for the AlGaAs system some years ago; the waveguide theories available seem adequate, but there is as yet little refractive index data available. Hence there is a place for theoretical models of refractive indices in new laser materials, perhaps based on those available for AlGaAs (134, 155).

(iv) *Degradation problems.* To select only two particular areas where theoretical study seems appropriate, consider (a) self-pulsing observed in degraded lasers (156), and (b) effects of facet coating in improving laser lives (143). Whilst the former effect has been tentatively attributed to microscale absorption centres (157), an extremely large absorption cross section is necessary to explain the observed pulsations via repetitive Q-switching (158). As regards (b), the problem is to explain why facet coating is effective when half-wave films are used (159, 160) which do not intentionally change the facet reflectivity or limit the optical field intensity. An improved understanding of these phenomena may shed further light on the mechanisms responsible for laser degradation.

It is anticipated that theoretical work on the topics listed here and on related subjects will continue to play an important part in semiconductor laser development.

ACKNOWLEDGEMENTS

I am indebted to J. Buus for a preprint of Reference 140, and to the U.K. Science Research Council for the award of a research fellowship.

REFERENCES

1. See, for example, W. R. Frensley and H. Kroemer, *Phys. Rev.*, B, **16**, 2642 (1977), and references therein.
2. C. M. Garner, Y. D. Shen, J. S. Kim, G. L. Pearson, W. E. Spicer, J. S. Harris, Jr., and D. D. Edwall, *J. Appl. Phys.*, **48**, 3147 (1977).
3. B. N. Sharapov, *Soviet Physics-Semiconductors*, **4**, 948 (1970); also *Fiz. Tekh. Poluprov.*, **4**, 1121 (1970).
4. R. D. Burnham, P. D. Dapkus, N. Holonyak, Jr., D. L. Keune, and H. R. Zwicker, *Solid State Electronics*, **13**, 199 (1970).

5. K. A. Shore, *Quarterly Report No. 2 on Post Office Contract No. 541509* (1972).
6. B. W. Hakki, *J. Appl. Phys.*, **45**, 288 (1974).
7. F. Stern, in *Laser Handbook*, ed. F. T. Arechi and E. O. Schulz-DuBois (North-Holland, Amsterdam, 1974) pp. 425–440.
8. K. A. Shore and M. J. Adams, *Optical and Quantum Electronics*, **8**, 269 (1976).
9. D. L. Rode, *J. Appl. Phys.*, **45**, 3887, (1974).
10. A. R. Goodwin, J. R. Peters, M. Pion, G. H. B. Thompson, and J. E. A. Whiteaway, *J. Appl. Phys.*, **46**, 3126 (1975).
11. K. A. Shore and M. J. Adams, *App. Phys.*, **9**, 161 (1976).
12. B. W. Hakki, *J. Appl. Phys.*, **44**, 5021 (1973).
13. B. W. Hakki, *J. Appl. Phys.*, **46**, 2723 (1975).
14. W. P. Dumke, *Solid State Electron.*, **16**, 1279 (1963).
15. B. W. Hakki, *J. Appl. Phys.*, **46**, 292 (1975).
16. I. Ladany, *J. Appl. Phys.*, **48**, 1935 (1977).
17. N. Chinone, *J. Appl. Phys.*, **48**, 3237 (1977).
18. A. S. Logginov and V. E. Solovev, *Phys. stat. sol.* (a), **41**, 371 (1977).
19. P. A. Kirkby, A. R. Goodwin, G. H. B. Thompson, and P. R. Selway, *IEEE Jnl.*, **QE-13**, 705 (1977).
20. J. Buus and M. Danielsen, *IEEE Jnl.*, **QE-13**, 669 (1977).
21. H. S. Sommers, Jr., and D. O. North, *J. Appl. Phys.*, **48**, 4460 (1977).
22. H. Statz, C. L. Tang, and J. M. Lavine, *J. Appl. Phys.*, **35**, 2581 (1964).
23. K. Konnerth and C. Lanza, *Appl. Phys. Letts.*, **4**, 120 (1964).
24. D. M. Eagles, *J. Phys. Chem. Solids*, **16**, 76 (1960).
25. E. O. Kane, *Phys. Rev.*, **131**, 79 (1963).
26. B. I. Halperin and M. Lax, *Phys. Rev.*, **148**, 722 (1966).
27. For a review, see E. Göbel and M. H. Pilkuhn, *J. de Physique*, **35**, C3-191 (1974).
28. W. W. Anderson, *IEEE Jnl.*, **QE-1**, 228 (1965).
29. H. Kogelnik and V. Ramaswamy, *Appl. Opt.*, **13**, 1857 (1974).
30. I. Hayashi, M. B. Panish, and F. K. Reinhart, *J. Appl. Phys.*, **42**, 1929 (1971).
31. M. J. Adams, *Optics Commun.*, **23**, 105 (1977).
32. G. H. B. Thompson and P. A. Kirkby, *IEEE Jnl.*, **QE-9**, 311 (1973).
33. G. H. B. Thompson, G. D. Henshall, G. E. A. Whiteaway, and P. A. Kirkby, *J. Appl. Phys.*, **47**, 1501 (1976).
34. H. C. Casey, Jr., M. B. Panish, W. O. Schlosser, and T. L. Paoli, *J. Appl. Phys.*, **45**, 322 (1974).
35. H. C. Casey, Jr., S. Somekh, and M. Ilegems, *Appl. Phys. Letts.*, **27**, 142 (1975).
36. A. Aiki, M. Nakamura, J. Umeda, A. Yariv, A. Katzir, and H. W. Yen, *Appl. Phys. Letts.*, **27**, 403 (1975).
37. D. R. Scifres, W. Streifer, and R. D. Burnham, *Appl. Phys. Letts.*, **29**, 23 (1976).
38. G. H. B. Thompson, D. F. Lovelace, and S. E. H. Turley, *Solid-State and Electron Devices*, **2**, 12 (1978).
39. M. Cross and M. J. Adams, *Solid-State Electron.*, **15**, 919 (1972).
40. W. O. Schlosser, *B.S.T.J.*, **52**, 887 (1973).
41. F. R. Nash, *J. Appl. Phys.*, **44**, 4696 (1973).
42. D. D. Cook and F. R. Nash, *J. Appl. Phys.*, **46**, 1660 (1975).
43. P. S. Epstein, *Proc. Nat. Acad. Sci.*, **16**, 627 (1930).

44. S. N. Stolyarov, *Sov. Jnl. of Quantum Electron.*, **2**, 144 (1972); also *Kvant. Elektr.*, **2**, 69 (1972).
45. M. Osinski, *Opt. and Quantum Electron.*, **9**, 361 (1977).
46. D. Marcuse, *IEEE Jnl.*, **QE-9**, 1000 (1973).
47. W. W. Rigrod, J. H. McFee, M. A. Pollack, and R. A. Logan, *J.O.S.A.*, **65**, 46 (1975).
48. W. Streifer, R. D. Burnham, and D. R. Scifres, *IEEE Jnl.*, **QE-12**, 494 (1976).
49. M. J. Adams, *Opt. and Quantum Electron.*, **10**, 17 (1978).
50. E. M. Conwell, *Appl. Phys. Letts.*, **23**, 328 (1973).
51. J. R. Carruthers, I. P. Kaminov, and L. W. Stulz, *Appl. Optics*, **13**, 2333 (1974).
52. H. A. Haus and R. V. Schmidt, *Appl. Optics*, **15**, 774 (1976).
53. H. Kogelnik, in *Integrated Optics*, ed. T. Tamir (Springer-Verlag, N.Y., 1975) pp. 13–81.
54. D. F. Nelson and J. McKenna, *Jnl. of Appl. Phys.*, **38**, 4057 (1967).
55. O. V. Bogdankevich, V. S. Letokhov, and A. F. Suchkov, *Sov. Phys.-Semiconductors*, **3**, 566 (1969); also *Fiz. Tekh. Poluprov.*, **3**, 665 (1969).
56. K. Unger, *Ann. Physik*, **19**, 64 (1967).
57. J. A. Arnaud and W. Mammel, *IEEE Trans.*, **MTT-23**, 927 (1975).
58. M. Abramowitz and I. A. Stegun, *Handbook of Mathematical Functions*, Dover, N.Y., 1965.
59. T. Tsukada, *J. Appl. Phys.*, **45**, 4899 (1974).
60. R. D. Burnham and D. R. Scifres, *Appl. Phys. Lett.*, **27**, 510 (1975).
61. P. A. Kirkby and G. H. B. Thompson, *J. Appl. Phys.*, **47**, 4578 (1976).
62. H. Yonezu, Y. Matsumoto, T. Shinohara, I. Sakuma, T. Suzuki, K. Kobayashi, R. Lang, Y. Nannichi, and I. Hayashi, *Japan. Jnl. Appl. Phys.*, **16**, 209 (1977).
63. P. A. Kirkby, A. R. Goodwin, and R. S. Baulcomb, presented at the conference 'Semiconductor Injection Lasers and their Applications', Cardiff, Wales, 1977.
64. A. Yariv and M. Nakamura, *IEEE Jnl.*, **QE-13**, 233 (1977).
65. Y. Nannichi, *J. Appl. Phys.*, **37**, 3009 (1966).
66. R. Ulbrich and M. H. Pilkuhn, *IEEE Jnl.*, **QE-6**, 314 (1970).
67. C. H. Gooch, in *Proceedings of the International Symposium on GaAs* (Institute of Physics, London, 1966), pp. 62–67.
68. S. Hasuo and T. Ohmi, *Japan. Jnl. Appl. Phys.*, **13**, 1429 (1974).
69. W. Streifer, R. D. Burnham, and D. R. Scifres, *IEEE Jnl.*, **QE-13**, 403 (1977).
70. D. R. Scifres, R. D. Burnham, and W. Streifer, *Appl. Phys. Letts.*, **31**, 112 (1977).
71. H. Namizaki, *IEEE Jnl.*, **QE-11**, 427 (1975).
72. M. J. Adams and P. T. Landsberg, in *Gallium Arsenide Lasers*, ed. C. H. Gooch (Wiley, London, 1969), pp. 5–79.
73. H. Hillbrand and P. Russer, *Electron. Letts.*, **11**, 372 (1975).
74. Y. Suematsu, S. Akiba, and T. Hong, *IEEE Jnl.*, **QE-13**, 596 (1977).
75. M. J. Adams, *Opto-electronics*, **5**, 201 (1973).
76. T. Ikegami and Y. Suematsu, *Electronics and Communications in Japan*, **53-B6**.
77. W. Harth, *Electron. Letts.*, **9**, 532 (1973).
78. W. Harth, *A.E.U.*, **29**, 149 (1975).
79. G. Arnold and P. Russer, *Appl. Phys.*, **14**, 255 (1977).
80. P. M. Boers, M. T. Vlaardingerbroek, and M. Danielsen, *Electron. Letts.*, **11**, 206 (1975).
81. W. Harth and D. Siemsen, *A.E.U.*, **30**, 343 (1976).
82. J. Angerstein and D. Siemsen, *A.E.U.*, **30**, 477 (1976).
83. K. Kajiyama, S. Hata, and S. Sakata, *Appl. Phys.*, **12**, 209 (1977).

84. Y. Suematsu and K. Furuya, *Trans. IECE*, **E60**, 467 (1977).
85. T. Kobayashi and S. Takahashi, *Japan. Jnl. Appl. Phys.* **15**, 2025 (1976).
86. J. Buus, M. Danielsen, P. Jeppesen, F. Mengel, M. Moeskjaer, and V. Ostoich, *Proceedings of the Second European Conference on Optical Fibre Communication* (1976), pp. 231–239.
87. T. Ozeki and T. Ito, *IEEE Jnl.*, **QE-9**, 388 (1973).
88. T. Ozeki and T. Ito, *IEEE Jnl.*, **QE-9**, 1098 (1973).
89. T. P. Lee and R. M. DeRosier, *Proc. IEEE*, **62**, 1176 (1974).
90. J. G. Farrington and J. E. Carroll, *Proceedings of the First European Conference on Optical Fibre Communications* (1975), pp. 135–137.
91. M. Danielsen, *IEEE Jnl.*, **QE-12**, 657 (1976).
92. K. Kobayashi, R. Lang, and K. Minemura, *Japan. Jnl. Appl. Phys.*, **15**, 281 (1976).
93. P. Russer, *A.E.U.*, **29**, 231 (1975).
94. R. Lang and K. Kobayashi, *IEEE Jnl.*, **QE-12**, 194 (1976).
95. P. Russer, G. Arnold, and K. Petermann, *Proceedings of the Third European Conference on Optical Fibre Communication* (1977), pp. 139–141.
96. Y. Suematsu and T. Hong, *IEEE Jnl.*, **QE-13**, 756 (1977).
97. D. E. McCumber, *Phys. Rev.*, **141**, 306 (1966).
98. H. Haug, *Phys. Rev.*, **184**, 338 (1969).
99. D. J. Morgan and M. J. Adams, *Phys. stat. sol.* (a), **11**, 243 (1972).
100. T. L. Paoli, *Appl. Phys. Letts.*, **24**, 187 (1974).
101. T. L. Paoli, *IEEE Jnl.*, **QE-11**, 276 (1975).
102. G. Guekos, H. Jäckel and K. F. Schmid, *Electron. Letts.*, **12**, 64 (1976).
103. H. Jäckel, *Electron. Letts.*, **12**, 289 (1976).
104. H. Jäckel and G. Guekos, *Opt. and Quantum Electron.*, **9**, 233 (1977).
105. R. Lang, K. Minemura, and K. Kobayashi, *Electron. Letts.*, **13**, 228 (1977).
106. M. Danielsen, J. Buus, F. Mengel, K. Mortensen, and K. Stubkjaer, *Proceedings of the Third European Conference on Optical Communications*, (1977), pp. 142–144.
107. T. Ito, S. Machida, K. Namata, and T. Ikegami, *IEEE Jnl.*, **QE-13**, 574 (1977).
108. W. R. Lange, *Electron. Letts.*, **14**, 7 (1978).
109. H. C. Casey, Jr., and F. Stern, *J. Appl. Phys.*, **47**, 631 (1976).
110. F. Stern, *J. Appl. Phys.*, **47**, 5382 (1976).
111. See, for example, E. O. Goebel, O. Hildebrand, and K. Löhnert, *IEEE Jnl.*, **QE-13** 848 (1977).
112. B. W. Hakki and T. L. Paoli, *J. Appl. Phys.*, **46**, 1299 (1975).
113. J. C. Dyment, J. E. Ripper, and T. H. Zachos, *J. Appl. Phys.*, **40**, 1802 (1969).
114. P. Garel-Jones and J. C. Dyment, *IEEE Jnl.*, **QE-11**, 408 (1975).
115. R. Keller, R. Salathé, and T. Tschudi, *IEEE Jnl.*, **QE-8**, 783 (1972).
116. T. Kobayashi and Y. Furukawa, *Japan. Jnl. Appl. Phys.*, **14**, 1981 (1975).
117. R. Keller, C. Voumard, and H. Weber, *Appl. Phys. Letts.*, **26**, 50 (1975).
118. C. J. Hwang, *Appl. Phys. Letts.*, **30**, 167 (1977).
119. H. Yonezu, I. Sakuma, K. Kobayashi, T. Kamejima, M. Ueno, and Y. Nannichi, *Japan. Jnl. Appl. Phys.*, **12**, 1585 (1973).
120. W. B. Joyce and R. W. Dixon, *J. Appl. Phys.*, **46**, 855 (1975).
121. H. Kressel and H. Mierop, *J. Appl. Phys.*, **38**, 5419 (1967).
122. C. D. Dobson and F. S. Keeble, *Proceedings of the International Symposium on Gallium Arsenide* (London, The Institute of Physics and the Physical Society, 1967), pp. 68–71.

123. R. S. Broom, *IEEE Jnl.*, **QE-4**, 135 (1968).
124. W. Nakwaski, *Electron. Technology* (Poland), **9**, 2 (1976).
125. F. D. Nunes, N. B. Patel, and J. E. Ripper, *IEEE Jnl.*, **QE-13**, 675 (1977).
126. See, for example, M. J. Adams and B. Thomas, *IEEE Jnl.*, **QE-13**, 580 (1977), and references therein.
127. C. H. Henry and D. V. Lang, *Phys. Rev.*, **B15**, 989 (1977).
128. For a review, especially with reference to GaAs, see P. T. Landsberg and M. J. Adams, *J. Luminescence*, **7**, 3 (1973).
129. H. Kressel and M. Ettenberg, *J. Appl. Phys.*, **47**, 3533 (1976).
130. For a recent review, see T. Kimura and K. Daikoku, *Opt. and Quantum Electron.* **9**, 33 (1977).
131. H. G. Unger, *A.E.U.*, **25**, 539 (1971).
132. J. J. Hsieh, J. A. Rossi, and J. P. Donnelly, *Appl. Phys. Letts.*, **28**, 709 (1976).
133. G. D. Pettit and W. J. Turner, *J. Appl. Phys.*, **36**, 2081 (1965).
134. M. J. Adams and M. Cross, *Solid State Electronics*, **14**, 865 (1971).
135. G. H. B. Thompson, *Optoelectronics*, **4**, 257 (1972).
136. T. L. Paoli, *IEEE Jnl.*, **QE-13**, 662 (1977).
137. M. Osinski, private communication.
138. For a recent account, together with references to the relevant literature, see G. B. Hokker and W. K. Burns, *Applied Optics*, **16**, 113 (1977).
139. T. Rozzi, T. Itoh, and L. Grun, *Radio Science*, **12**, 543 (1977).
140. J. Buus, to be published.
141. M. J. Adams, *Electron. Letts.*, **13**, 236 (1977).
142. H. Kogelnik, *Appl. Optics*, **4**, 1562 (1965).
143. H. Kressel and I. Ladany, *R.C.A. Rev.*, **36**, 230 (1975).
144. J. E. Carroll, S. G. Eldon, and G. H. B. Thompson, *Electron. Letts.*, **12**, 564 (1976).
145. K. Otsuka, *IEEE Jnl.*, **QE-13**, 520 (1977).
146. M. Cross, *Phys. stat. sol.* (a), **16**, 167 (1973).
147. P. R. Selway, P. A. Kirkby, A. R. Goodwin, and G. H. B. Thompson, *S.S.E.D.*, **2**, 38 (1978).
148. T. Ikegami, *Proceedings of the First European Conference on Optical Fibre Communication* (1975), pp. 111–113.
149. G. Arnold, F.-J. Berlec, and R. Petschacher, *Proceeding of the Third European Conference on Optical Fibre Communication* (1977), pp. 136–138.
150. P. R. Selway and A. R. Goodwin, *Electron. Letts.*, **12**, 25 (1976).
151. J. E. Ripper, F. D. Nunes, and N. B. Patel, *Appl. Phys. Letts.*, **27**, 328 (1975).
152. K. Aiki, M. Nakamura, T. Kurada, and J. Umeda, *Appl. Phy. Letts.*, **30**, 649 (1977).
153. L. Figueroa and S. Wang, *Appl. Phys. Letts.*, **31**, 45 (1977).
154. L. Figueroa and S. Wang, *Appl. Phys. Letts.*, **32**, 55 (1978).
155. M. A. Afromowitz, *Solid State Commun.*, **15**, 59 (1974).
156. T. L. Paoli, *IEEE Jnl.*, **QE-13**, 351 (1977).
157. D. Kato, *Appl. Phys. Letts.*, **31**, 588 (1977).
158. M. J. Adams, *Phys. stat. sol.* (a), **1**, 143 (1970).
159. I. Ladany, M. Ettenberg, H. F. Lockwood, and H. Kressel, *Appl. Phys. Letts.*, **30**, 87 (1977).
160. Y. Shima, N. Chinone, and R. Ito, *Appl. Phys. Letts.*, **31**, 625 (1977).

Chapter 16
Laser Modes as Eigenfunctions of an Operator Equation*

J. E. Ripper, M. D. Campos, M. A. A. Pudensi

16.1 INTRODUCTION

In most semiconductor laser mode theories (1–27), the laser is regarded as a waveguide, and the mirrors are assumed to reflect an exact image of the incident beam, at most attenuated by the mirror not being totally reflective. This is a reasonable approximation if the mirrors are perpendicular to the waveguide and the waveguide has lateral dimensions of several wavelengths. The approximation breaks down for thin double heterostructures or for non-perpendicular mirrors (28).

In this chapter we propose a new method of arriving at an approximate solution of mode problems which cannot be treated by the traditional methods. Basically the idea is to treat the laser mode as an eigenfunction of an operator equation so that the mathematical methods developed to treat the wave equations in quantum mechanics can be used as tools to solve the equation.

16.2 THE TRADITIONAL METHOD

We shall use as an example the mode solution given by Zachos and Ripper (1) for the stripe geometry homostructure semiconductor laser. We shall repeat here the main results of that paper since they are going to be useful later in this work.

It was assumed that the refractive index did not depend on the longitudinal or z-direction (perpendicular to the mirrors) and could be expanded in series:

$$n = \bar{n} \cdot \left[1 - \left(\frac{x}{x_e}\right)^2 - \left(\frac{y}{y_e}\right)^2 + \text{higher order terms}\right]^{1/2} \quad (16.1)$$

---
* This work was partially supported by Telecomunicações Brasileiras S/A and Fundação de Amparo à Pesquisa do Estado de São Paulo, Brazil.

where x is the transverse direction perpendicular to the junction and the origin is chosen at the maximum of the refraction index to eliminate first order terms. It was shown that the higher order terms could be neglected and that the waveguide modes are given by:

$$E_{mn}(x, y, z) = A_{mn} X(x) Y(y) \exp(\pm i k \gamma_{mn} z) \qquad (16.2)$$

where A_{mn} is a constant and

$$X_m(x) = H_m\left(\sqrt{2\frac{x}{\omega_{0x}}}\right) \exp\left[-\left(\frac{x}{\omega_{0x}}\right)^2\right] \qquad (16.3)$$

$$Y_n(y) = H_n\left(\sqrt{2\frac{y}{\omega_{0y}}}\right) \exp\left[-\left(\frac{y}{\omega_{0y}}\right)^2\right] \qquad (16.4)$$

and

$$\gamma_{mn} = \bar{n}\left[1 - \frac{2m+1}{\bar{n}x_e k} - \frac{2n+1}{\bar{n}y_e k}\right]^{1/2} \qquad (16.5)$$

The function H_m is the Hermite polynomial of order m, and the beam widths ω_{0x}, ω_{0y} are given by

$$\omega_{0x} = \sqrt{\frac{\lambda x_e}{\pi \bar{n}}}, \qquad \omega_{0y} = \sqrt{\frac{\lambda y_e}{\pi \bar{n}}} \qquad (16.6)$$

The mirrors were assumed to be totally reflective and to reflect the same field distributions that reached them or in effect just changing the sign of exponent in (16.2). This imposes a boundary condition

$$k_{\gamma_{mnq}} = \frac{q\pi}{L} \qquad (16.7)$$

where q is an integer and L is the distance between the two mirrors. Thus the mode solution becomes

$$E_{mnq} = 2A_{mn} X_m(x) Y_n(y) \cos\left(\frac{q\pi}{L} z\right) \qquad (16.8)$$

Obviously, this method could not take into account non-uniform mirror reflectivity or inhomogeneities of the laser medium.

16.3 THE OPERATOR EQUATION

Let us consider an electromagnetic wave whose field distribution is $E(x, y, z)$ and is propagating in the z-direction in a medium which can have a gain. After travelling a distance dz the field distribution is $E'(x, y, z+dz)$, and we can define an operator $dG(z)$ by the equation

$$E'(x, y, z+dz) = dG(z)E(x, y, z) \tag{16.9}$$

After travelling from z to a point z_1 the field distribution will be given by

$$E''(x, y, z_1) = \prod|_z^{z_1} dG(z) E(x, y, z) \tag{16.10}$$

where the symbol $\prod|_z^{z_1} dG(z)$ stands for the product:

$$dG(z_1) \cdot dG(z_1 - dz) \ldots dG(z+dz) \cdot dG(z) \tag{16.11}$$

A laser is an oscillator where the electromagnetic wave travels a closed path returning to the same point. Assuming that we arbitrarily choose the point 1 of coordinates (x_1, y_1, z_1) as a starting (and finishing) point and we start with an arbitrary wave distribution $E(x_1, y_1, z_1)$, we shall have after a round trip:

$$E'(x_1, y_1, z_1) = G_{11} E(x_1, y_1, z_1) \tag{16.12}$$

where

$$G_{11} = \prod|_1^1 dG(x, y, z) \tag{16.13}$$

If the field distribution $E(x_1, y_1, z_1)$ were a laser cavity mode, then after a round trip the field would reproduce itself except for a proportionality constant. This can actually be used as the definition of a laser mode. Thus, a laser mode is an eigenfunction of the operator equation

$$G_{11} E^1 = g E^1 \tag{16.14}$$

It is interesting to note that if a different starting point 2 were chosen, we would have a different eigenfunction but the same eigenvalue. It is easy to see that

$$G_{11} = G_{12} G_{21} \tag{16.15}$$

Thus Equation 16.14 becomes

$$G_{12} G_{21} E^1 = g E^1 \tag{16.16}$$

Let us operate at the left with the operator G_{21}:

$$G_{21} G_{12} (G_{21} E^1) = g(G_{21} E^1) \tag{16.17}$$

Since:

$$G_{21} G_{12} = G_{22}, \tag{16.18}$$

the field

$$E^2 = G_{21} E^1 \tag{16.19}$$

is an eigenfunction of the operator G_{22} with the same eigenvalue g.

The physical interpretation is obvious, in general the field of the laser mode is different at different points of the cavity and, knowing the field (the eigenfunction) at point 1, we can find it at point 2 using Equation 16.19. The eigenvalue g corresponds to the round trip gain of the laser mode and does not depend on the starting point.

Thus in equilibrium the threshold condition can be defined by

$$g = 1 \tag{16.20}$$

16.4 THE IDEALIZED HOMOSTRUCTURE LASER

To show a possible application of the eigenfunction method let us go back to the problem treated in Section 16.2. First, let us keep the assumption of totally reflective mirrors. As shown in Figure 16.1, let us call 1 the point

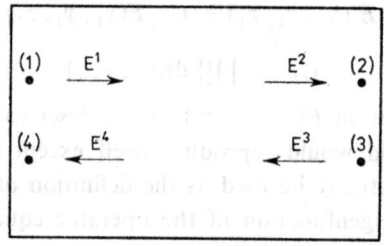

Figure 16.1 Schematic view of the fields obtained with the action of the operators G_b and G_m. For example: $E^2 = G_b E^1$ and $E^3 = G_m E^2$

just behind mirror A, 2 just in front of mirror B, 3 just behind mirror B and 4 just in front of mirror A.

By symmetry,

$$G_{21} = G_{43} = G_b \tag{16.21}$$

and

$$G_{32} = G_{14} = G_m \tag{16.22}$$

Thus the laser equation is given by

$$G_{11} E^1 = G_m G_b G_m G_b E^1 = g E^1 \tag{16.23}$$

or because of the symmetry

$$G_m G_b E^1 = g^{1/2} E^1 \tag{16.24}$$

Let us now analyse each operator separately. Let us first consider the bulk. Basically we are back to the waveguide problem. If the field distribution E

is supposed to be the solution given by Equation 16.2 we have

$$G_b E_{mn} = [\exp(\alpha_{mn} L + ik\gamma_{mn} L)] E_{mn} \quad (16.25)$$

where L is the length of the laser and α_{mn} is the gain per unit length of the laser.

If we now assume that the mirror is totally reflective and reflects exactly the same field distribution, we have

$$G_m E = E \quad (16.26)$$

regardless of what the field distribution E is.

Solving Equations 16.24–16.26 and applying the threshold condition $g = 1$ we get again Equation 16.7 and $\alpha_{mn} = 0$.

Let us now assume a dielectric mirror with a constant reflectivity R independent of the wave shape.

$$G_m E = R[\exp(i\psi)] E \quad (16.27)$$

where ψ is a phase delay introduced by the dielectric mirror.

In this case the solution of Equation 16.24 will lead to

$$\alpha_{mn} = \frac{1}{L} \ln\left(\frac{1}{R}\right) \quad (16.28)$$

$$k\gamma_{mn} L + \psi = q\pi \quad (16.29)$$

16.5 THE 'REAL' HOMOSTRUCTURE LASER-THEORY

In the previous section the eigenfunction method was illustrated with a problem which could as easily be treated by conventional methods. Let us now tackle a problem which cannot. In reality, a dielectric mirror does not have a constant reflectivity; for a plane wave the reflectivity depends on polarization and on the incidence angle. For a non-plane wave, such as the Hermite–Gaussian waves, the mirror actually does not reflect exactly the same mode distribution since the mode can be considered as a combination of plane waves, each with a different reflectivity.

Since in the direction parallel to the junction (y) the mode dimensions are much bigger than the wavelength, we can neglect its contribution to the mirror reflectivity; in this direction the plane wave approximation is a reasonable one. The same is not true in the direction perpendicular to the junction (x) where confinement reduces the dimensions considerably. Thus, to simplify the equations we shall treat the problem as a two-dimensional (x–z) one, since it was shown by Zachos and Ripper (1) that the mode equation can be separated in its x and y component.

If the mirror reflection does not have exactly the same contribution of the incident wave; then we can write

$$G_m E_{mnq} = E'_{mnq} \tag{16.30}$$

or from (16.2)

$$G_m A_{mnq} X_m Y_n \exp(\pm ik\gamma_{mn} z) = A_{mnq} X'_m Y_n \exp(\pm ik\gamma_{mn} z) \tag{16.31}$$

thus, ignoring the y and z dimensions, we can write

$$G_m X_m = X'_m \tag{16.32}$$

which means that the reflectivity variation is considered only in the x-direction.

Since it can be proved that the set of Hermite–Gaussian functions X_i is a complete set,* we can rewrite (16.32) as

$$G_m X_m = \sum_i a_{mi} X_i \tag{16.33}$$

where a_{mi} are constants to be determined.

Ignoring the y and z dependences, Equation 16.25 can be rewritten as

$$G_b X_m = g_m X_m \tag{16.34}$$

where

$$g_m = \exp(\alpha_m L) \tag{16.35}$$

and the full mode equation 16.24 written in a matrix form becomes

$$G_m G_b \begin{bmatrix} X_0 \\ X_1 \\ \vdots \\ X_i \\ \vdots \end{bmatrix} = \begin{bmatrix} g_0 a_{00} & g_0 a_{01} & \cdots & g_0 g_{0i} & \cdots \\ g_1 a_{10} & g_1 a_{11} & \cdots & g_1 a_{1i} & \cdots \\ \cdots\cdots\cdots\cdots\cdots\cdots\cdots\cdots \\ g_i a_{i0} & g_i a_{i1} & \cdots & g_i a_{ii} & \cdots \\ \cdots\cdots\cdots\cdots\cdots\cdots\cdots\cdots \end{bmatrix} \begin{bmatrix} X_0 \\ X_1 \\ \vdots \\ X_i \\ \vdots \end{bmatrix} \tag{16.36}$$

We can always find a rotation operator R such that the matrix above is diagonalized:

$$R \begin{bmatrix} g_0 a_{00} & g_0 a_{01} & \cdots & g_0 a_{0i} & \cdots \\ g_1 a_{10} & g_1 a_{11} & \cdots & g_1 a_{1i} & \cdots \\ \cdots\cdots\cdots\cdots\cdots\cdots\cdots\cdots \\ g_i a_{i0} & g_i a_{01} & \cdots & g_i a_{ii} & \cdots \\ \cdots\cdots\cdots\cdots\cdots\cdots\cdots\cdots \end{bmatrix} R^{-1} = \begin{bmatrix} b_0 & 0 & 0 & \cdots & 0 & \cdots \\ 0 & b_1 & 0 & \cdots & 0 & \cdots \\ 0 & 0 & b_2 & \cdots & 0 & \cdots \\ \cdots\cdots\cdots\cdots\cdots\cdots \\ 0 & 0 & 0 & \cdots & b_i & \cdots \end{bmatrix} \tag{16.37}$$

---

* It is easy to show that in analogy to the solutions of the Schrödinger equation, the solutions of any mode wave equations form a complete set.

Laser Modes as Eigenfunctions

This rotation operator is Hermitian with the operators G_m and G_b so that (16.35) can be rewritten as:

$$G_m G_b \begin{bmatrix} \bar{X}_0 \\ \bar{X}_1 \\ \bar{X}_2 \\ \vdots \\ \bar{X}_i \\ \vdots \end{bmatrix} = \begin{bmatrix} b_0 & 0 & 0 & \ldots & 0 & \ldots \\ 0 & b_1 & 0 & \ldots & 0 & \ldots \\ 0 & 0 & b_2 & \ldots & 0 & \ldots \\ \multicolumn{6}{c}{\dotfill} \\ 0 & 0 & 0 & \ldots & b_i & \ldots \\ \multicolumn{6}{c}{\dotfill} \end{bmatrix} \begin{bmatrix} \bar{X}_0 \\ \bar{X}_1 \\ \bar{X}_2 \\ \vdots \\ \bar{X}_i \\ \vdots \end{bmatrix} \qquad (16.38)$$

where

$$R \begin{bmatrix} X_0 \\ X_1 \\ X_2 \\ \vdots \\ X_i \\ \vdots \end{bmatrix} = \begin{bmatrix} \bar{X}_0 \\ \bar{X}_1 \\ \bar{X}_2 \\ \vdots \\ \bar{X}_i \\ \vdots \end{bmatrix} \qquad (16.39)$$

From (16.38) we see that the equations are decoupled and we can write:

$$G_m G_b \bar{X}_\gamma = b_\gamma \bar{X}_\gamma \qquad (16.40)$$

so $\bar{X}_\gamma$ are the eigenfunctions in gain operator equations and correspond to the mode distribution of the laser in the x-direction. Note that Equation 16.39 actually corresponds to a linear combination of the unperturbed Hermite–Gaussian modes. We can see here the similarity with the infinite order perturbation method used in quantum mechanics. Actually it should be noted that although in theory the matrices involved are of infinite dimensions, the convergence is very fast, because the gain coefficients g_i are very small except for very low orders i of the modes.

In practice, in solving the mode equations the problem is to determine the coefficients a_{im} of Equation 16.33.

In order to determine the mirror reflectivity we have first to transform the wavefunction X_m into a combination of plane waves. This can be done by Fourier transform, noting that except for a scaling factor the Hermite–Gaussian functions transform into themselves, thus

$$X_m(x) = \int F_m(k_x) \exp(ik_x x) \, dk_x \qquad (16.41)$$

where

$$F_m(k_x) = H_m\left(\frac{\sqrt{2}}{2}\omega_{0x}k_x\right)\exp\left[-\left(\frac{\omega_{0x}k_x}{2}\right)^2\right](i)^m \qquad (16.42)$$

For each of the plane waves $e^{ik_x x}$ the mirror reflectivity is:

$$R_{\text{TE}}(k_x) = \frac{\bar{n}\cos\theta - \sqrt{1-\bar{n}^2\sin^2\theta}}{\bar{n}\cos\theta + \sqrt{1-\bar{n}^2\sin^2\theta}} \qquad (16.43)$$

for the TE polarization (*E* parallel to the junction plane), and

$$R_{\text{TM}}(k_x) = \frac{\bar{n}\sqrt{1-\bar{n}^2\sin^2\theta} - \cos\theta}{\bar{n}\sqrt{1-\bar{n}^2\sin^2\theta} + \cos\theta} \qquad (16.44)$$

for the TM polarization (*E* perpendicular to the junction plane), where we define

$$\theta = \tan^{-1}\frac{k_x}{k_z} \qquad (16.45)$$

Thus X'_m as defined in Equation 16.32 can be given by

$$X'_m(x) = \int R(k_x) F_m(k_x) e^{ik_x x} dk_x \qquad (16.46)$$

If we assume X_m to be normalized,

$$a_{mi} = \int_{-\infty}^{+\infty} X'_m(x) X_i(x) dx \qquad (16.47)$$

The condition

$$b_m = \pm 1 \qquad (16.48)$$

defines the threshold condition for the *m*th order mode. In general, the zero order mode will have the lowest threshold, because its field is more concentrated near the gain region. Thus g_i becomes rapidly smaller as i increases and because of this the diagonalization of the matrix can be performed in terms of relatively few components.

It should be noted that as we introduce a mirror characteristic in which the eigenfunctions of G_b are not eigenfunctions of G_m, the eigenfunctions of the total gain $G_m G_b G_m G_b$ are not eigenfunctions of either G_m or G_b. Physically, this means that the change in intensity and field distribution introduced by the mirror is exactly cancelled by the change introduced by the bulk amplification. Another point to be noted is that gain guiding is automatically taken into account by the way G_b is calculated.

16.6 THE REAL HOMOSTRUCTURE LASER-COMPUTER CALCULATIONS

In order to calculate the modes of a homostructure, some assumptions on the device had to be made. First, the index of refraction was assumed to obey Equation 16.1 with $x_e = 1.7$ μm and $n = 3.6$ which was the average value found by Zachos and Ripper (1). This leads to $\omega_{0x} = 1.12$ μm taking a wavelength of 0.84 μm. To calculate the bulk gain, we assume for simplicity a region of constant loss C_2 outside it. Thus

$$\alpha_m = \frac{\sqrt{2}}{\omega_{0x}} \left[\int_{-\infty}^{-d/2} (-C_2) X_m^2 \, dx + \int_{-d/2}^{+d/2} C_1 X_m^2 \, dx + \int_{d/2}^{\infty} (-C_2) X_m^2 \, dx \right] \quad (16.49)$$

as $X_m \left(\dfrac{\sqrt{2}}{\omega_{0x}} x \right)$ is normalized, we get

$$\alpha_m = -C_2 + 2(C_1 + C_2) \frac{\sqrt{2}}{\omega_{0x}} \int_0^{d/2} X_m^2 \, dx \quad (16.50)$$

The following values were used for the constants:

$$L = 380 \text{ μm}, \quad d = 2.30 \text{ μm}, \quad C_2 = 125 \text{ cm}^{-1} \quad (16.51)$$

C_1 was assumed to be an adjustable gain parameter linked to the current necessary to satisfy the threshold condition (16.48). It should be noted that C_1 and C_2 refer to field amplitude gain; so these values are half of the usually found in the literature which refer to intensity gain.

Using these conditions we solved the problem with the aid of a digital computer (PDP-10), using a (10×10) matrix. The convergence is quite good and in fact little difference is found in the zeroth order mode if a (6×6) matrix is used.

For TE mode (polarization parallel to the junction), to achieve threshold $(b_0^2 = 1.000)$ we need:

$$C_1 = 20.74 \text{ cm}^{-1} \quad (16.52)$$

$$\bar{X}_0 = X_0 + 0.0061 X_2 + 0.0001 X_4 \quad (16.53)$$

For the same gain C_1 in the active region, the overall gain of the first order mode would be

$$b_1^2 = 0.112 \qquad (16.54)$$

For the TM mode, the corresponding results are

$$C_1 = 20.96 \text{ cm}^{-1}, \qquad \overline{X}_0 = X_0 - 0.0063X_2 - 0.002X_4, \qquad b_1^2 = 0.108 \qquad (16.55)$$

Taking the values of $\lambda = 0.83$ μm, $\omega_{0y} = 3$ μm and $S = 8.0$ μm for the stripe width, the following values are obtained for TE modes in the y-direction

$$\overline{y}_0 = y_0 - 0.0010y_2, \qquad b_0^2 = 1, \qquad b_1^2 = 0.520, \qquad C_1 = 16.11 \text{ cm}^{-1} \qquad (16.56)$$

It can be seen that there is no significant difference in the gain necessary to achieve threshold.

Finally, it should be noted that the field distributions found correspond to the reflected mode right behind the mirror and are not constant throughout the laser cavity.

16.7 THE DOUBLE HETEROSTRUCTURE LASER

Double heterostructure lasers in general have fields much more confined than homostructure laser. There we expect that the mirror reflectivity plays a more significant role. In fact it is known that the TE predominance is caused by the higher mirror reflectivity even though the way the problem is normally solved, by assuming the laser to be a waveguide whose modes are reflected with the same shape by a mirror whose reflectivity is averaged over the Fourier components, is obviously incorrect.

Even though the refraction index cannot be approximated by Equation 16.1 in the case of the heterostructure laser,* we can get a good idea of the physical picture by assuming the parabolic approximation but forcing a much bigger confinement.

We have carried out the calculation with the following data

$$\omega_{0x} = 0.25 \text{ μm}, \qquad d = 0.5 \text{ μm}, \qquad \bar{n} = 3.6$$
$$L = 380 \text{ μm}, \qquad C_2 = 125 \text{ cm}^{-1} \qquad (16.57)$$

* A correct calculation for double heterostructure lasers is being carried out and the results will be published shortly.

and have obtained the following results at threshold ($b_0^2 = 1.0000$). For the TE mode

$$C_1 = 18.70 \text{ cm}^{-1}$$

$$\overline{X}_0 \simeq 0.9985 X_0 - 0.1701 X_2 - 0.0252 X_4$$
$$+ 0.0204 X_6 + 0.0272 X_8 + 0.0198 X_{10} \qquad (16.58)$$

$$b_1^2 = 0.142$$

and for the TM mode

$$C_1 = 28.68 \text{ cm}^{-1}$$

$$\overline{X}_0 \simeq 0.9812 X_0 + 0.6475 X_2 - 0.0017 X_4$$
$$- 0.1815 X_6 - 0.1686 X_8 - 0.0928 X_{10} \qquad (16.59)$$

$$b_1^2 = 0.014$$

As expected we can see that the gain necessary to achieve threshold is significantly larger for TM than for TE modes. This result is consistent with the experimental measurements of the difference in threshold of these two polarizations (30).

Let us analyse the results. First we see that the gain necessary to operate the laser in the TM mode is slightly higher ($\sim 1\%$) than the TE mode. This is not experimentally detectable and is probably overwhelmed by strain effects introduced during the fabrication and/or bonding of the laser (29-30). Experimentally, we find both TE and TM operating lasers and in many instances lasers operating on different polarization. The favouring of the TE mode over the TM is caused by the fact that the mirror reflectivity is larger for the mode Fourier components not normal to the mirror ($k_x \neq 0$) thus leading to an overall smaller mirror loss. The difference is small, because the mode size is still several times larger than the wavelength inside the device. There is also a slight difference in the TE and TM wavefunctions. The fact that these wavefunctions are composed only of even modes is to be expected because of the fact that the zeroth order mode is symmetric even after the perturbation.

Finally the difference of overall gains (a factor of 10) between the zeroth and first order modes is certainly overemphasized by the assumption of a very sharp boundary between the gain and the lossy region. This can be verified by changing slightly the dimension of the gain region, choosing $d = 2.60$ μm and maintaining the same values for $\bar{n}$ and x_s. As expected the zeroth order mode is significantly affected, the gain C_1 is necessarily reduced to 17.80 cm^{-1} for the TE mode and 18.02 for the TM mode gain. However, for the first order mode the loss is greatly reduced and when $b_0^2 = 1.000$, $b_1^2 = 0.261$ for TE and 0.252 for TM, b_1^2 is greater for TE, because the non-normal Fourier components are more important for the first order modes than for the zeroth order one.

In conclusion we find that for the heterostructure, although the classical method does not find the exact wavefunction, the difference between the gains b_0^2 and b_1^2 is even smaller in the y-direction.

It is interesting to note that the mode field distribution is significantly altered by the mirror reflectivity, as can be seen in Figure 16.2. This result

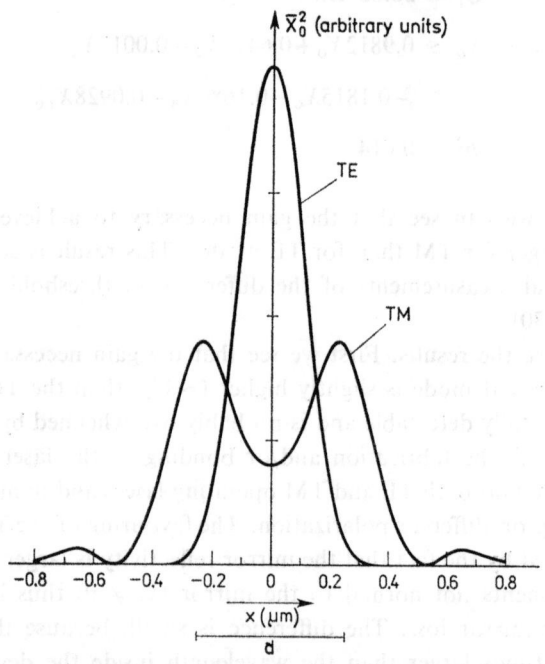

Figure 16.2 Plot of the field intensity as a function of the position coordinate x perpendicular to the junction plane for a laser with an active region of width $d = 0.5$ μm. TE and TM correspond respectively to electric fields parallel and perpendicular to the junction plane

has not been observed experimentally because the resolution of optical system in air is not sufficient to measure it. To observe it we would have to use an optical system with a medium of high refraction index. One possibility is the system illustrated in Figure 16.3. In this system a laser with a metallized mirror has a GaAs microlens behind of which can have a resolution of about 0.25 μm. This system is then coupled to a conventional optical system. The

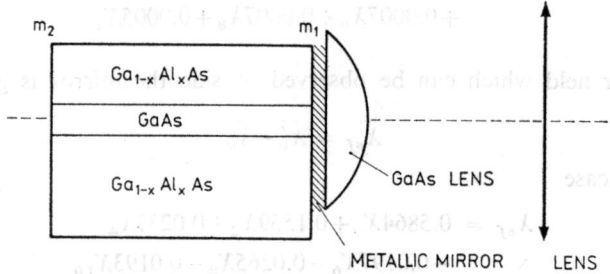

Figure 16.3 As referred to in the text in Section 16.7, a GaAs lens is coupled to a DH laser through a metallic film so as to obtain a high resolution image of the optical field. The lens behind the device is a part of the conventional optical measurement system

mirror has to be metallized so that reflectivity is assured, because the GaAs lens would neutralize the dielectric mirror reflectivity. Obviously, the mirror cannot be observed; a reflectivity of the order of 95% to 98% seems a good compromise being a good approximation of a totally reflective mirror for the laser cavity and allowing enough light out to be observed. Obviously, the experiment would correspond to a slightly different problem than the one solved above. The field observed E_0 would correspond to the solution of equations

$$E_0 = (1 - G_m) E_1 \qquad (16.60)$$

$$G_{m_1} G_b G_{m_2} G_b E_1 = E_1 \qquad (16.61)$$

where G_b and G_{m_2} have the same definitions of the problem above and

$$G_{m_1} E = RE \qquad (16.62)$$

for any field distribution with R being the reflectivity of the metallic mirror.

Another point to stress is that the field distribution calculated and given in Equations 16.53, 16.55, 16.58 and 16.59 correspond to those just behind the mirror; (E_1 or E_3 in Figure 16.1). To calculate the field in front of the mirror we would have to apply the operator G_b which would be equivalent to multiplying each component i by g_i (see Equation 16.34). Thus, if we pick the TE mode for $d = 0.5$ μm as given by Equation 16.58, the field in front of the mirror would be given by:

$$\bar{X}'_0 = 1.5849 X_0 - 0.0142 X_2 - 0.0015 X_4 \\ + 0.0007 X_6 + 0.0007 X_8 + 0.0005 X_{10} \tag{16.63}$$

The near field which can be observed outside the mirror is given by

$$\bar{X}_{nf} = \bar{X}'_0 - \bar{X}_0 \tag{16.64}$$

or in this case

$$\bar{X}_{nf} = 0.5864 X_0 + 0.1559 X_2 + 0.0237 X_4 \\ - 0.0197 X_6 - 0.0265 X_8 - 0.0193 X_{10} \tag{16.65}$$

If we compare the differences between the field $\bar{X}_0$ given by Equation 16.58 with $\bar{X}'_0$ given by Equation 16.63 it is easy to see that X'_0 deviates much less from the Gaussian field distribution X_0. The physical interpretation of the fact is that G_b includes a very strong gain guiding effect so that even though the field entering the bulk behind a mirror reflection is quite different from the Gaussian field distribution, the concentration of gain in the centre region tends to guide the field transforming it into a nearly Gaussian one. The non-uniform mirror reflectivity exactly cancels the effect, restoring the $\bar{X}_0$ distribution.

16.8 'TILTED MIRROR' SEMICONDUCTOR LASER

It was shown by Frescura et al. (28) that fabricating a laser with mirrors non-perpendicular to the laser stripe, the non-linearity in the laser output characteristics (the so-called 'kink') can be pushed to higher powers or completely eliminated.

Obviously, the traditional mode calculation methods cannot tackle this problem. Since the mirror is not perpendicular to the laser mode propagation direction, it obviously cannot reflect in that direction the same field distribution it has received. Again the eigenfunctions of G_b are not eigenfunctions of G_m and thus cannot be of the whole problem.

Following the lines of the previous problems treated in this chapter we are determining the gain necessary to achieve threshold and the field distribution as a function of the angle between the stripe and the mirrors. The results of this calculations will be published shortly.

16.9 CONCLUSIONS

We have shown that defining a laser mode as an eigenfunction of an operator equation makes it possible to solve many problems which cannot be solved by conventional methods. Examples were given. The method is powerful enough to tackle a number of different problems not only in semiconductor lasers but in any lasers where the modes cannot be assumed to be determined by either a waveguide confined by perfect plane mirrors (usual in semiconductor lasers) or by a cavity with uniform medium inside (usual in gas lasers). The method can also be expanded to allow for the gain operator to be dependent of the optical field to include saturation effects, spatial hole burning, etc.

ACKNOWLEDGMENTS

The authors would like to thank R. I. Bossi for experimental support, for some data used in the calculations, and some helpful discussions. They are grateful to H. K. E. Liesenberg for the help in running the computer program.

REFERENCES

1. T. H. Zachos and J. E. Ripper, 'Resonant modes of GaAs junction lasers', *IEEE J. Quantum Electronics*, **QE-5**, 29 (Jan. 1969).
2. T. L. Paoli, J. E. Ripper and T. H. Zachos, 'Resonant modes of GaAs junction lasers. II: High-injection level', *IEEE J. Quantum Electronics*, **QE-5**, 271 (June 1969).
3. J. C. Dyment, 'Hermite–Gaussian mode patterns in GaAs junction lasers', *Appl. Phys. Lett.*, **10**, 84 (Feb. 1967).
4. J. E. Ludman and K. M. Hergenrother, 'Far-field pattern of injection lasers and dielectric gradient in the inversion layer', *Solid State Electron.*, **9**, 863 (1966).
5. D. F. Nelson and J. McKenna, 'Electromagnetic modes of anisotropic dielectric waveguides at p–n junctions', *J. Appl. Phys.*, **38**, 4057 (Sept. 1967).
6. L. A. Rivlin and V. S. Shil'dyaev, 'Mode interaction in an injection semiconductor laser', *ZhTF Pis'ma*, **6**, 5, 659 (Sept. 1967).
7. N. E. Byer and J. K. Butler, 'Optical field distribution in close-confined laser structures', *IEEE J. Quantum Electronics*, **QE-6**, 291 (June 1970).
8. J. K. Butler, H. S. Sommers, Jr., and H. Kressel, 'High-order transverse cavity modes in heterojunction diode lasers', *Appl. Phys. Lett.*, **17**, 403 (Nov. 1970).
9. F. K. Reinhart, I. Hayashi, and M. B. Panish, 'Mode reflectivity and waveguide properties of double-heterostructure injection lasers', *J. Appl. Phys.*, **42**, 4466 (Oct. 1971).
10. J. K. Butler, 'Theory of transverse cavity mode selection in homojunction and heterojunction semiconductor diode lasers', *J. Appl. Phys.*, **42**, 4447 (Oct. 1971).

11. M. Cross and M. J. Adams, 'Wave-guiding properties of stripe geometry double-heterostructure injection lasers', *Solid State Electron.*, **15**, 919 (1972).
12. J. K. Butler and H. Kressel, 'Transverse mode selection in injection lasers with widely space heterojunction', *J. Appl. Phys.*, **43**, 3403 (Aug. 1972).
13. T. Ikegami, 'Reflectivity of mode at facet and oscillation mode in double-heterostructure injection lasers', *IEEE J. Quantum Electronics*, **QE-8**, 470 (June 1972).
14. P. A. Kirkby and G. H. B. Thompson, 'The effect of double heterojunction waveguide parameters on the far field emission patterns of lasers', *Opto-electronics*, **4**, 323 (1972).
15. H. S. Sommers, Jr., 'Experimental properties of injection lasers: Modal distribution of laser power', *J. Appl. Phys.*, **44**, 1263 (March 1973).
16. M. Cross, 'Selection mechanisms of transverse modes in semiconductor injection lasers', *IEEE J. Quantum Electronics*, **QE-9**, 517 May (1973).
17. H. Yonezu, K. Kobayashi, and I. Sakuma, 'Threshold current and lasing transverse mode in a GaAs-Al$_x$Ga$_{1-x}$As double-heterostructure laser', *Japan. J. Appl. Phys.*, **12**, 1593 (Oct. 1973).
18. W. H. Weber and K. F. Yeung, 'Waveguide and luminescent properties of thin-film Pb-salt injection lasers', *J. Appl. Phys.*, **44**, 4991 (1973).
19. F. R. Nash, 'Mode guidance parallel to the junction plane of double-heterostructure GaAs lasers', *J. Appl. Phys.*, **44**, 4696 (1973).
20. B. W. Hakki and C. J. Hwang, 'Mode control in GaAs large-cavity double-heterostructure lasers', *J. Appl. Phys.*, **45**, 2168 (1974).
21. J. E. Ripper, F. D. Nunes, and N. B. Patel, 'Filaments in semiconductor lasers', *Appl. Phys. Lett.*, **27**, 328 (1975).
22. F. D. Nunes, N. B. Patel, and J. E. Ripper, 'A theory of long time delays and internal Q-switching in GaAs junction lasers, *IEEE J. Quantum Electronics*, **QE-13**, 675 (1977).
23. P. A. Kirkby, A. R. Goodwin, G. H. B. Thompson, and P. R. Selway, 'Observations of self-focusing in stripe geometry semiconductor lasers and the development of a comprehensive model of their operation', *IEEE J. Quantum Electronics*, **QE-13**, 705 (1977).
24. S. J. Chua and B. Thomas, 'Spatial, spectral and temporal resolved measurements of semiconductor injection lasers, Part I. Homostructure lasers: Mode formation', *IEEE J. Quantum Electronics*, **QE-13**, 652 (1977).
25. T. L. Paoli, 'Waveguiding in a stripe-geometry junction laser', *IEEE J. Quantum Electronics*, **QE-13**, 662 (1977).
26. J. Buus, M. Danielsen, 'Carrier diffusion and higher order transversal modes in spectral dynamics of the semiconductor laser', *IEEE J. Quantum Electronics*, **QE-13**, 669 (1977).
27. J. E. Ripper and T. L. Paoli, 'Mode configurations in second-order mode-locked lasers', *IEEE J. Quantum Electronics*, **QE-8**, 74 (1972).
28. B. L. Frescura, C. J. Hwang, H. Luechinger, and J. E. Ripper, 'Suppression of output nonlinearities in double-heterostructure lasers by use of misaligned mirrors', *Appl. Phys. Lett.*
29. J. E. Ripper, N. B. Patel, and P. Brosson, 'Effect of uniaxial pressure on the threshold current double-heterostructure GaAs lasers', *Appl. Phys. Lett.*, **21**, 124 (1972).
30. N. B. Patel, J. E. Ripper, and P. Brosson, 'Behaviour of threshold current and polarization of stimulated emission of GaAs injection lasers under uniaxial stress', *IEEE J. Quantum Electronics*, **QE-9**, 338 (1973).

Chapter 17

Investigation of Optical Gain and its Saturation Behaviour in Semiconductor Lasers

M. H. Pilkuhn

17.1 INTRODUCTION

Optical gain is one of the most fundamental parameters of semiconductor lasers. It determines the laser performance below and above threshold. Basic properties like threshold current density, output power, mode and modulation behaviour are directly or indirectly determined by the gain and its dependence on wavelength as well as current density.

It is not surprising that a large number of theoretical (1–9) and experimental studies (10–26) have been devoted to the optical gain and its variation with laser structure, doping level of the active layer, temperature, etc.

In general, one has to distinguish between the 'unsaturated gain' at low photon density where spatial and energy carrier distributions are not affected, and the 'saturated gain' at high photon densities.

In the latter case of saturation effects, the following possibilities exist: (i) homogeneously broadened lines, i.e. gain saturation for all photon energies of the gain spectrum, and (ii) inhomogeneously broadened lines where saturation depends strongly on photon energy. In the first case, spatial hole burning and in the second case, both spectral and spatial hole burning can lead to a multimode behaviour of lasers.

In this chapter only some of the basic features of optical gain in semiconductor lasers will be discussed. First, experimental methods of gain measurement will be explained. Then the unsaturated gain will be discussed, in particular, its dependence on wavelength, temperature, and the doping level. Finally, recent results on gain saturation will be mentioned.

No attempt at a complete review will be made.

17.2 EXPERIMENTAL TECHNIQUES

17.2.1 Analysis of threshold data

Measurement of threshold current density, j_t, as a function of laser cavity length, L, (10, 11) or mirror reflectivity, R, gives information about the spectral maximum of the optical gain coefficient, g_m, at threshold:

$$\Gamma g_m = \frac{1}{L}\ln\left(\frac{1}{R}\right) + \alpha_e \tag{17.1}$$

(α_e = internal losses, Γ = confinement factor). This is simply the condition where internal losses, α_e, and mirror losses, $\frac{1}{L}\ln\left(\frac{1}{R}\right)$, are balanced by the gain. If the propagating laser mode extends spatially beyond the active layer width, d, the mode gain coefficient has to be modified by a 'confinement' or 'filling factor' (27), Γ, which is the ratio of light intensity within to that outside the active layer. In order to get experimental values of the maximum gain, the relation between g_m and threshold current density, $j_{t'}$, must be known. The most simple approximation is $g_m = \beta j_t$ (or βj_t^2) which historically was the first to provide information about gain coefficients (10, 11).

A theoretical consideration of the dependence of gain on current density by Stern (6) has led to the improved approximation

$$g_m = \beta'\left(\frac{\eta}{d}j_t - \gamma\right) \tag{17.2}$$

(η = internal quantum efficiency, d = width of active layer).

If g_m is expressed in [cm^{-1}], the coefficients β' and γ should have the following values: $\beta' = 5 \times 10^{-2}$ cm^{2}A^{-1}; $\gamma = 4.5 \times 10^3$ A/cm^{2}µm, and this relation is in reasonable agreement with experimental absorption data (28). In a detailed analysis of threshold data of double-heterostructure (DH)-lasers, the effect of leakage currents (29) and stripe width (30) has to be taken into account.

Finally, the internal losses, α_e, in Equation 17.1 may be analysed in a complex way by free carrier losses in the active layer, $\Gamma\alpha_{fc}$, and outside the active layer $(1-\Gamma)\alpha_{fco}$, scattering losses, α_s, and coupling losses α_c (29). The sum of all the losses determines the maximum gain coefficient. A serious limitation of the analysis of threshold data is that only one value of the gain, namely at the spectral maximum and at the threshold current density, is obtained. For a better understanding of lasers, the gain should be known as a function of frequency (wavelength) and current density (excitation level).

17.2.2 Analysis of Fabry–Perot resonances

In this method the intensity modulation depth of oscillations in the spontaneous emission spectra due to Fabry–Perot resonances is measured (16, 17). Figure 17.1 shows a schematic experimental arrangement according to Hakki

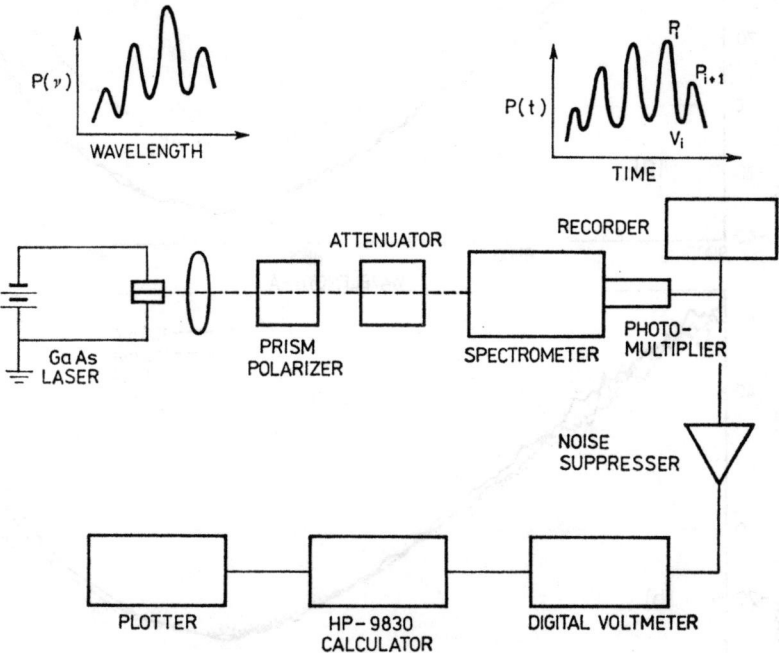

Figure 17.1 Experimental arrangement for gain measurements by analysing Fabry–Perot resonances, according to Hakki and Paoli (17)

and Paoli (17) from which unsaturated gain spectra as a function of current density can be obtained. Essentially, the ratio of the peak intensity, P_i, to the minimum valley intensity, V_i, of Fabry–Perot resonances is measured as a function of wavelength (converted into time). The average depth of modulation, r_i, is obtained by averaging two consecutive peaks (compare insert in Figure 17.1):

$$r_i = \frac{1}{2}\frac{P_i + P_{i+1}}{V_i}$$

The mode gain is then obtained from the modulation depth:

$$\Gamma g_i = \frac{1}{L}\ln\left(\frac{\sqrt{r_i}+1}{\sqrt{r_i}-1}\right) + \frac{1}{L}\ln R \tag{17.3}$$

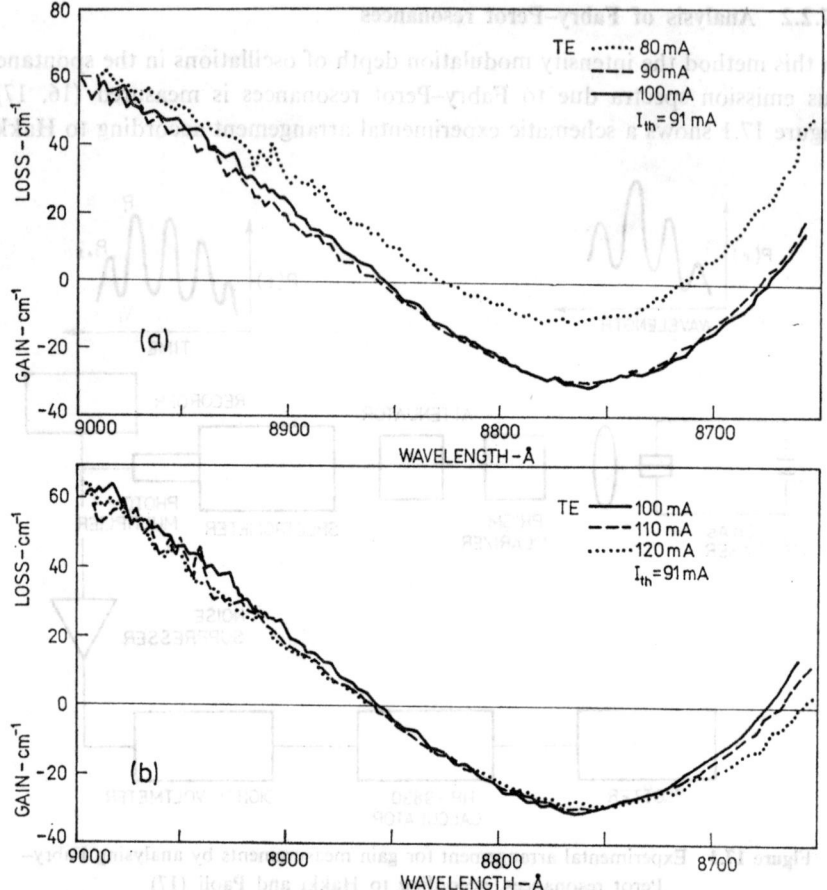

Figure 17.2 Gain spectra (TE polarization) for GaAs DH lasers at different currents (17). Threshold current I_{th} = 91 mA, room temperature

The results of Hakki and Paoli (17) for a GaAs DH-laser are shown in Figure 17.2, where the gain (loss) coefficient has been plotted for the TE mode as a function of wavelength and for different currents. It may be noticed that the gain spectra rapidly saturate above threshold. However, in this method it is difficult, in principle, to obtain gain spectra above threshold. The problem is that a laser oscillator exhibits a non-uniform light intensity profile along the laser axis (31, 32). If the carrier diffusion length is short compared with the laser length, the carrier density and the gain coefficient will also exhibit a profile along the laser axis (25). The analysis of Fabry–Perot resonances above threshold leads to spatially averaged gain spectra only.

17.2.3 Transmission experiments

A straightforward measurement of optical gain is possible in a simple transmission experiment (33, 34) in which the transmission (or absorption) coefficient of a narrow light beam traversing the active layer is measured as a function of wavelength, intensity, etc.:

$$I = I_0 R_1 R_2 e^{(g-\alpha_e)L} \qquad (17.4)$$

(I_0 = incident intensity, R_1, R_2 = reflection coefficients, I = transmitted intensity).

Recently, tunable dye lasers have been developed for the near infrared wavelength range (35) and used for transmission experiments in GaAs. In this way, the gain spectrum of optically pumped high purity GaAs layers

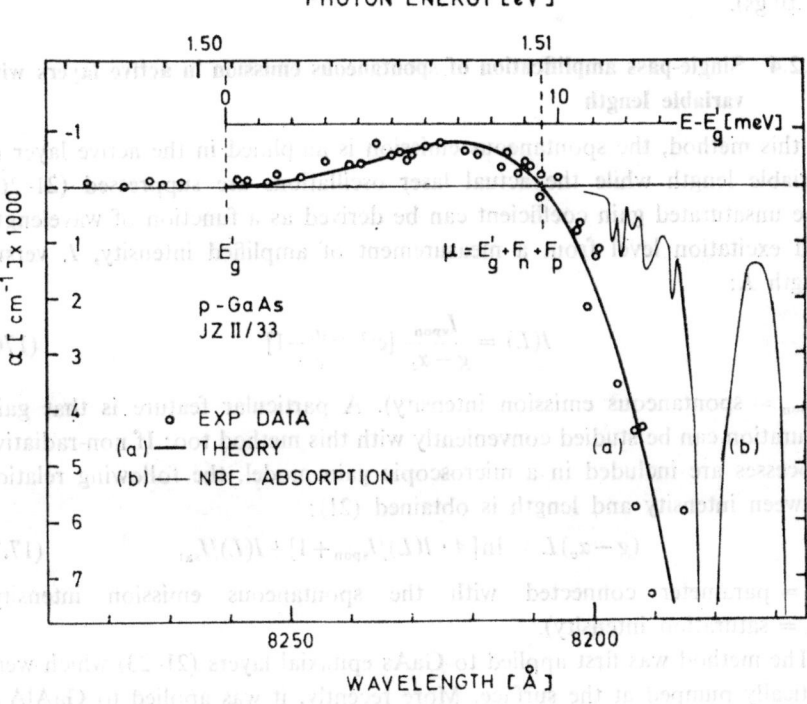

Figure 17.3 Gain and absorption in optically pumped GaAs ($N_a - N_d$ (77 K) = 6.6 × × 10^{13} cm^{-3}) derived from transmission data (two tunable laser experiments), according to Hildebrand et al. (36, 37). Curve (a) refers to high excitation (1 MW/cm^2). The solid line is a theoretical fit, the circles are experimental data. Curve (b) is a conventional high resolution transmission experiment at low excitation density for comparison. The bath temperature is 2 K

could be measured (36–37), and a result is shown in Figure 17.3 (after to Hildebrand et al. (36, 37)). A high power dye laser was used to pump the GaAs; the transmission coefficient was measured with a second tunable dye laser. Since the internal losses are very low, gain starts at the (reduced) bandgap, E'_g (see Figure 17.3), and changes into absorption at the chemical potential $\mu = E'_g + F_n + F_p$ (F_n, F_p = quasi-Fermi levels). This is an experimental demonstration of the semiconductor laser condition (38) for band-to-band transitions

$$E_g < h\nu < E'_g + F_n + F_p \tag{17.5}$$

Although the transmission experiments are convenient for basic studies of optically pumped GaAs layers, they are difficult to apply to DH-laser structures in which laser oscillations have to be suppressed (e.g. by antireflective coatings).

17.2.4 Single-pass amplification of spontaneous emission in active layers with variable length

In this method, the spontaneous emission is amplified in the active layer of variable length while the actual laser oscillations are suppressed (21–26). The unsaturated gain coefficient can be derived as a function of wavelength and excitation level from a measurement of amplified intensity, I, versus length L:

$$I(L) = \frac{I_{\text{spon}}}{g - \alpha_e} [e^{(g-\alpha_e)L} - 1] \tag{176}$$

(I_{spon} = spontaneous emission intensity). A particular feature is that gain saturation can be studied conveniently with this method too: If non-radiative processes are included in a microscopic gain model, the following relation between intensity and length is obtained (21):

$$(g - \alpha_e)L = \ln[A \cdot I(L)/I_{\text{spon}} + 1] + I(L)/I_{\text{sat}} \tag{17.7}$$

(A = parameter connected with the spontaneous emission intensity, I_{sat} = saturation intensity).

The method was first applied to GaAs epitaxial layers (21–23) which were optically pumped at the surface. More recently, it was applied to GaAlAs–GaAs–DH-lasers (24, 26). In this case, a light source transparent to the GaAlAs-layer must be used. Most commercially available dye lasers allow only investigations of DH-lasers with large energy gap GaAlAs. With the advent of suitable high power near infrared dye lasers (35) practical DH-laser structures can be studied.

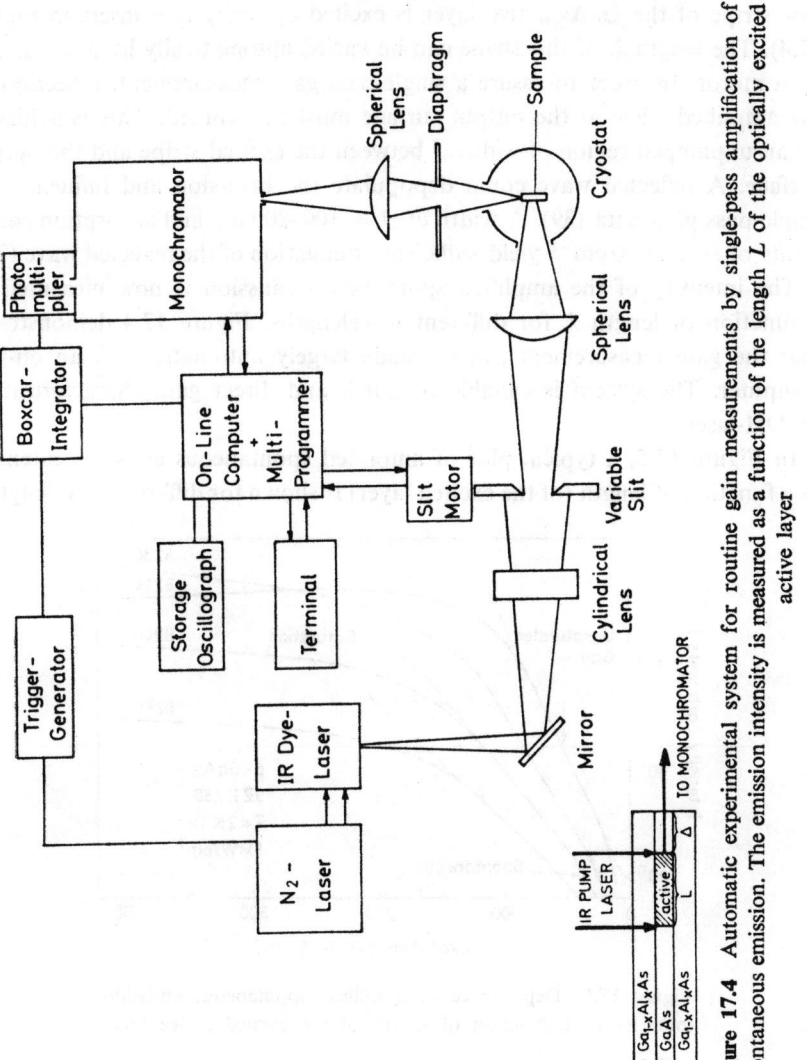

Figure 17.4 Automatic experimental system for routine gain measurements by single-pass amplification of spontaneous emission. The emission intensity is measured as a function of the length L of the optically excited active layer

An experimental system which can be used for quick routine gain measurements in DH-laser structures is shown in Figure 17.4: The beam of a nitrogen-laser pumped IR dye laser is imaged on the sample in such a way that a narrow stripe of the GaAs active layer is excited optically (see insert in Figure 17.4). The length L of this stripe can be varied automatically by a slit driven by a motor. In order to assure a single-pass gain measurement, reflection of the amplified wave at the output surface must be avoided. This is achieved by an unpumped region of width Δ between the excited stripe and the output surface. A reflected wave could depopulate the inversion and influence the single-pass gain data (39). A width of $\Delta = 100\text{--}200$ µm and absorption coefficients of several $10\,\text{cm}^{-1}$ yield sufficient attenuation of the reflected wave (26).

The intensity of the amplified spontaneous emission is now measured as a function of length L for different wavelengths. Figure 17.4 demonstrates that the gain measurement can be made largely automatic with an on-line computer. The system is capable of quick and direct gain characterization of DH-lasers.

In Figure 17.5, a typical plot of amplified spontaneous emission intensity as a function of length (of the excited layer) is shown for different wavelengths.

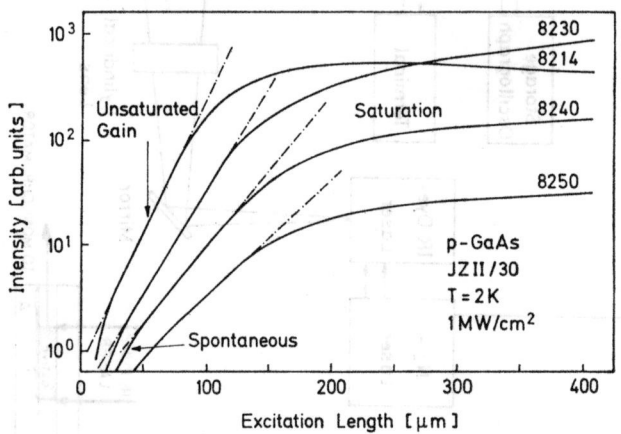

Figure 17.5 Dependence of amplified spontaneous emission intensity as a function of length of the excited active layer

In a logarithmic intensity scale, the three regions of (i) spontaneous emission (without amplification), (ii) unsaturated gain, and (iii) saturation effects can be clearly distinguished. For example, the strong increase of saturation effects towards short wavelengths (25) becomes immediately visible in an intensity-length plot such as the one in Figure 17.5.

In summary, the single-pass amplification of spontaneous emission in layers of different lengths seems to be the most versatile and direct method to study gain and gain saturation in semiconductor lasers (even absorption or spontaneous emission, if desired). Quick and reliable characterization of the gain in DH-lasers is possible with automatic experimental systems.

17.3 UNSATURATED GAIN IN GaAs LASERS

The dependence of the optical gain coefficient on photon energy and excitation level has been calculated by Stern (6), and Figure 17.6 shows various gain spectra for 77 K and 297 K. The excitation level is expressed through

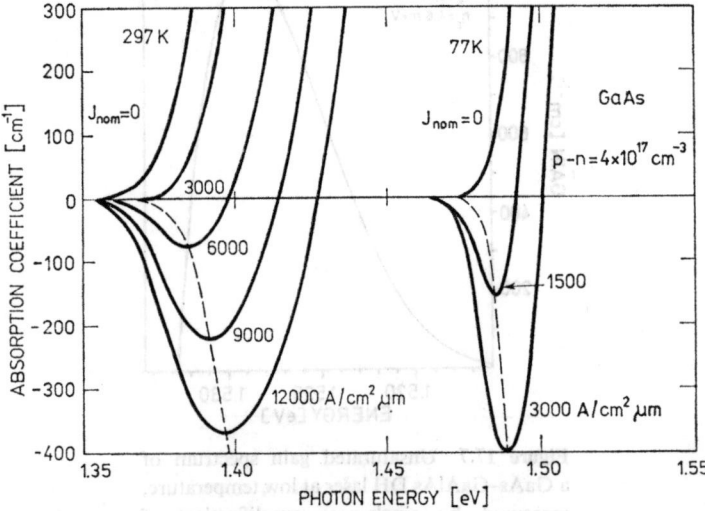

Figure 17.6 Calculated gain (absorption) coefficients at 77 K and 297 K as a function of photon energy and for different nominal current densities, after Stern (6)

a nominal current density $j_{nom} = j \cdot \eta/d$ as a parameter at the spectra. (d = width of the active region in µm, η = internal quantum efficiency.) It may be noted that the width of the gain spectra increases considerably with increasing temperature as observed also experimentally (26). The theoretical gain or absorption coefficient of Figure 17.6 should be compared with the experimental data obtained from transmission experiments as shown in Figure 17.3. In general, the agreement is very good; all curves have a characteristic shape with gain below the quasi-Fermi levels, and down to the energy gap, and absorption above the quasi-Fermi levels.

A satisfactory theoretical fit with the experimental data can in most cases be obtained by neglecting k-selection rules for the band-to-band recombination (23–26).

In the gain measurements due to single-pass amplification of spontaneous emission (compare Section 17.2.4), only the gain spectrum up to the quasi-Fermi levels is measured. A typical unsaturated gain spectrum for a DH-laser and at low temperature is shown in Figure 17.7. The real carrier tempera-

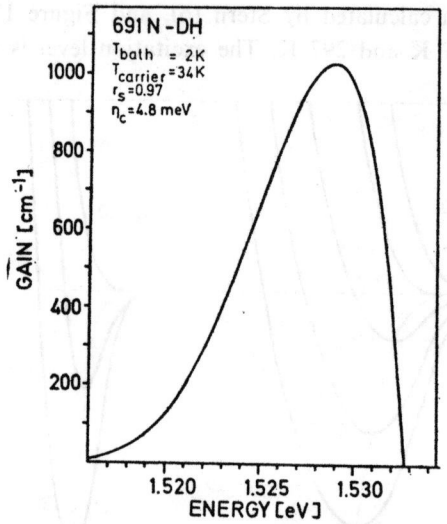

Figure 17.7 Unsaturated gain spectrum of a GaAs–GaAlAs DH laser at low temperature, measured by single-pass amplification of spontaneous emission. The active layer was undoped and was 1 μm thick (26)

ture is usually higher than the bath temperature and has to be determined from the high energy Boltzmann-tail of the emission. Note that the maximum gain can easily reach values of the order of 10^3 [cm^{-1}]. In DH-structures, the maximum gain values are usually higher than in optically excited bulk materials at the same excitation level by a factor of 2–3. This is due to improved confinement as well as lack of surface recombination.

Gain spectra of the type depicted in Figure 17.7 can be well described by recombination (without k-selection) between parabolic bands modified by

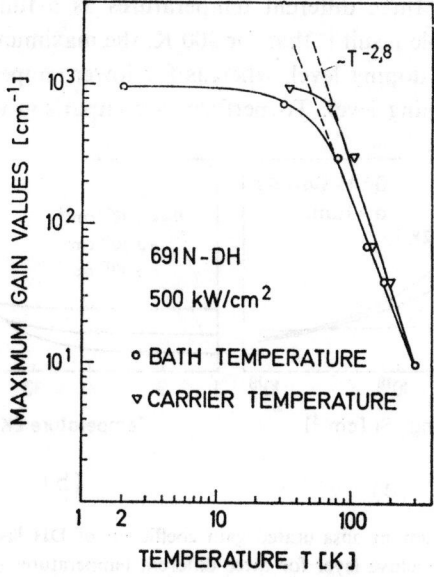

Figure 17.8 Variation of the unsaturated maximum gain coefficient with temperature for GaAs–GaAlAs DH lasers (undoped active layer). The excitation level was constant and equal to 500 kW/cm² (26)

tail states (1, 40, 41). The parameter r_s in Figure 7 describes carrier density, the parameter η_c the tail states (41).

Variations in unsaturated maximum gain with temperature (26) are depicted in Figure 17.8 at constant excitation level and for undoped active layers. A careful distinction between bath temperature and carrier temperature has to be made as just mentioned. In the temperature range 100 K < T < 300 K one gets (26):

$$g_{max} \sim T_{carrier}^{-2.8}$$

The temperature dependence of the gain should depend on the doping level, and Figure 17.9(b) depicts a theoretical calculation by Stern (42). The nominal current density which is necessary to reach a gain of 100 cm⁻¹ has been plotted as a function of temperature. At high temperature, e.g. 300 K, high doping levels should be favourable and reduce the nominal current density; at low temperatures the reverse should be true.

Unsaturated gain measurements as a function of temperature and doping have been recently made by Weber (43): In Figure 17.9(a) the maximum gain is plotted for three different temperatures as a function of Si-doping level. The remarkable result is that for 300 K, the maximum gain is practically independent of the doping level, whereas for lower temperatures it increases with decreasing doping level. To perform a comparison with the theoretical

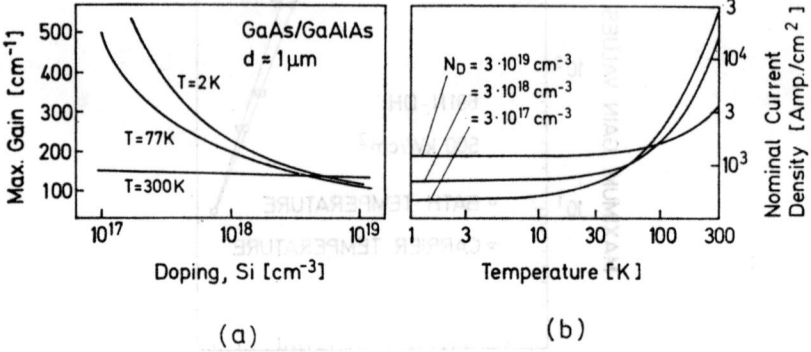

Figure 17.9 (a) Maximum unsaturated gain coefficient of DH lasers as a function of Si-doping level of the active layer for three different temperatures (43). Excitation level was 500 kW/cm². (b) Theoretical dependence of nominal current densities on temperature for different doping levels, after Stern (42)

results of Figure 17.9(b) this has to be translated into a nominal current density-temperature plot. Apparently, the 'crossover' of the curves in Figure 17.9(b) occurs at a higher temperature than theoretically predicted, namely at about 300 K, where the gain is independent of the doping level.

Gain spectra at high doping level can be quite complex and include band-to-acceptor in addition to band-to-band transitions, of course, modified by tail states. Recent results by Weber (43) are depicted in Figure 17.10.

The picture shows unsaturated gain spectra at a bath temperature of 77 K which corresponds to a carrier temperature of about 128 K. The laser structures were DH-lasers with the following Si-doping levels: from top to bottom the spectra correspond to $N_a - N_d = 5.0 \times 10^{18}$, 3.5×10^{18}, and 1.1×10^{18} cm^{-3}. In all cases, the gain due to band-to-band transitions is the strongest, and the following effects may be noticed: (i) The gain maximum shifts slightly to lower energy with increasing doping level, which is the familiar 'gap shrinkage effect'. (ii) The gain spectrum becomes considerably broader with increasing doping level.

Furthermore, a structure on the low energy side develops due to gain in the band-acceptor transitions. This additional gain builds up systematically with increasing doping level and has its maximum at about 1.46 eV. In

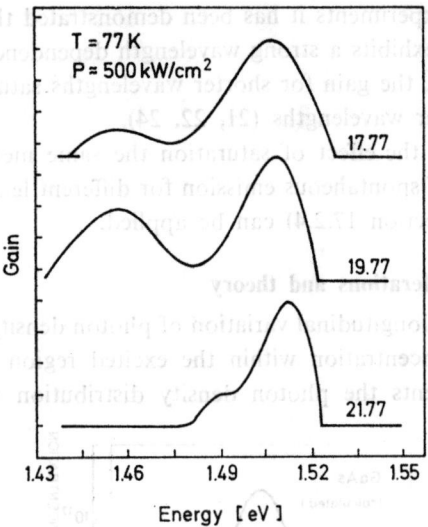

Figure 17.10 Unsaturated gain spectra of DH lasers at 77 K bath temperature and for different high Si-doping levels of the active layer (43): Sample 17.77: 5×10^{18} cm^{-3}; 19.77: 3.5×10^{18} cm^{-3}; 21.77: 1.1×10^{18} cm^{-3}. Excitation level: 500 kW/cm^2

Figure 17.10, it is most pronounced for sample 19.77 with $N_a - N_d = 3.5 \times 10^{18}$ cm^{-3}. Finally, at the highest doping levels, the band-acceptor gain spectrum seems to merge with the band-to-band gain.

Certainly, the details of unsaturated gain spectra at high doping levels, in particular the influence of tail states and impurity bands, still require more experimental work. The method of single-pass amplification of spontaneous emission is the most convenient one to consider these problems.

17.4 SATURATION EFFECTS

Saturation of the band-to-band laser emission has been treated in several papers (8, 17–20, 44) and the effect of gain saturation on the longitudinal as well as on the transverse gain profile has been studied. However, so far in all of these calculations only the effect of saturation on the maximum gain or on the integrated gain has been considered, and no calculations on the wavelength dependence of gain saturation are available.

In a variety of experiments it has been demonstrated that gain saturation in semiconductors exhibits a strong wavelength dependence, and that within the gain bandwidth, the gain for shorter wavelengths saturates more rapidly than that for longer wavelengths (21, 22, 24).

In order to study the effect of saturation the same method of measuring the amplification of spontaneous emission for different lengths of the excited region (compare Section 17.2.4) can be applied.

17.4.1 Basic considerations and theory

In Figure 17.11 the longitudinal variation of photon density, chemical potential, and carrier concentration within the excited region are depicted. The broken line represents the photon density distribution without saturation

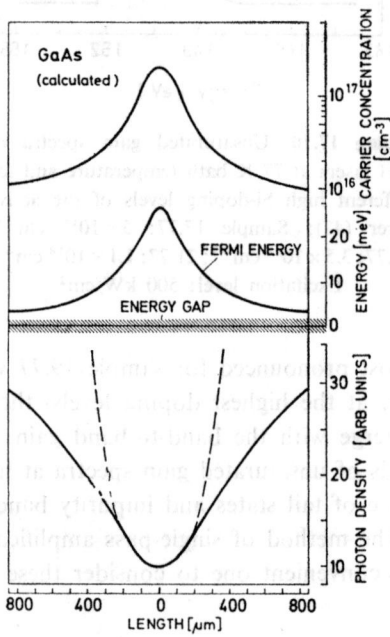

Figure 17.11 Longitudinal variation of photon density, Fermi energy and carrier concentration within the excited region of a semiconductor laser (25)

effects: $\varrho \sim \cosh(x)$, the solid line shows the distribution with saturation. As expected, the curves for the photon density, ϱ, and the carrier density, n, agree with those calculations which treat saturation effects integrally (8, 44).

In general, the spectral variation of the gain with changing photon density should be included in the description. The effect of wavelength dependence on gain saturation can already be discussed qualitatively by means of a curve for the shape of the chemical potential; saturation will be the strongest on the high-energy side of the laser spectrum, because the variation of the chemical potential shifts the energy at which gain turns into absorption.

Diffusion currents due to spatial variations of the quasi-Fermi levels in the longitudinal direction should also have an influence but are usually neglected. The effect of diffusion currents on the transverse light distribution of stripe geometry lasers has been discussed in great detail by Hakki (18-20). Though for the longitudinal photon flux distribution, diffusion might be important to reduce the effect of spatial hole burning in Fabry-Perot lasers (18), it does not wash out the photon flux distribution shown in Figure 17.11 appreciably, because the length of the excited region is large compared with the diffusion length.

A theoretical treatment of the wavelength dependence of gain saturation has recently been published by Göbel et al. (25). The amplification of the light emitted within the active region of a semiconductor was calculated in a self-consistent way similar to the theory of Cross and Oldham (39). The theoretical treatment can be applied to gain experiments, especially to the dependence of light intensity on the length of the excited region, and it yields stimulated emission spectra for different lengths of the active region, i.e. for different levels of saturation.

In the theory, the active layer is separated into infinitesimally small regions with index i, and the change of light intensity is given by

$$\Delta \varrho_{(E)}^i \sim [\Omega r_{\text{spon}}^i(E) + r_{\text{stim}}^i(E) \varrho_{(E)}^{i-1}] \Delta x \qquad (17.8)$$

In Equation 17.8, ϱ is the photon density, $r_{\text{spon}}(E)$ is the spontaneous emission rate, $r_{\text{stim}}(E)$ is the stimulated emission rate, Δx is the width of the region i, and Ω is the actual solid angle. The equation can be integrated iteratively along the lasing axis; however, it must be considered that r_{spon}^i and r_{stim}^i depend implicitly on the photon density ϱ^i. This procedure yields the spectral and the total light output of the laser which now, of course, includes the effect of the wavelength dependent saturation, because the carrier concentration profile is calculated self-consistently and the spectral shape of the gain curve (which is essentially r_{stim}^i) and of the spontaneous emission is taken into account. The influence of the gain saturation on the spectral output can be studied in this way just by varying the length L of the excited region.

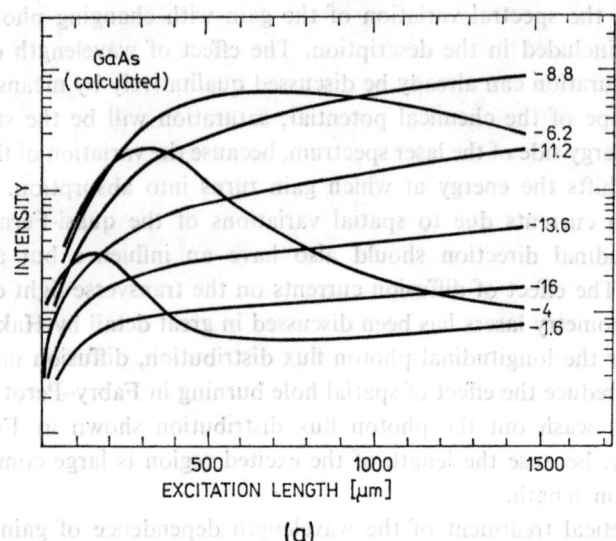

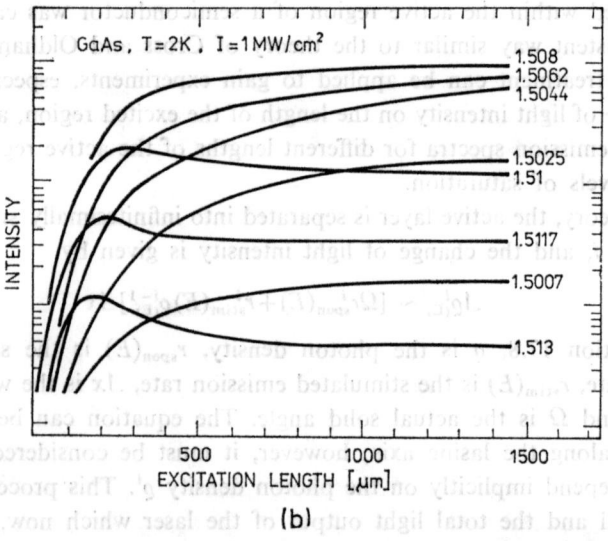

Figure 17.12 Amplified intensity as a function of the length of the excited region for different photon energies of the band-to-band laser transition. (a) Calculated curves: energies (in meV) written on the right-hand side of the curves are relative to the unperturbed bandgap energy. Parameters: $T_{carrier} = 40$ K, $n = 8.7 \times 10^{17}$ cm^{-3}. (b) Experimental curves for p-type epitaxial GaAs layers. Excitation level 1 MW/cm^2 (25)

17.4.2 Comparison between experiment and theory

The calculated curves for the spectrally resolved light output as a function of the length of the excited region for constant excitation level is shown in Figure 17.12. The carrier temperature for the calculation was assumed to be 40 K. The curves can be characterized by two regions, namely the region of spontaneous and unsaturated stimulated emission at small lengths, and the region of saturated stimulated emission. It should be noted in Figure 17.12 that saturation does not occur uniformly but shows a strong wavelength dependence: At short wavelength, saturation occurs fast as the length of the excited region and, therefore, the photon density increases. For long wavelengths, gain saturation occurs at comparatively large lengths, i.e., at high light levels. For these long wavelengths, gain can even turn into absorption, which means a decrease of light output with increasing length, a rather surprising result.

The corresponding experimental results are shown in Figure 17.12(b) for GaAs layers at a bath temperature of 2 K. These results are in good qualitative agreement with the theoretical predictions of Figure 17.12(a).

The fact that saturation depends strongly on the emission wavelength has significant bearing on the emission spectra. This means that as soon as saturation becomes important, the spectral shape of the stimulated emission cannot be described by unsaturated gain curves.

In Figure 17.13, stimulated emission spectra (not gain spectra!) for different lengths of the excited region are depicted. Again, a comparison between calculated curves (Figure 17.13(a)) and experimental results (Figure 17.13(b)) is shown. The wavelength dependence of the gain saturation influences the stimulated emission spectra as follows: due to the change of gain into absorption for the short wavelength emission, the emission intensity on the high-energy side of the spectrum decreases as the length of the excited region increases. The fact that the short wavelength emission saturates more rapidly than the emission for the longer wavelengths causes a shift of the emission maximum to lower energies with increasing length. All the significant features of the experimental spectra are adequately described by the theoretical curves of Figure 17.13.

Qualitatively similar results are obtained for room temperature lasers. Emission spectra of GaAs–GaAlAs double-heterostructure lasers show the following effects due to wavelength dependent saturation (25): On the high-energy side, the emission intensity decreases, and the emission maximum shifts to lower energies as the length of the laser increases. For any practical laser device (possibly with the exception of DFB lasers) these changes in the

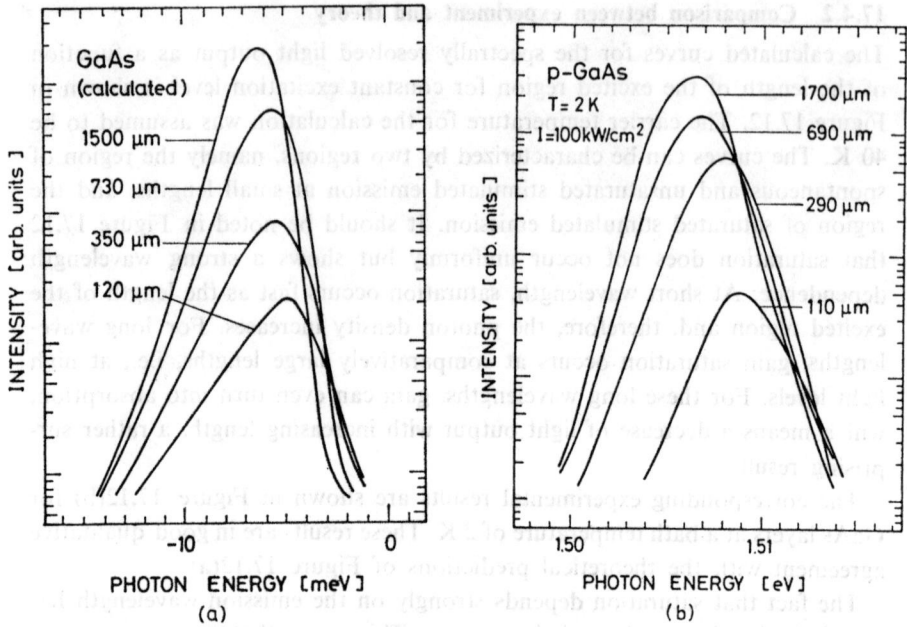

Figure 17.13 Stimulated emission spectra for different lengths of the excited layer. (a) Calculated curves: $T_{carrier} = 40$ K, $n = 1.7 \times 10^{17}$ cm^{-3}. (b) Experimental curves: same samples as in Figure 17.12, excitation level 100 kW/cm² (25)

effective gain profile must be considered, since they affect the modulation and amplification behaviour, and probably mode instabilities (45).

Finally, another consequence of the wavelength dependent gain saturation is that the position of the stimulated emission maximum is not simply determined by the change of the pump level (band filling) but depends strongly on the length of the active region and on the photon density within the active region, i.e., the amount and the sign of the shift depend on the level of saturation. This is demonstrated by the theoretical results of Figure 17.14((a) and (b)) (25), where emission spectra for different excitation levels are depicted. The excitation length for the curves in Figure 17.14(a) was assumed to be 85 μm whereas in the case of Figure 17.14(b) the excitation length was 1500 μm. Since for the shorter length one obtains unsaturated emission spectra, at least for the lower excitation levels, the shift of the emission is not influenced by the wavelength dependent gain saturation. The shift of the peak obtained for this case is somewhat weaker than proportional to $I_{pump}^{2/3}$, which would be expected in a simple band filling model. At the highest pump level in Figure

17.14(a) saturation already becomes important, and, because of the wavelength dependence of the saturation, the high-energy side saturates more rapidly. This would result in a shift to lower energies. Consequently, the net shift observed when the excitation power is raised is the sum of both contri-

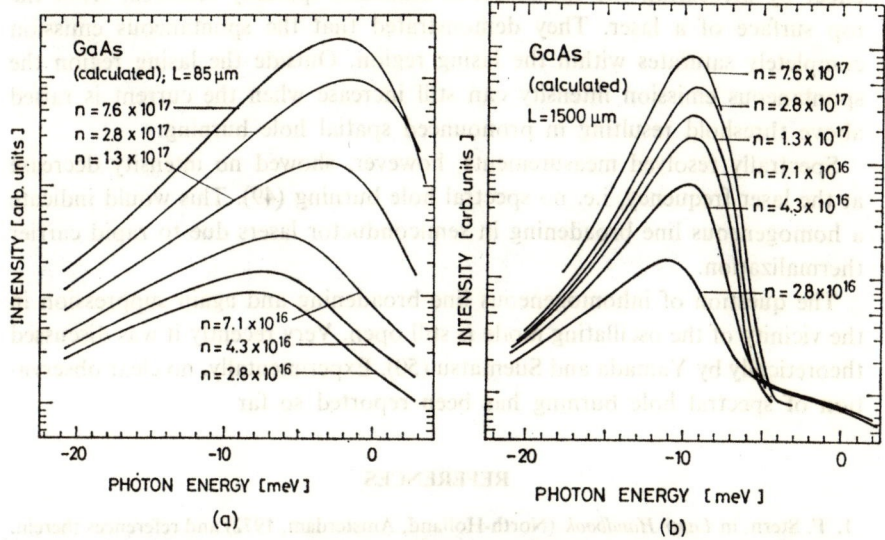

Figure 17.14 Calculated stimulated emission spectra for different excitation levels. (a) Length of the excited region is 85 μm, $T_{carrier} = 40$ K. (b) Length of the excited region is 1500 μm, $T_{carrier} = 40$ K (25)

butions. This means that for the case of saturation the position of the chemical potential is no longer determined by the pump level only, but also by the photon density. Actually, the effect of the photon density, which tends to lower the chemical potential because of the enhanced stimulated emission, can overcome the effect of the increased pump power, which of course would lift the chemical potential, resulting in a net shift of the peak emission to lower energies as shown in Figure 17.14(a).

The effect of the saturation on the peak shift naturally gets even more pronounced at the larger laser length of 1500 μm as shown in Figure 17.14(a). In this case of extreme saturation the peak position is nearly independent of the excitation power and again it might even shift to lower energies with increasing pump level. This basically reflects the fact that in the case of extreme saturation the position of the chemical potential is fixed. This can also be observed by different experiments, namely the saturation behaviour of the spontaneous emission in semiconductor lasers (46–49).

The results on the saturation of spontaneous emission will be briefly discussed in this last section: The saturation behaviour should depend strongly on spatial uniformity of the laser action and on mode stability (46, 47, 49). Recently, Nakamura et al. (49) reported a direct observation of the saturation effect by measuring the spontaneous emission, spatially resolved, from the top surface of a laser. They demonstrated that the spontaneous emission completely saturates within the lasing region. Outside the lasing region the spontaneous emission intensity can still increase when the current is raised above threshold resulting in pronounced spatial hole burning.

Spectrally resolved measurements, however, showed no intensity decrease at the laser frequency, i.e. no spectral hole burning (49). This would indicate a homogeneous line broadening in semiconductor lasers due to rapid carrier thermalization.

The question of inhomogeneous line broadening and again suppression in the vicinity of the oscillating mode is still open. Very recently it was discussed theoretically by Yamada and Suematsu (50). Experimentally, no clear observation of spectral hole burning has been reported so far.

REFERENCES

1. F. Stern, in *Laser Handbook* (North-Holland, Amsterdam, 1972) and references therein.
2. F. Stern, *IEEE J. Quantum Electron.*, **QE-9**, 290 (1973).
3. C. J. Hwang, *Phys. Rev.*, **B 2**, 4117 (1970).
4. R. F. Kazarinov, *Sov. Phys.-Semicond.*, **7**, 525 (1973).
5. M. J. Adams, *Solid-State Electron.*, **12**, 661 (1969).
6. F. Stern, *J. Appl. Phys.*, **47**, 5382 (1976).
7. H. Haug, *Phys. Rev.*, **184**, 338 (1969).
8. R. Lang, *IEEE J. Quantum Electron.*, **QE-10**, 825 (1974).
9. Y. Nishimura, *J. Quantum Electron.*, **QE-9**, 1011 (1973).
10. M. H. Pilkuhn, *phys. stat. sol.*, **25**, 9 (1968).
11. M. H. Pilkuhn, H. Rupprecht, and S. Blum, *Solid-State Electron.*, **7**, 905 (1964).
12. A. R. Goodwin and G. H. B. Thompson, *IEEE J. Quantum Electron.*, **QE-6**, 311 (1970).
13. I. Hayashi, M. B. Panish, and F. K. Reinhart, *J. Appl. Phys.*, **42**, 1929 (1971).
14. M. Ettenberg and H. Kressel, *J. Appl. Phys.*, **43**, 1204 (1972).
15. G. H. B. Thompson and P. A. Kirkby, *IEEE J. Quantum Electron.*, **QE-9**, 311 (1973).
16. B. W. Hakki and T. L. Paoli, *J. Appl. Phys.*, **44**, 4113 (1973).
17. B. W. Hakki and T. L. Paoli, *J. Appl. Phys.*, **46**, 1299 (1975).
18. B. W. Hakki, *J. Appl. Phys.*, **44**, 5021 (1973).
19. B. W. Hakki, *J. Appl. Phys.*, **45**, 288 (1974).
20. B. W. Hakki, *J. Appl. Phys.*, **46**, 292 (1975).
21. K. L. Shaklee and R. F. Leheny, *Appl. Phys. Lett.*, **18**, 475 (1971);
    K. L. Shaklee, R. F. Leheny, and R. E. Nahory, *J. Lumin.*, **7**, 284 (1973).
22. E. Göbel and M. H. Pilkuhn, *J. Phys. (Paris)*, **35**, C3, 191 (1974).

23. E. Göbel, H. Herzog, M. H. Pilkuhn, and K. H. Zschauer, *Solid-State Commun.*, **13**, 719 (1973);
    E. Göbel, *Appl. Phys. Lett.*, **24**, 492 (1974).
24. J. Bakker and G. A. Acket, *IEEE J. Quantum Electron.*, **QE-13**, 567 (1977).
25. E. Göbel, O. Hildebrand, and K. Löhnert, *IEEE J. Quantum Electron.*, **QE-13**, 848 (1977).
26. O. Hildebrand, E. Göbel and K. Löhnert, *Appl. Phys.*, **15**, 149 (1978).
27. D. Marcuse, *Light Transmission Optics*, van Nostrand, 1972, p. 305.
28. H. C. Casey, Jr., and F. Stern, *J. Appl. Phys.*, **47**, 631 (1976).
29. H. C. Casey, Jr., *J. Appl. Phys.*, in press (1978).
30. W. T. Tsang, to be published.
31. R. Ulbrich and M. H. Pilkuhn, *IEEE J. Quantum Electron.*, **QE-6**, 314 (1970); *Appl. Phys. Lett.*, **16**, 516 (1970).
32. Y. Nannichi, *J. Appl. Phys.*, **37**, 3009 (1966).
33. J. R. Crowe and R. M. Craig, Jr., *Appl. Phys. Lett.*, **4**, 57 (1964).
34. M. J. Coupland, K. G. Hambleton, and C. Hilsum, *Phys. Lett.*, **7**, 231 (1963).
35. O. Hildebrand, *Opt. Comm.*, **10**, 310 (1974);
    K. M. Romanek, O. Hildebrand, and E. Göbel, *Opt. Comm.*, **21**, 16 (1977).
36. O. Hildebrand, B. O. Faltermeier, and M. H. Pilkuhn, *Solid State Commun.*, **19**, 841 (1976).
37. O. Hildebrand, E. Göbel, and M. H. Pilkuhn, *Proc. XIIIth Int. Conf. Phys. Semicond. Rome*, (1976), p. 942.
38. M. G. A. Bernard and G. Duraffourg, *phys. stat. sol.*, **1**, 699 (1961).
39. P. S. Cross and W. G. Oldham, *IEEE J. Quantum Electron.*, **QE-11**, 190 (1975).
40. G. Lasher and F. Stern, *Phys. Rev.*, **133**, A 553 (1963).
41. E. O. Kane, *Phys. Rev.*, **131**, 79 (1963).
42. F. Stern, *Phys. Rev.*, **148**, 186 (1966).
43. H. Weber, *Diplom-Thesis*, Stuttgart, 1978.
44. W. W. Rigrod, *J. Appl. Phys.*, **36**, 2487 (1965).
45. M. J. Adams, *Electron. Lett.*, **13**, 236 (1977).
46. T. L. Paoli, *IEEE J. Quantum Electron.*, **QE-9**, 267 (1973).
47. P. Brosson, J. E. Ripper, and N. B. Patel, *IEEE J. Quantum Electron.*, **QE-9**, 273 (1973).
48. T. L. Paoli and P. A. Barnes, *Appl. Phys. Lett.*, **28**, 714 (1976).
49. M. Nakamura, K. Aiki, and J. Umeda, *Appl. Phys. Lett.*, **32**, 322 (1978).
50. M. Yamada and Y. Suematsu, *J. Appl. Phys.*, to be published.

Chapter 18

Stripe Geometry Heterojunction Injection Lasers and the Effect of Optical and Carrier Confinement

G. H. B. Thompson

18.1 TYPES OF STRIPE LASER

In many applications of semiconductor heterostructure lasers an optical source is needed whose width in the plane of the heterostructure is not much greater than its thickness in the perpendicular direction. For instance, in communications applications a source width of 5–10 µm combined with a thickness of 0.5 µm is a suitable compromise for both supplying sufficient output power and for launching the light satisfactorily into a graded index multimode optical fibre.

Providing suitable optical confinement over widths as great as this is difficult, particularly if lasing is to be restricted to a zero order transverse mode. For instance, in a dielectric guide of 10 µm the dielectric step must be less than 0.002 to give single mode operation.

In the majority of stripe lasers the light is confined by restricting the current flow so that the optical gain is mainly supplied in a narrow area. A convenient means is to provide a 'current window' in the structure which is not co-planar with the active layer. This can be done in a variety of ways. The simplest design is the oxide insulated stripe laser illustrated in Figure 18.1(a) (1, 2). Other designs include the shallow mesa stripe (3, 4), the planar stripe (5) and derivatives (6, 7, 8) in which a reverse-biassed p–n junction is incorporated to block current flow, and the shallow proton bombarded stripe (9), illustrated in Figure 18.1(c), in which an insulating region is created. Because the 'current window' and the active layer are not co-planar there is a certain amount of current spreading in all these devices.

There is a second class of stripe lasers in which the injection current is very precisely restricted to a particular width and in which various additional means may be incorporated to restrict sideways diffusion of carriers in the

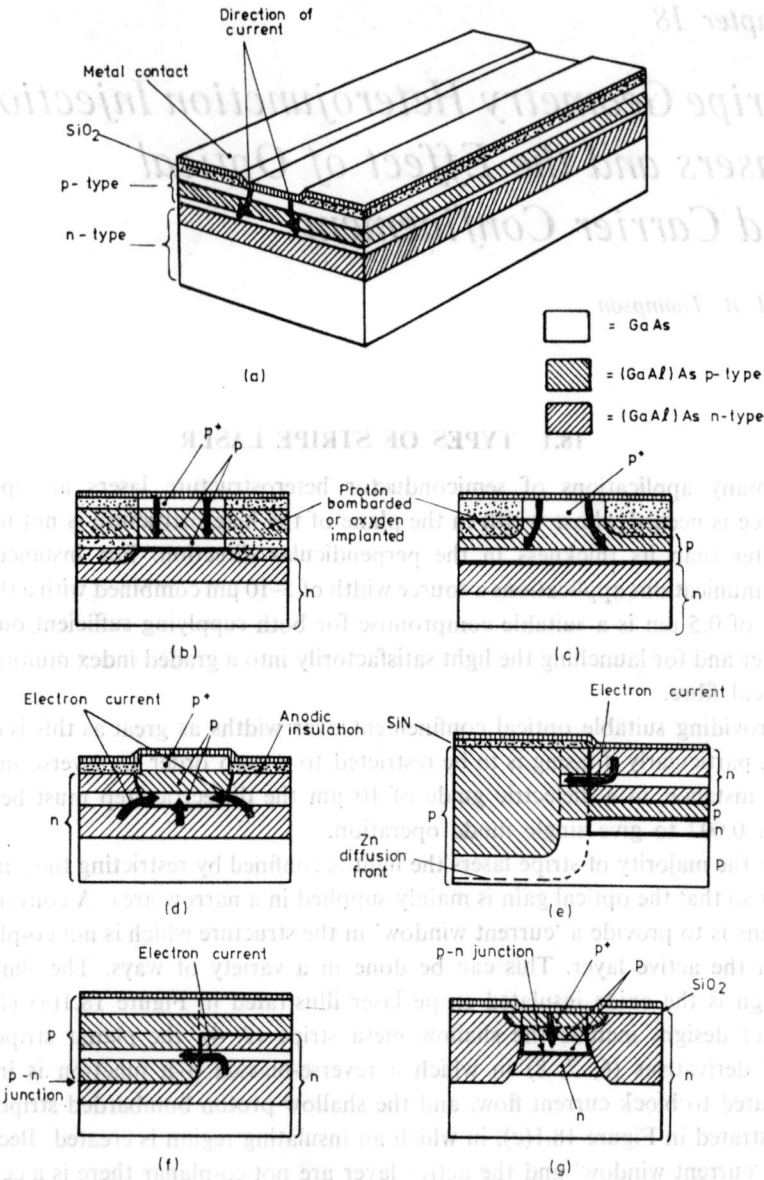

Figure 18.1 Various configurations of stripe geometry laser. (a) Oxide insulated stripe contact laser; (b) deep bombarded or implanted stripe; (c) shallow bombarded stripe; (d) twin transverse junction (or deep diffused) stripe laser; (e) transverse single-heterojunction laser; (f) transverse homojunction laser; (g) buried heterostructure stripe laser

active layer itself. The simplest is the deep proton bombarded (9) or oxygen implanted device (10) of Figure 18.1(b), where there is no restriction to lateral diffusion in the active layer. Similarly in the transverse homojunction laser (11, 12) of Figure 18.1(f) there is no restriction to lateral diffusion, which in fact largely determines the width of the lasing filament. In contrast the twin transverse junction laser (13, 14) of Figure 18.1(d) and the transverse single-heterojunction laser (15) of Figure 18.1(e) incorporate heterojunctions, and/or p–n junctions with selective injection of electrons, to confine the transverse distribution. Finally the buried heterostructure laser (16) of Figure 18.1(g) uses a pair of heterojunctions for confinement in the junction plane in the same way as in the perpendicular plane.

In the latter group of devices, with the exception of the proton bombarded laser the confinement behaviour is similar to that which is well known for the various heterostructure types in the perpendicular plane. We will only deal briefly with this. In the first group of lasers and in the deep proton bombarded laser the principles of optical and carrier confinement are less well known and we will deal with these in greater detail.

18.2 CARRIER CONFINEMENT IN STRIPE LASERS

We will consider the lateral confinement of carriers in stripe lasers before we deal with the lateral confinement of light, because in the considerable number of stripe lasers designs where there is no deliberate optical waveguide the lateral distribution of injected carriers also determines the width of the optical distribution.

18.2.1 Methods of carrier confinement

In simple stripe lasers where there is no positive barrier to prevent transverse current flow, edge leakage becomes significant for stripe widths of less than 30 μm and for widths less than 10 μm it can account for more than half the total current. Some positive confinement becomes very worthwhile for widths below 10 μm, and for widths below 2 μm strong confinement by heterostructure barriers is appropriate.

Four methods of carrier confinement are used in stripe lasers. In the first method, used in stripe contact lasers where only the current is constrained (see Figures 18.1(a) and (b)), the carrier confinement is simply provided by the additional resistance presented to that part of the current which takes

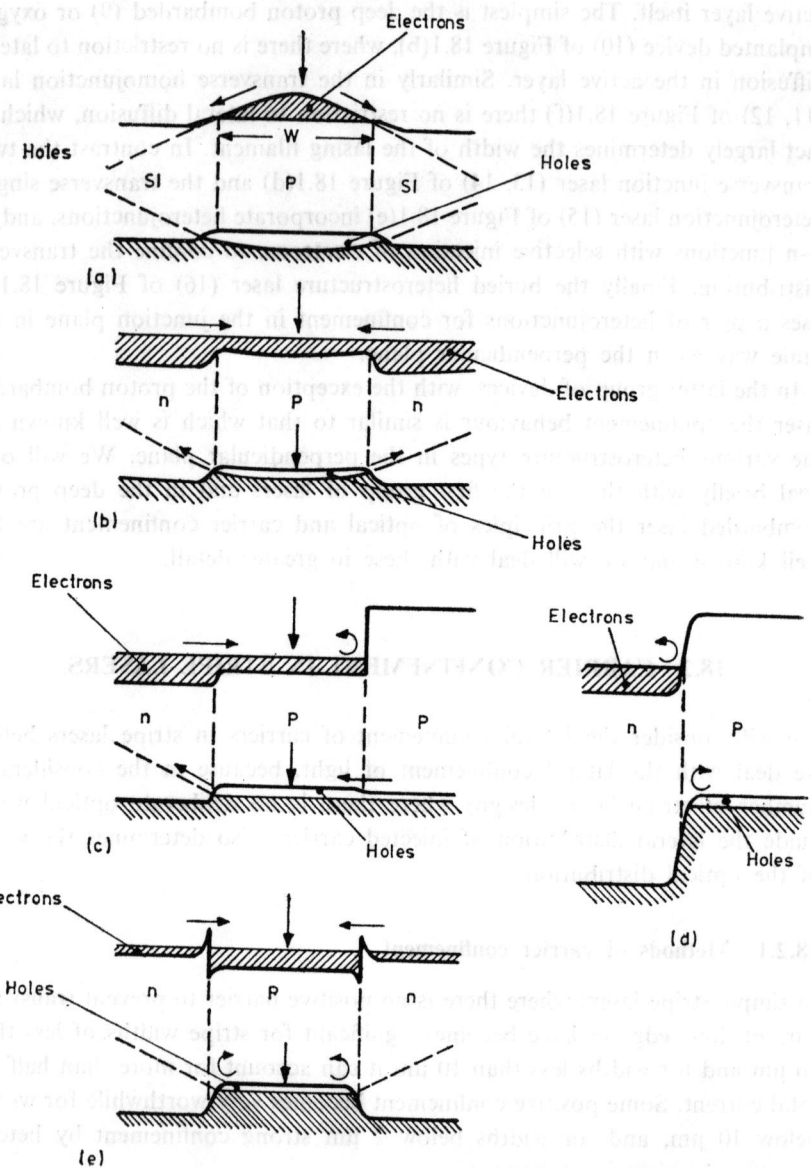

Figure 18.2 Band diagram showing effectiveness of carrier confinement in various types of stripe laser. (a) Deep proton bombarded and oxygen implanted lasers; (b) twin transverse junction laser; (c) transverse single heterojunction laser; (d) barrier to by-pass injection of carriers in structures (b), (c) and (e) provided by p–n junction in passive layer; (e) buried heterostructure laser

the longer path to the edge of the inverted part of the active layer. The effectiveness of the constraint depends on the resistivity and geometry of the layers.

In the second method, employed in deep proton bombarded and oxygen implanted lasers, the course of the current flow is determined by the geometry of the semi-insulating regions, (see Figure 18.1(c)), so that the entire current is directed to injecting carriers into the prescribed section of the active layer. However, once the carriers are injected they are capable of diffusing sideways, and can enter the semi-insulating regions without restriction. Here they can, in general, recombine at an enhanced rate. The distribution of minority carriers in the direction parallel to the junction is illustrated in Figure 18.2(a). Minority electrons injected across the p–n junction, as indicated by the vertical arrow, flow down the concentration gradient into the semi-insulating regions on either side. The confining barrier for majority holes at the semi-insulating region boundary is largely removed by the net negative charge injected into the region so that holes flow in at an equal rate to the electrons to take part in the recombination. The diffusion process intercepts a substantial proportion of the injection current which flows within a diffusion length of the stripe edge.

The third method of carrier confinement is by a transverse p–n junction barrier. This is used both in the twin transverse junction laser (Figure 18.1(d)) and in the transverse single heterojunction laser (Figure 18.1(e)). The band structure in the active layer parallel to the main heterostructure is illustrated for the twin transverse junction laser in Figure 18.2(b) and for the transverse single heterostructure laser in Figure 18.2(c). The main electron current that is injected normally to the heterostructure is indicated by the vertical arrow. The p–n junction or junctions which traverse the active layer act as reasonable confinement barriers for majority holes and as additional sources of injection of minority electrons. They perform the same function as in an orthodox single heterostructure laser and require heavy n-doping to do this most effectively. Since the p–n junction must by continuous through the cross section of the structure, it must traverse the higher band-gap layers of the heterostructure outside the stripe section. In this region it presents a useful block to the flow of shunt current, as indicated in Figure 18.2(d). With the same voltage applied as in the active layer a considerable potential barrier remains which reduces the injection current by several orders of magnitude.

The fourth method of carrier confinement is by the use of a heterojunction, as employed in the buried heterostructure laser of Figure 18.1(d). In this structure identical methods are used for carrier confinement in the two transverse directions, and the overall confinement efficiency approaches 100%. The distribution of carriers in the plane parallel to the main heterostructure is illustrated in Figure 18.2(e). The transverse heterojunctions at both edges provide good confinement of holes in the central region and also an additional source for electron injection.

18.2.2 Current spreading in stripe contact lasers

The current spreading in a stripe laser, as it affects the threshold current and the incremental efficiency, is a complex process in which the behaviour both underneath and outside the stripe contact must be taken into consideration. Also in practice the distinction between current spreading and transverse diffusion becomes somewhat blurred in processes which take place in the active layer. Various simplified approaches can be made. One such treatment has been presented by D'Asaro (17) and Dumke (18) which deals particularly with the edge leakage current that flows through the active layer outside the area of the stripe. A simple linearized analysis allows the total leakage current and its transverse distribution outside the stripe to be obtained in terms of the characteristics of the p–n junction and the resistivity and thickness of the heterostructure layers. This serves as a useful introduction to the general problem of transverse carrier flow in stripe lasers. However a further question which must be answered is where under the stripe contact does the leakage

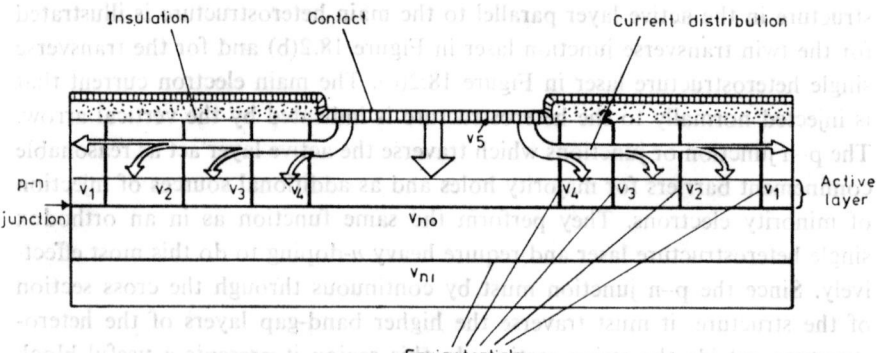

Figure 18.3 Simplified diagram of equipotentials and current distribution between contact and active layer in stripe contact laser

current originate, since if it is close to the edges and at no expense to the centre region, then it does not affect the threshold of the laser but only the width of the optical output. The analysis of external leakage current must therefore be followed by an investigation into the behaviour under the stripe.

Stripe Geometry Heterojunction Injection Lasers

For the present purpose the potential distribution over the cross section of the laser in the layers adjacent to the contact (or 'current window') is assumed to take the simple form illustrated in Figure 18.3. The n-side boundary of the active layer is assumed to be a plane of equipotential V_{no}. This is a reasonable assumption when the conductivity of the n-passive layer is relatively high, which is normal. The equipotentials in the thin layer on the p-side (v_1, etc.) are taken to be perpendicular to the p–n junction in the region outside the contact, with purely transverse current flow. The analysis of the lateral distribution of the voltage outside the contact is then reduced to the solution of a one-dimensional differential equation relating the voltage gradient along the x-direction to the current taken by the p–n junction as follows:

$$d^2V/dx^2 = (J_0/gt)\exp(eV/nkT) \tag{18.1}$$

where g is the conductivity of the p-layer, t is the combined thickness of the p-layers (including the active layer if p-doped), V is the voltage on the p-side of the active layer at position x measured with respect to the voltage at $x = 0$, J_0 is the current density at $x = 0$, and $\exp(eV/nkT)$ gives the relative current-voltage dependence of the laser junction below threshold. Solving this equation shows that the current density J through the active layer drops off with distance x from the contact according to the relation

$$J = 2J_0/(x/l_s+2^{1/2})^2 \tag{18.2}$$

where l_s is a spreading length given by

$$l_s = (gtnkT/J_0 e)^{1/2} \tag{18.3}$$

The total current/unit length, J_{fringe}, which flows into the fringe region on both sides of the stripe is given by

$$J_{\text{fringe}} = 2.2^{1/2} l_s J_0 \tag{18.4}$$

Measurements show that the value of the spreading length l_s is of the order of 2–5 μm, which means that half the current leaks out sideways in stripes of around 10 μm width.

18.2.3 Carrier loss by diffusion

Lateral spreading of current in stripe lasers occurs in the active layer as well as in the passive layer adjacent to the contact. If the active layer is of the same doping type as the passive layer (usually p-type), the current spreading takes place by the same ohmic process as in the p-passive layer. However, if the carriers are injected in appreciable concentration into the active layer, as occurs if the active layer doping is low or of the opposite type to the passive layer, then the movement of injected carriers takes place by diffusion.

In stripe lasers where deep proton bombardment or oxygen implantation has produced an insulating region intersecting the active layer, diffusion provides the only mechanism for the lateral flow of carriers; the diffusion is ambipolar if both types of carrier are present in equal concentrations. Hakki (19) has investigated this behaviour in both shallow and deep proton bombarded lasers (Figures 18.1(c) and 18.1(b), respectively). In the shallow bombarded lasers concerned, the observed distribution of carriers (as deduced from measurements of spontaneous emission) is consistent with a diffusion length of 6 μm, both within and outside the stripe. In the deep proton bombarded devices the characteristics of the active layer at the edge of the stripe are changed by the bombardment and in particular the carrier recombination time is reduced. This produces a high recombination velocity for carriers and reduces their concentration at the boundary to around 25% of their peak concentration at the centre of the stripe. Blum et al. (10) have investigated the behaviour in deep oxygen implanted lasers. The dominant effect of the oxygen is to reduce the carrier mobility rather than the recombination time, so reducing the lateral flow of carriers and maintaining a higher concentration under the stripe.

The diffusion profile of the injected carriers, both within and outside the region where the injection current flows, has been analysed by Hakki (19). In the general case it is necessary to take into account the different values of the diffusion constant and the diffusion length in the central and outer regions, which we represent respectively by D_1, l_1 and D_2, l_2, and in particular the ratio $D_1 l_2/D_2 l_1$ which we represent by ζ. ζ is unity when the active layer characteristics are uniform. The carrier concentration n is given as a function of position x from the centre line of the stripe, of width w, by:

$$n = \begin{cases} n_0(1 - A\cosh(x/l_1)) & \text{for } -w/2 < x < w/2 \\ n_0 B\exp(-x/l_2) & \text{for } x < -w/2 \text{ and } w/2 < x \end{cases} \quad (18.5)$$

where n_0 is the injected carrier concentration which would be produced by the input carrier density J if there were no sideways diffusion. The coefficients A and B are given by:

$$A = \{\cosh(w/2l_1) + \zeta\sinh(w/2l_1)\}^{-1}$$
$$= \exp(-w/2l) \quad \text{when } \zeta = 1 \quad (18.6)$$
$$B = A\zeta\sinh(w/2l_1)\exp(w/2l_2)$$
$$= \sinh(w/2l) \quad \text{when } \zeta = 1$$

with $\zeta = D_1 l_2/D_2 l_1 = (D_1/\tau_1)^{1/2}/(D_2/\tau_2)^{1/2}$

Various profiles of injected carrier concentration are illustrated in Figure 18.4 for different values of $w/2l_1$ and ζ. The profiles given in (a), (b) and (c) for $\zeta = 1$ show the effect of varying the stripe width w (or l_1). When the stripe is wider than about five diffusion lengths the distribution of carrier concentration over the centre region of the stripe is very flat and the maximum concentration at the centre line approaches within 10% of the value in a broad contact laser at the same current density. Under such circumstances (see Figure 18.4(c)) the carrier concentration at the edge of the stripe rises to approximately $\zeta/(1+\zeta)$ of its value at the centre. The total current per unit width which is lost by sideways diffusion corresponds to that flowing into a width of the stripe which is a fraction $2/(1+\zeta)$ of the diffusion length under the stripe. This value of effective width is about half of that calculated for current spreading in Equation 18.4 (depending on the exact value of ζ).

The current lost at the edges of the stripe causes a considerable lowering of the injected carrier concentration within a diffusion length of the edge but a much smaller lowering at the centre, particularly for stripe widths greater than four diffusion lengths. Hence provided the lasing filament is confined to the centre region of the stripe the lost current at the edge has little effect on the threshold current of the wider lasers. In this instance, therefore, the previously derived treatment of D'Asaro and Dumke is not applicable although, as will be described later, circumstances arise in which the lost current in wider lasers becomes important.

When the stripe is less than a few diffusion lengths wide (see Figure 18.4(a)), the peak carrier concentration at the centre of the stripe drops considerably below the value n_0 for a wide stripe, and continues to diminish as the stripe width is further reduced. The height of the peak is also affected by the value of ζ, and this dependence becomes increasingly significant as the width of the stripe is narrowed.

Figures 18.4(d) and (e) show the carrier concentration profiles for $w/l_1 = 2$ and for values of ζ of both 0.25 and 4. Hakki has found that ζ is reduced to about 0.25 in proton bombarded lasers as a result of a large reduction in the carrier recombination rate in the proton bombarded regions (19). In contrast the results of Blum *et al.* (10) for oxygen implanted lasers are reasonably consistent with a value of ζ of 4 (see Section 18.4.2), indicating a large reduction in the diffusion constant in the bombarded region with little change in the recombination rate. Figures 18.4(d) and (e) show that the peak of the carrier concentration distribution is appreciably increased for $\zeta = 4$ compared with $\zeta = 1$ and appreciably reduced for $\zeta = 0.25$. When the stripe width is further narrowed to less than about one diffusion length the height

504 G. H. B. Thompson

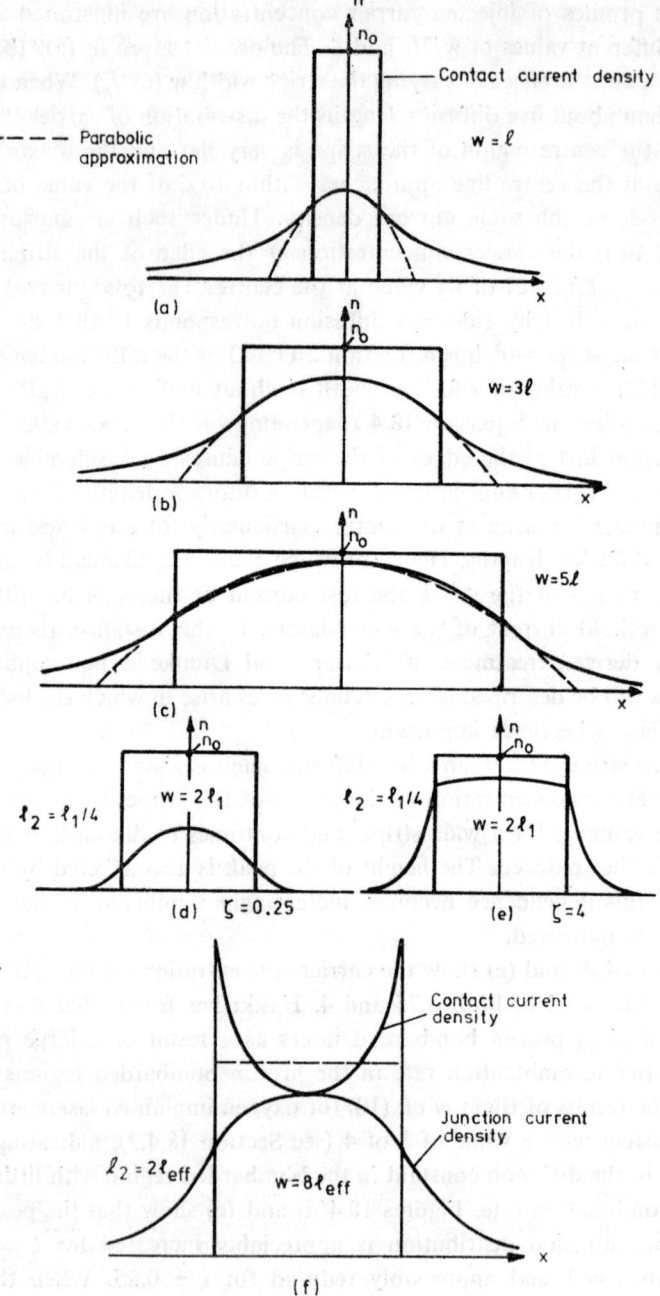

Figure 18.4 (see opposite page)

of the peak becomes directly proportional to ζ. These results show that deep proton bombarded lasers with $\zeta = 0.25$ become relatively unsatisfactory for stripe widths of less than about 1.5 diffusion lengths, whereas oxygen implanted lasers with $\zeta = 4$ become particularly suitable in respect of current confinement.

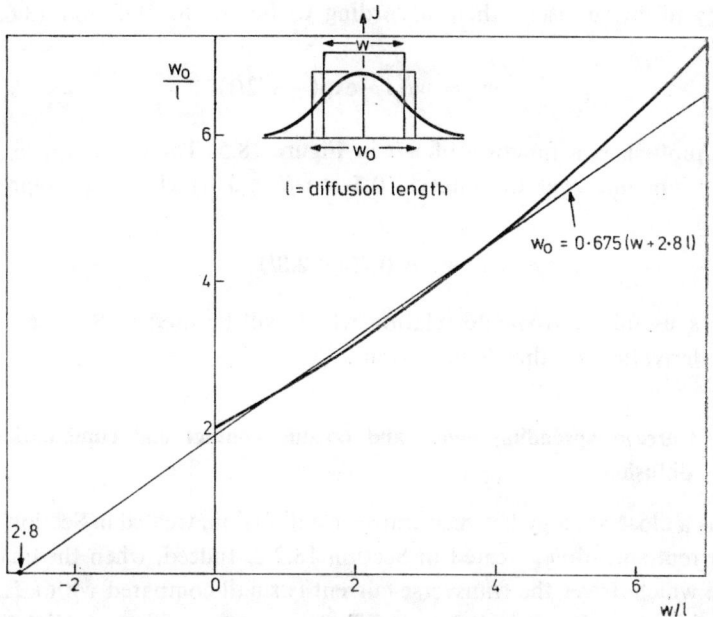

Figure 18.5 Effective stripe width relates current density at centre of stripe to current supplied under conditions of lateral carrier diffusion

Figure 18.4 Theoretical lateral distribution of injected carrier concentration or non-lasing recombination current density in stripe lasers. Profiles determined by diffusion. (a), (b) and (c) material uniform within and outside stripe and width of stripe varying from one to five diffusion lengths; (d) example of deep proton bombarded stripe laser. Diffusion length outside stripe reduced by factor of 4 owing to reduction in τ. Stripe width equal to two internal diffusion lengths. (e) Example of deep oxygen implanted stripe laser. Diffusion length reduced by factor of 4 outside stripe owing to reduction in diffusion coefficient. Stripe width equal to two internal diffusion lengths. (f) Example of lateral distribution of junction current with non-uniform contact current density under circumstances where contact is closely coupled to active layer

To describe the peak value n_m of the carrier concentration at the centre of the stripe it is convenient to define an effective width w_0 over which the input carrier I, if perfectly constrained, would produce a uniform carrier concentration of n_m. In the case where the characteristics of the active layer are uniform across the width ($\zeta = 1$), which is approximately true in the majority of stripe lasers, then, according to Equations 18.5 and 18.6, w_0 is given by:

$$w_0 = w/(1-\exp(-w/2l)) \qquad (18.7)$$

w_0/l is plotted as a function of w/l in Figure 18.5. The curve approximates to a straight line over the interval $0.5 < w/l < 5$ which can be represented by

$$w_0 = 0.7(w+2.8l) \qquad (18.8)$$

This is a useful approximate relation which will be used in Section 18.4.1.2 in the derivation of threshold current.

18.2.4 Current spreading under and outside contact and combination with diffusion

There is a close analogy between transverse diffusion, treated in Section 18.2.3, and current spreading, treated in Section 18.2.2. Indeed, when the change in voltage which drives the transverse current is small compared with kT, Equation 18.1 can be written precisely as a diffusion equation with $\exp(eV/nkT)$ in the RHS replaced by $(1+eV/nkT)$. In any situation where current spreading and transverse diffusion occur in combination it is reasonable, as an approximation, to treat the two processes as one and modify the values of the diffusion constants appropriately. The behaviour under the contact is generally of most significance since this is where the light in most stripe lasers is mainly concentrated (see next section). It is the region where the smallest changes in voltage occur and hence where the replacement of $\exp(eV/nkT)$ by its linear approximation causes the least inaccuracy.

Up to this point the transverse flow of carriers has been treated in the same one-dimensional manner both underneath and outside the contact. Underneath the contact this is a very considerable approximation and it is less true for current flow in the passive layer than it is for diffusion in the active layer. The conditions under the contact differ from those outside in that the direction of major current flow is normal (y-direction) rather than transverse (x-direction) to the junction. Any variation in voltage at the junction causes a change in the voltage drop between the junction and the contact and hence, in addition to inducing some lateral flow of current, changes the current density normal to the junction. This new factor affects the transverse distribution of injected carriers. By introducing a further tendency to stabilize the carrier concentration it has the effect of reducing the effective diffusion length.

Figure 18.4(e) shows a typical distribution of the contact current density and the junction current density under conditions where the contact is closely coupled. This applies to the case where the diffusion length under the stripe is half that outside ($\zeta = 2$) and $\frac{1}{8}$ of the stripe width, and for a spacing between the contact and the active layer which gives a voltage drop of about $nkT/3$, e.g. 15 mV. The figure shows that the junction current density at the centre of the stripe is very nearly equal to the contact density at the same position, and that the extra current that flows is confined almost entirely to the two ears of additional current density at each edge of the stripe.

18.2.5 Carrier confinement in buried heterostructure stripe laser

In the buried heterostructure laser the functions of channelling the current to the active region and confining the injected carriers within the same region are combined in one structure.

The active region consists of a filament of lower bandgap material surrounded on all four sides by (GaAl)As material of higher bandgap as illustrated in the cross section of Figure 18.1(g). In the structure shown the p–n junction encircles the active region, itself p-doped, on three sides, providing three corresponding sources of electron injection. The fourth side provides the source of hole injection. After injection both carrier types are well confined by the heterojunction barriers and can accumulate in the active region up to concentrations well beyond those in the injecting regions.

After encircling the active layer the p–n junction continues through the higher bandgap material on both sides of the stripe until it reaches the SiO_2 insulation at the surface of the wafer. Relatively little by-pass injection current passes through these sections of the p–n junction because the potential barrier which must be surmounted in the higher bandgap material is greater. The leakage current density concerned is of the same order as the very small portion that results from leakage of injected carriers across the confining barriers of the active region.

In practice the buried heterstructure is found to give very good carrier confinement. This is indicated by the excellent results which have been achieved with measured threshold current as low as 7 mA in devices of 1.5 μm width and 300 μm length with an active layer thickness of 0.2 μm (4). The corresponding threshold current density is comparable to that in a broad contact laser.

18.3 OPTICAL CONFINEMENT IN STRIPE LASERS

Three processes are important in considering optical confinement in stripe lasers: dielectric waveguiding, so-called gain-guiding and self-focusing. Dielectric guiding as in the buried heterostructure follows well-known principles and will not be treated further. Gain-guiding has a particularly significant bearing on the behaviour of conventional stripe lasers and a physical description of the process will be presented. Self-focusing effects, which have been tentatively suggested in the past (20, 21) have recently been shown also to contribute significantly to the behaviour of stripe lasers (22). This process will also be described.

18.3.1 Gain-guiding

This section attempts to interpret gain-guiding in physical terms and to establish the fundamental principles which lie at the basis of the behaviour. Two models of the gain-guide, one with abrupt steps between the gain and loss regions and one where the gain/loss characteristic is continuously graded over the laser cross section, illustrate slightly different aspects of the phenomenon.

18.3.1.1 Abrupt steps between gain and loss

Figure 18.6 gives a perspective illustration of a typical two-dimensional wave which is generated in a gain-guiding situation, where an elongated region of uniform gain is surrounded by regions which are either transparent or optically lossy. The overall wave motion creates a herring-bone pattern. The spreading V-shaped waves in the outer transparent or lossy regions are linked together at their apex in the gain region by curved sections of the wavefront. The outer waves die away in amplitude with distance from the guide axis, and fan out like the wake from a ship.

Energy is propagated in the direction normal to the wavefronts. In the centre stripe the majority of the power flux is propagated along the axis of the guide. A proportion, however, moves out to the edges of the guide and is fed into the diverging waves in the 'wake'. The magnitude of the proportion

Stripe Goemetry Heterojunction Injection Lasers

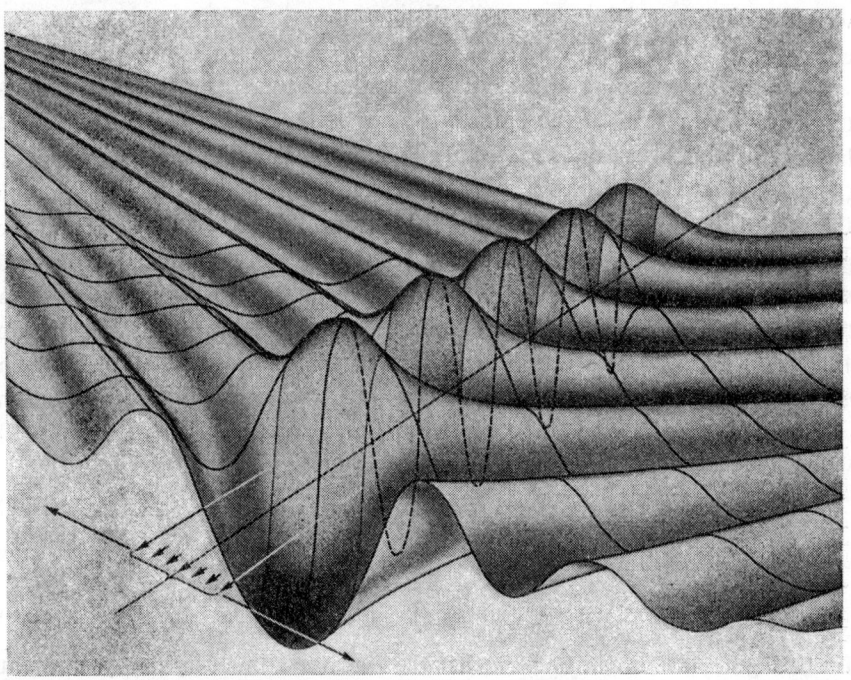

Figure 18.6 Three-dimensional representation of shape of optical wave over length and breadth of a gain-guided mode in a stripe geometry laser

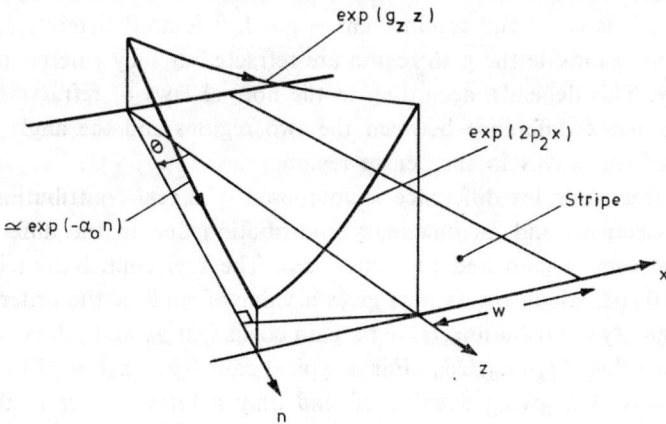

Figure 18.7 Diagrammatic representation of relation between longitudinal gain coefficient, lateral decay coefficient and emergent angle of leaky waves in gain-guided mode of stripe laser

which leaks out determines the dissipation of the structure, and the effectiveness of the gain-guiding.

The way the diverging waves die away with distance from the axis controls the transverse spread of the light, and may be regarded as an indication of how well the light is confined. Figure 18.7 shows how the coefficient $2p_2$ for the lateral fall-off in optical intensity is related to the gain coefficient g_z along the axis of the centre region, the angle θ between the axis of the laser and the propagation direction of the diverging plane waves in the outer region, and the loss coefficient of these waves. This loss coefficient differs negligibly from the loss coefficient α_0 for plane waves in the outer region if θ is small. Figure 18.7 gives a diagrammatic representation of the complete relation between the transverse propagation constant (p_1+jp_2), the longitudinal propagation constant $(\beta_z+jg_z/2)$ and the plane wave propagation constant $(\mu\beta_0-j\alpha_0/2)$ which is given by

$$(p_1+jp_2)^2 = (\mu\beta_0-j\alpha_0/2)^2 - (\beta_z+jg_z/2)^2 \tag{18.9}$$

where μ is the refractive index and $\beta_0 = 2\pi/\lambda$. θ is related to p_1 by:

$$\sin\theta \simeq p_1/\mu\beta_0 \tag{18.10}$$

From Figure 18.7 it can be seen that the transverse intensity decay constant $2p_2$ is given approximately by:

$$2p_2 \simeq (g_z+\alpha_0)/\sin\theta \tag{18.11}$$

The angle θ is very important in determining the transverse decay constant, and when it is small the confinement is good. θ is itself determined by the way the waves inside the gain region are refracted as they emerge across the boundary. This depends, according to the normal laws of refraction, on the refractive index difference between the two regions and the angle of propagation of the waves in the centre region.

The refractive index difference is composed of a real contribution due to injected electrons and an imaginary contribution due to the difference between the central gain and the outer loss. The real contribution is of the order of 0.002, which on its own gives a value of $\sin\theta$ of the order of 0.03. The imaginary contribution, taking a gain coefficient g_0 and a loss coefficient α_0, has a value $j(g_0+\alpha_0)/2\beta_0$. For a typical case $((g_0+\alpha_0) = 150$ cm^{-1}) we find $\delta\mu = 0.001j$, giving a value of $\sin\theta$ only a little smaller at 0.02 than that suggested for the real part of the refractive index. Putting these values back into Equation 18.11 and taking a value of $(g_z+\alpha_0)$ of 100 cm^{-1} gives a transverse decay length of the order of a few microns.

18.3.1.2 Continuous gain/loss profile

In a more realistic model of the stripe laser the gain should be taken to vary continuously across the transverse section, peaking at the centre and falling away to a region of loss at the edges. This gain distribution may be accompanied by a similar continuous variation of the refractive index, but with a minimum value at the centre, giving a distribution that tends to spread the wave rather than confine it. If the distributions are parabolic a simple theoretical analysis can be made (23). Here we give a descriptive treatment of the behaviour.

In the distributed model there are no abrupt boundaries between regions of gain and loss that can give rise to specific reflections, and there is no positive source of optical confinement. The spread of the lasing mode is determined largely by diffraction, augmented by any anti-focusing effects associated with the refractive index profile. The amount of diffraction is related to the width of the beam. The width of the beam can be taken to be not much greater than the width of the gain region, (in certain circumstances, as explained below, it can be considerably less). The way that such a beam would spread by diffraction in the absence of gain can be estimated using reasonable assumptions for the shape of its intensity distribution and the curvature of the wavefront. The proportion of the beam which spreads outside the gain region over the length of the laser can then be regarded as a measure of the diffraction loss and can be used as a criterion for the performance of the stripe laser.

We will assume the transverse intensity distribution to be Gaussian, which is true for a parabolic gain profile, and set its width, to the $1/e^2$ points of intensity equal to s_{eff}. Let us initially assume that the wavefront is plane. Such a source produces a far-field distribution which is also Gaussian with an angular half width $\theta_{1/2}$ (measured between the centre line and the direction where the intensity is reduced by $1/e^2$) given by

$$\sin\theta_{1/2} = 2\lambda/\pi\mu s_{\text{eff}} \tag{18.12}$$

where μ is the refractive index. The effective width of the beam at a distance L from the source is approximately $(2L\sin\theta_{1/2} + s_{\text{eff}})$. A convenient criterion of a beam which is adequately directional for laser operation is that it should not spread to more than double its size over the length L of the laser. On this basis Equation 18.12 shows that the minimum allowable value of the optical width is given by:

$$(s_{\text{eff}})_{\min} = (2L\lambda/\pi\mu)^{1/2} \tag{18.13}$$

For a laser of length, say, 300 μm this equation indicates that the beam width should be at least 7 μm.

The transverse distribution of the light generated in the lasing mode of the stripe laser does not, of course, expand as it propagates from one end of the resonator to the other. It settles to such a form, along the whole length of the laser, that the effect of the gain distribution in enhancing the centre and attenuating the edges of the optical distribution just counteracts the spreading effect of diffraction, and the wave propagates along the laser axis unchanged in shape but with the necessary amplification. The equilibrium optical distribution is Gaussian but the wavefront is curved. If there is no refractive index profile to cause anti-focusing, then it can be shown that the equilibrium distribution corresponds precisely, in both shape and curvature of wavefront, to that in a Gaussian beam at the point in the transition region between the near- and far-field where the beam has expanded by a factor of $2^{1/2}$, compared with its width at the waist. The angle $\theta_{1/2}$ at which the beam diverges at this point is a factor of $1/2^{1/2}$ less than the final diffraction angle but precisely equal to the angle given in terms of s_{eff} in Equation 18.12. Hence no adjustment need be made to Equation 18.13 to take account of the curvature of the wavefront. If a refractive index profile that causes anti-focusing is present the curvature of the wavefront for a given beam width is increased. This increases the value of $\sin\theta_{1/2}$ or alternatively means that for given diffraction loss a larger beam width must be used. The minimum acceptable value of s_{eff} is then greater than that given by Equation 18.13 and it is necessary to use the exact analysis of a parabolic gain profile to find its value.

The width of the beam in the laser is determined by the width of the gain profile and the magnitude of the diffraction. If the width of the optical distribution is small, the diffraction loss is large and the majority of the gain is taken up in counteracting this loss rather than in imparting overall amplification to the wave.

In a precise treatment of the gain/loss equilibrium the diffraction loss is an unsatisfactory quantity, because it requires the definition of a boundary which the escaping energy must cross before it is defined as a loss. Instead we will consider absorption loss. The magnitude per unit length of this loss can be obtained by an appropriate integration of the product of optical intensity and loss coefficient in the lossy region.

To produce a net amplification of the wave, the gain contribution over the width of the stripe where the gain is positive must be greater than the loss contribution over the remaining width where the gain is negative.

Expressed more precisely this requires that the integral of the optical-intensity/ gain product in the centre region be greater than that of the optical-intensity/ loss product in the outer regions. This condition can in principle be reached in all reasonable circumstances, however narrow the gain region, by a sufficient increase in the current density. An exact treatment shows that when the gain and loss profile can be approximated to a parabola, the maximum distance the light can penetrate into the lossy region is such that the effective optical width s_{eff}, measured to the $1/e^2$ points, is twice the width of the region

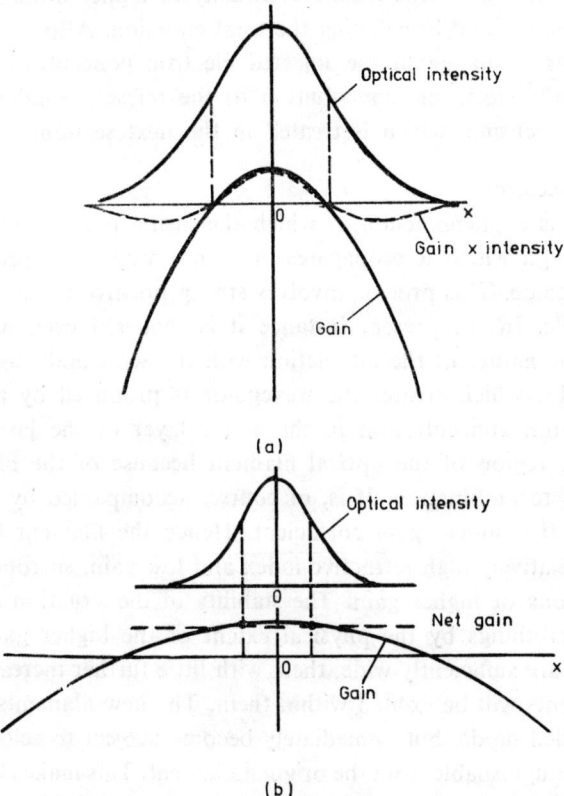

Figure 18.8 Corresponding distributions of gain and optical intensity in gain-guided modes. (a) Narrow gain distribution; (b) wide gain distribution

where the gain is positive. The losses then exactly compensate the gain. The corresponding distribution of gain and optical intensity are illustrated in Figure 18.8(a).

A different situation arises when the gain region is wide. The diffraction loss of the laser mode is small and the lasing emission is generated over a width where the gain barely exceeds the end losses of the laser. Because of the curvature at the peak of the gain profile this can correspond to only a fraction of the width over which the gain is positive. This behaviour is illustrated in Figure 18.8(b).

The situation becomes more complicated, however, as the laser is driven harder, because then the gain profile is affected by the stimulated emission, and its shape changes. This results eventually in higher order lateral modes reaching threshold and broadening the total emission. Also a change in gain profile implies a change in the injected electron concentration and hence a change in the electronic contribution to the refractive index. This is the cause of self-focusing, which is treated in the next section.

18.3.2 Self-focusing

Self-focusing is a phenomenon in which the lasing beam interacts with the medium through which it propagates in such a way as to provide its own optical waveguide. This process involves strong positive feedback, and tends to be unstable. In the present instance it is rendered even more unstable because of the nature of the interaction with the semiconductor. The rise in refractive index which creates the waveguide is produced by a drop in the injected electron concentration in the active layer of the laser. This drop occurs in the region of the optical filament because of the high local rate of stimulated recombination. It is, of course, accompanied by a decrease in the value of the optical gain coefficient. Hence the filament finds itself in a region of relatively high refractive index and low gain, surrounded on both sides by regions of higher gain. The stability of the situation is determined amongst other things by the physical extent of the higher gain regions. If these regions are sufficiently wide, then, with little further increase in current, further filaments will be excited within them. The new filaments are initiated in a gain-guided mode, but immediately become subject to self-focusing and are then indistinguishable from the original filament. This makes the behaviour in a wide laser relatively complicated. However in a sufficiently narrow laser the behaviour is simpler. The regions of higher gain bordering the filament are not then wide enough to support either a gain-guided mode or a higher order transverse mode within the original filament and it is possible to obtain a stable zero order self-focused filament over an appreciable range of current.

In narrow stripe lasers the self-focusing action has to counteract an initial carrier distribution which has a considerable peak at the centre of the stripe and which produces anti-focusing. As the current is increased above threshold the current taken by the lasing flattens the peak and initially there may be a slight increase in the width of the optical distribution. However, after a sufficient increase in current the curvature at the peak of the carrier distribution becomes inverted and self-focusing starts. This improves the optical guiding and the filament contracts in width. The optical losses also become less and the injected carrier concentration at the centre of the distribution can therefore dip appreciably below the level in the neighbouring regions before the gain drops below the lower level now required. This means that the current available for stimulated emission increases and that a greater optical output is produced than in the previous condition.

However as the drive current is further increased the filament continues to narrow, and the inward diffusion of injected carriers is insufficient to prevent the carrier concentration in the two peaks which border the filament from increasing in height. This diverts current from the lasing emission and reduces the incremental efficiency. The overall effect of the self-focused operation on the light/current characteristic is therefore to enhance the output in the current range over which the focusing is first initiated but to follow this with a region of reduced slope where the filament narrows and the spontaneous output increases. Eventually a first order mode is produced.

18.4 THRESHOLD CURRENT IN STRIPE LASERS

This section deals with the threshold current of stripe lasers where there is no deliberate optical or carrier confinement in the lateral direction. The treatment applies to those lasers in which the width of the lasing filament is determined by the natural distribution of injected carriers which results from lateral current spreading and carrier diffusion.

It is helpful to consider the increase in current density in narrow stripes in two parts. First, we can consider the effect of the sideways spreading of the injected carrier concentration. This reduces the peak carrier concentration at the centre of the stripe to below the value that would apply in a broad stripe laser at the same injected current density. The current density must be raised to counteract this effect. Secondly, we can consider the effect of the gain-guiding process which determines how far the filament spreads from the centre line of the stripe. Because the filament spreads into regions of lower gain the net modal gain is somewhat less than the peak value of the gain at the centre. Hence to reach threshold the current density must be further increased so that the net gain reaches the same value as in a broad laser. In the process the peak gain exceeds that value by an appropriate amount which increases as the laser gets narrower.

The major effect on threshold current in all except the narrowest stripe lasers comes from the first of the above mechanisms,—the lateral spreading of carriers. Two ways of treating the lateral spreading of carriers have been suggested. D'Asaro (17) and Dumke (18) have shown that there is an effective lateral leakage current that flows outside the edge of a stripe laser, as described in Section 18.2.2, which corresponds to the current which would flow through some particular additional width of stripe at the current density supplied by the contact. The implication is that this represents an additional current which must be added to the normal current flowing into the stripe to give the increased value of the threshold.

Hakki has used a different approach, as described in Section 18.2.3, considering the concentration of carriers injected at the centre of the stripe and comparing it with that which would occur at the same current density but without sideways diffusion. In general a much smaller additional current is required to raise this concentration to its original value than is required to compensate the total lateral leakage current at the edges of the laser. However, as discussed in Section 18.2.4, the contact is not necessarily a constant current source. If it is sufficiently closely coupled to the active layer, with negligible series resistance, additional current can be drawn from the outer edges of the contact so as to maintain the injected carrier concentration at a constant level over the whole width and to provide the extra current in the lateral leakage component. This causes the behaviour to revert precisely to that obtained by D'Asaro and Dumke. However, in general, the contact resistance is sufficient to severely limit the amount of additional current drawn and the behaviour approximates more closely to the lateral diffusion model of Hakki. We will use the diffusion model in the analysis which follows, although to take account of possible interaction with the contact we will be prepared to use an effective value for the diffusion length beneath the contact which is less than that outside.

To complete the diffusion model it is necessary to consider the gain-guiding process which determines how the modal gain is derived from the non-uniform gain distribution at the centre of the stripe. Various parameters affect the gain-guiding behaviour. However, they can all be lumped together into a single parameter which we call the *characteristic gain-guide width* s_1. The increase in current density required to raise the central gain peak to its threshold value depends on the ratio of the actual filament width to the gain-guide width. For example, a stripe laser which gives a filament width of $2s_1$ requires twice the current density that is necessary in a very wide laser.

The treatment presented in the following sections describes the range of circumstances in which the threshold-vs.-stripe-width relation can be represented by a carrier-diffusion plus gain-guiding model, using in most instances only the two variable parameters mentioned above, namely the effective diffusion length and the characteristic gain-guide width.

18.4.1 Gain-guiding as a function of stripe width

To analyse the behaviour of the gain-guided mode in a stripe laser we need first to find the transverse distribution of gain and dielectric constant in terms of the distribution of injected carriers and then to find the net gain of the guided mode in terms of the magnitude of the distribution.

We amalgamate the distribution of gain and dielectric constant produced by the injected carriers into a distribution of complex dielectric constant $\varepsilon_1 + j\varepsilon_2$ where the imaginary part ε_2 can be shown to be related to the local gain coefficient g by

$$\varepsilon_2 \simeq g\varepsilon_1^{1/2}/\beta_0 \qquad (18.14)$$

where $\beta_0 = 2\pi/\lambda$. We will assume that both ε_1 and ε_2 depend approximately linearly on injected carrier concentration so that we may write

$$\varepsilon_2 = b\varepsilon_1 \qquad (18.15)$$

where b is approximately constant and normally has a value around -1.5.

Analysis of the optical guiding produced by the complex dielectric profile when the overall distribution is parabolic shows that the light in the lowest order transverse mode has a Gaussian transverse distribution. We will set the effective width of the distribution to the $1/e^2$ intensity points equal to s_{eff}. The following relation can be shown to exist between s_{eff} and the relative gain distribution g/g_z normalized to the mode gain g_z

$$g/g_z - 1 = 8(Bs_0)^2 \{s_{\text{eff}}^{-2} - 16(x/s_{\text{eff}}^2)^2\} \qquad (18.16)$$

where

$$B^2 = \pm (b^2 + 1)^{1/2} - b \qquad (18.17)$$

the positive sign of the two alternatives applying when the gain peaks at the centre, and where s_0 is a characteristic length given by

$$s_0 = (\beta_0 \varepsilon_1^{1/2} g_z)^{-1/2} \qquad (18.18)$$

which is normally of the order of 3 μm.

To express g in terms of the injected carrier distribution as given in Equation 18.5 for a particular stripe width we will assume that there is a linear relation between gain and carrier concentration of the form

$$g \propto n - n_t \tag{18.19}$$

where n_t is the injected carrier concentration at which the gain is zero. If the injected carrier concentration required to give the mode gain g_z under uniform conditions is n_{th} then we may write the following relation for $g/g_z - 1$

$$g/g_z - 1 = (n/n_{th} - 1)/\beta \tag{18.20}$$

where

$$\beta = (1 - n_t/n_{th}) \tag{18.21}$$

Using Equation 18.5 we therefore obtain the following transverse distribution of gain

$$g/g_z - 1 = \{(n_0/n_{th})(1 - A\cosh(x/l_1) - 1)\}/\beta \tag{18.22}$$

Equation 18.22 is in the form of a hyperbolic cosine whilst Equation 18.16 is parabolic. To match the two sufficiently accurately we fit the curves at three points, viz. the central peak and the points on each side where $g = g_z$. This approximation underestimates the gain when it is greater than average and overestimates it when it is less than average. Hence it tends to cancel out the discrepancies. Matching the two distributions gives the following two equations for n_0/n_{th} and s_{eff} in terms of A.

$$(1 - A)n_0/n_{th} = 1 + 8s_1^2/s_{eff}^2 \tag{18.23a}$$

$$= 1 + A(n_0/n_{th})(\cosh(s_{eff}/4l_{eff}) - 1) \tag{18.23b}$$

where

$$s_1 = \beta^{1/2} B s_0 = [\beta\{(b^2+1)^{1/2} - b\}/\beta_0 \varepsilon_1^{1/2} g_z]^{1/2} \tag{18.24}$$

s_1 is a useful quantity for describing the gain-guiding process to which we referred at the beginning of this section and which we have called the *characteristic gain-guide width*.

Although there is no analytic solution to these equations for s_{eff} in terms of A there is a simple solution for A, and hence w, (see Equation 18.6), in terms of s_{eff}. For convenience we write this relation in terms of a normalized filament width x, a normalized stripe width w' and a normalized diffusion length l' all expressed in terms of the characteristic gain-guide width s_1 and given by

$$x = s_{eff}/s_1, \qquad w' = w/s_1, \qquad l' = l_{eff}/s_1 \tag{18.25}$$

The required solution is

$$A = [1 + (1 + x^2/8)\{\cosh(x/4l') - 1\}]^{-1} \quad (18.26)$$

The normalized stripe width w' can then be obtained in terms of A, according to Equation 18.6 by

$$w' = 2l'\ln[\{1/A + (1/A^2 + \zeta^2 - 1)^{1/2}\}/(1+\zeta)]$$
$$= 2l'\ln(1/A) \quad \text{when} \quad \zeta = 1 \quad (18.27)$$

18.4.1.1 Filament width

The relations of Equations 18.25, 18.26 and 18.27 give a series of curves of normalized optical filament width x against normalized stripe width w', with normalized diffusion length l' and ζ as variable parameters. An example is shown in Figure 18.9 of a set of curves of s_{eff}/l_{eff} (i.e. x/l') against w/l_{eff}

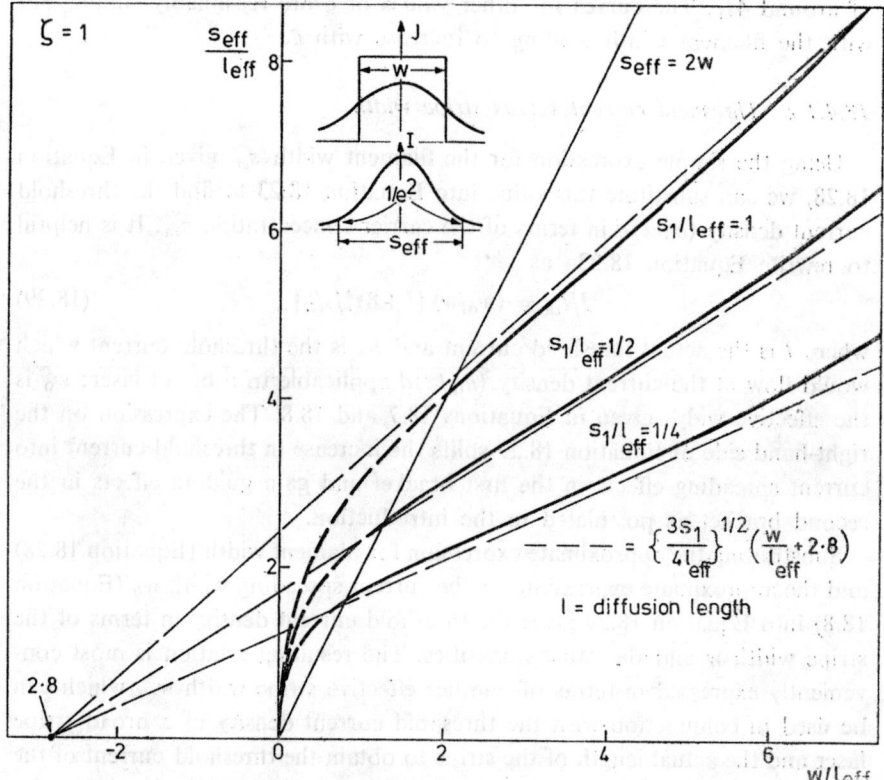

Figure 18.9 Theoretical relation between filament width and stripe width in gain-guided modes in a stripe laser with diffusion length as a parameter

(i.e. w'/l') for various values of l' with $\zeta = 1$ (i.e. applying when there is no discontinuity in the carrier transport characteristics at the edge of the stripe). Over the range of w'/l' from 1 to 6 the curves can be fitted quite well by straight lines (as pointed out in one instance by Hakki (24)) which have the approximate form

$$s_{\text{eff}} = A_1(\beta^{1/2}Bs_0/l_{\text{eff}})^{1/2}(w + B_1 l_{\text{eff}}) \tag{18.28}$$

where A_1 and B_1 are reasonably independent of l' but depend on ζ. For $\zeta = 1$ $A_1 \simeq (3/4)^{1/2}$ and $B_1 \simeq 2.8$.

Examination of the curves for $\zeta = 1$ shows that the filament width measured to the $1/e^2$ intensity points is of the order of the stripe width, being somewhat greater for narrow stripes and somewhat less for broad stripes, the difference in both instances being somewhat more marked when the diffusion length is greater. The division between 'broad' and 'narrow' occurs at a stripe width of around $4s_1$. The curves for other values of ζ are reasonably similar but with the filament width tending to increase with ζ.

18.4.1.2 Threshold current versus stripe width

Using the simple expression for the filament width s_{eff} given in Equation 18.28, we can substitute this value into Equation 18.23 to find the threshold current density $(n_0/\tau)d$ in terms of the carrier concentration n_{th}. It is helpful to rewrite Equation 18.23a as

$$I/I_{\text{th}} = (w_0/w)\,[1 + 8s_1^2/s_{\text{eff}}^2] \tag{18.29}$$

where I is the actual threshold current and I_{th} is the threshold current which would flow at the current density $(n_{\text{th}}/\tau)d$ applicable to a broad laser; w_0 is the effective width given in Equations 18.7 and 18.8. The expression on the right-hand side of Equation 18.29 splits the increase in threshold current into current spreading effects in the first bracket and gain guiding effects in the second bracket as postulated in the introduction.

Substituting the approximate expression for filament width (Equation 18.28) and the approximate expression for the current spreading width w_0 (Equation 18.8) into Equation 18.29 gives the threshold current density in terms of the stripe width w and the other quantities. The resulting relation is most conveniently expressed in terms of another effective stripe width w_{eff} which can be used in conjunction with the threshold current density of a broad stripe laser and the actual length of the stripe to obtain the threshold current of the stripe laser. If we assume that n_{th} and β are constant, then w_{eff} is given by

$$w_{\text{eff}} = 0.7s_2(y + 8/y) \tag{18.30}$$

where

$$y = (w + B_1 l_{\text{eff}})/s_2, \quad B_1 \simeq 2.8 \quad (18.31)$$

$$s_2 = (4 l_{\text{eff}} s_1 / 3)^{1/2}$$

with the other quantities as in Equation 18.25.

Equation 18.30 is a rough approximation, which is reasonably valid over the range of stripe widths between $0.5 l_{\text{eff}}$ and $5 l_{\text{eff}}$, and when the carrier transport properties do not change between the region under the stripe and the region outside. It provides a useful means of analysing the relevant experimental results and of sorting out the contributions of the various parameters, particularly when backed up by other independent experimental evidence.

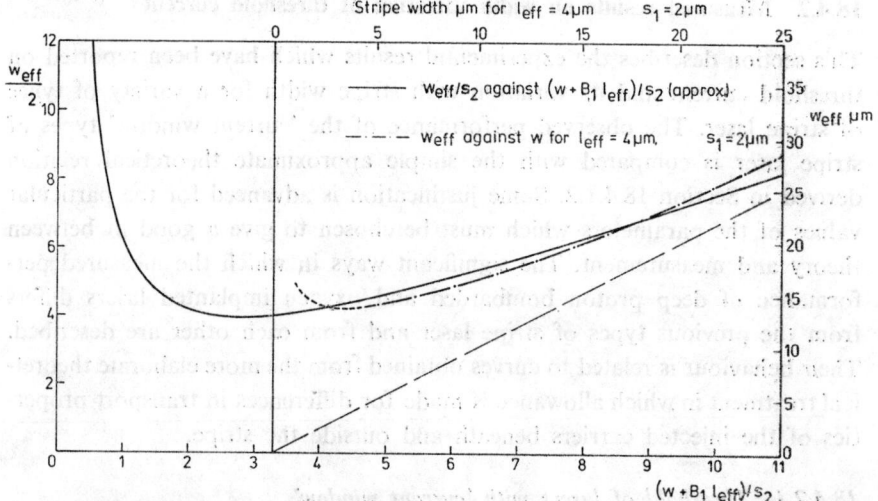

Figure 18.10 Universal approximate theoretical curve giving increase in threshold current versus width for stripe lasers with uniform properties within and outside stripe. Plotted in terms of effective width w_{eff}. More exact specific curve also given for comparison

Figure 18.10 gives a plot of the normalized effective stripe width w_{eff}/s_2 in Equation 18.30 against the quantity y. To relate y more directly to the width of the stripe we need to know the value of the diffusion length l_{eff}. With this information a simple construction can be used on the plot to obtain a direct relation between the actual and effective stripe width. The vertical axis, together with the scale on the horizontal axis, is shifted to the right so that the two axes intersect at the point $y = B_1 l_{\text{eff}}/s_2$. This provides a direct plot of w_{eff}/s_2 against w/s_2. Absolute scales are indicated as an example on the figure for w_{eff} against w with $l_{eff} = 4$ μm and $s_1 = 2$ μm. Also illustrated is the exact curve which applies for these values as calculated using Equations 18.26, 18.27 and 18.7. It can be seen that the approximate curve represents the exact curve reasonably well. It should also be noted that in the region where the curves diverge at narrow stripe widths, the 'exact' curve is no longer reliable because the optical filament spreads appreciably outside the stripe.

In general the approximate curve, in conjunction with the shift of axis to allow for the diffusion length, provides a reasonably universal relation between e_{eff} and w for any pair of values of the parameters s_2 and l_{eff} which is satisfactory over the range of validity for w/l_{eff} quoted above.

18.4.2 Measured results on width variation of threshold current

This section describes the experimental results which have been reported on threshold current and its variation with stripe width for a variety of types of stripe laser. The observed performance of the 'current window' types of stripe laser is compared with the simple approximate theoretical relation derived in Section 18.4.1.2. Some justification is advanced for the particular values of the parameters which must be chosen to give a good fit between theory and measurement. The significant ways in which the measured performance of deep proton bombarded and oxygen implanted lasers differs from the previous types of stripe laser and from each other are described. Their behaviour is related to curves obtained from the more elaborate theoretical treatment in which allowance is made for differences in transport properties of the injected carriers beneath and outside the stripe.

18.4.2.1 *Threshold of lasers with 'current window'*

The results of threshold current versus width measurements on 'current window' types of stripe lasers are given in Figure 18.11. This shows a plot of w_{eff} against w for a variety of stripes reported by various workers. These include the internally striped s diffused lasers of Takusagawa et al. [6], the MBE embedded stripe lasers of Lee and Cho (25), the planar stripe lasers of Yonezu et al. (5), the internally striped lasers of Burnham and Scrifes (8), and the shallow proton bombarded stripe lasers of Dyment et al. (26) and Rossi et al. (27), each type comprising devices of various lengths. It is interesting to note the similarity that exists in the performance of a large range of devices, and particularly of the three sets of devices that give the least increase in threshold current density compared with broad contact lasers.

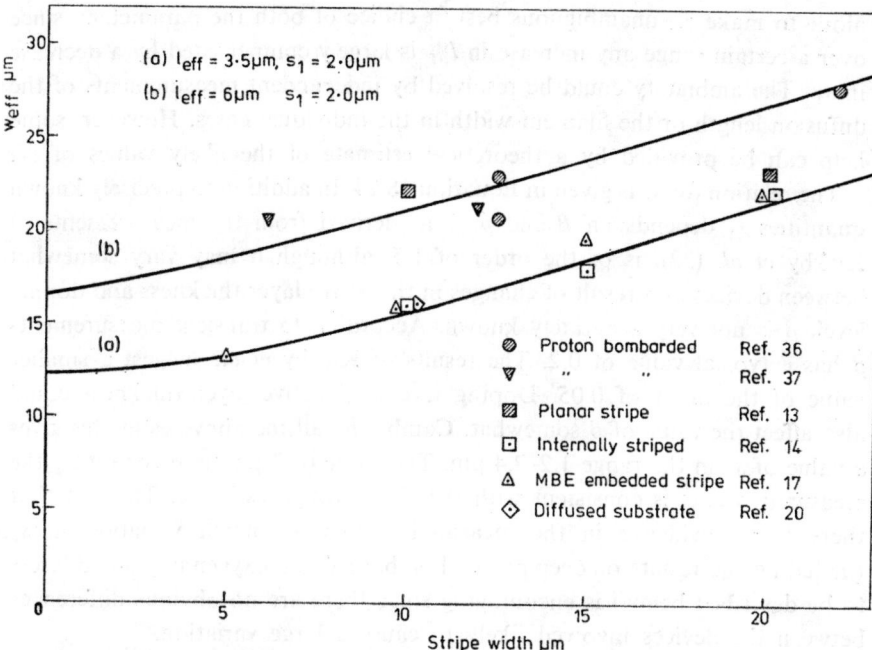

Figure 18.11 Observed value of effective width w_{eff} (derived from increase in threshold current density) for various types of stripe laser with 'current window', plotted against stripe width. Theoretical curves for $s_1 = 2$ μm, and diffusion lengths of 3.5 and 6 μm, also given

The similarity in the latter instance applies despite a considerable variation between different devices in the thickness and doping level of the active layer (0.18–0.4 μm in the thickness of the active layer and doping between 10^{17} cm^{-3} n-type and 5×10^{17} cm^{-3} p-type). The common factor between the best results is probably that in these lasers the passive layers contribute little extra current spreading beyond that already occurring in the active layer (including diffusion) by virtue of the low doping of the relevant layers and their total thickness of 1 μm or less.

We also show two theoretical curves in Figure 18.11 based on the approximate Equation 18.30 which give a good fit with the measured results using appropriate values of the parameters l_{eff} and s_1. The theoretical curves (a) and (b) apply respectively for effective diffusion lengths of 3.5 μm and 6 μm and for a value of the characteristic gain guide width s_1 of 2 μm in both curves. The fit to the measured points is excellent for curve (a) and satisfactory for curve (b). However, it is not possible on the basis of the measured points alone to make an unambiguous best-fit choice of both the parameters, since over a certain range any increase in l_{eff} is largely compensated by a decrease in s_1. The ambiguity could be resolved by independent measurements of the diffusion length or the filament width in the individual cases. However, some help can be provided by a theoretical estimate of the likely values of s_1.

The relation for s_1 is given in Equation 18.24. In addition to precisely known quantities s_1 depends on B and β. B as derived from the measurements of Kirkby et al. (22), is of the order of 1.5, although it may vary somewhat between devices as a result of changes in the active layer thickness and doping level. β is not very accurately known. According to transient measurements β has a typical value of 0.2. The results of Kirkby et al. suggest a smaller value of the order of 0.05. Doping level and active layer thickness could also affect the value of β somewhat. Combining all the above estimates gives a value of s_1 in the range 1.2–2.4 µm. The value of 2 µm used for fitting the measured results is consistent with the above rough estimate. The fact that there is no evidence in the measured results for much variation in s_1 (including the results on deep proton bombarded and oxygen implanted lasers to be described below) is encouraging since there are no obvious differences between the devices involved likely to cause a large variation.

Although s_1 can be expected to vary somewhat between lasers of different length and the contribution to s_1 from B and β can be expected to be somewhat dependent on the thickness and doping level of the active layer, it seems reasonable to conclude from these results that the predominant cause of variation in w_{eff}/w in actual devices is variation in the effective diffusion length l_{eff}. Variations in the thickness and doping level of the p-passive layer probably constitute the major factors controlling l_{eff} although variation in the doping level of the active layer when it reaches a level comparable with the concentration of minority carriers injected at threshold, must also be taken into consideration.

18.4.2.2 Threshold of lasers with deep semi-insulating region

We will now deal with stripe lasers in which there is a change in the properties of the semiconductor across the boundary edge of the stripe. First let us consider the observed performance characteristics of these lasers. Figure 18.12 shows the measured values of w_{eff} versus w for deep proton bombarded lasers as obtained by Dyment et al. (26) and Rossi et al. (27) and for deep oxygen implanted lasers obtained by Blum et al. (10). Measurements on shallow proton bombarded lasers by the first two sets of workers are also included for comparison. The results show that the deep bombardment and

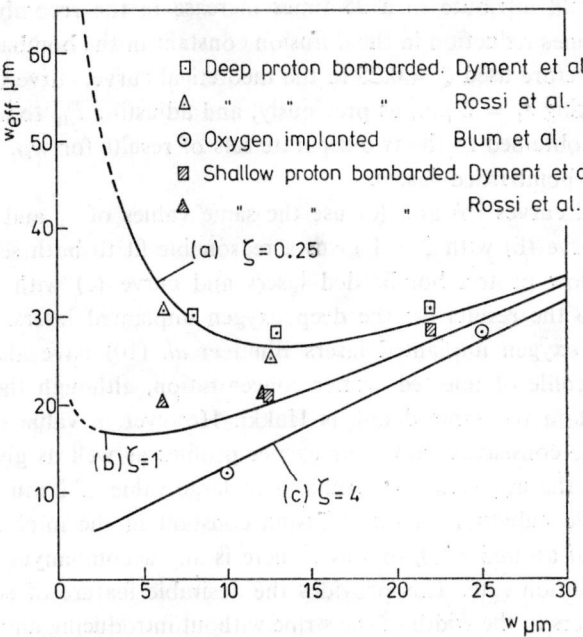

Figure 18.12 Theoretical relation between effective width w_{eff} (derived from increase of current density) and actual width of stripe lasers with various values of ζ and with $l' = 2$. Measured results for both deep and shallow proton bombarded and oxygen implanted lasers also shown, taking $s_1 = 2$ μm and diffusion length = 5 μm

deep implantation produce lasers with significantly different performance from each other as well as from the control group. Deep proton bombardment causes an appreciable increase in w_{eff} and in threshold current, which becomes rapidly greater as the stripe width is reduced below 10 μm. In contrast the deep oxygen implantation causes an appreciable reduction in w_{eff} below that of the control group, also particularly for widths less than 10 μm. It therefore seems that this form of fabrication would be particularly suitable for narrow stripes.

Theoretical curves of w_{eff} against w are also given in Figure 18.12. Values of ζ considerably differing from unity are used to obtain these curves, which require an exact solution of Equations 18.23, 18.26 and 18.27. By examination of the transverse distribution of injected carriers in deep proton bombarded lasers Hakki deduced a value of ζ of around 0.25 for these lasers which he could attribute to a 25 times increase in the recombination rate and an 0.7 times reduction in the diffusion constant in the bombarded region. We have therefore used $\zeta = 0.25$ in the theoretical curve, curve (a) in Figure 18.12. By taking $s_1 = 2$ μm, as previously, and adjusting l_{eff} to around 5 μm a good fit is obtained to the two separate sets of results for w_{eff} against w in deep proton bombarded lasers.

Theoretical curves (b) and (c) use the same values of s_1 and l_{eff} while ζ is varied. Curve (b) with $\zeta = 1$ gives a reasonable fit to both sets of results for the shallow proton bombarded lasers and curve (c) with $\zeta = 4$ gives a good fit to the results for the deep oxygen implanted lasers. In the case of the deep oxygen implanted lasers Blum et al. (10) have also measured the lateral profile of injected carrier concentration, although they have not interpreted it in the same detail as Hakki. However, a value of around 4 appears to be consistent with their carrier profiles as well as giving a satisfactory fit to the w_{eff} versus w results. This large value of ζ must arise from a considerable reduction in the diffusion constant in the implanted region, by a factor of around 1/20, or less if there is any accompanying increase in the recombination rate. This provides the desirable feature of confining the injected carriers to the width of the stripe without introducing any appreciable recombination at the edges.

REFERENCES

1. J. C. Dyment, *Appl. Phys. Lett.*, **10**, 84 (1967).
2. J. E. Ripper, J. C. Dyment, L. A. D'Asaro, and T. L. Poole, *Appl. Phys. Lett.*, **18**, 155 (1971).
3. T. Tsukada, R. Ito, H. Nakashima, and O. Nakada, *IEEE J. Quant. Electron.*, **QE-9**, 356 (1973).
4. S. Iida and Y. Watanabe, *Japan J. Appl. Phys.*, **13**, 1249 (1974).
5. H. Yonezu, I. Sakuma, K. Kobayashi, T. Kamejima, M. Ueno, and Y. Nannichi, *Japan J. Appl. Phys.*, **12**, 1585 (1973).
6. M. Takusagawa, H. Nishi, S. Osaka, M. Morimoto, H. Imai, H. Takanashi, and T. Misugi, *Proceedings of Electron Devices Conference*, Washington 1975.
7. K. Itoh, M. Inoue, and I. Teramoto, *IEEE J. Quant. Electron.*, **QE-11**, 421 (1975).
8. R. D. Burnham, D. R. Scifres, J. C. Tramontana, and A. S. Alimonda, *IEEE J. Quant. Electron.*, **QE-11**, 418 (1975).
9. J. C. Dyment, L. A. D'Asaro, J. C. North, B. I. Miller, and J. E. Ripper, *Proc. IEEE (Lett.)*, **60**, 726 (1972).
10. J. M. Blum et al., *IEEE J. Quant. Electron.*, **QE-11**, 413 (1975).
11. H. Namizaki, H. Kan, M. Ishii, and A. Ito, *J. Appl. Phys.*, **45**, 2785 (1974).
12. H. Namizaki, *IEEE J. Quant. Electron.*, **QE-11**, 427 (1975).
13. H. Yonezu et al., *Japan J. Appl. Phys.*, **16**, 209 (1977).
14. G. H. B. Thompson, D. F. Lovelace, and S. E. H. Turley, To be published.

15. W. Susaki, T. Tanaka, H. Kan, and M. Ishii, *IEEE J. Quant. Electron.*, **QE-13**, 587 (1977).
16. T. Tsukada, *J. Appl. Phys.*, **45**, p. 4899 (1974).
17. L. A. D'Asaro, *J. Luminescence*, **7**, 310 (1973).
18. W. P. Dumke, *Solid-State Electron.*, **16**, 1279 (1973).
19. B. W. Hakki, *J. Appl. Phys.*, **44**, 5021 (1973).
20. A. K. Jonscher and M. H. Boyle, *Proceedings of IPPS Symposium on GaAs*, Reading (1966), p. 78.
21. G. H. B. Thompson, *Opto-electronics*, **4**, 257, (1972).
22. P. A. Kirkby, A. R. Goodwin, G. H. B. Thompson, and P. R. Selway, *IEEE J. Quant. Electron.*, **QE-13**, 705, (1977).
23. D. D. Cook and F. R. Nash, *J. Appl. Phys.*, **46**, 1660, (1975).
24. B. W. Hakki, *J. Appl. Phys.*, **46**, 2723, (1975).
25. T. P. Lee and A. Y. Cho, *Appl. Phys. Lett.*, **29**, 164, (1976).
26. J. C. Dyment, C. J. Hwang, and A. R. Hartman, unpublished results quoted in Reference 19.
27. J. Rossi and J. P. Donelly, Unpublished 1976.

Chapter 19

The Possibilities of Influencing the Spectral Behaviour of Injection Lasers

H. Bachert

19.1 INTRODUCTION

When in the early 60's many teams investigated the first injection lasers, they found rather poor properties: it was necessary to cool the lasers at liquid nitrogen temperatures, continuous operation was rarely observed, the near and far field distributions were inhomogeneous and so was the spectral behaviour. The aim of this chapter is to discuss the ways of influencing the spectral properties of injection lasers. Inbuilt methods, such as high doping p-active regions, special two-dimensional waveguides, DFB or DBR-structures and others will not be touched on because of the multiplicity of technological problems involved—a subject requiring separate treatment. We concentrate on methods closely connected with external systems which we want to distinguish in composite cavity resonators, coupled resonators, and resonators where the laser only acts as an active medium without its own resonance behaviour. The aim of this kind of investigations is to construct an injection laser with properties similar to those of gaseous lasers. Another field is to investigate the special properties of the gain and loss mechanism in injection lasers, the kind of line broadening, etc., i.e. application of the above-mentioned techniques to investigate lasers themselves.

We will show that it is possible to operate injection lasers at least with spectral properties given essentially only by an external system with coherence lengths of metres, wavelengths adjustable by about 200 Å with powers up to watts if they are excited with high current pulses. We applied these methods to GaAs–AlGaAs single heterostructure lasers but it is known that other materials act in a very similar way.

19.2 GAIN AND LOSS IN INJECTION LASERS

The emission properties of injection lasers are considerably affected by the spectral gain distribution and its dependence on the pump current density. If we consider the threshold condition for a Fabry–Perot laser

$$R_1 R_2 \exp(g - \alpha_1) 2l = 1 \tag{19.1}$$

we observe laser action for a net gain $g - \alpha_1 \geq 0$, where g is the gain per unit length, α_1 represents internal losses of the medium between the two mirrors with reflectivities R_1 and R_2, and l is the distance between the mirrors.

Obviously, changes of the gain and losses related to the spectral emission energy $h\nu$ strongly influence this threshold condition.

At first, let us consider the dependence between gain and emission energy $g \equiv g(h\nu)$, the gain profile. Following Lasher and Stern (1), and Thompson and Popov (2) we can write

$$g(h\nu) = \frac{B}{h\nu} \int_{-\infty}^{+\infty} \varrho_c(E_c) \varrho_v(E_v) [f_c(E_c) - f_v(E_v)] dE_c \tag{19.2}$$

where B is a constant containing the matrix element for the transition probability, $E_v = E_c - h\nu$, with E_v and E_c being the valence and conduction band energies, respectively, $f_c(E_c)$ and $f_v(E_v)$ are the electron occupation probabilities for the conduction and valence bands, and $\varrho_c(E_c)$, $\varrho_v(E_v)$ are the corresponding densities of states. General solutions of (19.2) are complicated but simple cases can be calculated. If we consider the band tail at the conduction band to be exponential with the quasi Fermi level E_{fc} located within it and the levels at the valence band to be separated from E_v by E_{fc}, we can find a solution (2)

$$g(E) = C \exp(-E/E_0) \tanh(E/2kT) \tag{19.3}$$

where E is taken from the energy $E_{fc} - E_{fv}$, g is zero by definition from the laser condition $h\nu < E_{fc} - E_{fv}$, E_0 is the energy where the density of states in the conduction band tail is lowered to $1/e$ and C is considered to be a constant depending on the transition probability and the doping levels. Figure 19.1(a) shows the variation of g with the emission energy. This very simple expression cannot fit all types of injection lasers (Homo-, SH- and DH-structures) or doping concentrations, but gives a sufficient impression of the principal dependence.

Rewriting (19.1) in

$$j \sim g = \alpha_1 + \frac{1}{l} \ln(R_1 R_2)^{1/2} \tag{19.4}$$

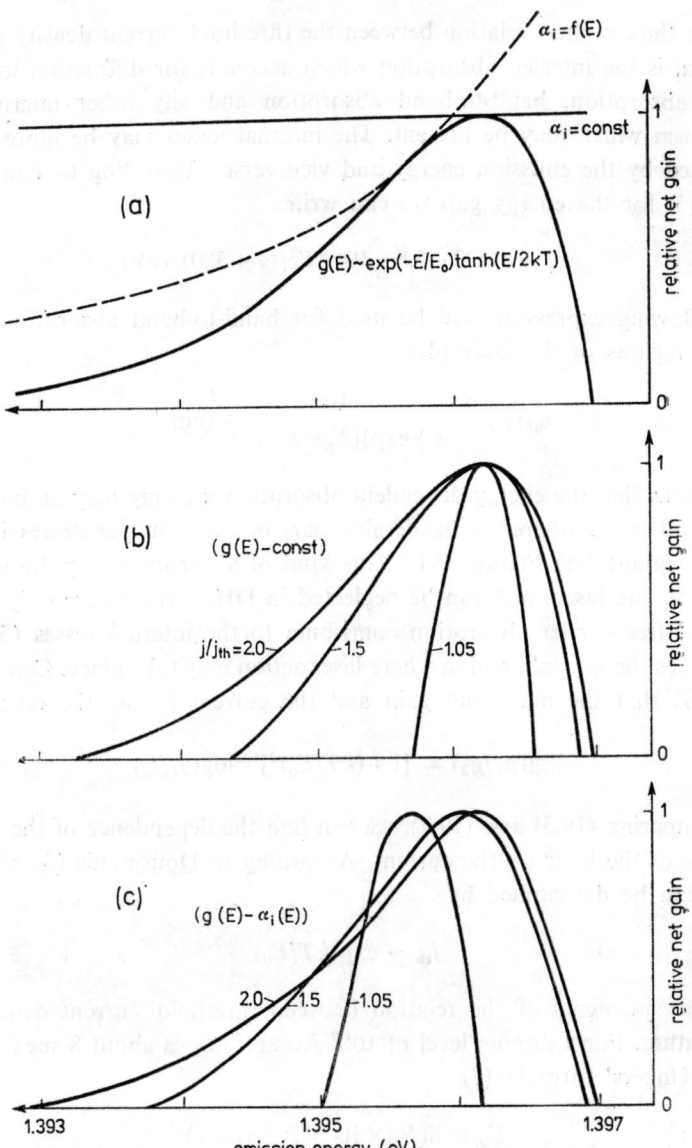

Figure 19.1 (a) Variation of gain as a function of emission energy from Equation 19.3 with energy dependent and independent losses, respectively; (b) normalized net gain as a function of emission energy for constant losses; (c) normalized net gain as a function of emission energy for energy dependent losses. The parameter for (b) and (c) is the current overdrive factor i/i_{th}

we have the common relation between the threshold current density and the losses. α_i is the internal absorption which accounts for diffraction loss, free carrier absorption, band-to-band absorption and any other internal loss mechanism which may be present. The internal losses may be more or less influenced by the emission energy and vice versa. According to Panish and Casey (3) for the energy gap we can write

$$E_g = 1.522 - 5.8 \times 10^{-4} T^2/(T+300) \text{ (eV)} \tag{19.5}$$

The following expression will be used for band-to-band absorption in the passive regions of the laser (4):

$$\alpha_{bb}(E) = \frac{10^4}{1+\exp[(E_g-E)/E_0]} \text{ (cm}^{-1}) \tag{19.6}$$

This means that the energy dependent absorption α_{bb} may play an important role in all lasers where a considerable part of the light penetrates into the passive regions (see Figure 19.1). This kind of absorption may dominate in homostructure lasers and can be neglected in DH-lasers, where only diffraction and free carrier absorption contribute to the internal losses (5). $g - \alpha_i$ determines the net gain region where laser action may take place. Calculations show (7) that the maximum gain and the current follow the relationship

$$\log(g_1/g_2) = [1+(kT/E_0)^2]^{1/2} \log(j_1/j_2) \tag{19.7}$$

and comparing (19.3) and (19.7), we can find the dependence of the spectral net gain of the laser on the current. According to Dousmanis (6), the value of E_0 can be determined by

$$j_{th} \sim \exp(kT/E_0) \tag{19.8}$$

from measurements of the relation between threshold current density and temperature. For a doping level of 10^{18} Atcm^{-3}, E_0 is about 8 meV according to Ungers' formula (7)

$$E_0 \approx 8(N_D \times 10^{-18})^{1/2} \text{ (meV)} \tag{19.9}$$

It should be noted that E_0 in DH-lasers does not represent the bandtail parameter but an effective experimental value. Figure 19.1(b) shows the net gain of injection lasers as a function of emission energy for different current levels using energy independent losses. This assumption does not seem to be valid especially in homojunction and SH-lasers, because a considerable part of the guided light meets passive regions near the absorption edge and the total losses should be energy-dependent (8, 9). Constant and energy dependent losses have been qualitatively shown in Figure 19.1(a) and the influence on the net gain can be seen in Figures 19.1(b) and (c) for calculations with and without a considerable influence of energy dependent losses. Recently published results of Sommers (5) show that for LOC-lasers there is almost no influence of the emitted wavelength on the internal losses such as is expected in the case of DH-lasers. Our own results on homojunction and SH-lasers indicate such an influence.

The width of the gain profile seems to explain why a junction laser emits a rather broad spectrum. Figure 19.2 shows an emission spectrum of an

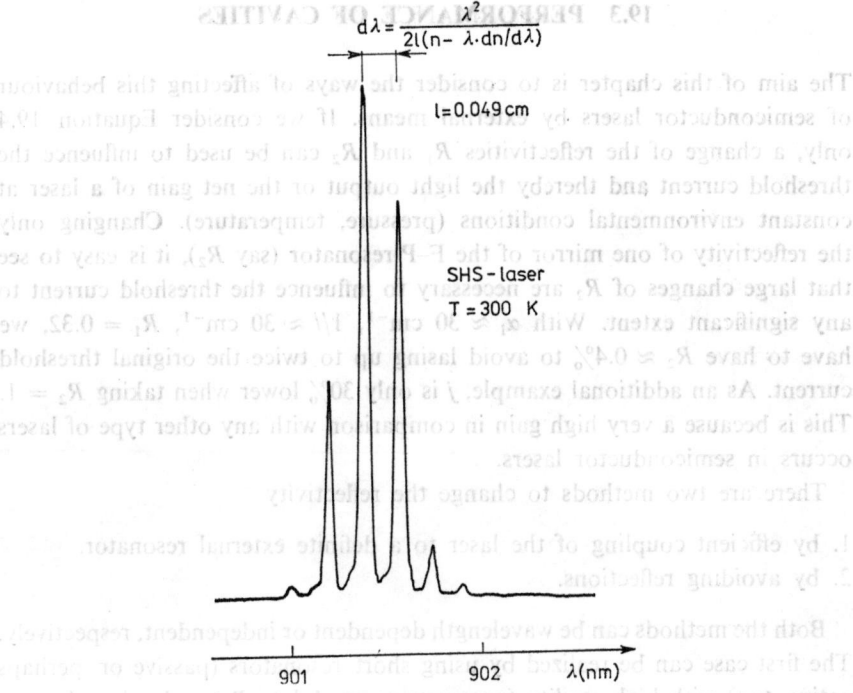

Figure 19.2 Emission spectrum with longitudinal modes separated as given in Equation 19.10

injection laser. It is easy to see that the emitted spectrum covers only a tenth of the calculated net gain even at high excitation levels. This means that only a limited number of lines can be emitted simultaneously. Otherwise one would predict a homogeneous broadened line in a semiconductor laser, because the lifetime of excited carriers is lower in the case of stimulated emission than in spontaneous emission. Homogeneous line broadening would favour only one small emission line. The observed coexistence of a number of lines (or modes) can be explained by assuming that inhomogeneous and homogeneous line broadening are both generally present in semiconductor lasers. For more information about this fact see for instance Wright et al. (10). It should be mentioned, however, that there are many of additional explanations for multimode operation.

19.3 PERFORMANCE OF CAVITIES

The aim of this chapter is to consider the ways of affecting this behaviour of semiconductor lasers by external means. If we consider Equation 19.4 only, a change of the reflectivities R_1 and R_2 can be used to influence the threshold current and thereby the light output or the net gain of a laser at constant environmental conditions (pressure, temperature). Changing only the reflectivity of one mirror of the F-P-resonator (say R_2), it is easy to see that large changes of R_2 are necessary to influence the threshold current to any significant extent. With $\alpha_i \approx 30$ cm^{-1}, $1/l \approx 30$ cm^{-1}, $R_1 = 0.32$, we have to have $R_2 \approx 0.4\%$ to avoid lasing up to twice the original threshold current. As an additional example, j is only 30% lower when taking $R_2 = 1$. This is because a very high gain in comparison with any other type of lasers occurs in semiconductor lasers.

There are two methods to change the reflectivity

1. by efficient coupling of the laser to a definite external resonator,
2. by avoiding reflections.

Both the methods can be wavelength dependent or independent, respectively. The first case can be realized by using short resonators (passive or, perhaps active, too) with high quality (necessary to modulate R_2) and a length comparable to the laser resonator, so that mode spacings of both resonators

$$d\lambda = \frac{\lambda^2}{2l(n - \lambda_c \, dn/d\lambda)} \qquad (19.10)$$

are of the same order of magnitude. The dispersion term is of the order of -1.5 for GaAs and can be taken to be 0 for most materials used as passive resonators. The reflectivity has maxima at a length $l = k\lambda/2$ of the passive resonator, where k is an integer. Practical aspects are given in the experimental chapters.

The second case can be realized by antireflection coatings (AR) or polishing an oblique angle between 10° and 15° (the maximum is Brewster's angle $\alpha_B = \arctan(l/n) = 15.5°$ for GaAs).

The minimum reflectivity is found for AR-coatings with a refractive index of $n = (n_{GaAs})^{1/2} \approx 1.9$ and a layer thickness of $\lambda/4$. The most often used AR-coating is made by evaporating SiO ($n \approx 1.9$), but sometimes multilayers are also applied. The aim of such coatings is to increase the laser threshold or the gain of the laser without exciting internal longitudinal modes so hat it is possible to change the effective R_2' by external means. This leads to wavelength dependent losses:

$$\alpha(\lambda) = \alpha_i + \frac{1}{l} \ln[R_1 R_2'(\lambda)]^{-1/2} \qquad (19.11)$$

where $R_2' = R_2 + R_{2\text{ext}}(\lambda)$ with R_2 being the reflectivity of the coated side of the laser (as an optimum near to 0) and $R_{2\text{ext}}(\lambda)$—a part consisting of complicated functions containing the optical quality of the external system, the yield of the optical feedback and the wavelength dispersive properties.

Obviously, the relation between R_2 and $R_{2\text{ext}}$ determines the properties of the whole system.

19.3.1 Decrease of the internal reflectivity R_2

The procedure used most frequently to minimize R_2 is evaporation of SiO in high vacuum. An exact thickness control is most essential because of the sensitivity of j_{th} on R_2. Using Equation 19.7 and $\alpha_i = 30 \text{ cm}^{-1}$, $1/l = 30 \text{ cm}^{-1}$, $E_0 = 8$ meV, $T = 300$ K, we find a change of the gain compared to that of the uncoated laser as given in Figure 19.3. Having a threshold gain for the uncoated laser of about 65 cm^{-1} the small change of R_2 from 0.1 to 0.2% lowers the possible gain by about 230 cm^{-1}! It is therefore necessary to control the reflectivity R_2 very carefully during deposition. The methods used are:

1. Thickness control of $d = (\lambda/4) \cdot n_{SiO}$. It often gives poor results, because the required accuracy is about 30 Å if we want to have a minimum value of R_2 of about 0.003 (see for instance experimental and theoretical results in References 11–13).

2. Sommers (5) evaporated SiO deposited to a thickness exceeding $\lambda/4$ and thinned by successive etchings in a 10% solution of HF. The effect was controlled by using the expression (14)

$$\frac{P_2}{P_1} = \left(\frac{R_1}{R_2}\right)^{1/2} \left(\frac{1-R_2}{1-R_1}\right) \qquad (19.12)$$

where P_1 and P_2 are the light intensities on both sides of the operated laser. Yamamoto (11) used similar effects when measuring the luminescence intensity during evaporation.

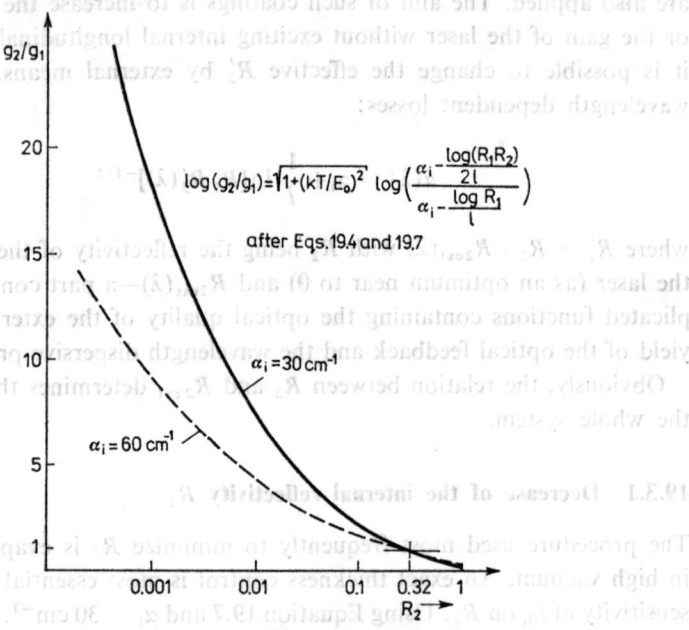

Figure 19.3 Relative gain at threshold $g(R_2)/g$ ($R_2 = 0.32$) versus R_2 with internal losses α_i as a parameter

3. During evaporation the threshold current can be measured and the deposition must be immediately stopped when the maximum is reached. We operate the laser at twice its threshold before evaporation and stop the process when lasing does not occur.

None of the three kinds of thickness or reflectivity control can prevent an altering of the layer as it comes from the ambient during laser investigations, such as deposition of vapours (especially in vacuum at liquid nitrogen temperatures), degradation when operating the laser (power densities of up to 10^7 Wcm^{-2} can be reached) or the simple influence of temperature.

In 1967 Mohn et al. (15) showed how to avoid reflections on one side of the laser-cavity. They polished the lasers at an angle near Brewster's giving zero reflection of the vertically polarized light. The lasers were operated in an external cavity equipped with spherical mirror. Mohn et al. were able to show that the mode spacing of the long resonator determines the spectral properties of the system. The results we obtained with such lasers for various kinds of external resonators are shown in the course of this chapter. It should be noted that it is neither necessary nor convenient to meet Brewster's angle, because a much smaller angle (10–12° has proved to be optimal) also avoids the reflection of light back into the active layer. Of course, the efficiency of external systems (expressed by R_2' in (19.11)) will be lower than in cavities with AR-coated lasers. Two factors could be responsible:

1. Far from Brewster's angle and (or) from completely vertical polarization of the light the normal part of about 30% of the emitted light is reflected and thus lost for the external system.
2. The experimental set-up is more complicated in that the laser and the external system have different axes causing difficulties with alignment and again with the part of light which can feed back with the laser which has a divergence of normally 30–50° perpendicular to the junction.

Both factors imply that a great part of the light with inconvenient polarization will be refracted and cannot contribute to R_2'. Otherwise we found an increase of the threshold up to three times that of the normal laser indicating a very small R_2 (which defines now only a fraction of the light reflected in % of the active layer) of the order of 10^{-4} and therefore permitting an extremely high gain which can more than compensate for the higher losses of the external system. At least it should be noticed that this system is very insensitive to ambient influences but degradation of mirror 2 during experiments can lead to a back scattering in per cents of the active layer thus lowering the threshold of the laser again.

19.3.2 Various combinations of lasers and external cavities

Figure 19.4 shows a number of typical combinations of lasers and resonators. They can be divided into three groups:

1. *Resonators consisting of uncoated lasers combined with passive resonators of comparable length* (Figure 19.4(a))

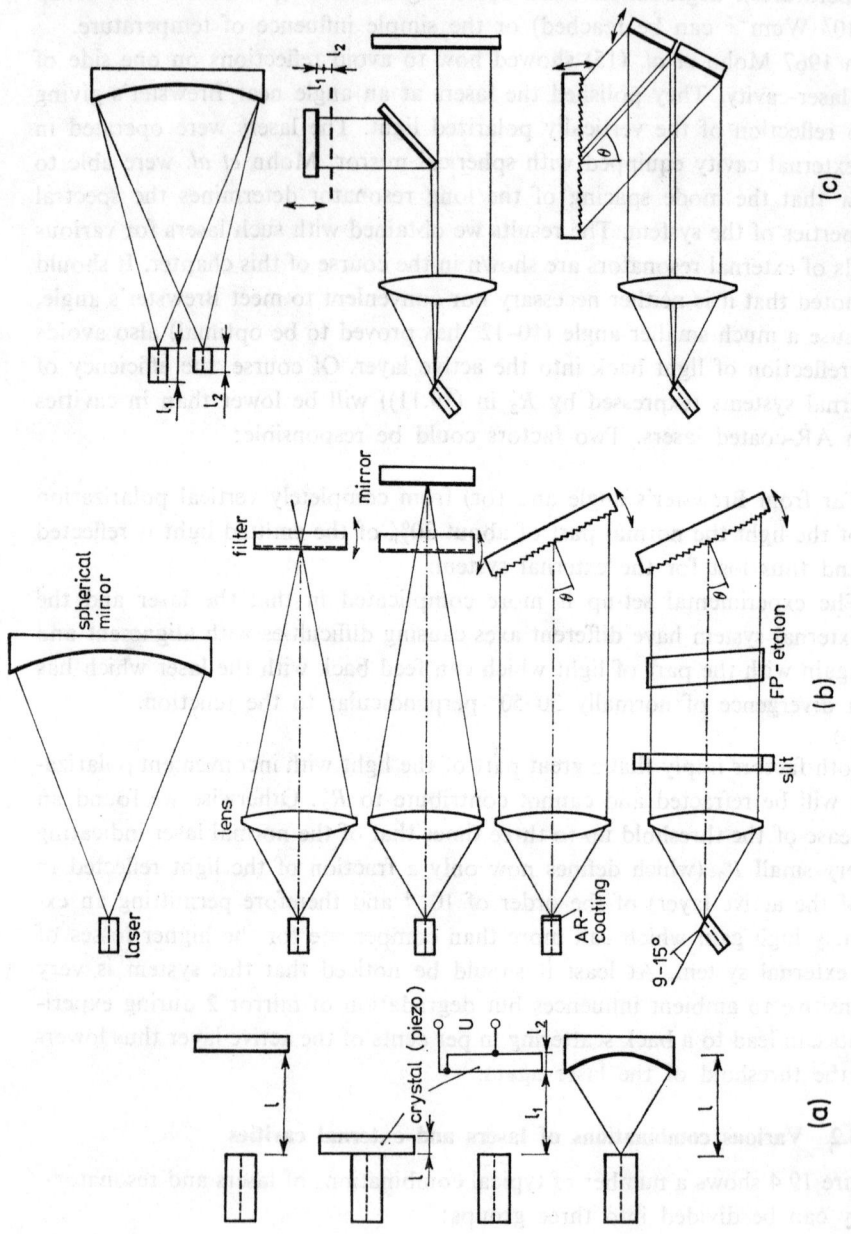

Figure 19.4 Various combinations of lasers and external cavities (a), (b), (c)—group 1, group 2 and group 3 cavities, respectively (see text)

A passive resonator can be made by adjusting a plane mirror parallel to the laser endface or a glass plate (a material of higher refractive index is better) directly stuck to one end of the laser. The quality of these closely coupled resonators determines the spectral properties of the system. Short spherical mirrors belong to the same group as well as various modifications of composed passive resonators which can be tuned by changing their lengths, e.g., using piezo effects ($LiNbO_3$ should be mentioned as an example for a transparent material). Such composite cavity injection lasers have been discussed in a number of publications which appeared between 1969 and 1972 (16–19). We will show some experimental results in Section 19.4. The main disadvantage of group 1 systems is the low possible current and therefore power range in which they can be operated efficiently because the change of R_2 given by the external system is too small to suppress the original laser properties in a wider excitation range. It must be mentioned, however, that there are possibilities to improve these systems (20).

2. In the second group (Figure 19.4(b)) we summarize *systems with resonators having a length much larger than the laser cavity itself* to avoid interactions between modes of the external and internal cavities. This is to be expected if the length of the external resonator gives a mode distance comparable to the spectral width of a normal internal mode (in pulse regime not less than 0.01 nm). The above-mentioned condition is fulfilled by cavities longer than 5 cm. Group 2 systems should again be divided into cavities where R_2 lies near R_1, and those where R_2 is much less than R_1. Initial experiments have been made with uncoated lasers and a non-selective light reflecting system (the upper in Figure 19.4(b)). The other arrangements given show different wavelength selective properties and should be classified as dispersive cavities. This group will be more extensively dealt with in Section 19.4.

3. This group is a combination of group 1 and group 2 systems as shown in Figure 19.4(c). It includes the incorporation of coupled lasers with similar and different spectral properties. The results will also be given in Section 19.4.

19.3.3 Optical losses of external systems

In order to construct an efficient system of group 2 or 3 we have to determine its main components concerning the losses as given by Equation 19.11. The cavities consist of three components:

1. a laser or active medium with a reflectivity R_1 at one side and R_2 at the side of the external system,

2. a more or less efficient imaging system (lens or spherical mirror),
3. a dispersive or non-dispersive reflecting system.

Inside this cavity additional selective elements can be positioned, such as interferometers, gases (for instance in intracavity spectroscopy), beam splitters, etc. Following a very informative paper of Bogatov et al. (21) we find for $R'_2(\lambda)$:

$$R'_2(\lambda) = R_2 + B^* \frac{(1-R_2)^2}{\int f_2(x,y) \, dx \, dy} \int f(-x,-y) \varphi(x,y,\lambda) \, dx \, dy \quad (19.13)$$

where $f(x, y)$ determines the intensity distribution of the emitted light of the laser at the laser mirror 2 (the x-direction lies in the plane of this mirror transverse to the active region and y is perpendicular to that direction), $\varphi(x, y, \lambda)$ gives the intensity distribution reflected back to the emitting surface, and B^* is the part of the emitted light which can reach the lens of the external system. The minus signs in $f(-x, -y)$ are related to the reflection, because it turns the direction of the electric field. Assuming an intensity distribution which is constant in the y-direction (this should be valid for broad contact lasers with $w \geq 50$ μm), we find simplified equations

$$f(x, y) \to f(x) = \exp(-x^2/\sigma) \quad (19.14)$$

with $\sqrt{\sigma}$ the halfwidth of light intensity on the emitting surface in the x-direction which can be approximated by the active layer thickness if it amounts to more than about 1 μm, and

$$\varphi(x,y,\lambda) \to \frac{R_3 T^2}{\sqrt{\pi \sigma_1}} \iint f(x) f(x') \exp\left[-\frac{x - x' - \frac{dx}{d\lambda}(\lambda - \lambda_0)^2}{\sigma}\right] dx \, dx' \quad (19.15)$$

where R_3 is the reflectivity of the dispersive or non-dispersive reflector at wavelength λ, T is the transmission of the imaging system including all media inside the external resonator, $dx/d\lambda$ is the linear dispersion of the external resonator, λ_0 is the central wavelength of the resonator, σ_1 is the diameter of the reflected image of the active region, and x' represents the x-axis for the reflected beam. Using (19.14) and (19.15), we can approximate $R'_2(\lambda)$ by

$$R'_2(\lambda) = AB^*(1-R_2)^2 R_3 T^2 \sqrt{\frac{2\sigma}{2\sigma + \sigma_1}} \exp\left[-\frac{(dx/d\lambda)^2(\lambda-\lambda_0)^2}{2\sigma + \sigma_1}\right] \quad (19.16)$$

with a factor A near 1 resulting from the small mistake which has been made by assuming $\varphi(x, y, \lambda) \to \varphi(x, \lambda)$. Then the linewidth of a laser which is operated in an external cavity will be given by

$$\Delta\lambda = \sqrt{\frac{2\sigma+\sigma_1}{(\mathrm{d}x/\mathrm{d}\lambda)^2}} \tag{19.17}$$

where σ_1 includes the quality of the imaging system used with

$$\sigma_1 \approx (\lambda/S)^2 \tag{19.18}$$

(S the aperture of the optics). This means that a dispersive system will allow a minimum linewidth if the aperture is large. Now let us present some conclusions valid for most type 2 and 3 resonators:

1. The lasers should have thick active regions, because otherwise $\sigma_1 \gg \sigma$ and R'_2 will be small.
2. The reflectivity R_2 has to be made as low as possible (see Section 19.3.1) in order to give a high R'_2/R_2 ratio and thus to avoid an influence of the internal cavity (if not desired).
3. The emission must be homogeneous (near field) for allowing A in (19.15) to be near 1.
4. The angular distribution of the emitted light should be small and consist of the zero order transverse mode producing a far field pattern that can be collected with reasonable efficiency by an optics of higher aperture thus giving a large B^* in (19.16).
5. The transmission of the optical system has to be high. Lenses should be antireflection coated for the wavelength of the laser. Many camera lenses seem to have coatings only for the visible range. In general, specially made lenses and some microscope lenses are far more efficient.

19.3.4 Adjustment of external resonators

Composite cavity resonators (group 1) are rather easy to adjust considering the fact that plane mirrors are exactly parallel to the laser mirror or that the axes of curved mirrors correspond to that of the laser, respectively. The main problem is to control the length of the resonators if dielectric materials directly connected with the laser are not used. All adjusting elements have to consist of materials with a negligible expansion coefficient (such as invar). Otherwise the ambient temperature can be used for exact adjustment. A simple calculation shows that a change of the resonator length of 0.1 μm gives 1 nm wavelength change ($\mathrm{d}\lambda = \lambda \mathrm{d}l/l$) for one longitudinal mode. It is most profitable to observe the spectrum and not the intensity on the backside of the laser because the overall efficiency of a composite cavity changes only a little in comparison with that of the uncoupled laser.

The following routine is generally used to adjust the various types of large cavity external resonators (groups 2 and 3). The main tool in the experiments is an infrared image converter (or a TV camera) positioned at the opposite end of the laser. A He–Ne-laser sighting beam can be used to define and align the optical axes of the components but with some experience this is not necessary.

1. A diaphragm on the lens or near the laser can be used to find the reflected beam, if the laser is excited.
2. After the diaphragm is removed the image at the converter shows the silhouette of the laser. Of course, the laser mounting must allow undisturbed observation of both sides. If the axis of the monitor coincides with that of the laser, it is possible to adjust the lens also in that axis. Eventually step 1 must be repeated.
3. The imaging system (lens or spherical mirror) is now shifted towards or away from the laser up to a position where the silhouette of the laser completely covers the screen of the image converter. Following that the lens is adjusted transverse to the active region of the laser and normally the coupling between the laser and external cavity takes place.
4. Exact adjusting is achieved if the lens is so positioned that the threshold current of the laser-resonator system shows a minimum. The threshold can be measured at the screen of the image converter, when the far field is observed. Optimum coupling can also be determined by measuring the power output.
5. Then the emitted light is focused at the entrance slit of a high resolution spectrograph which permits further control of the system and investigations. For high resolution measurements a Fabry–Perot interferometer can also be used.

Instead of the simple image converter also an infrared microscope can be used showing the exact position of the reflected beam transverse to and in the plane of the active region. For visible lasers converters are naturally not necessary.

19.4 EXPERIMENTAL RESULTS

Most experiments have been done with a view to achieving single mode (or, preferably, small line) operation from injection lasers, thus improving the spectral properties of that type of lasers. The comparatively broad gain

19.4.1 Composite cavity lasers

Figure 19.5(a) shows the influence of a short external resonator on the losses of injection laser with a given gain (17). The conditions for proper adjustment have been given in 19.3.4. In order to achieve single mode operation up to high excitation levels it is favourable to use high quality passive resonators giving a halfwidth of the loss minima less than $2\Delta h\nu_1$. The length

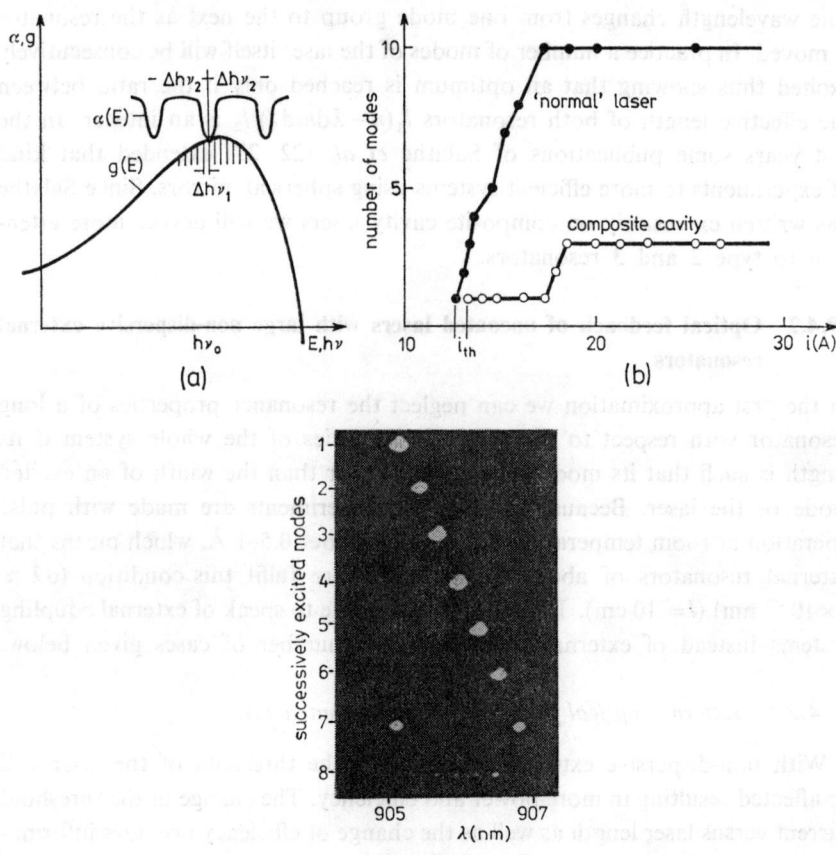

Figure 19.5 The effect of composite cavity injection lasers on (a) the gain and loss relation; (b) the number of excited modes, and (c) the spectra when changing the length of the passive resonator (laser length 0.0335 cm, $T = 300$ K)

l_2 should be small enough for only one peak of the net gain $g(E)-\alpha(E)$ to give laser action. With a SiC platelet, Eliseev et al. (16) found single mode operation up to 50% above threshold. Figure 19.5(b) shows the different number of excited modes of a normal and a composite cavity laser as a function of current level obtained in such experiments. More extensive experiments with plane adjustable resonators have shown that the efficiency of such systems is rather low. The threshold current can only be influenced by about 5% and therefore the power in a single mode will not be very high. Figure 19.5(c) shows some experimental results obtained with an adjustable air resonator consisting of one laser endface and a silvered sliding mirror. The wavelength changes from one mode group to the next as the resonator is moved. In practice a number of modes of the laser itself will be consecutively excited thus showing that an optimum is reached only if the ratio between the effective length of both resonators $l_1(n-\lambda dn/d\lambda)/l_2$ is an integer. In the last years some publications of Salathe et al. (22, 23) extended that kind of experiments to more efficient systems using spherical mirrors. Since Salathe has written extensively on composite cavity lasers we will devote more attention to type 2 and 3 resonators.

19.4.2 Optical feedback of uncoated lasers with large non-dispersive external resonators

In the first approximation we can neglect the resonance properties of a long resonator with respect to the spectral properties of the whole system if its length is such that its mode distance is smaller than the width of an excited mode of the laser. Because most of the experiments are made with pulse operation at room temperature this value is about 0.5–1 Å, which means that external resonators of about 10 cm and more fulfil this condition ($d\lambda \approx 4 \times 10^{-3}$ nm) ($l=10$ cm). It thus seems possible to speak of external coupling systems instead of external resonators in a number of cases given below.

19.4.2.1 External optical feedback of an uncoated laser

With non-dispersive external cavities only the threshold of the laser will be affected resulting in more power and efficiency. The change of the threshold current versus laser length as well as the change of efficiency provides information about the performance of the external system. In view of the poor knowledge of the properties of the lenses often used in geometries for which they were not intended this creates an excellent opportunity for determining empirically the best combinations of lenses, mirrors and other components. By

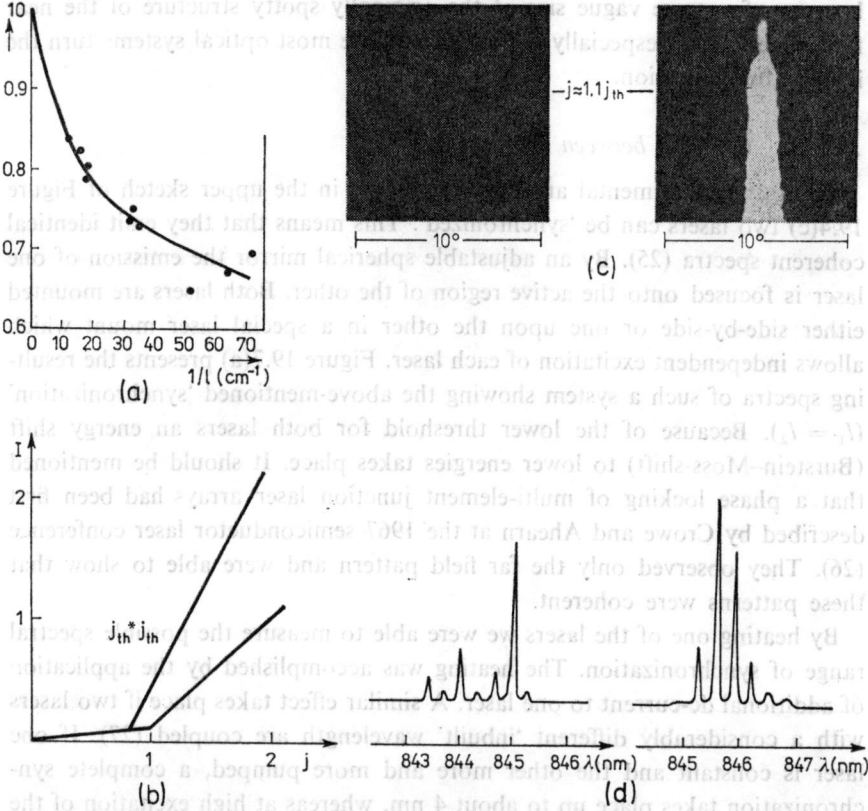

Figure 19.6 Influence of an external system on the properties of injection lasers. (a) decrease of threshold current as a function of the reciprocal length of the laser; (b) change of differential quantum efficiency; (c) improvement of the far field in the plane of the active region; (d) improvement of the spectra

measuring the power emitted from the opposite side of the laser (Equation 19.11) it is possible to determine an effective R_2' if one assumes $\varphi(x, y, \lambda)$ to be constant (see Equation 19.13). Figure 19.6 shows a number of results which normally can be observed by coupling a laser with an external cavity. Figure 19.6(a) gives the change of the threshold current density as a function of reciprocal laser length (9, 24), Figure 19.6(b) shows the increase of power efficiency, 19.6(c)—the change in the far field distribution, and 19.6(d) the change of the laser spectrum. Figures 19.6(c) and (d) show that improvement of the spatial and spectral properties of the laser often takes place as a result of a considerably more homogeneous field distribution in the case of coupling because of a more vague size of the originally spotty structure of the near field. The latter is especially important because most optical systems turn the image after reflection.

19.4.2.2 Coupling between different lasers

Using the experimental arrangement shown in the upper sketch of Figure 19.4(c) two lasers can be 'synchronized'. This means that they emit identical coherent spectra (25). By an adjustable spherical mirror the emission of one laser is focused onto the active region of the other. Both lasers are mounted either side-by-side or one upon the other in a special laser mount which allows independent excitation of each laser. Figure 19.7(a) presents the resulting spectra of such a system showing the above-mentioned 'synchronization' ($l_1 = l_2$). Because of the lower threshold for both lasers an energy shift (Burstein–Moss-shift) to lower energies takes place. It should be mentioned that a phase locking of multi-element junction laser arrays had been first described by Crowe and Ahearn at the 1967 semiconductor laser conference (26). They observed only the far field pattern and were able to show that these patterns were coherent.

By heating one of the lasers we were able to measure the possible spectral range of synchronization. The heating was accomplished by the application of additional dc-current to one laser. A similar effect takes place if two lasers with a considerably different 'inbuilt' wavelength are coupled (27). If one laser is constant and the other more and more pumped, a complete synchronization takes place up to about 4 nm, whereas at high excitation of the second laser a part of the original spectrum remains. Many other experiments in this area have been made, especially with a view to investigating amplification characteristics (see Reference 28) or the change of threshold current when pumping one laser with more or less light from another.

Now let us consider the coupling between lasers of different length. If we assume no influence of the long connecting cavity, it is possible to define the distance between longitudinal modes which can oscillate in both lasers. This case is nearly identical to that of composite cavity lasers. We find a difference of the selected modes

$$\Delta\lambda_{\text{sel}} = \frac{\lambda^2}{2\Delta l(n - \lambda \, \mathrm{d}n/\mathrm{d}\lambda)} \tag{19.19}$$

where $\Delta l = |l_1 - l_2|$. The ratio $l_{1,2}/\Delta l$ should be an integer if optimal selection is desired. Figure 19.7(b) shows the interaction of two lasers with $l_1 = 440$ μm and $l_2 = 340$ μm which results in a spectrum with a mode distance

Influencing the Spectral Behaviour of Injection Lasers 547

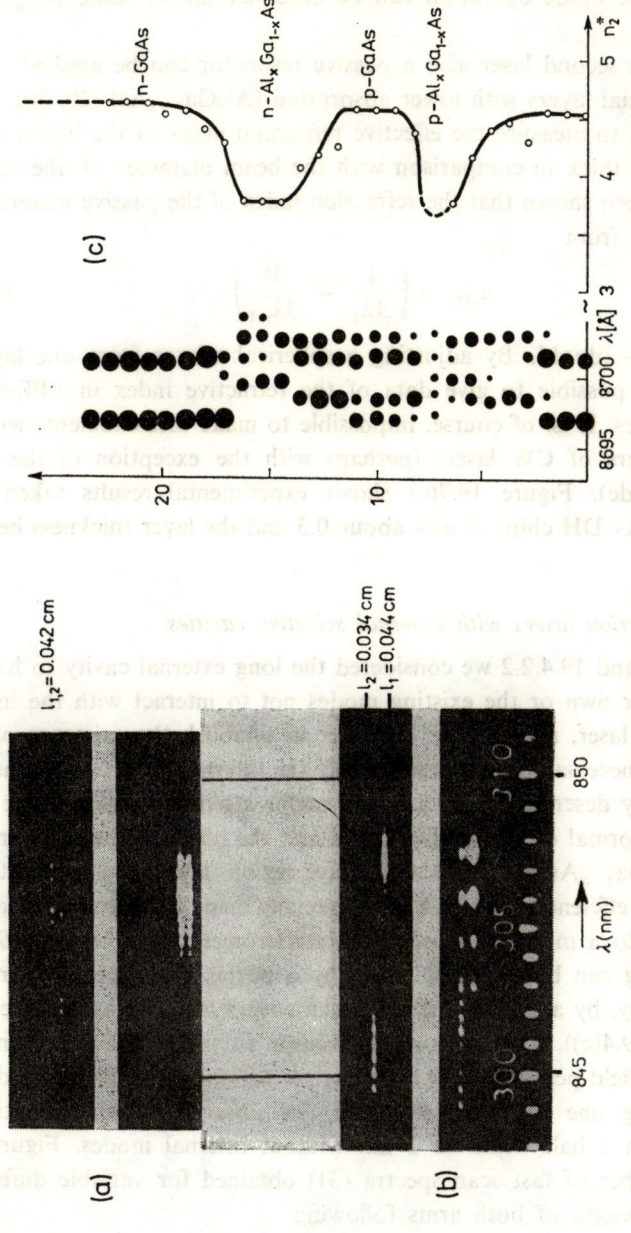

Figure 19.7 (a) Spectral 'synchronisation' of two uncoated injection lasers of equal length; (b) selection of longitudinal modes by a passive resonator shown on the right-hand side if two injection lasers of different length are coupled; (c) selection of longitudinal modes by a passive resonator. The measured change of the effective index of refraction is shown on the right-hand side $Al_xGa_{1-x}As$-layers.

of 0.75 nm in accordance with Equation 19.19. Using two lasers with very small Δl, single mode operation can be observed up to some 10% above threshold.

Instead of a second laser also a passive resonator can be applied. For it we used epitaxial layers with lower absorption ($Al_xGa_{1-x}As$) (29, 30). There it was possible to measure the effective refraction index of the layers if they were relatively thick in comparison with the beam diameter of the focused laser. It has been shown that the refraction index of the passive material can be determined from

$$n_{eff_2} = \left(\frac{1}{\Delta\lambda_1} - \frac{1}{\Delta\lambda_{sel}}\right) \quad (19.20)$$

with $n_{eff} = n - \lambda dn/d\lambda$. By adjusting a spherical mirror from one layer to the next it is possible to give data of the refractive index in LPE-double heterostructures. It is, of course, impossible to make measurements with the very thin layers of CW lasers (perhaps with the exception of the active layer waveguide). Figure 19.7(c) shows experimental results taken from an $Al_xGa_{1-x}As$ DH chip; X was about 0.3 and the layer thickness between 3 and 5 µm.

19.4.2.3 Injection lasers with external selective cavities

In 19.4.2.1 and 19.4.2.2 we considered the long external cavity to have no modes of their own or the existing modes not to interact with the internal modes of the laser, respectively. In order to establish the existence of such modes it was necessary to suppress largely the internal modes. This has been done in a way described in 19.3.1. We prefer lasers with an oblique angle between the normal of one polished face and the plane of the active region. We used $Al_xGa_{1-x}As$ SH-lasers with active regions about 2 µm thick to give high coupling effciences (see 19.3.3). A specially made $f/1.3$ lens sends a collimated beam to a modified Michelson interferometer (see Figure 19.4(c)). Beam splitting can be managed either by a normal halfsilvered mirror or, more efficiently, by a silvered mirror which covers only the half of the beam (see Figure 19.4(c)). This is possible because in the zero order transverse mode the far field perpendicular to the active region is due mainly to diffraction. Blocking one interferometer arm, we observe a nearly continuous spectrum with a half width of 2 nm without internal modes. Figure 19.8 shows a number of fast scan spectra (31) obtained for variable differences between the length of both arms following

$$\Delta\lambda_{sel} = \lambda^2/2\Delta l. \quad (19.21)$$

Influencing the Spectral Behaviour of Injection Lasers

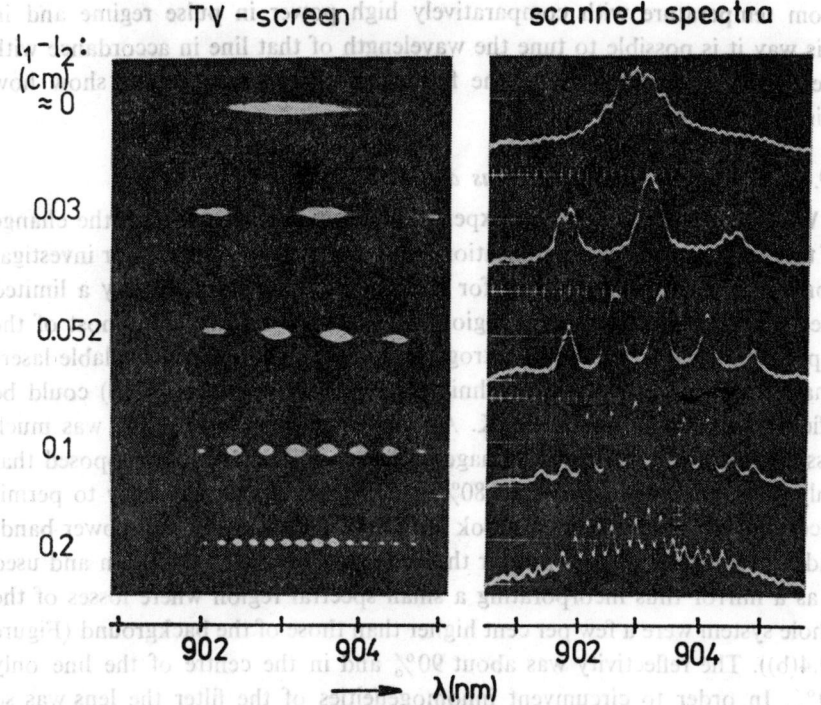

Figure 19.8 Spectra obtained by coupling an injection laser with a modified Michelson interferometer

The nominal length of one arm was 50 cm. This beating between two independent resonators 50 cm long shows that modes of the external resonator do exist. It is thus clear that any change of the spectral properties in an external cavity is reproduced in the laser. The width of a selected mode in comparison with the distance between modes again depends on the quality of the whole resonator and therefore on the efficiencies of lenses, mirrors, splitters, etc. Additional selective means incorporated in a long resonator (such as F–P interferometers) could produce very small lines.

19.4.3 The application of dispersive external cavities

Several experiments performed up to 1969 have shown that semiconductor asers can be efficiently influenced by external means as described in 19.4.1 and 19.4.2. Evaluations of the effective reflection coefficients and the influence of dispersive elements have been given in Equation 19.16. Thus, dispersive external cavities should allow the laser to emit a small spectral line also at room temperature with comparatively high power in pulse regime and in this way it is possible to tune the wavelength of that line in accordance with the broad gain distribution. The following experimental results show how this task has been solved.

19.4.3.1 Use of optical filters as dispersive elements

When in 1969 a number of experiments were carried out with the change of threshold current, synchronization between two lasers and similar investigations, it was necessary to look for a medium which reflected only a limited spectral range into the active region of the lasers. At that time most of the experiments had to use liquid nitrogen cryostates, because the available lasers (made by the Zn^- diffusion technique or by homostructure LPE) could be efficiently operated only at 80 K. At room temperature coupling was much less efficient and difficult to manage, although obtainable. We supposed that only reflectivities of more than 80% would give enough efficiency to permit successful experiments. So we took a dielectric filter with a half-power bandwidth of approximately 1 nm for the desired wavelength of 850 nm and used it as a mirror thus incorporating a small spectral region where losses of the whole system were a few per cent higher than those of the background (Figure 19.4(b)). The reflectivity was about 90% and in the centre of the line only 30%. In order to circumvent inhomogeneities of the filter the lens was so adjusted that the beam was focused to its surface. (This kind of adjustments is the most stable to couple injection lasers with an external system.) The result of this experiment was surprising: The spectrum changed very strongly in two groups of modes (first published in 1970 (32), see also (33)). This behaviour is illustrated in Figure 19.9(a). The next step was to move the filter a few millimeters towards the lens thus avoiding coupling with the light reflected from the background. The transmitted light was then focused onto an additional plane mirror. According to Equation 19.15 the transmission T_f (with our filters about 0.7) determines the spectral behaviour of the system. We find for a nearly Gaussian transmission

$$T_f = 0.7\exp\left[-4\ln2 \cdot \left(\frac{\lambda - \lambda_0}{\Delta\lambda}\right)^2\right]$$

with $\Delta\lambda = 1$ nm a halfwidth for the twice transmitted part of the light: $\Delta\lambda_f = 0.7$ nm. In order to measure the single mode operation it was necessary to make lasers with a mode distance in the same order. We prepared lasers with $l_1 = 0.012$ cm giving $\Delta\lambda_1 = 0.625$ nm. Although the effective reflectivity R_2 was certainly below 0.40 in comparison with the 0.32 of the uncoupled

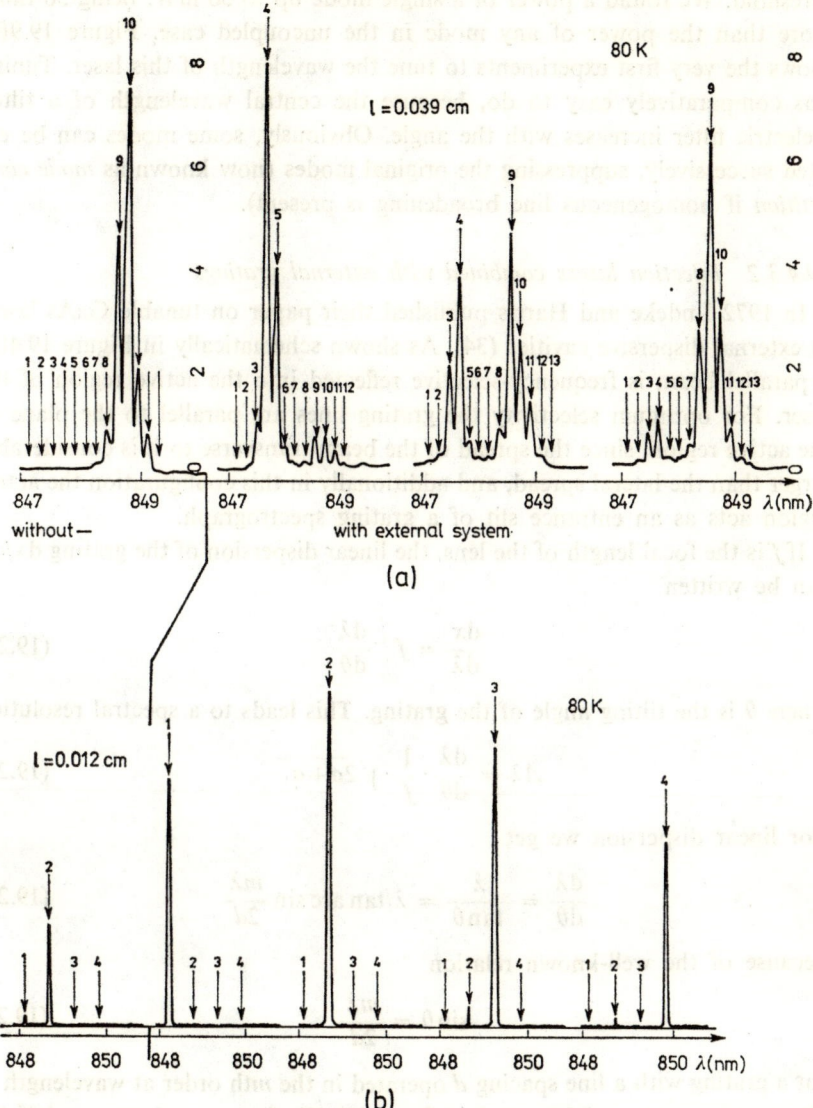

Figure 19.9 Use of a transmission filter to influence the spectra of an injection laser: (a) the filter is used as a selective mirror; (b) tuning of a single mode by a combination of transmission filter and mirror

laser, it was possible to find single mode operation up to 10% above threshold. We found a power of a single mode up to 50 mW, being 30 times more than the power of any mode in the uncoupled case. Figure 19.9(b) shows the very first experiments to tune the wavelength of this laser. Tuning was comparatively easy to do, because the central wavelength of a tilted dielectric filter increases with the angle. Obviously, some modes can be excited successively, suppressing the original modes (now known as *mode competition* if homogeneous line broadening is present).

19.4.3.2 Injection lasers combined with external gratings

In 1972 Ludeke and Harris published their paper on tunable GaAs lasers in external dispersive cavities (34). As shown schematically in Figure 19.4(b) a parallel beam is frequency-selective reflected into the active region of the laser. For optimum selectivity the grating lines are parallel to the plane of the active region, since the spread of the beam transverse to it is considerably larger than the lateral spread, and additionally in this configuration the active region acts as an entrance slit of a grating spectrograph.

If f is the focal length of the lens, the linear dispersion of the grating $dx/d\lambda$ can be written

$$\frac{dx}{d\lambda} = f \left/ \frac{d\lambda}{d\theta} \right. \tag{19.22}$$

where θ is the tilting angle of the grating. This leads to a spectral resolution

$$\Delta\lambda = \frac{d\lambda}{d\theta} \cdot \frac{1}{f} \cdot \sqrt{2\sigma + \sigma_1} \tag{19.23}$$

For linear dispersion we get

$$\frac{d\lambda}{d\theta} = \frac{\lambda}{\tan\theta} = \lambda/\tan\arcsin\frac{m\lambda}{2d} \tag{19.24}$$

because of the well-known relation

$$\sin\theta = \frac{m\lambda}{2d} \tag{19.25}$$

for a grating with a line spacing d operated in the mth order at wavelength λ. For our lenses ($s \approx 0.75$) we find from (19.18) that $\sigma_1 \approx 1$ μm, and if the active layer thickness σ is also 1 μm, we find

$$\Delta\lambda \approx \sqrt{3} \cdot \frac{\lambda}{fS\tan\arcsin\frac{m\lambda}{2d}} \tag{19.26}$$

Influencing the Spectral Behaviour of Injection Lasers 553

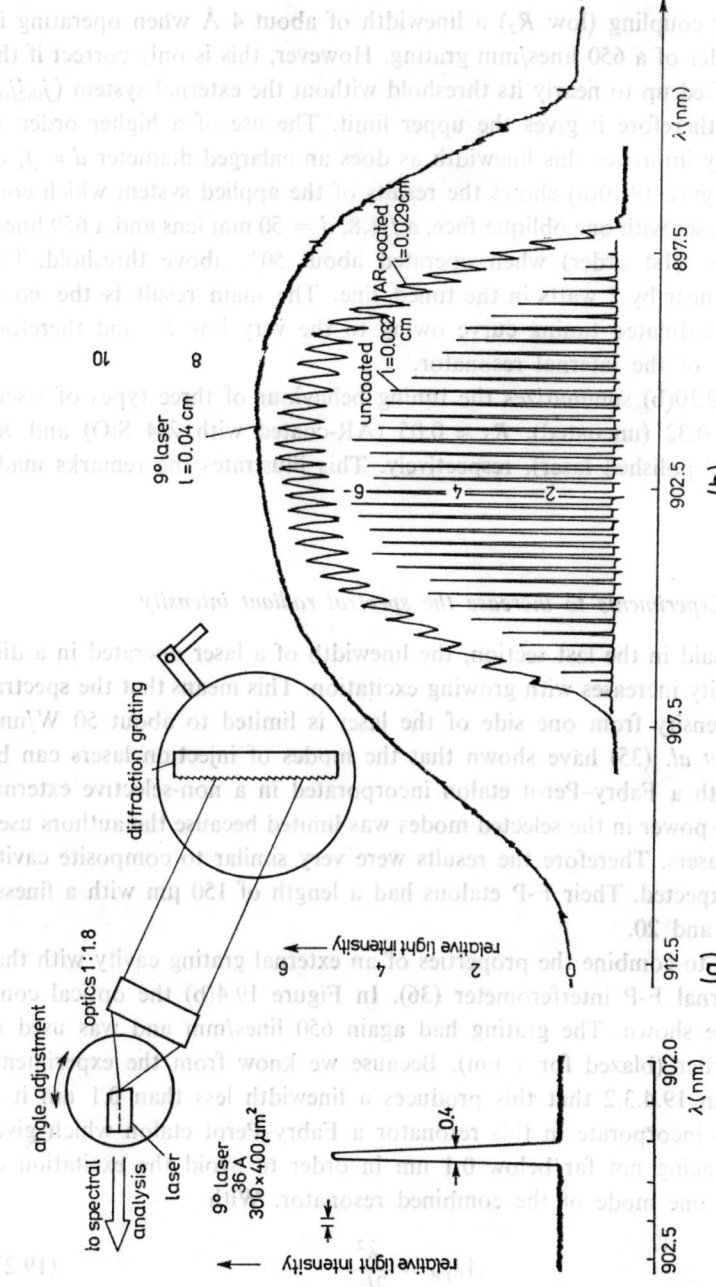

Figure 19.10 Intensity of a tuned line as a function of emission wavelength: (a) experimental setup; (b) tuning curves for uncoated, AR-coated and polished lasers. The lasers have been operated below their uncoupled threshold

where f is the focal length and S is the aperture of the lens. Thus we expect for efficient coupling (low R_2) a linewidth of about 4 Å when operating in the first order of a 650 lines/mm grating. However, this is only correct if the laser is excited up to nearly its threshold without the external system ($j_{th}/j_{th_0} \approx 2$), and therefore it gives the upper limit. The use of a higher order m considerably improves this linewidth as does an enlarged diameter $d = f_s$ of the lens. Figure 19.10(a) shows the results of the applied system which consisted of a laser with one oblique face, a $f/1.8$, $d = 50$ mm lens and a 650 lines//mm grating (1st order) when operated about 50% above threshold. The power was near by 2 watts in the tuned line. The main result is the completely unmodulated tuning curve owing to the very low R_2 and therefore low quality of the internal resonator.

Figure 19.10(b) summarizes the tuning behaviour of three types of lasers with $R_2 = 0.32$ (uncoated), $R_2 \approx 0.05$ (AR-coated with $\lambda/4$ SiO) and $R_2 < 0.02$ (10° polished laser), respectively. This illustrates the remarks made in 19.3.1.

19.4.3.3 Experiments to increase the spectral radiant intensity

As was said in the last section, the linewidth of a laser operated in a dispersive cavity increases with growing excitation. This means that the spectral radiant intensity from one side of the laser is limited to about 50 W/nm. Voumard et al. (35) have shown that the modes of injection lasers can be selected with a Fabry–Perot etalon incorporated in a non-selective external cavity. The power in the selected modes was limited because the authors used uncoated lasers. Therefore the results were very similar to composite cavity lasers as expected. Their F-P etalons had a length of 150 μm with a finesse between 1 and 20.

We tried to combine the properties of an external grating cavity with that of an external F-P interferometer (36). In Figure 19.4(b) the optical components are shown. The grating had again 650 lines/mm and was used in the first order (blazed for 1 μm). Because we know from the experiments described in 19.4.3.2 that this produces a linewidth less than 0.1 nm it is possible to incorporate in this resonator a Fabry–Perot etalon which gives a mode spacing not far below 0.1 nm in order to avoid the excitation of more than one mode of the combined resonator. With

$$\Delta\lambda_{FP} = \frac{\lambda^2}{2l_{FP}} \qquad (19.27)$$

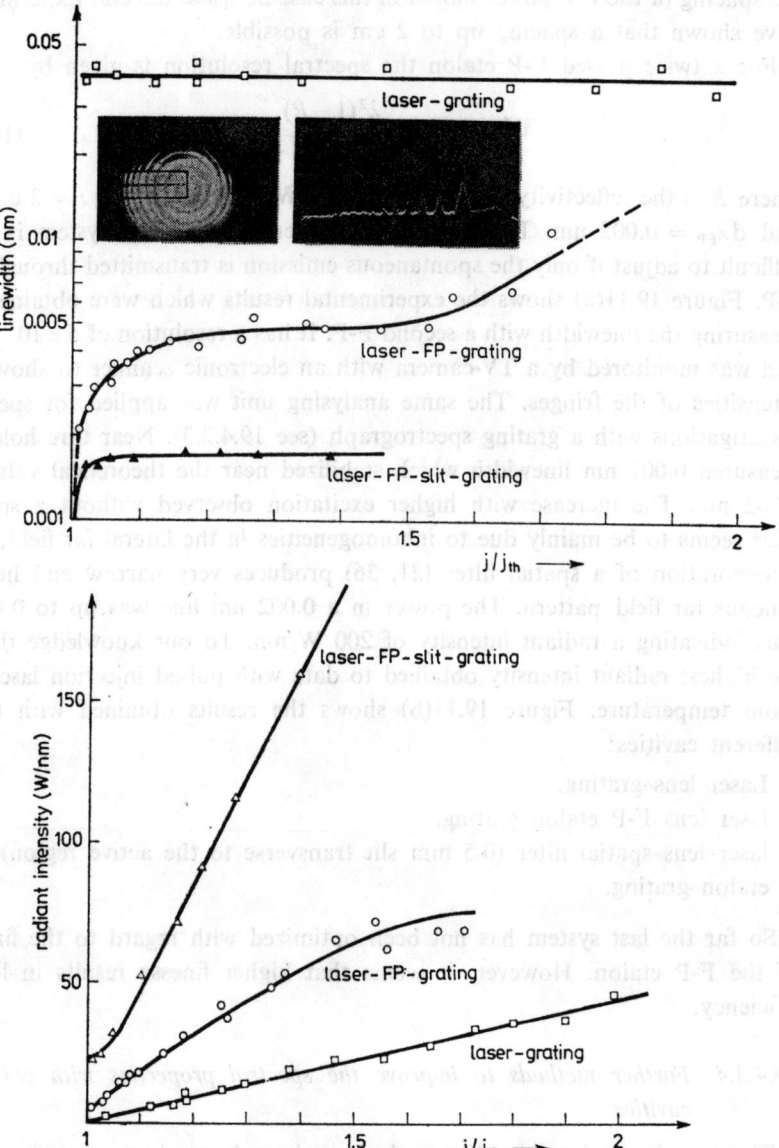

Figure 19.11 Incorporation of a Fabry–Perot etalon in a dispersive external resonator. (a) The halfwidth of the emitted line as a function of current overdrive $(i - i_{th})/i_{th}$; (b) the radiant power density of one line for different cavities versus current overdrive

the spacing of the F-P plates should in this case be $l_{FP} \approx 0.5$ cm. Experiments have shown that a spacing up to 2 cm is possible.

For a twice passed F-P etalon the spectral resolution is given by

$$d\lambda_{FP} = \frac{\lambda^2(1-R)}{2l_{FP}\pi\sqrt{R}\sqrt{1+\sqrt{2}}} \tag{19.28}$$

where R is the reflectivity of the F-P plates. With $R = 0.6$ and $l = 2$ cm we find $d\lambda_{FP} = 0.002$ nm. The low R was chosen, because the system is very difficult to adjust if only the spontaneous emission is transmitted through the F-P. Figure 19.11(a) shows the experimental results which were obtained by measuring the linewidth with a second F-P. It has a resolution of 5×10^{-4} nm and was monitored by a TV-camera with an electronic scanner to show the intensities of the fringes. The same analysing unit was applied for spectral investigations with a grating spectrograph (see 19.4.2.3). Near threshold we measured 0.001 nm linewidth which stabilized near the theoretical value of 0.002 nm. The increase with higher excitation observed without a spatial filter seems to be mainly due to inhomogeneities in the lateral far field. The incorporation of a spatial filter (21, 36) produces very narrow and homogeneous far field pattern. The power in a 0.002 nm line was up to 0.4 W, thus indicating a radiant intensity of 200 W/nm. To our knowledge this is the highest radiant intensity obtained to date with pulsed injection lasers at room temperature. Figure 19.11(b) shows the results obtained with three different cavities:

1. Laser–lens–grating,
2. laser–lens–F-P etalon–grating,
3. laser–lens–spatial filter (0.5 mm slit transverse to the active region)–F-P etalon–grating.

So far the last system has not been optimized with regard to the finesse of the F-P etalon. However, it seems that higher finesse results in lower efficiency.

19.4.3.4 Further methods to improve the spectral properties with external cavities

The experimental results discussed above have been obtained with reflection gratings in first order ($m = 1$). Since for a given focus and aperture of the lens only m determines the linewidth, it is convenient to use high orders. Bogatov et al. (21) calculated $\Delta\lambda = 0.02$ nm for a grating with $d = 10$ µm operating in the twentieth order (see Equation 19.24). Similar values were obtained by Sommers (5) with a 625/mm ($d = 1.6$ μm) grating in the third order. The main problem is to get gratings which are blazed for 0.9 μm in a higher order (reflectivity !).

A feature of transmission gratings (5) is that the lens that projects the beam on the grating also forms the external beam. Owing to its lower reflectivity the coupling efficiency is low but the external efficiency can be very high. Thus a metallized back facet of the laser can be used to improve the efficiency further. However, this configuration can in principle also be used with reflection gratings, because a higher amount of the light will be reflected from the grating in the $m = 0$ order (like a mirror) producing a highly parallel beam.

A third method is based on the coupling of different F-P etalons. One of these should have high finesse and low distance l_{FP} between the etalon plates thus giving a spacing between adjacent modes of more than 5 nm. A 50 μm air F-P gives a $\Delta\lambda_{FP} = 8$ nm for $\lambda = 0.89$ μm (see Equation 19.27). With $R = 0.95$ it follows from (19.26) that $d\lambda_{FP} = 0.07$ nm. A second F-P etalon of larger distance l_{FP} (see 15.4.3.3) could then give a linewidth of less than 10^{-3} nm.

19.4.3.5 Double-grating experiments

The introduction of a beam splitter with a second grating into the collimated beam path of a normal grating cavity allows the laser to oscillate in two different wavelengths depending on the adjustment of the gratings (37, 38). This technique is useful in studying the interaction of spectral modes in the laser. For more efficiency we used two independently adjustable fully silvered mirrors where the beam covers one half of one mirror as described in 19.4.2.3 (see Figure 19.4(c)) and one grating only. Figure 19.12(a) shows the intensity of one line at a fixed wavelength when the second is tuned. The compensation of modes is very strong and depends on its intensity relation. Thus a more intensive mode takes the power from the weaker mode in the range shown. This is always smaller than the normal tuning curve and seems to be very similar to the range of mode competition often observed by coupling a grating with an uncoated laser (10, 39). It indicates a limited range in which modes cannot coexist, and can be explained by a not completely homogeneously broadened line. The results of double grating experiments show a variety of effects. One of them has been described by Bogatov et al. (37) as anomalous interaction of spectral modes due to stimulated scattering of the laser radiation by dynamic inhomogeneities of the electron density. Rossi et al. (38) pointed out that time delay effects (i.e. the time between the onset of the current and light pulses) are also responsible for some dis-

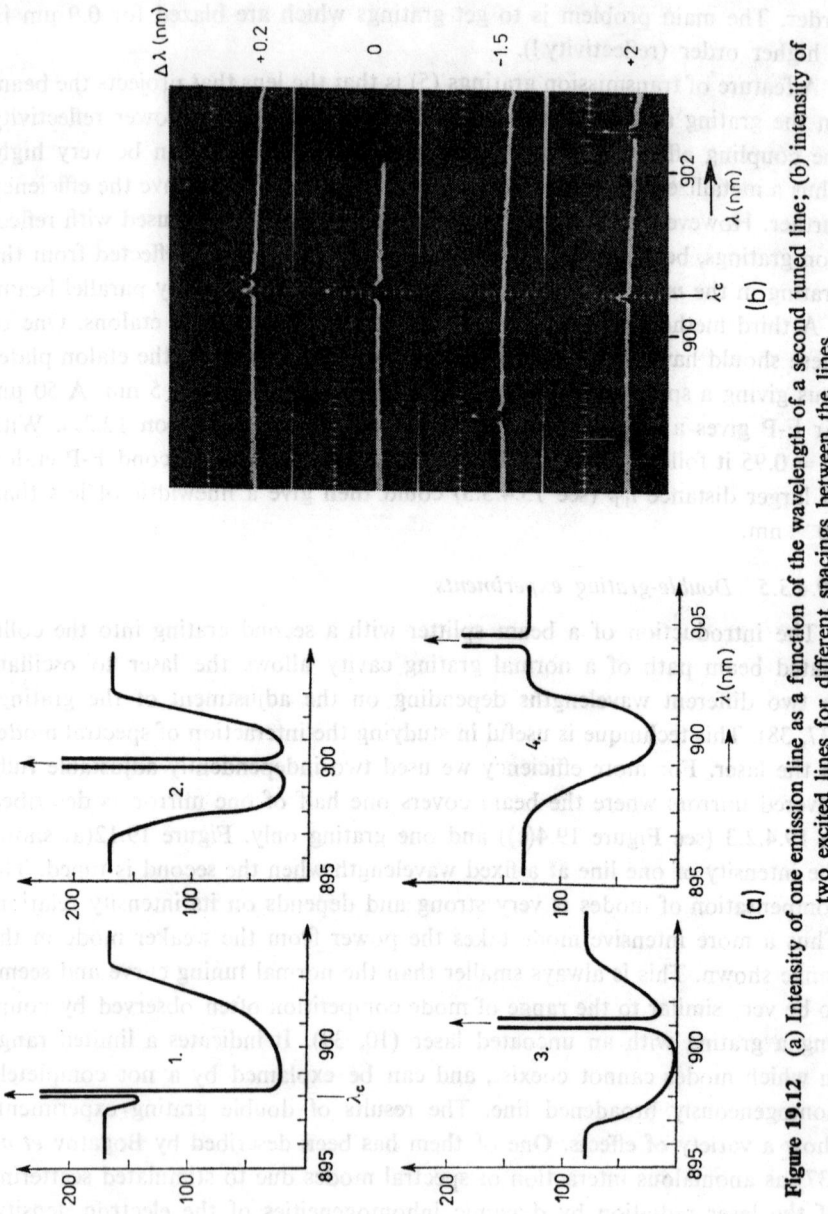

Figure 19.12 (a) Intensity of one emission line as a function of the wavelength of a second tuned line; (b) intensity of two excited lines for different spacings between the lines

continuities in the tuning behaviour, because a high-energy line (with normal delay) can saturate the absorber, thereby allowing a low-energy line to oscillate. The diffraction loss for a lower energy line suppresses the laser effect as a result of a breakdown in light guiding.

Rossi *et al.* (40) have shown that the time delay depends on the lasing wavelength, because a breakdown in guiding should first affect the longest wavelength which does not have the gain needed to overcome the additional diffraction loss. These results also account for the relative abrupt slope of the tuning curves (Figures 19.10(a) and (b)) if one compares it with the rather flat gain distribution on the longer wavelength side (Figure 19.1). The tuning range is limited for a given length of the current pulse because of the delay between light and current pulse.

Finally, it should be noted that in double-grating experiments we often measure another type of anomalous interaction of two excited lines. This is shown in Figure 19.12(b). If the position of the tuned line approaches the stable line up to less than 0.5 nm, suddenly both lines appear again with nearly the same intensities. At present we cannot offer any satisfactory explanation, but it seems that the equal carrier lifetime for both lines indicating an inhomogeneously broadened recombination line could be responsible (if one follows Wright *et al.* (10)). Similar results have been obtained by coupling two independently tuned lasers.

19.5 CONCLUSIONS

We have shown that injection lasers can be efficiently operated in external selective cavities. In pulse regime at room temperature spectral power densities of more than 100 W/nm, a tuning range of more than 20 nm and powers of 5 W for 0.1 nm lines have been obtained permitting a number of applications in spectroscopy, optical communications, measuring devices and semiconductor research. The lasers used can serve as model substances with properties transferable to any other material for required wavelengths. If the semiconductor is the only active material of the laser, the properties of such a laser should be comparable with many gaseous lasers with some advantages like pumping, high efficiency, tunability, high modulation frequencies and, last but not least, a very compact and stable mount. If the development of LOC-DH structures results in low threshold CW lasers, it could be expected that they replace some types of gaseous lasers. Continuously tuned lines with powers up to 100 nW CW and coherence length of meters should be possible.

The problem of the beam collimation has been touched upon only briefly. If spatial filters in external cavities and proper lenses were used, it might not be a serious problem.

Experiments, especially spectral and temporal measurements, with double grating cavities which are still in progress may answer some questions about the nature of recombination lines in semiconductor lasers. The observed localized gain suppression in the vicinity of an excited line (Figure 19.12(a)) could be ascribed to be homogeneous line width of the transitions.

REFERENCES

1. G. Lasher, F. Stern, *Phys. Rev.*, **137**, A553 (1964).
2. Yu. M. Popov, G. M. Strachovski, N. N. Shiukin, *FTP*, **3**, 803 (1969); G. H. B. Thompson, unpublished results (1969).
3. M. B. Panish, H. C. Casey, Jr., *JAP*, **40**, 163 (1969).
4. W. W. Anderson, *IEEE J. of Quantum Electron.*, **QE-1**, 228 (1965).
5. H. S. Sommers, Jr., *RCA Review*, **38**, 33 (1977).
6. G. C. Dousmanis, H. Nelson, D. L. Staebler, *APL*, **5**, 174 (1964).
7. K. Unger, *Z. f. Physik*, **207**, 332 (1967).
8. H. Bachert, S. Raab, *phys. stat. sol.*, **29**, K175 (1968).
9. H. Bachert, P. G. Eliseev, A. Keiper, S. Raab, *phys. stat. sol.* (a), **8**, 477 (1971).
10. P. D. Wright, J. J. Coleman, N. Holonyak, Jr., M. J. Ludowise, G. E. Stillman, J. A. Rossi, *JAP*, **47**, 3580 (1976).
11. T. Yamamoto, K. Kawamura, *Proc. IEEE*, **12**, 1967 (1966).
12. E. J. Walker, A. E. Michel, *JAP*, **35**, 2285 (1964).
13. R. Vuilleumier, N. E. Collins, J. M. Smith, J. C. S. Kim, H. Raillard, *Proc. IEEE*, **55**, 141 (1967).
14. M. Ettenberg, H. S. Sommers, Jr., H. Kressel, H. F. Lockwood, *APL*, **18**, 571 (1971).
15. E. Mohn, R. F. Broom, C. Deutsch, J. Hatz, *Phys. Lett.*, **24A**, 561 (1967).
16. P. G. Eliseev, I. Ismailov, M. A. Manko, V. P. Strachov, *ZETF-letters*, **9**, 594 (1969).
17. Yu. M. Popov, N. N. Shuikin, *FTP*, **4**, 45 (1970).
18. P. G. Eliseev, M. A. Manko, *Short Notes FIAN*, Moscow, **4**, 47 (1970).
19. H. Bachert, P. G. Eliseev, S. Raab, *Z. prikl. spectr.*, **16**, 810 (1972).
20. O. V. Bogdankevich, S. A. Darsnek, P. G. Eliseev, *Semiconductor lasers*, Izdatelstvo 'Nauka', Moscow, 1976, p. 114.
21. A. P. Bogatov, P. G. Eliseev, B. N. Sverdlov, *Trudy FIAN*, **91**, 75 (1977).
22. R. Salathe, C. Voumard, H. Weber, *phys. stat. sol.* (a), **23**, 675 (1974).
23. C. Voumard, R. Salathe, H. Weber, *Optics Comm.*, **13**, 130 (1975).
24. H. Bachert, *Dissertation*, Karl-Marx-Universität Leipzig, 1971.
25. H. Bachert, P. G. Eliseev, M. A. Manko, S. Raab, *Z. prikl. spektr.*, **13**, 232 (1970).
26. J. W. Crowe, W. E. Ahearn, *IEEE J. of Quantum Electron.*, **QE-4**, 169 (1968).
27. H. Bachert, A. Keiper, S. Raab, *Wiss. Zt. d. KMU Leipzig*, **20**, 261 (1971).
28. J. W. Crowe, W. E. Ahearn, *IEEE J. of. Quantum Electron.*, **QE-2**, 283 (1966).
29. S. Raab, A. Keiper, H. Bachert, P. G. Eliseev, *Patent DWP*, 111469 (1974).

30. A. Keiper, *Dissertation*, Humboldt Universität Berlin, 1974.
31. H. Bachert, A. P. Bogatov, Ch. A. Dshalovov, P. G. Eliseev, A. Keiper, M. A. Manko, *Kvant. Elektr.* (to be published).
32. D. Ackermann, H. Bachert, P. G. Eliseev, A. Keiper, M. A. Manko, S. Raab, *Lasers and their Applications*, Dresden, 1970, **12**, 719 (1970).
33. D. Ackermann, P. G. Eliseev, A. Keiper, M. A. Manko, S. Raab, *Kvant. Elektr.*, **1**, 85 (1971).
34. R. Ludeke, E. P. Harris, *APL*, **20**, 499 (1972).
35. C. Voumard, R. Salathe, H. Weber, *Appl. Phys.*, **7**, 123 (1975).
36. H. Bachert, A. P. Bogatov, P. G. Eliseev, *Lasers and their Applications*, Dresden, 1977, K68.
37. A. P. Bogatov, P. G. Eliseev, B. N. Sverdlov, IEEE J. of Quantum Electron., **QE-11**, 510 (1975).
38. J. A. Rossi, J. J. Hsieh, H. Heckscher, *IEEE J. of Quantum Electron.*, **QE-11**, 538 (1975).
39. T. L. Paoli, J. E. Ripper, A. C. Morosini, N. B. Patel, *IEEE J. of Quantum Electron.*, **QE-11**, 525 (1975).
40. J. A. Rossi, H. Heckscher, G. E. Stillman, S. R. Lhinn, *APL*, **23**, 254 (1973).

Chapter 20

Coherence of the Radiation Emitted by Semiconductor Lasers

M. A. Herman

20.1 INTRODUCTION

Injection lasers are promising in many respects as light sources for optical information processing or transmitting systems. They are compact in size, easy to handle and small in power consumption. However, it is well known that there still remain many difficulties which must be overcome before injection lasers can be put into widespread practical use.

The possibility of applying injection lasers in coherent optics, mainly in computer holographic memories, as well as in optical fibre communication systems has led to investigations on the coherence of the radiation emitted by these light sources. From the point of view of the above-mentioned applications the requirements for coherence of the radiation emitted by injection lasers are especially severe. The development of statistical and quantum optics observed in the last decades has permitted a thorough and deep understanding of the statistical properties of the radiation emitted by partially coherent light sources, a class of sources, to which also the injection lasers belong.

The modern concept of coherence is based on investigations of statistical fluctuations of optical electromagnetic fields. Optical coherence phenomena may be said to be manifestations of correlations between these statistical field fluctuations. In the statistical description of an electromagnetic field, optical disturbances at a fixed point in space are treated as random functions of time. Correlation functions of any given order correspond to appropriate statistical moments.

In the case of fields generated by thermal sources of radiation, the complete statistical description is contained in the second-order correlation function. With the development of lasers, other types or radiation of different statistical nature have been studied, and higher-order correlation functions have been used.

Two main groups of problems can be distinguished in coherence investigations.

The first group concerns the coherence properties of free fields and the methods of measuring these properties.

The second group concerns the origin and growth of coherence within the source itself.

From the point of view of design and manufacture of injection lasers the second group of coherence problems is, of course, much more important. This is connected with the hopes to control the coherence properties of the injection laser by optimizing the technology and the construction of this device. However, the theoretical and experimental difficulties posed by the second group of problems are considerable, and somebody who wanted to investigate the coherence phenomena concerning junction lasers would have to start with the first group of coherence problems. This also applies to our lecture. We will start with a brief recollection of the basic definitions of the theory of partial coherence and with the description of measurement methods most frequently used in coherence investigations. Detailed information on the modern concept of coherence in optics is available in many original papers (1–5) review articles (6–8) as well as in monographs and handbooks (9–13).

In view of this, we can confine ourselves only to the basic definitions and results that will be used later on in this chapter.

20.2 SECOND-ORDER COHERENCE OF ELECTROMAGNETIC RADIATION

In the case of electromagnetic fields generated by semiconductor junction lasers, which are characterized by relatively low coherence properties, the second-order theory of coherence is sufficient for discussing the existing experimental results. Moreover, most frequently only the scalar approximation of this theory may be used.

20.2.1 Elementary coherence concepts and definitions

The *mutual coherence function*, the basic quantity in the scalar theory of coherence, is defined as

$$\Gamma(P_1, P_2, \tau) = \langle V(P_1, t+\tau) V^*(P_2, t) \rangle \tag{20.1}$$

where $V(P_1, t)$ and $V(P_2, t)$ are the field values (in complex representation) at space-points P_1 and P_2 and the sharp brackets denote the averaging operation over time.

The normalized form $\gamma(P_1, P_2, \tau)$ of the mutual coherence function defined above is expressed by the formula

$$\gamma(P_1, P_2, \tau) = \frac{\Gamma(P_1, P_2, \tau)}{\sqrt{\Gamma(P_1, P_1, 0)\Gamma(P_2, P_2, 0)}} \qquad (20.2)$$

and is termed the *complex degree* of coherence.

Time averaging in formula 20.1 is a useful procedure when the radiation field is composed of a finite number of periodic components of fixed phase and amplitude such as, for example, the field of an intensity stabilized multimode CW junction laser. In such a case $V(P, t)$ is defined as an analytic signal associated with a Cartesian component of the electric vector E of the e–m field. The advantages of dealing with coherence problems in terms of analytic signals (1) instead of real field quantities (2) stem from the following considerations (9):

— the analytic signal involves only positive frequencies;
— its modulus is the envelope of the real function with which it is associated;
— there is a particularly simple relation between real and imaginary parts of this signal;
— the analytic signal can be represented as a linear transform of the real function with which it is associated;
— it provides a very convenient framework for extracting the quasi-monochromatic approximations which form the basis for most practical problems involving coherence theory.

There are three essentially equivalent methods of associating an analytic signal $V(P, t)$ with a given real function $V^r(P, t)$ of the real variable t.

In the first method the analytic signal is defined in terms of the Fourier transform, in the second—in terms of the Hilbert transform, and in the third—in terms of the Dirac functions (9).

According to the first method, most frequently used in practical applications, the analytic signal may be expressed by the formula

$$V(P, t) = \int_0^\infty \hat{V}(P, f) e^{-i2\pi ft} df \qquad (20.3)$$

where $\hat{V}(P, f)$ is equal to the doubled value of the Fourier transform of the real function $V^r(P, t)$.

It is important to note that the mutual coherence function $\Gamma(P_1, P_2, \tau)$ similarly as the functions V occurring in Equation 20.1, is also an analytic signal. Thus, the following equations are valid for it:

$$\Gamma(P_1, P_2, \tau) = \int_0^\infty \hat{\Gamma}(P_1, P_2, f) e^{-i2\pi f t} df \qquad (20.4)$$

where the Fourier transform with respect to τ is

$$\hat{\Gamma}(P_1, P_2, f) = \begin{cases} \int_{-\infty}^{+\infty} \Gamma(P_1, P_2, \tau) e^{i2\pi f t} d\tau > 0 \\ 0 \qquad \text{if } f < 0 \end{cases} \qquad (20.5)$$

The function $\hat{\Gamma}(P_1, P_2, f)$ is termed the *mutual power spectrum*.

In some cases concerning junction laser radiation problems (single mode operation) it is appropriate to specify the radiation as a random process generated by an ensemble of systems with identical macroscopic properties. The radiation in such a random process is then described by functions of the form

$$P_s\{V_1^r(P_1, t_1), V_2^r(P_2, t_2), ..., V_s^r(P_s, t_s)\} dV_1^r(P_1, t_1), ..., dV_s^r(P_s, t_s) \qquad (20.6)$$

defined as the probability that a given radiation of the system will have a value between $V_1^r(P_1, t_1)$ and $V_1^r(P_1, t_1) + dV_1^r(P_1, t_1)$ at position P_1 and time $t_1, ...$ and a value between $V_s^r(P_s, t_s)$ and $V_s^r(P_s, t_s) + dV_s^r(P_s, t_s)$ at position P_s and time t_s. In the 'ensemble of systems' we consider V^r to be a real function. In the following equations we omit the r superscript for the sake of notational convenience. The analytic signal is here considered only in connection with time averages. From P_s we can calculate moments of the form

$$M(V_1, ..., V_s) = \underset{s\text{-fold integral}}{\int ... \int} V_1 V_2 ... V_s P_s(V_1, V_2, ..., V_s) dV_1 ... dV_s \qquad (20.7)$$

These moments may be calculated directly from measurements as

$$M(V_1, ..., V_s) = \lim_{N \to \infty} \frac{1}{N} \left(\sum_{i=1}^{N} V_{1i} V_{2i} ... V_{si} \right) \qquad (20.8)$$

where i denotes which system in the ensemble is being considered.

It should be noted that P_s may be calculated only when all the moments of P_s are known. These moments are in general functions of time. However, if all P_s, and hence all moments, are independent of absolute time, the process is called a *stationary random process*. When the random process is stationary, one often assumes an ergodic type hypothesis and equates the infinite time average of any system i with the ensemble average. That is, one assumes that

$$\lim_{N \to \infty} \frac{1}{N} \left(\sum_{i=1}^{N} V_{1i} \ldots V_{si} \right) = \langle V_{1i} \ldots V_{si} \rangle |_{i \text{ is fixed}} \quad (20.9)$$

We shall assume this equivalence to be true in all the stationary processes we consider. In such cases, the temporal variation of V for any system is termed a *stationary random series*. It should be noted here that the mutual coherence function defined by formula 20.1 is of course simply a second-order moment. As a consequence of assumption 20.9 the moment $\Gamma(P_1, P_2, \tau) = \langle V(P_1, t+\tau) V^*(P_2, t) \rangle$ has, however, meaning for either a random stationary time series or a time variation made up of periodic components (9).

In principle, there is no reason to assume that in random processes the second moment $\Gamma(P_1, P_2, \tau)$ characterizes all physical processes of interest. Indeed, to characterize completely a physical process, P_s which includes all moments of V as $s \to \infty$ is needed. The assumption is often made, however, that the random processes may be described by Gaussian statistics, and in this case, all higher moments are determined from $\Gamma(P_1, P_2, \tau)$.

Although all the information concerning the second-order coherence of the optical fields are contained in the mutual coherence function, in many cases it proves helpful to separate as much as possible those coherence effects which arise from the finite spatial extent of the primary source of the radiation from those which arise from the finite spectral width of the radiation. A complete separation of these effects is, of course, impossible, but a partial separation which is useful for many problems is possible. The appearance of these two effects can be explained by the following argument. Radiation of finite spectral width Δf cannot be coherent since its various Fourier components are not correlated. Similarly, an extended source cannot radiate coherently due to the lack of correlation between its various spatial elements. The two effects are called *temporal coherence* and *spatial coherence effects*, respectively. Temporal coherence is represented by

$$\Gamma(P_1, P_1, \tau) = \langle V(P_1, t+\tau) V^*(P_1, t) \rangle$$

and can be measured using the Michelson two-beam interferometer. Spatial coherence, on the other hand, is represented by

$$\Gamma(P_1, P_2, 0) = \langle V(P_1, t) V^*(P_2, t) \rangle$$

and can be measured, for example, using the Young two-pinhole interferometer. In the first case the dependence of the correlation on the parameter

τ is exhibited, with the points P_1 and P_2 coincident and fixed; in the second case the dependence on the position of the two points is exhibited while the time delay $\tau \ll 1/\Delta f$ is kept fixed.

It is a basic property of the mutual coherence function $\Gamma(P_1, P_2, \tau)$, in terms of which the degree of coherence was defined, that in vacuum this function obeys two wave equations (6)

$$\nabla_j^2 \Gamma = \frac{1}{c^2} \frac{\partial^2 \Gamma}{\partial \tau^2} \qquad (j = 1, 2) \qquad (20.10)$$

where ∇_j^2 denotes the Laplacian operator with respect to the coordinates of the point P_j. With the help of Equation 20.10 one may study the distribution of second-order coherence throughout an optical field.

20.2.2 Measurement procedure

Suppose that a beam of quasimonochromatic light from small source S is divided into two beams in a Michelson interferometer, and that the two beams are superposed after a path delay $\Delta l = c\Delta t$ (c velocity of light) has

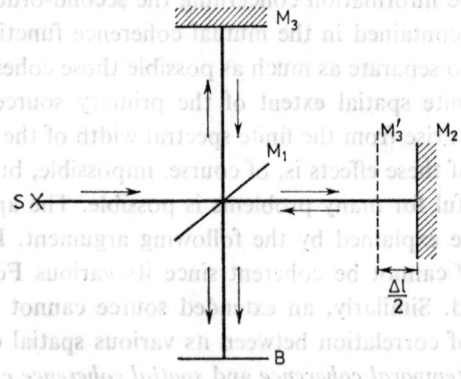

Figure 20.1 Light beams in the Michelson interferometer; S, light source; M_1, half-dividing glass plate; M_2, movable mirror; M_3, fixed reference mirror; B, observation screen

been introduced between them (Figure 20.1). In the region of superposition, interference fringes that can be observed on screen B are formed if Δl is sufficiently small. The usual measure of the sharpness of interference fringes

Coherence of the Radiation

is the so-called *visibility* of the interference pattern defined after Michelson by

$$\mathscr{V}(r) = \frac{I_{\max} - I_{\min}}{I_{\max} + I_{\min}} \quad (20.11)$$

$I_{\max}$ and $I_{\min}$ represent here the intensity maxima and minima in the immediate neighbourhood of a point $P(r)$ in this interference pattern. The visibility of the interference fringes or the fringe contrast on screen B depends on the path difference Δl or on the corresponding time delay Δt between the two light beams in the interferometer. So, the appearance of the fringes of definite visibility on screen B is said to be a *manifestation* of temporal coherence between the two beams, or what is equivalent, of temporal coherence of the light radiated by the source S. In general, interference fringes will only be observed if $\Delta t \Delta f \ll 1$, where Δf is the effective band-width of the light. The time $(\Delta t)_c \sim 1/\Delta f$ is called the *coherence time* of the light and the corresponding path $c(\Delta t)_c$ the *coherence length* (6). From the point of view of practical applications it is usual to define the coherence length or the coherence time as the path difference Δl or the time delay Δt corresponding to the visibility $\mathscr{V} = 0.5$ of the interference pattern (14, 15). Let us next consider the Young type interference experiment used for measuring the spatial coherence of the investigated light source. The idea of this experiment can be seen in the schematic diagram shown in Figure 20.2. S represents a light

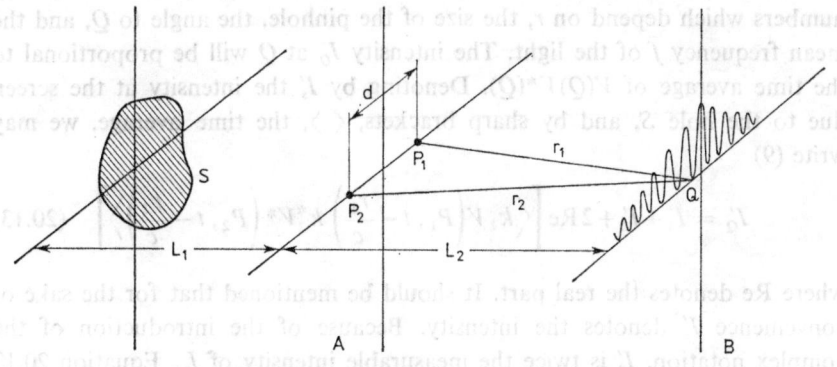

Figure 20.2 Schematic diagram illustrating Young's interference experiment; S, light source; A, screen with two pinholes P_1 and P_2; B, observation screen; Q, a representative point on the interference pattern

source of finite spatial extent and finite but narrow spectral width, and P_1 and P_2 are two pinholes separated by distance d on screen A, at distance L_1 from the source. Interference fringes are observed in the intensity pattern on screen B at distance L_2 from screen A. The visibility of the fringes is increased by decreasing the size of the primary source S. For a fixed source size, the visibility of the fringes is varied by changing the separation d of the pinholes. This variation in visibility with pinhole separation is referred to as the effect of spatial coherence. Examining the quality of the fringes as we move away from the centre of the interference pattern, we note that the visibility diminishes with the increasing path difference r_1-r_2, eventually vanishing completely. The decrease in fringe visibility with increasing path difference is often regarded as an illustration of imperfect time coherence. Thus, when measuring spatial coherence the path difference r_1-r_2 should be as small as possible.

Let $V(P_1, t)$ and $V(P_2, t)$ be suitable chosen complex representations of the field disturbances at P_1 and P_2, respectively, in Figure 20.2. Then, assuming, for example, that we are dealing with a Cartesian component of the electric vector, V is propagated in free space by wave equation 20.10. Since this equation is a linear one, we may represent the disturbance at Q as a superposition of the contributions from P_1 and P_2, thus

$$V(Q, t) = k_1 V\left(P_1, t-\frac{r_1}{c}\right) + k_2 V\left(P_2, t-\frac{r_2}{c}\right) \quad (20.12)$$

where the k_i are propagators independent of time. They are pure imaginary numbers which depend on r, the size of the pinhole, the angle to Q, and the mean frequency $\bar{f}$ of the light. The intensity I'_Q at Q will be proportional to the time average of $V(Q)V^*(Q)$. Denoting by I'_s the intensity at the screen due to the hole S, and by sharp brackets, $\langle \rangle$, the time average, we may write (9)

$$I'_Q = I'_1 + I'_2 + 2\text{Re}\left[\left\langle k_1 V\left(P_1, t-\frac{r_1}{c}\right) k_2^* V^*\left(P_2, t-\frac{r_2}{c}\right)\right\rangle\right] \quad (20.13)$$

where Re denotes the real part. It should be mentioned that for the sake of convenience I'_s denotes the intensity. Because of the introduction of the complex notation, I'_s is twice the measurable intensity of I_s. Equation 20.13 may be rewritten if we note that k_1 and k_2 are independent of time and if we set $t_s = r_s/c$ ($s = 1, 2$). Consequently, we obtain

$$I'_Q = I'_1 + I'_2 + 2k_1 k_2^* \text{Re}[\langle V(P_1, t-t_1)V^*(P_2, t-t_2)\rangle] \quad (20.14)$$

The term inside the brackets is simply the complex cross correlation function of the two disturbances. This function depends on time difference $\tau = t_1 - t_2$

only and if we have $\tau \Delta f \ll 1$ (where Δf is the spectral width of the radiation), which holds for injection lasers, we may rewrite Equation 20.14 as

$$I'_Q = I'_1 + I'_2 + 2k_1 k_2^* \operatorname{Re} \Gamma(P_1, P_2, \tau) \qquad (20.15)$$

Let us note that

$$\Gamma(P_s, P_s, 0) = \langle V(P_s, t) V^*(P_s, t) \rangle \qquad (20.16)$$

is simply the intensity at point P_s ($s = 1, 2$). With this in mind, however, we may normalize $\Gamma(P_1, P_2, \tau)$ in Equation 20.15 and noting that

$$k_1 k_2^* \sqrt{\Gamma(P_1, P_1, 0) \Gamma(P_2, P_2, 0)} = \sqrt{|k_1|^2 \Gamma(P_1, P_1, 0) |k_2|^2 \Gamma(P_2, P_2, 0)} \qquad (20.17)$$

we may rewrite Equation 20.15 as

$$I'_Q = I'_1 + I'_2 + 2\sqrt{I'_1 I'_2} \operatorname{Re}[\gamma(P_1, P_2, \tau)] \qquad (20.18)$$

Here we have used the definition of $\gamma(P_1, P_2, \tau)$ as given by Equation 20.2. If we perform Young's interference experiment under the best conditions, that is, if $I'_1 = I'_2 = I'$ with the path differences $r_1 - r_2$ very small, and if we write γ in the form

$$\gamma(P_1, P_2, \tau) = |\gamma(P_1, P_2, \tau)| e^{i\Phi(P_1, P_2, \tau)} \qquad (20.19)$$

we may rewrite Equation 20.18 as

$$I'_Q = 2I'[1 + |\gamma(P_1, P_2, \tau)| \cos \Phi(P_1, P_2, \tau)] \qquad (20.20)$$

From this equation, using $I'_{Q\max} = 2I'(1+|\gamma|)$, $I'_{Q\min} = 2I'(1-|\gamma|)$ and Equation 20.3, one gets the visibility of the fringes as

$$\mathscr{V} = |\gamma(P_1, P_2, \tau)| \qquad (20.21)$$

Thus the modulus of the complex degree of coherence is equal to the visibility of the interference fringes in the centre of screen B and can be measured easily. However, when measuring the modulus of the complex degree of coherence at $\tau = 0$, one usualy uses Equation 20.18 in the following modified form:

$$|\gamma(P_1, P_2, 0)| = \frac{I_{Q\max} - I_1(Q) - I_2(Q)}{\sqrt{I_1(Q) I_2(Q)}} \qquad (20.22)$$

where $I_{Q\max}$ is the measured maximal intensity in the interference pattern at point Q in the centre of screen B, $I_1(Q)$ and $I_2(Q)$ are the measured intensities of the light reaching the point Q through the pinhole P_1 only, or P_2 only, respectively.

20.2.3 Review of experimental data concerning second order coherence of radiation emitted by junction lasers

The first reports concerning experimental investigations of second order-coherence effects of injection lasers radiation date back to the years 1966–67 (16, 17). In these reports the spatial coherence of pulsed GaAs homojunction lasers operating at liquid nitrogen temperatures was estimated. The lasers radiated coherently for a primary source area of breadth equal to 35 μm. The coherence was independent of diode current in the range between the

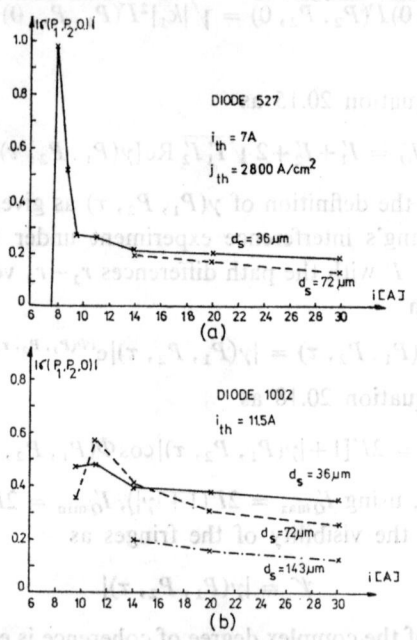

Figure 20.3 Complex degree of coherence of the e-m radiation emitted by pulsed GaAs homojunction laser as a function of the laser current and the distance d_s between two testpoints on the radiating surface of the p–n junction. Measurements have been performed at 77 K for two laser diodes (18)

threshold and the double value of the threshold. More extensive investigations of spatial coherence effects for homojunction pulsed GaAs lasers were performed by Büchtemann and Höhn (18) in 1969. The degree of spatial coherence

$|\gamma(P_1, P_2, 0)|$ of the radiation emitted by these laser diodes was determined with respect to two points p_1 and p_2 at the p–n junction of the diodes as a function of its current and the distance between these points using the Young two pinhole interferometer. The two pinholes P_1 and P_2 of screen A (Figure 20.2) of this interferometer were imaged by an optical system on the p–n junction of the investigated laser diode creating there two test points p_1 and p_2 of the primary radiation source. The radiation coming out from those two test points was investigated by measuring the respective intensities in the interference pattern on screen B (generated by pinholes P_1 and P_2) and by calculating from formula 20.22 the modulus of the complex degree of coherence. It should be noted that the degree of coherence of the radiation from the two test points p_1 and p_2 on the radiating surface of the source, and of the radiation from points P_1 and P_2 which are real optical images of the two test points, has the same value under the experimental conditions which makes it possible to neglect the imaging errors and the diffraction effects. These conditions were guaranteed by the optical system used. It was found that $|\gamma(P_1, P_2, 0)|$ had a maximum slightly above the threshold. In general it decreased as the distance $d_s = dK$ (where K is the magnification of the optical imaging system) between the test points increased (Figure 20.3)). Near threshold there was a strong spatial modulation of $|\gamma|$ and of the light intensity in the near field of the laser diode (Figure 20.4). To explain these phenomena, simultaneous existence of several transverse modes (19) and, at higher currents, of superradiation was proposed. Similar experiments concerning broad contact double heterojunction $GaAs-Al_xGa_{1-x}As$ pulsed lasers operating at room temperatures have been performed recently in the Institute of Physics of the Polish Academy of Sciences (20) using the Young two-pinhole interferometer shown schematically in Figure 20.5. The results shown in Figure 20.6 confirm the fact that for multimode DH lasers operating at room temperature the dependences of the complex degree of coherence $|\gamma|$ on diode current and on the distance between the two test points on the p–n junction are practically the same as for homojunction lasers operating at liquid nitrogen temperatures. The dependence of $|\gamma(P_1, P_2, 0)|$ on the angular distance between the two test points on the wave front of the radiation emitted by pulsed GaAs homojunction laser operating at 77 K was investigated by Bykovskii et al. (15) using a Fresnel biprism technique. The results are shown in Figure 20.7. The independence of $|\gamma|$ of angular distance in the plane perpendicular to the p–n junction is obvious, because the radiation is generated in this plane practically by one lasing spot (channel). The decrease of $|\gamma|$ with increasing angle distance observed in the plane parallel to the p–n

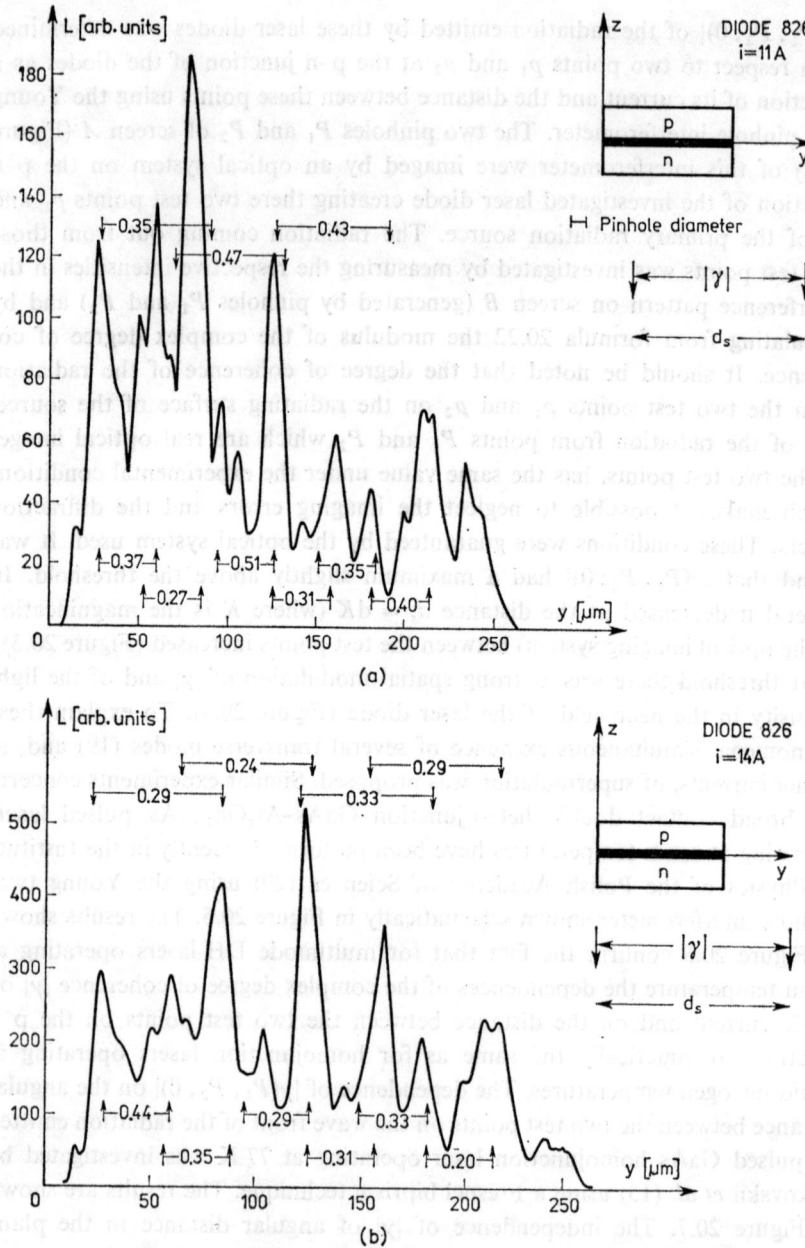

Figure 20.4 The spatial fluctuations of the radiation intensity in the near field pattern of the homojunction GaAs pulsed laser diode operating at 77 K and the complex degree of coherence for two values of the diode current (18)

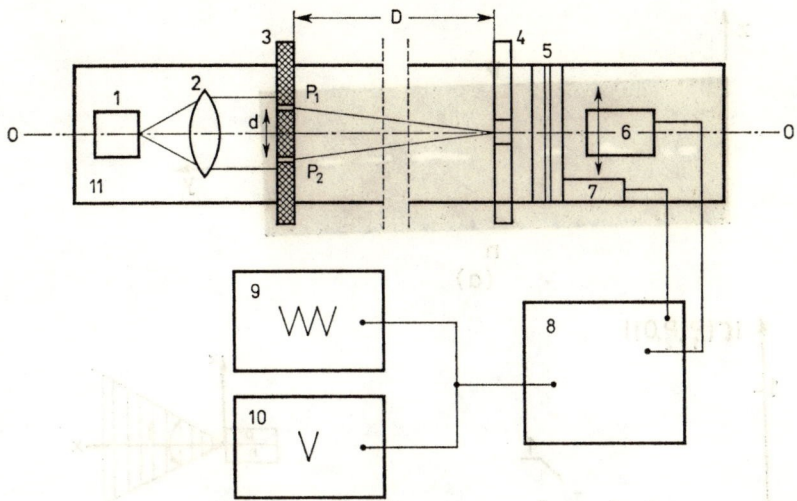

Figure 20.5 Schematic diagram of the Young interferometer used in Reference 20 for the measurement of the complex degree of coherence of DH lasers operating at room temperature; 1, laser diode; 2, collimator; 3, diaphragm with two pinholes; 4, diaphragm with a slit; 5, light choper; 6, photomultiplier; 7, reference photodetector; 8, lock-in nanovoltmeter; 9, X-T plotter; 10, digital voltmeter; 11, optical bench

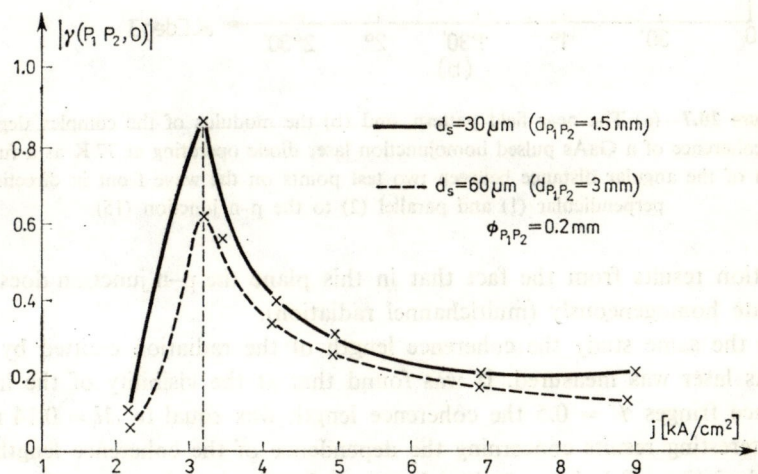

Figure 20.6 Modulus of the complex degree of coherence of the radiation emitted by pulsed DH multimode GaAs–Al$_x$Ga$_{1-x}$As lasers as a function of the laser current and the distance d_s between two testpoints on the radiating surface of the p–n junction (20)

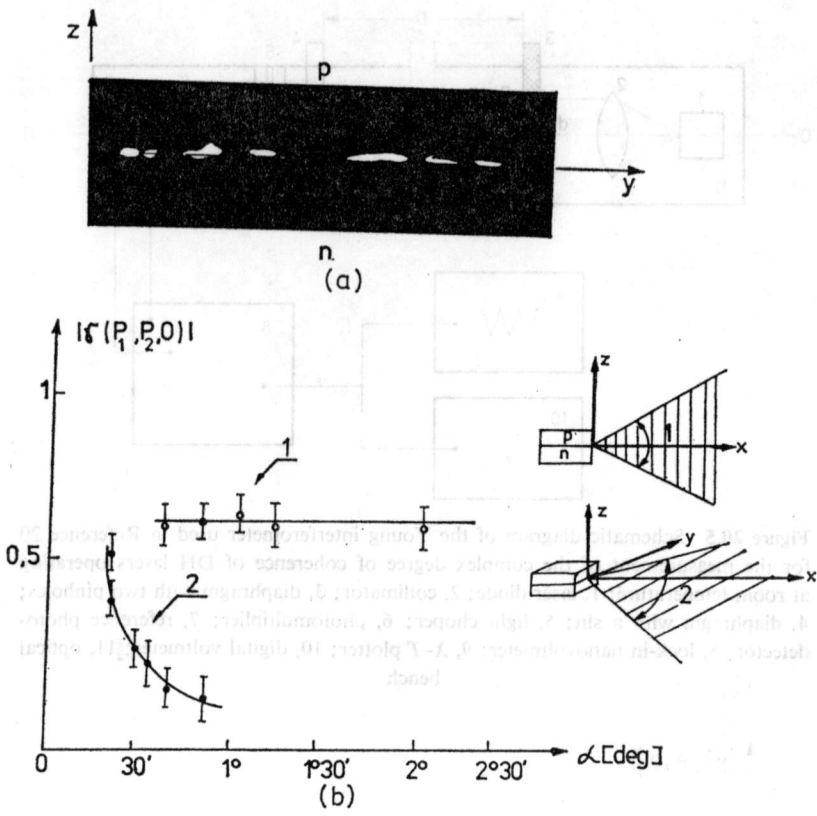

Figure 20.7 (a) The near field pattern, and (b) the modulus of the complex degree of coherence of a GaAs pulsed homojunction laser diode operating at 77 K as a function of the angular distance between two test points on the wave front in directions perpendicular (1) and parallel (2) to the p–n junction (15)

junction results from the fact that in this plane the p–n junction does not radiate homogeneously (multichannel radiation).

In the same study the coherence length of the radiation emitted by the GaAs laser was measured. It was found that at the visibility of the interference fringes $\mathscr{V} = 0.5$ the coherence length was equal to $\Delta l = 0.14$ mm.

Interesting results concerning the dependence of the coherence length on the deviation of lasing wavelength caused by the junction heating process during the current pulse has been reported in Reference 21. These results, shown in Figure 20.8 for different operation conditions of the investigated laser diodes, were obtained using the external Fabry–Perot interferometer

Coherence of the Radiation

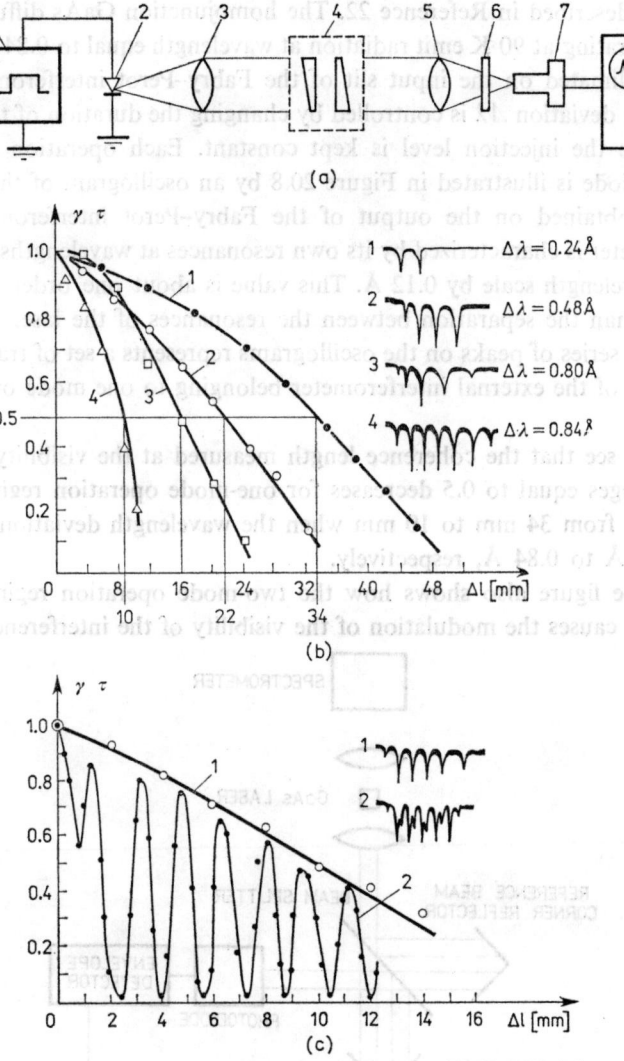

Figure 20.8 (a) Schematic diagram illustrating the external Fabry–Perot interferometer technique for investigations of the dependence of coherence length on lasing wavelength deviation (22); 1, pulse generator; 2, laser diode; 3, collimating lens; 4, Fabry–Perot interferometer; 5, focusing lens; 6, filter; 7, photomultiplier; 8, oscillograph. (b) Modulus of the degree of temporal coherence of a laser diode operating in a one-mode regime and oscillograms of the output signals of the Fabry–Perot interferometer (21). (c) Modulus of the degree of temporal coherence for lasers operating in one-mode (1) and two-mode (2) regimes (21)

technique described in Reference 22. The homojunction GaAs diffusion laser diodes operating at 90 K emit radiation at wavelength equal to 0.84 μm which is then collimated on the input slit of the Fabry–Perot interferometer. The wavelength deviation $\Delta\lambda$ is controlled by changing the duration of the current pulse while the injection level is kept constant. Each operation regime of the laser diode is illustrated in Figure 20.8 by an oscillogram of the radiated spectrum obtained on the output of the Fabry–Perot interferometer. The interferometer is characterized by its own resonances at wavelengths separated on the wavelength scale by 0.12 Å. This value is about one order of magnitude less than the separation between the resonances of the laser resonator. Thus, each series of peaks on the oscillograms represents a set of transmission resonances of the external interferometer belonging to one mode of the laser diode.

One can see that the coherence length measured at the visibility of interference fringes equal to 0.5 decreases for one-mode operation regime of the laser diode from 34 mm to 10 mm when the wavelength deviation increases from 0.24 Å to 0.84 Å, respectively.

The same figure also shows how the two-mode operation regime of the laser diode causes the modulation of the visibility of the interference fringes.

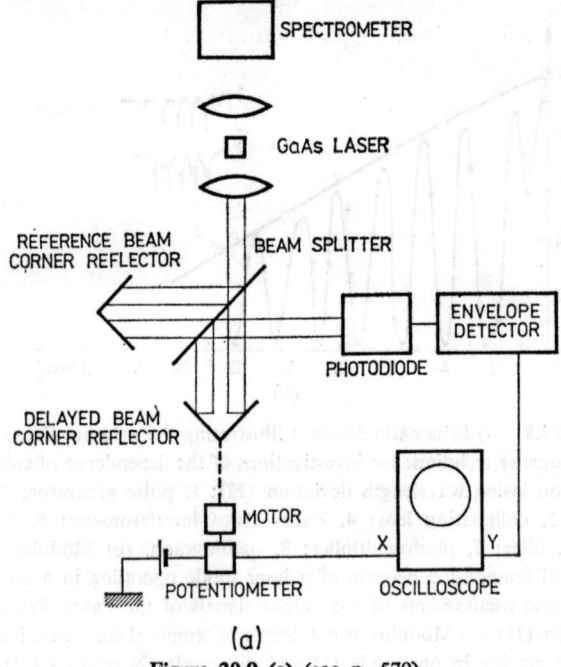

Figure 20.9 (a) (see p. 579)

Coherence of the Radiation

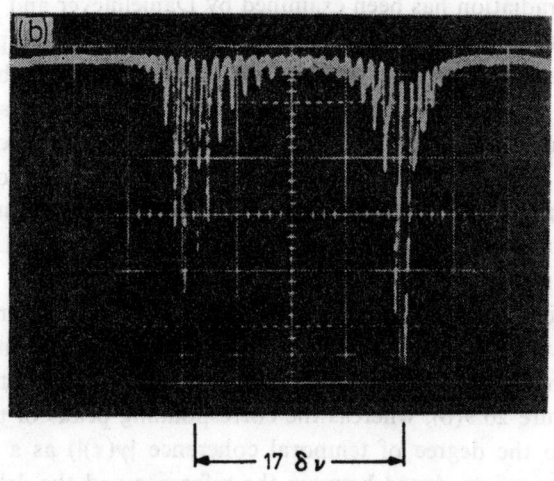

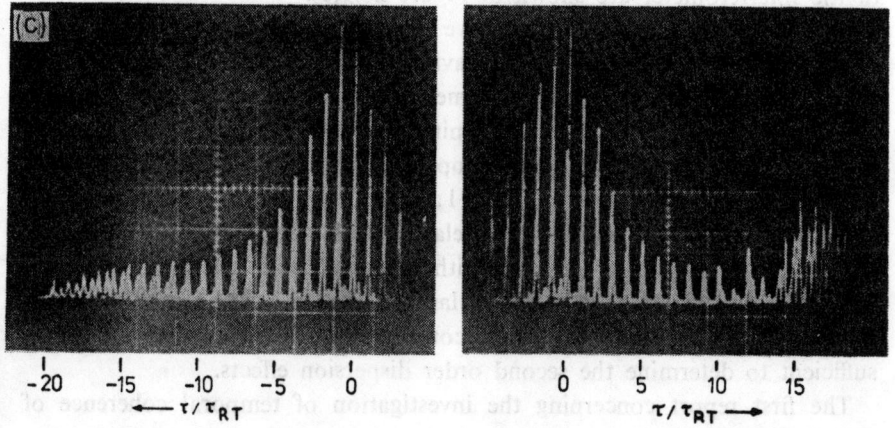

Figure 20.9 (a) Experimental arrangement showing the motor-driven interferometer and the use of corner reflectors to avoid optical feedback into the laser cavity. The magnitude of the coherence function is presented directly on the face of the storage oscilloscope as a function of the delay introduced by the moving mirror (24). (b) The oscillation spectrum at ~ 8400 Å consisting of two families of longitudinal modes and separated by 22.8 Å ($i/i_{th} = 2.2$) (24). (c) Correlation peaks as a function of the relative delay τ introduced between the reference and delayed beams of the interferometer (24)

The effect of dispersion (the dependence of the index of refraction of the lasing medium on lasing wavelength) on the temporal coherence properties of laser output radiation has been examined by Danielmeyer and Weber (23). This effect consists in shifting the maximum of the coherence peak on the scale of the relative delay time between the interfering beams, by the value proportional to the value of the refraction index of the dispersive medium and in diminishing the degree of temporal coherence of the laser radiation.

Experimental investigations concerning the dispersion effect on temporal coherence properties of a stripe geometry GaAs laser with a diffused junction operating CW at 80 K at current levels $i/i_{th} = 2.2$ have been used by Brackett (24) for studying the second-order dispersion effects (different time delays for signals at slightly different frequencies) occurring in this laser diode. The measurements were performed using a motor-driven Michelson interferometer shown in Figure 20.9(a). The oscillation spectrum of the investigated laser is shown in Figure 20.9(b), whereas the corresponding peaks of the visibility (proportional to the degree of temporal coherence $|\gamma(\tau)|$) as a function of the relative delay τ introduced between the reference and the delayed beams of the interferometer are shown in Figure 20.9(c).

The GaAs laser is itself a dispersive medium. Since the e–m fields in the GaAs travel back and forth in the cavity, the field amplitude in the output beam essentially repeats itself in a time equal to the round-trip of the laser cavity τ_{RT}. Because of this, the magnitude of the coherence function, as it appears on the face of the oscilloscope, consists of a series of correlation peaks at $\tau = k \cdot \tau_{RT}$, where $k = 0, \pm 1, \pm 2, \ldots$. For each value of k we can see in Figure 20.9(c) a successive correlation peak, corresponding to a greater length of the dispersive medium, with the result that the peak decays in amplitude and increases in width at larger values of τ. The measurement of either the amplitude or width of the correlation peaks as a function of τ are sufficient to determine the second order dispersion effects.

The first report concerning the investigation of temporal coherence of pulsed SH and DH GaAs–$Al_xGa_{1-x}As$ lasers operating at room temperature was published in 1974 (14).

For one-mode operation regime, the coherence length measured at visibility 0.5 was equal to 27 mm. This value coincides with the results obtained for one-mode homojunction lasers operating at 90 K.

For DH lasers operating in multimode regime the coherence length is smaller and does not exceed the value of a few millimeters (20). The schematic diagram of the Michelson interferometer used in (20) for investigations of temporal coherence properties of multimode DH lasers is shown in Figure

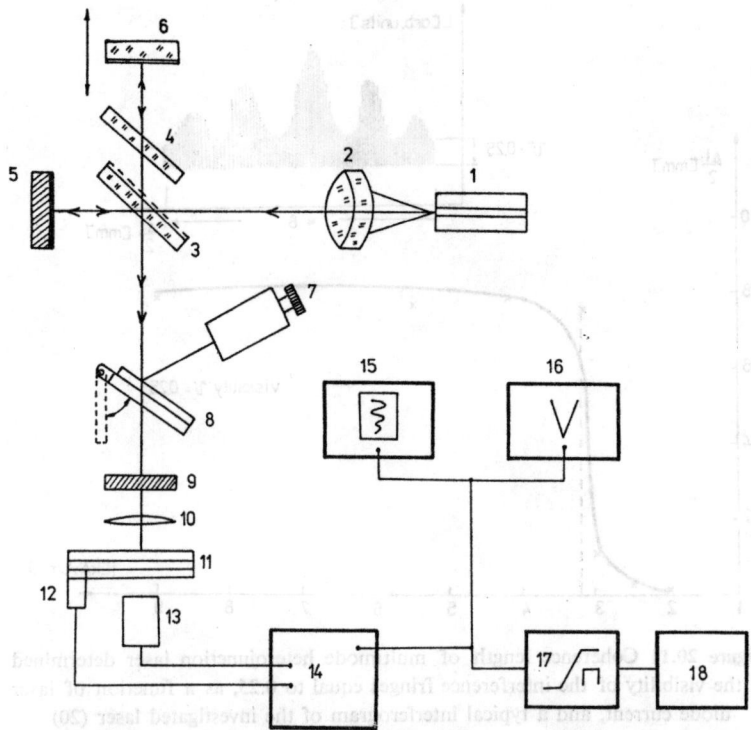

Figure 20.10 Schematic diagram of the Michelson interferometer used for coherence length measurement of multimode laser diodes (25); 1, laser diode; 2, collimator; 3, half-dividing glass plate; 4, compensating plate; 5, fixed reference mirror; 6, movable measuring mirror; 7, image converter; 8, movable observation mirror; 9, light filter; 10, focusing lens; 11, light choper; 12, reference photodetector; 13, photomultiplier; 14, lock-in nanovoltmeter; 15, X-T plotter; 16, digital voltmeter; 17, pulse forming unit with amplitude discriminator; 18, counting unit

20.10 (25). The results concerning the coherence length of multimode lasers, measured as a function of laser current are shown in Figure 20.11.

To sum up the experimental data presented above concerning the second-order coherence properties of the radiation emitted by junction lasers one may state that:

(a) the highest spatial coherence of laser radiation appears when the value of the laser current is equal to, or near, the threshold current—this seems to be caused by the fact that at these currents the laser radiates most homogeneously on the entire p–n junction surface;

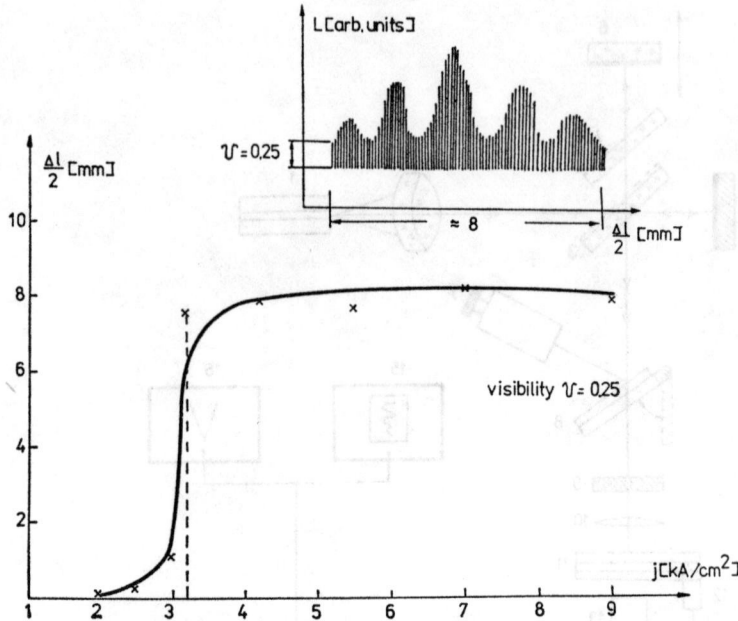

Figure 20.11 Coherence length of multimode heterojunction laser determined at the visibility of the interference fringes equal to 0.25, as a function of laser diode current, and a typical interferogram of the investigated laser (20)

(b) the highest temporal coherence of laser radiation appears when the laser operates in a one-mode regime—this is caused by the fact that at one-mode operation the spectral width of the radiation emitted is the lowest;
(c) even in the one-mode regime, the temporal coherence can be diminished, when the p–n junction of the laser diode is heated during the current pulse supplying the laser—this is caused by the fact that the wavelength of laser radiation deviates from its initial value, which is equivalent to a broadening of the spectral width of the radiation;
(d) the dispersion effects appearing in junction lasers give rise to a decrease of the temporal degree of coherence of the output radiation.

20.3 THE ROLE OF COHERENCE OF SEMICONDUCTOR LASERS RADIATION IN PRACTICAL APPLICATIONS OF THESE DEVICES

It has already been mentioned that semiconductor heterojunction lasers are promising as light sources for optical information processing and transmitting systems. Among different possible applications of semiconductor lasers in these systems, the most interesting and probably also the most important are the applications of laser diodes as light sources in holography and optical fibre communication.

Let us discuss briefly the main requirements concerning the coherence of laser diodes radiation from the point of view of these two kinds of applications.

20.3.1 Semiconductor lasers in holography

A hologram is a photographic recording of the interference pattern produced by two beams of light falling on a film. The requirement that the recorded interference fringes must be well defined to ensure a high quality of the hologram necessitates illumination from a source of high degree of spatial and temporal coherence. Accordingly, CW gas lasers are most commonly used as light sources for hologram production. In the reconstruction of developed holograms the requirement of coherence is not so severe, so that hologram reconstruction using injection laser is permissible. An analysis of the influence of insufficient coherence of the radiation emitted by a hologram reconstructing source on the quality of the reconstructed hologram was performed by Minami et al. (26). In that analysis no use was made of the coherence function approach, but the coherence properties were represented by $\Delta\lambda$, the width of the spectrum, and by Δd, the size of the light emitting region. The results have shown that the real image obtained from a hologram by reconstruction with a source of e-m radiation with insufficient coherence undergoes a lateral shift of Δx and Δy in the plane parallel to the hologram plane and a defocusing of Δz in the direction perpendicular to the hologram plane. Defocusing is caused by $\Delta\lambda$, and the lateral shift along the direction of the x axis (the direction perpendicular to the holographic carrier fringes) is a function of $\Delta\lambda$ and Δd_x, whereas the one along the direction of the y axis (the direction parallel to the holographic fringes) is a function of Δd_y only. These results have been confirmed experimentally using an optical system shown in Figure 20.12 with a GaAs-Al$_x$Ga$_{1-x}$As heterojunction laser diode as the light source. The injection laser current was varied. As its value becomes larger, so does the spectral width $\Delta\lambda$ of the emitted laser light, as well as the size Δd of the light-emitting region. The reconstructed images are shown in Figure 20.13. Figures 20.13(a), (b) and (c) are images reconstructed with this laser operating with I_{th}, $3.5 \times I_{th}$ and $5 \times I_{th}$, respectively, where I_{th} is the threshold current for lasing. Image degradation due to the defocusing and the lateral shift is most remarkable in 13(c), and the least in 13(a). Figures 20.13(d), (e) and (f) are also images reconstructed with the same laser diode

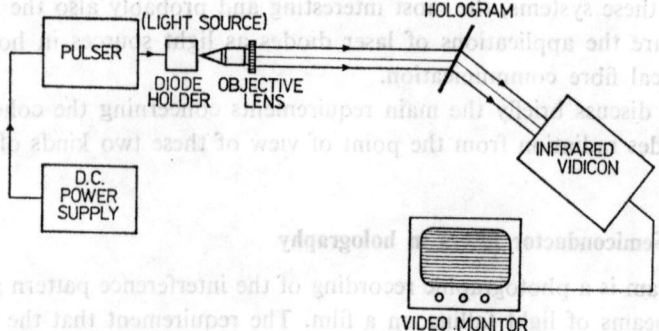

Figure 20.12 Optical system for holographic image reconstruction with an injection laser (26)

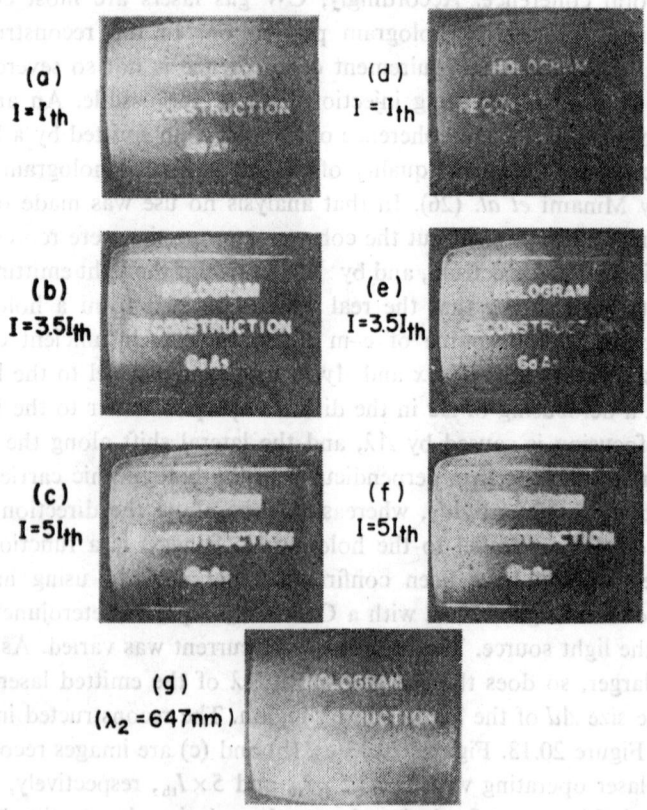

Figure 20.13 Holographic images reconstructed with a GaAs–$Al_xGa_{1-x}As$ heterojunction laser (a)–(f) and with a highly coherent Kr^+ laser lights (26)

with I_{th}, $3.5 \times I_{th}$ and $5 \times I_{th}$, respectively. But in these cases, the direction of the long side (along p–n junction) of the light-emitting region is rotated by 90 degrees as compared with the former cases where the direction of the long side is parallel to the carrier fringes of the hologram. Comparing these two groups of photographs, one can examine the effect of the size of the light-emitting region Δd (representing the spatial coherence of the source) independently of that of the spectral width $\Delta \lambda$ (representing the temporal coherence). It is remarkable that the main effect on image quality have: (a) temporal coherence and (b) spatial coherence in the direction perpendicular to the holographic carrier fringes. Recently, the technology of injection laser diodes has been developed to such a sophisticated level that the manufacturing of laser diodes emitting radiation of high spatial and temporal coherence has become possible. This applies to the buried stripe-geometry DH injection lasers that permits the mode structure of the emitted radiation to be controlled by the technology of the laser diode and which additionally can operate CW at room temperature. The optical arrangement used to make holograms with injection lasers is shown in Figure 20.14(a) (27). To increase the diameter of the laser beam from its small initial diameter, the beam is projected through a beam expander. Following the expander, a beam splitter deflects dawnward approximately 50% of the beam. This second beam is redirected by a flat-surface mirror to the second beam splitter. The main beam passes through the beam splitter, redirected by the mirror and is then combined by the second beam splitter with the other beam. The interbeam angle is made to be 0.5°, which corresponds to an average fringe frequency of 10 lines/mm in the hologram. The objects used in the experiments reported in Reference 27 were letters written on 1 mm thick glass plates. The size of the individual letters on the glass plate was 3 mm², a line width of ap-

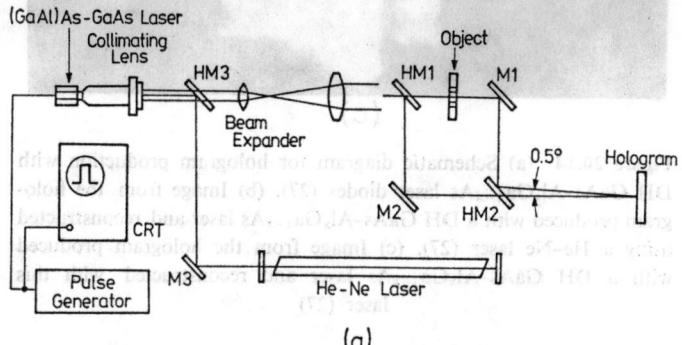

Fig. 20.14 (see p. 586)

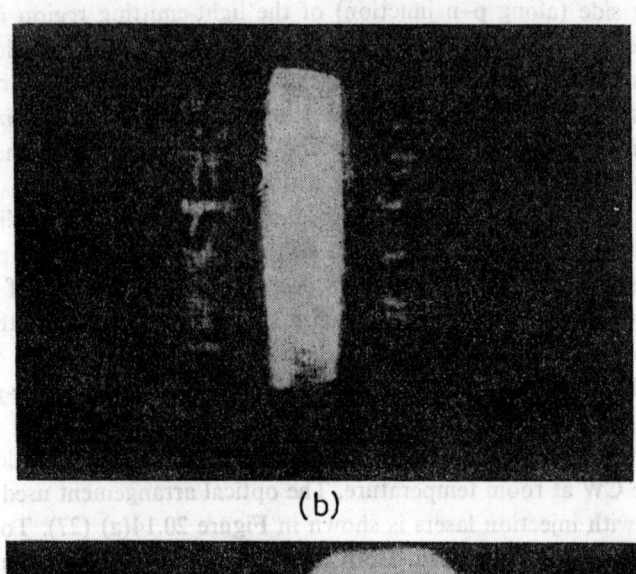

(b)

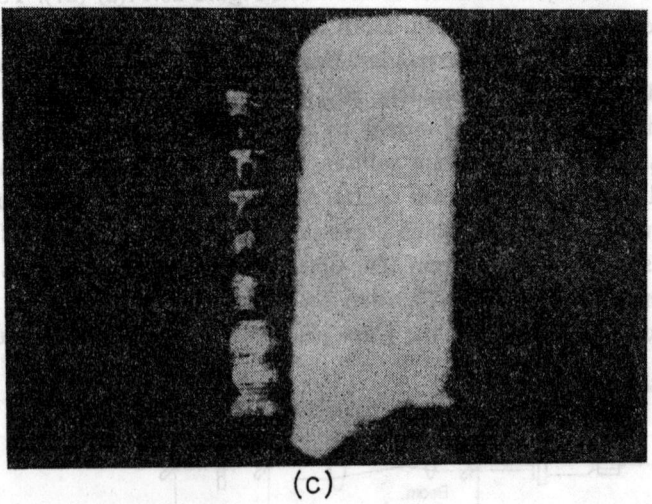

(c)

Figure 20.14 (a) Schematic diagram for hologram production with DH GaAs–Al$_x$Ga$_{1-x}$As laser diodes (27). (b) Image from the hologram produced with a DH GaAs–Al$_x$Ga$_{1-x}$As laser and reconstructed using a He–Ne laser (27). (c) Image from the hologram produced with a DH GaAs–Al$_x$Ga$_{1-x}$As laser and reconstructed with this laser (27)

proximately 0.5 mm. The object transparency was placed in either the upper or the lower beam at approximately 70 cm from the hologram. The other beam served as a reference beam. Operated in pulse mode (pulse width 150 ns, duty cycle 0.5%) at room temperature the laser diode radiated approximately 4 mW in peak output (through the collimating lens) with a coherence length of about 15 mm. The developed holograms were reconstructed with coherent light from He–Ne laser (Figure 20.14(b)) and with light of pulsed DH semiconductor laser used for production of this hologram (Figure 20.14(c)). The images reconstructed with these two different light sources but recorded on the hologram by the same DH laser have a similar quality.

A more detailed analysis of the problem of image reconstruction from holograms using injection lasers has been made by Morozov (28, 29), Samoylov et al. (30, 31, 32), and Popov (33). Some results of this analysis are shown in Figures 20.15 and 20.16. The problem of recording Fourier

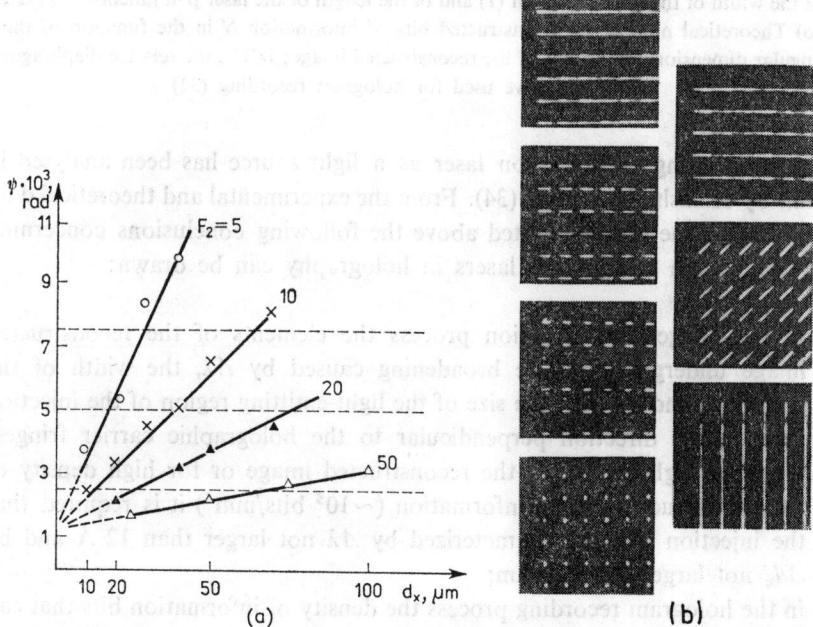

Figure 20.15 (a) The dependence of the angular dimensions ψ of a point of the reconstructed image on the linear dimension d_x of the light-emitting region of the laser p–n junction (32). (b) Photographs of some fragments of the reconstructed transparency consisting of 10^4 points on a 1 mm^2 hologram at different values of the length of the radiating laser p–n junction: 1, 80 μm; 2, 60 μm; 3, 40 μm; 4, 20 μm; and at different orientations of the laser p–n junction against the plane of holographic carrier fringes: a, parallel; b, 45°; c, perpendicular (32)

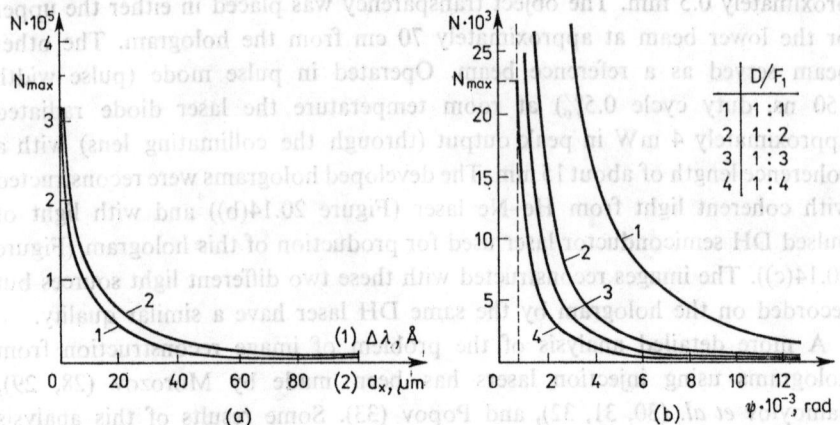

Figure 20.16 (a) The number of reconstructed bits of information N in the function of the width of the laser spectrum (*1*) and of the length of the laser p–n junction (*2*) (32). (b) Theoretical number of reconstructed bits of information N in the function of the angular dimensions of a point of the reconstructed image; D/F_1, the relative diaphragm of the objective used for hologram recording (32)

holograms using the injection laser as a light source has been analysed in detail by Kalashnikov *et al.* (34). From the experimental and theoretical data presented in the literature cited above the following conclusions concerning the application of injection lasers in holography can be drawn:

a. in the image reconstruction process the elements of the reconstructed image undergo a definite broadening caused by $\Delta \lambda$, the width of the spectrum, and by Δd_x, the size of the light-emitting region of the injection laser in the direction perpendicular to the holographic carrier fringes; thus, for high quality of the reconstructed image or for high density of the reconstructed bits of information ($\sim 10^5$ bits/mm²) it is required that the injection laser be characterized by $\Delta \lambda$ not larger than 12 Å and by Δd_x not larger than 20 μm;

b. in the hologram recording process the density of information bits that can be recorded on a hologram depends strongly on the temporal coherence of the laser radiation. A decrease of coherence length, e.g. to about 30% (160 μm) of its maximum value (450 μm), causes a decrease of the recorded bits density to about 20% (0.77×10^3 bits/mm²) of its maximum value (3.7×10^3 bits/mm²) (34). The negative influence of the insufficient spatial coherence of the injection laser on the density of information bits that can be recorded on a hologram can be compensated for by suitably selecting the magnification factor of the optical system forming the reference beam (34); thus, the insufficient spatial coherence of the injection laser does not play a critical role in this process.

20.3.2 Semiconductor lasers in optical communication systems

In order to evaluate channel capacities and transmission losses of optical communication system sit is often indispensable to study the statistical properties of partially coherent waves. This results from the fact that optical waves generated in semiconductor lasers and LEDs, the two kinds of sources most promising as light emitters for optical communication systems, are only partially coherent in space and time. The spectrum widths of these light sources are comparable to, or larger than, those of PCM signals in a high capacity and high speed optical communication channels (35), and, the temporal coherence of light pulses generated in these diodes has a considerable effect on the channel capacity. Furthermore, the actual optical fibres have index fluctuations, and these inhomogeneities lead to scattering losses and mode conversions in the fibres and decrease the coherence of guided waves (42). A comprehensive analysis of the fundamental propagation properties, the excitation characteristics, and the spatial and temporal coherence of partially coherent optical waves in random gradient fibres has been presented by Miyazaki (36) for incident light waves radiated from semiconductor lasers and from light emitting diodes. It has been shown theoretically that when the spatial coherence of optical sources exciting the graded-index fibre decreases, then in the same way the excitation coefficient of this fibre decreases too. This is especially distinct for lower modes of the fibre. Thus, lower modes are efficiently excited in the case of coherent laser beams while incident waves of low coherence such as light waves of LEDs, excite higher modes. An analysis of the transmission equations for the temporal correlation function of the electric field in the fibre has shown that gradient mode filterings for higher modes with lossy inhomogeneous claddings preserve the width of the input signal pulse (preserve the temporal coherence of this pulse) and protect the signal pulse from pulse broadenings caused by mode group delays and dispersions of higher modes through mode conversions and mixings due to index fluctuations.

The problem of fibre excitation by partially coherent sources for step index fibres has been analysed by Carpenter and Pask (37). In this analysis the exciting source is defined by the coherence properties of its radiation on the face of the waveguide. The source is assumed to be a circular disc of radius

d over the fibre entrance. The fibres studied are circular in cross section, with core radius ϱ and step refractive index profile. The source and fibre geometry is shown in Figure 20.17. In this case (for a system with circular

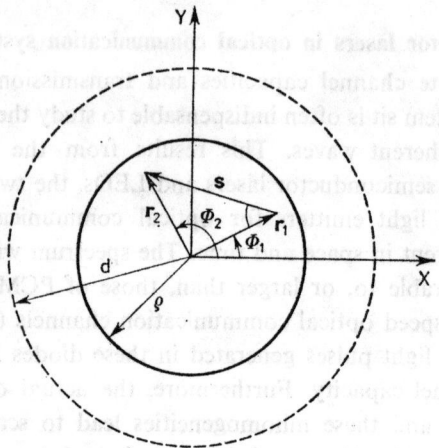

Figure 20.17 Source (radius d) and fibre (radius ϱ) geometry (37)

symmetry) the complex degree of coherence of the source radiation is defined by the formula

$$\gamma_{12} = \frac{2J_1(S/R_{\text{coh}})}{S/R_{\text{coh}}} \qquad (20.23)$$

where R_{coh} is a parameter called the *radius of coherence* of the source that determines the degree of coherence, $S = s/\varrho$ is the normalized distance between the source test points, and J_1 is the usual Bessel function.

The results of the theoretical analysis are shown in Figure 20.18. It can be seen (Figure 20.18(a)) that in general the power accepted by the fibre increases as the source becomes more coherent. An increase in the waveguide parameter

$$V = \left(\frac{2\pi\varrho n_1}{\lambda}\right) \sin\vartheta_c \qquad (20.24)$$

where $\cos\vartheta_c = n_2/n_1$ and n_1, n_2 are the core and cladding refractive indices respectively, improves the power acceptance ratio, particularly for highly incoherent sources.

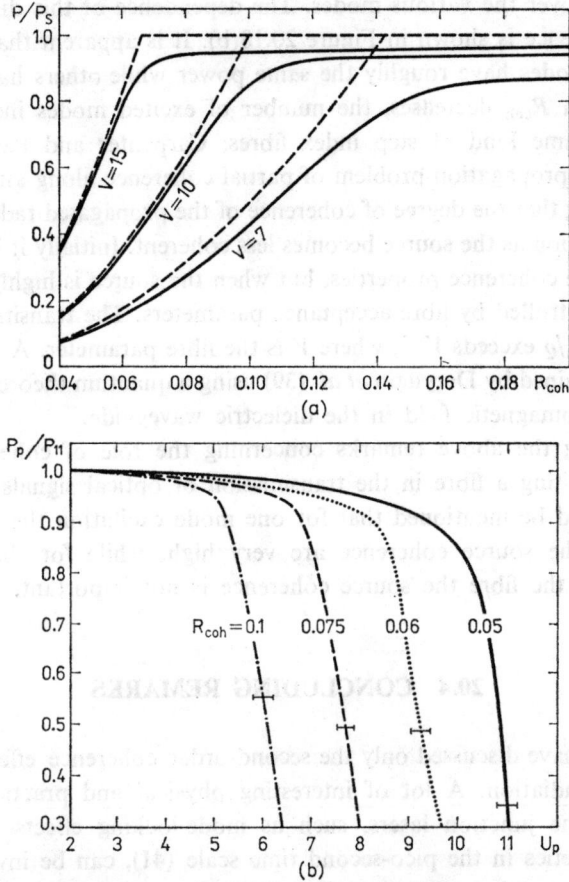

Figure 20.18 (a) Total power in the bound modes P normalized by the source output P_s as a function of the radius of coherence of the source R_{coh} for different values of the waveguide parameter V and for $D = \varrho/d = 1$ and $\vartheta_c = 0.1$. Continuous curves and dashed curves were obtained by different theoretical approximations (37). (b) Bound mode power P_p normalized by the power of the lowest order HE_{11} mode P_{11} as a function of the eigenvalue U_p of the appropriate mode field, for various degrees of coherence ($D = 1$, $V = 20$, $\vartheta = 0.1$). On each curve, the point where $1/R_{coh} = U_p$ is marked by a horizontal bar (37)

The behaviour of power propagating along an optical fibre depends on its distribution over the various modes. The dependence of this distribution on source coherence is shown in Figure 20.18(b). It is apparent that for a given R_{coh} some modes have roughly the same power while others have very little power. When R_{coh} decreases, the number of excited modes increases.

For the same kind of step index fibres, Carpenter and Pask have also analysed the propagation problem of partial coherence along an optical fibre (38), showing that the degree of coherence of the propagated radiation undergoes a transition as the source becomes less coherent. Initially it is determined by the source coherence properties, but when the source is highly incoherent, then it is controlled by fibre acceptance parameters. The transition occurs as the ratio R_{coh}/ϱ exceeds V^{-1}, where V is the fibre parameter. A similar result has been obtained by Deryugin et al. (39) using a quantum theoretical analysis of the electromagnetic field in the dielectric waveguide.

Concluding the above remarks concerning the role of coherence of the radiation exciting a fibre in the transmission of optical signals through the fibre it should be mentioned that for one mode excitation the requirements concerning the source coherence are very high, while for the multimode operation of the fibre the source coherence is not important.

20.4 CONCLUDING REMARKS

Thus far we have discussed only the second-order coherence effects of injection lasers radiation. A lot of interesting physical and practical problems concerning the junction lasers, such as mode-locking effects (40), or the radiation kinetics in the pico-second time scale (41), can be investigated by studying higher-order coherence effects. This is realised in practice by using a non-linear detection technique in Michelson interferometry (41), or by using the photon counting technique (8). It also seems reasonable to hope that coherence investigations of injection lasers will lead to a new method of investigating degradation phenomena in these lasers. The statistical properties of laser radiation should depend on the homogeneity of the radiation field in the active layer of the laser. Since degradation starts with dark spots and dark line defects of the lasing structure, the homogeneity of the radiation field is influenced by these degradation effects, so the coherence of the emitted radiation should also be influenced by degradation effects. Much more interesting physical problems are, however, connected with the origin and growth of the radiation coherence within the injection laser itself.

So far this problem has been analysed only for lasers where the atomic system could be considered as a two level system. In semiconductors the physical situation is more complicated, since the recombining carriers current populates energy bands with definite distribution probabilities. Consequently, one has to analyse the band-band transition problem.

To sum up our discussion of the radiation coherence of injection lasers the following arguments may be cited to justify the importance of the problem:

(i) A miniature coherent light source is required for holography (holographic computer memories, hologram reading and recording heads) and for optical communication systems.
(ii) It may be possible to control the coherence of laser radiation optimizing the technology of laser structures.
(iii) Application of coherence investigations to degradation problems and to problems concerning light generation dynamics in laser diodes seems to be possible.
(iv) Interesting scientific problems resulting from the following groups of questions
   (a) The source—how does coherence arise in the source, how can it be controlled by the source parameters and how can technology make this possible?
   (b) The free e-m field—how can the coherence of laser diode radiation be measured, which exploitation parameters of laser diode are related to the coherence of its radiation and how?

ACKNOWLEDGEMENTS

The author is deeply indebted to Professor A. Kujawski for stimulating discussions and helpful criticism.

REFERENCES

1. E. Wolf, *Proc. Roy. Soc.* A, **230**, 246 (1955).
2. A. Blanc-Lapierre, P. Dumontet, *Rev. d'Optique*, **34**, 1 (1955).
3. R. J. Glauber, *Phys. Rev.*, **130**, 2529 (1963).
4. R. J. Glauber, *Phys. Rev.*, **131**, 2766 (1963).
5. E. Cahill, R. J. Glauber, *Phys. Rev.*, **177**, 1857 (1968).
6. L. Mandel, E. Wolf, *Rev. Mod. Phys.*, **37**, 231 (1965).
7. G. J. Troup, R. G. Turner, *Rep. Prog. Phys.*, **37**, 771 (1974).
8. H. P. Baltes, *Appl. Phys.*, **12**, 221 (1977).

9. M. J. Beran, G. B. Parrent, Jr., *Theory of Partial Coherence*, Prentice-Hall Inc., Englewood Sliffs, New Jersey, 1964.
10. J. Perina, *Coherence of Light*, Van Nostrand Reinhold Comp., London, 1971.
11. J. F. Vinson, *Coherence Optique Classique et Quantique*, Dunod, Paris, 1969.
12. W. H. Louisell, *Quantum Statistical Properties of Radiation*, J. Wiley and Sons, New York, 1973.
13. R. Loudon, *The Quantum Theory of Light*, Clarendon Press, Oxford, 1973.
14. B. S. Vvedenskii, A. S. Logginov, K. S. Senatorov, *Kvantovaya Elektronika*, **1**, 1232, (1974).
15. Yu. A. Bykovskii, V. A. Elhov, A. I. Larkin, *Phys. Tekhn. Poluprov.*, **4**, 962 (1970).
16. M. I. Natan, *Appl. Optics*, **5**, 1522 (1966).
17. R. Vuilleumier, N. E. Collins, J. M. Smith, C. S. Kim, H. Raillard, *Proc. IEEE*, **55**, 1420 (1967).
18. W. Buchtemann, D. H. Hohn, *Optik*, **29**, 401 (1969).
19. J. C. Dyment, *Phys. Letters*, **10**, 84 (1967).
20. M. A. Herman, T. Bryśkiewicz, B. Wiktor, *Electron Technology*, **12**, 49 (1979)
21. Yu. Bykovskii, V. L. Velichanskii, V. A. Elhov, Yu. P. Zaharov, A. I. Larkin, V. A. Maslov, R. V. Riabova, D. M. Samoilovitch, V. L. Smirnov, *Dokl. AN USSR Fiz.*, **203**, 1027 (1972).
22. Yu. Bykovskii, V. L. Velichanskii, I. C. Goncharov, V. A. Maslov, V. V. Nikitin, *Fiz-Tekh. Poluprov.*, **5**, 498 (1971).
23. H. G. Danielmeyer, H. P. Weber, *Phys. Rev.*, **3A**, May (1971).
24. C. A. Brackett, *IEEE J. Quantum Electronics*, **QE-8**, 66 (1972).
25. M. A. Herman, B. Wiktor, J. Biernacki, A. Biernacki, *Pomiary Automatyka Kontrola*, **23**, 375 (1977).
26. M. Minami, Y. Unno, Y. Mizobuchi, *Appl. Optics*, **10**, 1629 (1971).
27. T. Sugaya, A. Iwamoto, *Optics Commun.*, **10**, 37 (1974).
28. V. N. Morozov, *Kvantovaya Elektronika*, **5**, 5 (1973).
29. V. N. Morozov, *Rozprawy Elektrotechn.*, 1979 (in press).
30. V. V. Nikitin, V. D. Samoylov, G. I. Semenov, *Kvantovaya Elektronika*, **1**, 7, (1974).
31. V. P. Karpelcev, V. D. Samoylov, *Kvantovaya Elektronika*, **1**, 167 (1974).
32. N. G. Basov, V. V. Nikitin, V. D. Samoylov, G. I. Semenov, *Optical Methods of Information Processing*, Nauka, Leningrad, 1974.
33. Yu. M. Popov, *Semiconductor Sources of Electromagnetic Radiation*, ed. M. A. Herman, Polish Sc. Publ., Warszawa, 1976.
34. S. P. Kalashnikov, I. I. Klimov, V. V. Nikitin, G. I. Semenov, *Kvantovaya Elektronika*, **4**, 1666 (1977).
35. Y. Miyazaki, *Archiv. Elektr. Übertr.*, **28**, 160 (1974).
36. Y. Miyazaki, *Opt. Quant. Electronics*, **9**, 153 (1977).
37. D. J. Carpenter, C. Pask, *Opt. Quant. Electronics*, **8**, 545 (1976).
38. D. J. Carpenter, C. Pask, *Optics. Commun.*, **22**, 99 (1977).
39. I. A. Deryugin, S. S. Abdullaev, A. T. Mirzaev, *Kvantovaya Elektronika*, **4**, 2173 (1977).
40. H. Bachert, P. G. Eliseev, M. A. Manko, V. P. Strahov, S. Raab, C. M. Thay, *IEEE J. Quant. Electron.*, **QE-11**, 507 (1975).
41. M. A. Manko, *Rozprawy Elektrotechniki*, 1979 (in press).
42. B. Crosignani, P. Di Porto, M. Bertolotti, *Statistical Properties of Scattered Light*, Academic Press, New York, 1975.

Part V
Optical Communication Systems

Chapter 21

Properties of Optoelectronic Devices for Optical Communication Systems

J. C. Dyment

21.1 INTRODUCTION

In this chapter, several properties of optoelectronic devices will be discussed with emphasis being placed on how these properties determine the overall performance of fibre optic transmission systems. In its simplest form, an optical communications system has three major components as illustrated in Figure 21.1: (1) a transmitter, which contains the light source (usually an

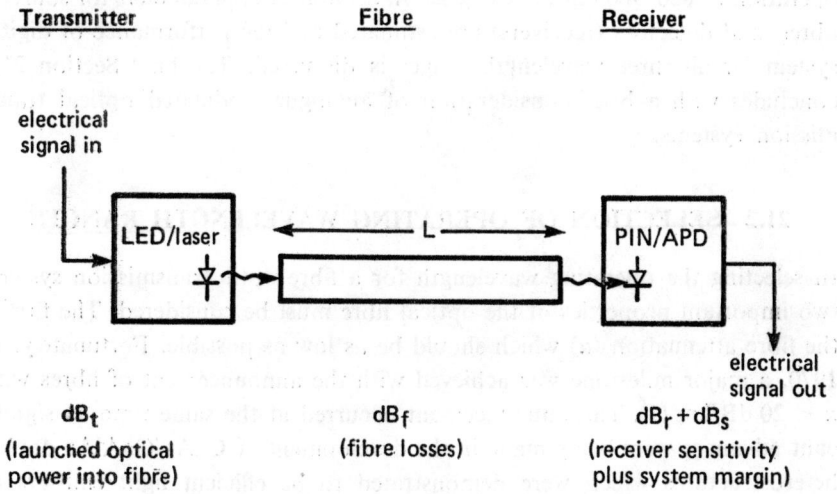

Figure 21.1 Schematic of fibre optic transmission system

LED or injection laser), (2) an optical fibre, which carries the light over a distance L to the receiver, and (3) a receiver, which contains a light detection device (usually a PIN diode or an avalanche photodiode (APD)). Several decisions must be made in selecting those devices to optimize the performance of a fibre optic transmission system. For example,

— what source-detector combinations are appropriate?
— what wavelength range?
— what modulation range and technique (analogue or digital)?
— what type of optical fibre (step or graded index, multimode or single mode)?

In general, there are several possible combinations of components which can be used to construct a desired system. However, often there is some overriding characteristic of a component which will dictate its choice over another. It is the purpose of this chapter to identify and discuss some of the considerations for selecting components.

We begin in Section 21.2 with a discussion of those optical fibre and detector properties which have led to the identification of three wavelength ranges for system operation. Then, in Section 21.3, we discuss material aspects of light sources appropriate for these three wavelength ranges. Section 21.4 presents a discussion of several properties of LED/laser light sources for operation at 800–900 nm wavelengths. In Section 21.5, parameters for sources, fibres, and detectors (receivers) are estimated and the performance of digital system for all three wavelength ranges is discussed. The final Section 21.6 concludes with a brief consideration of analogue modulated optical transmission systems.

21.2 SELECTION OF OPERATING WAVELENGTH RANGES

In selecting the operating wavelength for a fibre optic transmission system, two important properties of the optical fibre must be considered. The first is the fibre attenuation (α) which should be as low as possible. Fortunately, in 1970, a major milestone was achieved with the announcement of fibres with $\alpha < 20$ dB/km (1). This announcement occurred at the same time as significant advances were being made in the development of GaAs/GaAlAs double heterostructures which were demonstrated to be efficient light emitters at wavelengths $\lambda = 800$–900 nm (2). This wavelength range was matched quite well to a minimum in fibre attenuation. With improved fibre design and fabrication techniques, fibre losses have been steadily reduced so that, today, typical values of $\alpha < 5$ dB/km are found at $\lambda = 800$–900 nm (see Figure 21.2). Even lower fibre attenuations are measured at wavelengths beyond 1000 nm which is one of the reasons for considering longer wavelength light sources. A second fibre property, the chromatic dispersion, is another compelling reason for longer wavelength systems. Chromatic dispersion decreases with increasing wavelength and is near zero in the vicinity of 1300 nm (3).

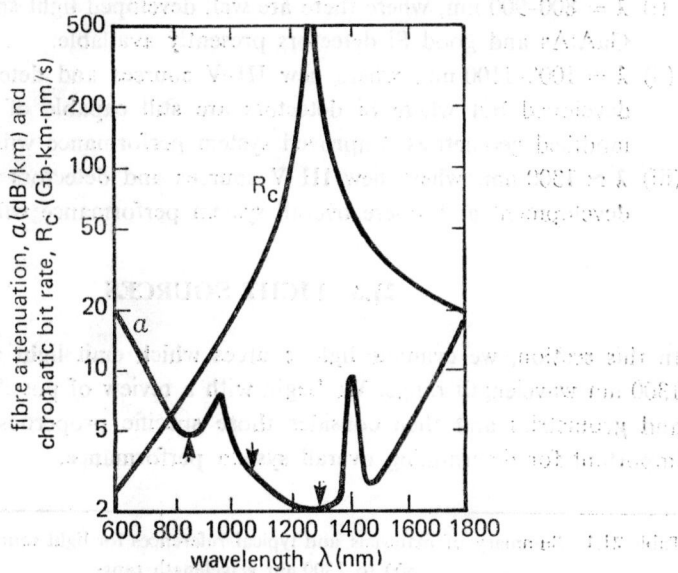

Figure 21.2 Typical values for fibre attenuation and chromatic bit rate as a function of wavelength. Arrows indicate the three wavelength regions discussed in this chapter for operation of fibre optic transmission systems

Thus, the corresponding chromatic bit rate (R_c) becomes very large for wavelengths near 1300 nm as illustrated in Figure 21.2.

Thus, from the fibre point of view, the two important properties of α and R_c both improve with wavelength and operating wavelengths of $\lambda \simeq 1300$ nm would appear to be optimal. Unfortunately, the availability of both light sources and detectors is quite limited for $\lambda \simeq 1300$ nm. At the present time, there are no commercial light sources at 1300 nm, although several laboratories have research programs to develop suitable alloys based on group III and group V elements (Section 21.3). Investigations into III–V materials for long wavelength detectors are also underway. These materials should have higher quantum efficiencies than Ge and also smaller leakage currents. Thus, detectors from III–V materials will have lower excess noise generated by multiplied dark current.

In view of the above discussion, three wavelength regions have been identified for optical transmission systems, as indicated on Figure 21.2 with arrows:

(i) $\lambda = 800$–900 nm, where there are well developed light sources of GaAs/GaAlAs and good Si detectors presently available.

(ii) $\lambda = 1000$–1100 nm, where new III–V sources and detectors are being developed but where Si detectors are still capable of operation with modified geometries. Improved system performance will occur.

(iii) $\lambda \simeq 1300$ nm, where new III–V sources and detectors are both under development and where overall system performance will be optimized.

21.3 LIGHT SOURCES

In this section, we evaluate light sources which emit light within the 800–1300 nm wavelength range. We begin with a review of possible source types and geometries and then consider those specific properties which will be important for determining overall system performance.

Table 21.1 Summary of materials and typical references for light sources emitting in the 800 to 1300 nm wavelength range

Light source	800–900 nm	1000–1100 nm	1300 nm
LEDs	GaAs/GaAlAs (4–6)	GaInAs (7) GaInAsP/InP (8) GaAsSb (9)	InGaAsP/InP (10)
Injection lasers	GaAs/GaAlAs (2, 4) GaAs/InGaP (11)	InGaAs/InGaP (12) InGaAsP/InP (13) GaAsSb/GaAlAsSb (14)	GaInAsP/InP (15)
YAG/LED	—	Nd:YAG fibres (16)	Nd:YAG fibres (17)

For LEDs and lasers, heterojunctions have been designated with two materials separated by a slash (/). The lower bandgap active layer is on the left of the slash while the wider bandgap confining layer is on the right.

21.3.1 Types of light sources

Three classes of solid state light source can be considered for the 800–1300 nm wavelength region. These are listed in Table 21.1 and include light emitting diodes (LEDs), injection lasers, and neodymium yttrium aluminum garnet

(Nd : YAG) crystals pumped by LEDs. Also included in Table 21.1 is a summary of the principal materials and typical references (2, 4–17) for each class. YAG/LED sources (16, 17) have operated in both of the longer wavelength ranges, although the research effort has been considerably smaller than for the III–V compounds used in p–n junction sources. The main advantage of YAG/LED lasers over the III–V compound sources is their very narrow output spectrum ($\simeq 0.2$ nm). However, disadvantages of these sources are that they are very inefficient, bulky, and probably will require external modulators. Accordingly, they are not seen to be as important in the near future for optical communications as are semiconductor diode sources, and they will not be further discussed here.

21.3.2 Materials for LED and laser sources

Intensive work is underway to develop LED and injection laser sources which utilize III–V compound semiconductor materials incorporating both ternary and quaternary alloys. Ternary alloys are formed by combining two of the

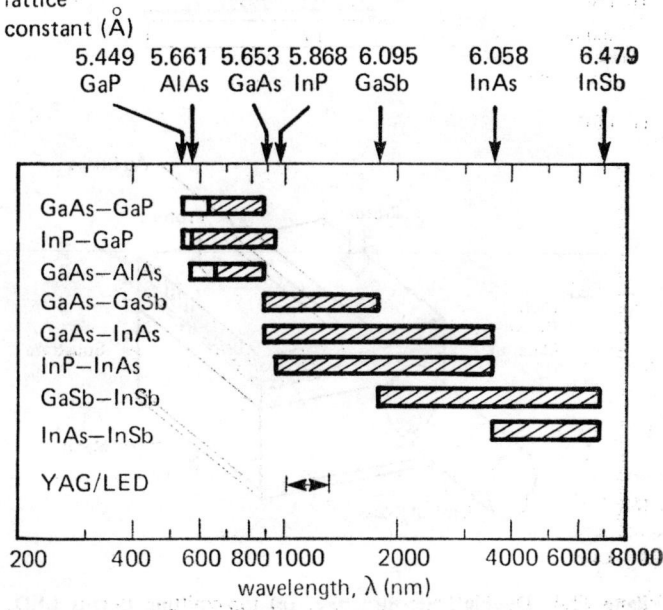

Figure 21.3 Room temperature wavelength ranges for several III–V ternary alloys plus YAG/LED lasers. Cross-hatching indicates the direct bandgap range employed for high efficiency. Lattice constants for binary materials are indicated

elemental binary compounds together. Because the composition is continuously variable it is possible to span a wide range of wavelengths with a single ternary alloy system. As an example, light sources formed from the ternary $GaAs_xSb_{1-x}$ span the range from 870 nm for pure GaAs ($x = 1$) to 1700 nm for pure GaSb ($x = 0$). This wavelength range, as well as the ranges for several other common ternary alloys, are shown in Figure 21.3.

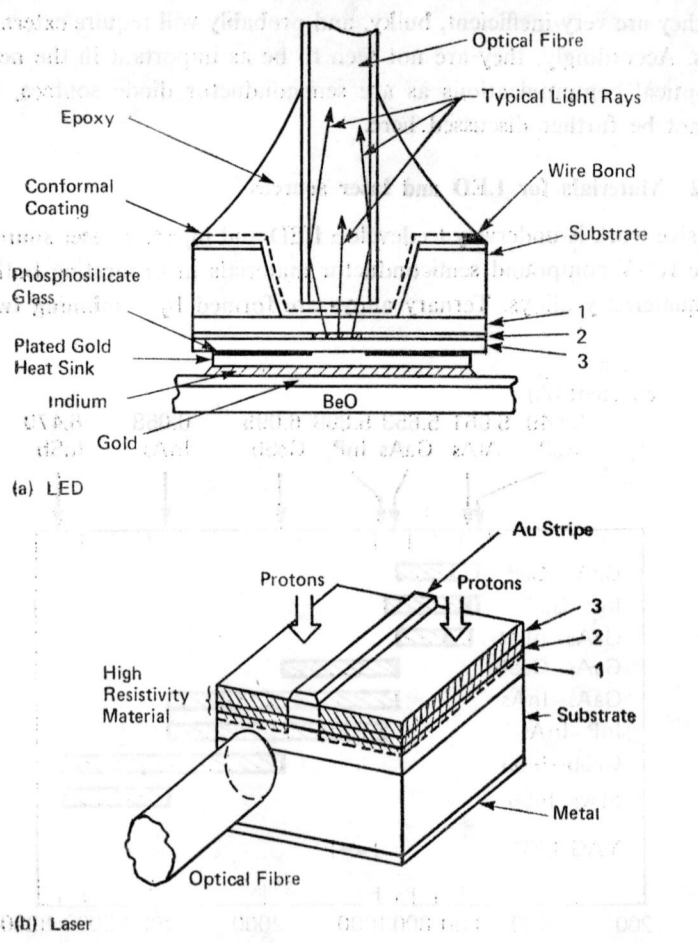

Figure 21.4 Double heterostructure; (a) top-emitting Burrus LED; (b) stripe geometry laser with proton bombardment current confinement. Layer 2 is the active one while layers 1 and 3 are wider bandgap confining layers. Some typical material combinations are listed in Table 21.1

For efficient light generation, the alloy composition must be chosen so that the energy band gap is direct; these regions are indicated by cross-hatching in Figure 21.3. In designs incorporating a double heterostructure (2, 4, 5), a further condition of lattice matching must also be fulfilled. Mismatch arises in double heterostructure designs when two dissimilar materials with differing lattice constants are placed adjacent to one another. For devices which incorporate heterojunctions, a slash (/) has been used in Table 21.1 to distinguish the materials employed for the central active region (left side of the slash) from the two adjacent confining layers (right side of the slash).

The close lattice matching achieved with GaAs and AlAs is the main reason for the extensive use of GaAlAs alloys for the 800–900 nm wavelength ranges. Longer wavelength materials such as GaInAs, GaAsSb, and InGaAsP require more careful attention to the lattice matching problem. For these, the difficulty is often one of preparing a ternary alloy on a binary substrate. For example, large lattice mismatch occurs for alloys such as GaInAs or GaAsSb on GaAs substrates. It is necessary to grow a compositionally graded layer between the substrate and the device layer in order to gradually change the lattice constant. The grading can be carried out either by step grade or by linear grade techniques (4). An added degree of flexibility is provided by quaternary alloys, such as GaInAsP, which allow independent control of energy bandgap and lattice constant over fairly wide energy ranges. Quaternary alloys have recently been used for the preparation of CW laser devices operating at room temperature (13).

21.3.3 Device configurations

Figure 21.4 illustrates typical LED and laser structures employing a double heterostructure design. For the Burrus design of Figure 21.4(a) light exits from the top of the LED in a direction perpendicular to the junction. Good coupling to fibres is achieved by close butting (6). Other forms of top emitting LEDs employ structures in which the substrate has not been etched; usually a lensed LED (18) or a lensed fibre (19) is used to effectively couple light. In another type of LED, the light exits from the side of the device (20). Side emitting LEDs are similar to the laser structure of Figure 21.4(b), except that the stripe contact extends only a short distance from the exit face.

Diode lasers usually employ some kind of current confinement to form a stripe geometry. The method of confinement by proton bombardment (21) illustrated in Figure 21.4(b) is just one of many stripe geometry structures which have been reported (2). Coupling to fibres is usually achieved by utilizing microlenses either on the fibre (22) or adjacent to the emitting facet (23).

21.4 PROPERTIES OF GaAlAs LIGHT SOURCES

In this section, we discuss those properties of GaAlAs LED and laser light sources which are believed to be the most important for optical communications systems. These discussions form the basis for assigning those values listed in Table 21.2 which have been assumed for the systems calculations of Section 21.5.

Table 21.2 Typical parameters for GaAlAs LED and laser light sources (λ = 800–900 nm)

Parameter	LED			Laser
Active layer carrier concentration (carriers/cm³)	4×10^{18}	8×10^{18}	1.4×10^{19}	5×10^{17}
Ext. efficiency, η (%)	7	4.6	2.3	Differential $\eta \simeq$ 15–20 (one end)
Typ. optical output, P_0	3	2	1	5–10 mW
	(Rad. Int. I_R (mW/sr) at 150 mA dc)			
Risetime, t_r (ns)	14	7	4	0.1
Modulation (MHz)	40	90	150	1000
	(3 dB optical bandwidths, f_3)			
Spectral width, $\Delta\lambda$ (nm)	40	45	45	1–2
Typical linearity (THD = Total Harmonic Distortion)	-35 dB (I_{dc} = 100 mA, I_p = 100 mA)			-40 dB (2.5 mW peak-peak optical modulation)
Reliability (h)	> 10^6			> 10^5

21.4.1 Device efficiency, power output, and rise times for LEDs

The carrier concentration in the active layer has been found to play a profound role on LED performance. Table 21.2 shows how device efficiency (η) and radiant intensity (I_R) vary for three different high radiance LEDs fabricated at our laboratories. These devices have Ge-doped active layers approximately 1 μm thick. It is to be noted that the efficiency decreases by a factor of about 3 as the carrier concentration increases from 4×10^{18} cm^{-3} to 1.4×10^{19} cm^{-3}. Over this same range, it is also found that LED risetime (t_r) decreases from 14 ns to 4 ns. The risetime (t_r) is the time for the optical output power to increase from 10% to 90% of its maximum value under square wave current excitation. Thus the three LEDs listed in Table 21.2 represent different options which are available for the trade-off between efficiency and risetime in these structures. At 4×10^{18} cm^{-3}, the LED has been optimized for total light output; at 1.4×10^{19} cm^{-3}, the structure has been optimized for speed of response. The middle LED, at 8×10^{18} cm^{-3}, represents a typical compromise between these two parameters which might be employed in a practical system.

21.4.2 Modulation properties of LEDs

If the LED is modulated with an angular frequency ω, it has been shown that the optical output power (P_0) will decrease as

$$P_0 = \frac{1}{\sqrt{1+\omega^2 \tau_{sp}^2}} \quad (21.1)$$

where τ_{sp} is the spontaneous lifetime of the injected minority carriers. The modulation characteristics of the LED are described by a 3 dB optical bandwidth defined as follows:

$$f_3 = 0.28/\tau_{sp} \quad (21.2)$$

This is obtained by determining the value of ω which results in $|P_0(\omega)/P_0(0)| = 1/2$. Thus, as LED efficiency decreases, f_3 values increase and the system designer has to decide what trade-offs can be made between optical output power and modulation rate expected for the LED. The f_3 values of Table 21.2 are for devices with fairly high capacitances, of typically several hundred picofarads. Further increases in 3 dB optical bandwidths can be achieved by utilizing techniques such as proton bombardment to reduce junction capacitance (24). In such cases, risetimes less than 2 ns have been obtained.

Both the reduction in η and increase in f_3 at high carrier concentration arise from smaller values of τ_{sp} which can be expressed in terms of τ_r and τ_{nr}, the radiative and non-radiative components, as $1/\tau_{sp} = 1/\tau_r + 1/\tau_{nr}$. At high carrier concentrations, the amount of Ge incorporated into the lattice is near the limit of solid solubility. Accordingly, many defects are introduced into the crystal in this regime which reduce the overall values of τ_{sp} by decreasing τ_{nr}.

21.4.3 Device efficiency, power output, and delay time for lasers

Laser properties are much less affected by the carrier concentration in the active layer. The value of 5×10^{17} cm^{-3} stated in Table 21.2 is typical of many values reported in the literature. There is little influence on external differential quantum efficiency of the laser which is typically 15–20% (one end) or in the vicinity of 0.2 mW/mA slope. Thus P_0 values of about 5–10 mW are achieved with current increases of 25–50 mA above I_{th}.

The 'risetime' of a laser can be considered to be faster than 0.1 ns, a limitation which is imposed by the detector employed for the measurement. A more meaningful quantity to measure is the laser 'delay time' which is the time between the application of a square current pulse and the onset of stimulated emission. It has been shown (25) that the time delay (t_d) is given by

$$t_d \simeq \tau_{sp} \ln\left[\frac{(I-I_0)}{(I-I_{th})}\right] = \tau_{sp} \ln\left[\frac{I_p}{I_p+(I_0-I_{th})}\right] \quad (21.3)$$

where I_{th} is the threshold current, I_0 is a prebias current, I_p is the pulsed current superimposed on I_0, and I is the total current given by $I = (I_0 + I_p)$. It is assumed that $I > I_{th}$ and $I_0 < I_{th}$. The second form of the expression shows that the delay is determined by the pulse amplitude I_p and the difference

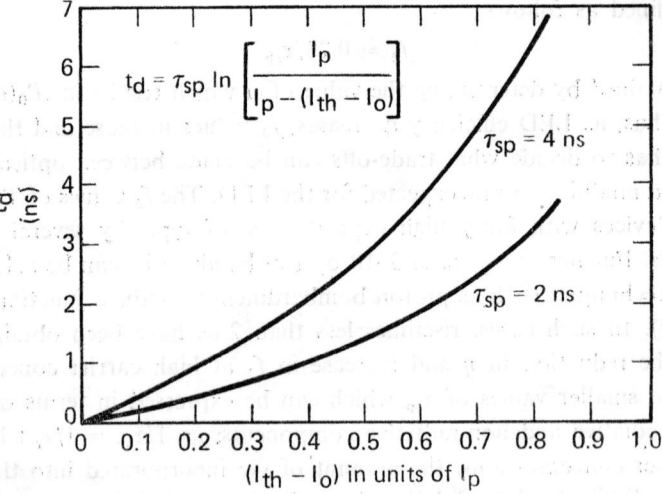

Figure 21.5 Values of laser time delay t_d as a function of parameter $(I_{th}-I_0)$ plotted from Equation 21.3 for two values of spontaneous lifetime (τ_{sp} = 2ns, 4ns). I_0 is pre-bias current (below I_{th}) and I_p is the superimposed pulse amplitude

between pre-bias and threshold currents. There is also a dependence on τ_{sp} and hence some influence on the active layer carrier concentration. Figure 21.5 shows a plot of Equation 21.3 for two values of τ_{sp} as a function of

($I_{th} - I_0$). In order to achieve values of t_d < 0.5 ns required for GHz modulation rates, ($I_{th} - I_0$) should be in the vicinity of $0.1 I_p$ to $0.25 I_p$. For example, at $I_p = 50$ mA, the laser must be pre-biased to within 5–13 mA of I_{th}.

21.4.4 Spectral widths of LEDs and lasers

The spectral width $\Delta\lambda$ is the full width measured between the points where the power output P_0 has decreased to 1/2 of its peak value. Typical values of $\Delta\lambda \simeq 45$ nm are appropriate for LEDs and $\Delta\lambda \simeq 1$–2 nm for lasers. In the case of lasers, it should be mentioned that $\Delta\lambda$ will depend strongly on the drive conditions and the extent to which the laser is driven above threshold. However, the $\Delta\lambda = 1$–2 nm value is typical of many lasers with a Fabry–Perot cavity. Considerably narrower $\Delta\lambda$ values have been achieved in some other laser structures which employ external cavities (26) or distributed feedback (27). However, both of these designs present some problems for the systems designer. In the former case, the laser is bulky and accurate cavity alignment must be maintained. In the latter case, the output power is low. Accordingly, our discussions will consider only the Fabry–Perot laser structure as shown in Figure 21.4(b).

21.4.5 Linearity of LEDs

Linearity of LEDs and lasers is of prime importance when analogue optical information is to be transmitted. A measure of linearity is provided by the total harmonic distortion (THD) as shown in Figure 21.6(a) for a Ge-doped LED at four different modulation levels. The data are taken at a fundamental frequency of 1000 Hz but no significant changes in distortion are observed up to 5 MHz. At a given modulation level, there is a minimum in the curve which occurs at a bias which does not vary more than 15% from device to device. With increasing modulation level, the minimum broadens and shifts to higher bias with an associated increase in THD for the LED. The distortion of Ge-doped LEDs is only weakly dependent on active layer carrier concentration and THD values typically in the range of −33 to −35 dB below the fundamental are observed at 100 mA dc bias, 100 mA peak-to-peak modulation currents. Figure 21.6(b) shows the 2nd and 3rd harmonic components of THD for typical Ge-doped LEDs. The minimum in THD is seen to result from a strong minimum occurring in the 2nd harmonic component.

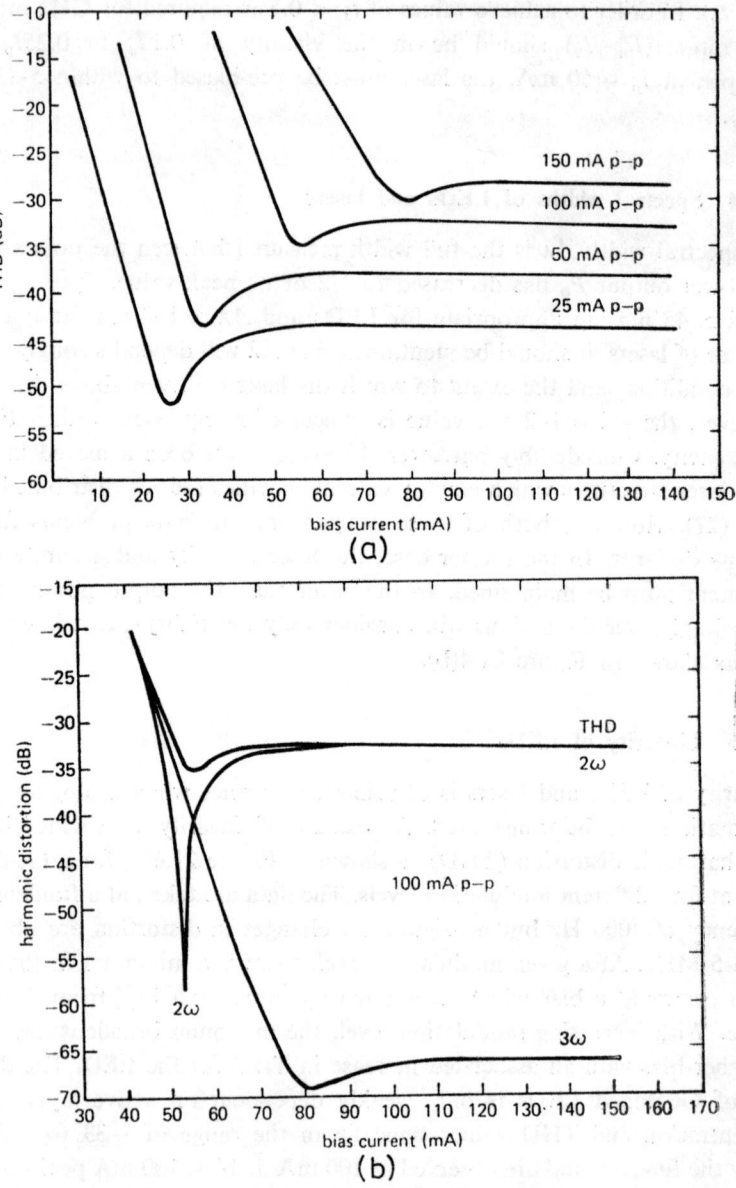

Figure 21.6 (a) Total harmonic distortion (THD) and (b) harmonic components of THD for a typical Ge-doped double heterostructure LED with a Burrus design (as in Figure 21.4 (a)). Plots are taken as a function of bias current with modulation levels as indicated

21.4.6 Possible linearization scheme for LEDs

Although the amount of distortion generated by LEDs is acceptable for single channel transmission over short distances (28), improved linearity is desired for multichannel optical signal transmission. In order to compensate for the inherent non-linearity of the LED, circuit design techniques can be

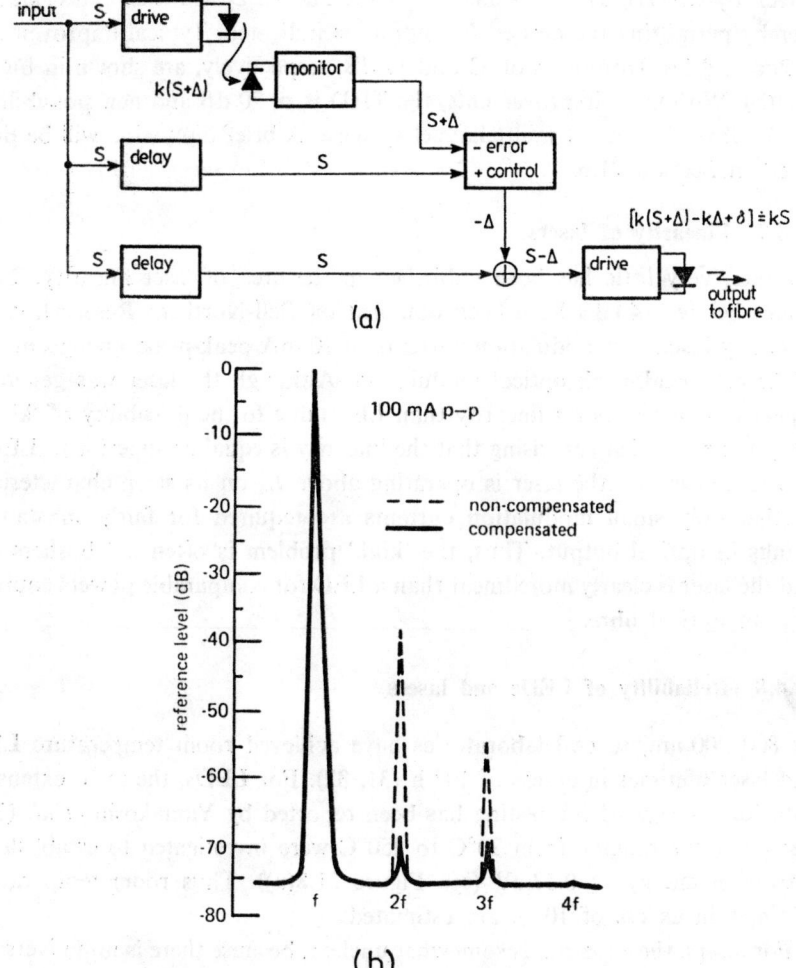

Figure 21.7 (a) Schematic diagram of quasifeedforward compensation scheme; (b) Frequency spectrum of original and compensated optical signal for quasifeedforward compensation of LEDs with a Burrus design. Compensation provides improvements in 2nd and 3rd order distortions of 33 dB and 17 dB, respecitvely

employed where necessary. A number of compensation techniques can be used (29) but one of the preferred schemes is known as 'quasifeedforward', as illustrated in Figure 21.7(a). Two equivalent LEDs are required. From the first LED, the drive signal S is distorted to $k(S+\Delta)$ where k describes the amplification factor and Δ is the induced non-linearity. The error signal is isolated, and then subtracted from S so that the second LED is being driven by $(S-\Delta)$. The non-linearity of the 2nd LED will then cancel out Δ, thereby permitting the desired kS output as indicated. Typical improvements in 2nd and 3rd harmonics of 33 and 17 dB, respectively, are shown in Figure 21.7(b). With these improvements, the THD is $\simeq$ 70 dB and new possibilities can be investigated for multichannel systems. A brief discussion will be presented in Section 21.6.

21.4.7 Linearity of lasers

To date, very little has been published in the area of laser linearity. THD values below -40 dB have been obtained on Bell-Northern Research stripe geometry lasers for modulation currents of 10 mA peak-peak, giving outputs of 2.5 mW peak-peak optical modulation. Although the laser was generally expected to have poorer linearity than LEDs due to the possibility of 'kinks' (30), it is somewhat surprising that the linearity is equal or superior to LEDs. The reason is that the laser is operating above I_{th} on its steep characteristic so that only small modulation currents are required for fairly substantial swings in optical output. Thus, the 'kink' problem is often not bothersome and the laser is clearly more linear than a LED for comparable powers coupled into an optical fibre.

21.4.8 Reliability of LEDs and lasers

At 800–900 nm, several laboratories have achieved room temperature LED and laser lifetimes in excess of 10^4 h (31, 32). For LEDs, the most extensive data for accelerated life testing has been reported by Yamakoshi et al. (31). Temperatures ranging from 20°C to 250°C were investigated to establish an activation energy of 0.57 eV (see Figure 21.8(a)). Thus room temperature lifetimes in excess of 10^6 h are estimated.

For lasers, the situation is somewhat unclear, because there is no universally accepted definition of laser lifetime. Bell Telephone Laboratories has assembled the most complete set of data in which laser lifetime is defined as the length of time for which the laser output power P_0 exceeds 1 mW (33). Numerous current adjustments are allowed. Using this definition, the distribution of

Properties of Optoelectronic Devices for Optical Communication Systems

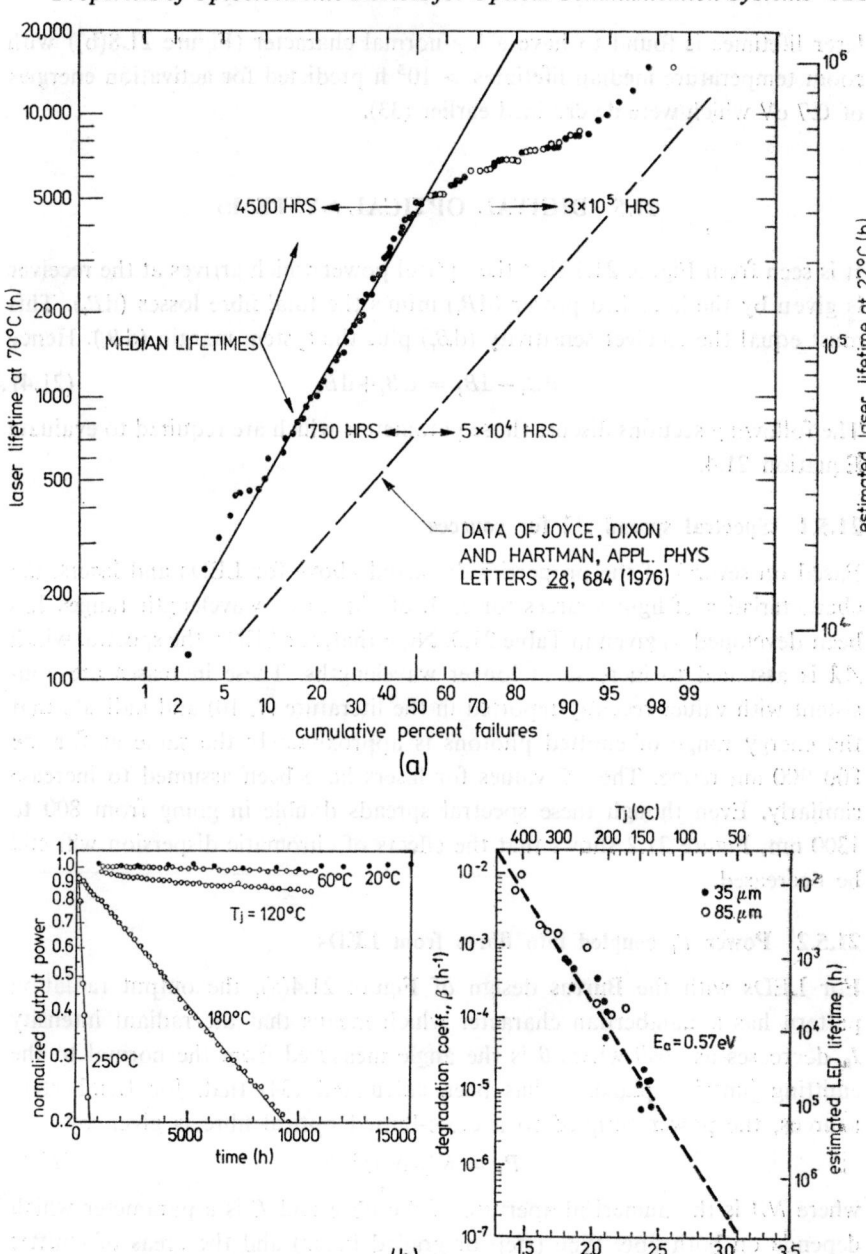

Figure 21.8 Accelerated life testing data for (a) LEDs and (b) lasers taken from (31) and (32), respectively. For LEDs, it is assumed that normalized output power varies with time t as $\exp(-\beta t)$ where β is a degradation coefficient depending on junction temperature (T_j) and activation energy (E_a) as $\beta = \beta_0 \exp(-E_a/kT_j)$. For lasers, the upper data show the lifetime results at 70°C for selected 'superior' crystals compared to more typical crystals (lower dashed line). Assuming $E_a = 0.7$ eV, median lifetimes at 22°C are predicted to be 3×10^5 h and 5×10^4 h, respectively, for these two crystal sets

laser lifetimes is found to have a log-normal character (Figure 21.8(b)) with room temperature median lifetimes $> 10^5$ h predicted for activation energies of 0.7 eV which were determined earlier (33).

21.5 DIGITAL OPTICAL SYSTEMS

It is seen from Figure 21.1 that the optical power which arrives at the receiver is given by the launched power (dB_t) minus the total fibre losses (dB_f). This must equal the receiver sensitivity (dB_r) plus the system margin (dB_s). Hence

$$dB_t - dB_f = dB_r + dB_s \qquad (21.4)$$

The following sections discuss those parameters which are required to evaluate Equation 21.4.

21.5.1 Spectral spread $\Delta\lambda$ for sources

Based on several of the properties discussed above for LEDs and lasers, the characteristics of light sources for each of the three wavelength ranges has been developed as given in Table 21.3. Note that, for LEDs, the spectral width $\Delta\lambda$ is assumed to increase at longer wavelengths. These increases are consistent with values recently reported in the literature (7, 10) and indicate that the energy range of emitted photons is approximately the same as for the 800–900 nm range. The $\Delta\lambda$ values for lasers have been assumed to increase similarly. Even though these spectral spreads double in going from 800 to 1300 nm, Figure 21.2 shows that the effects of chromatic dispersion will still be decreased.

21.5.2 Power P_f coupled into fibres from LEDs

For LEDs with the Burrus design of Figure 21.4(a), the output radiation pattern has a Lambertian character which means that the radiant intensity I_R decreases as $\cos\theta$ where θ is the angle measured from the normal to the emitting junction plane. It has been calculated (34) that, for Lambertian sources, the power coupled to a close-butted optical fibre is given by

$$P_f = KI_R(NA)^2 \qquad (21.5)$$

where NA is the numerical aperture of the fibre and K is a parameter which depends on both fibre type (step or graded index) and the areas of emitter and fibre core (A_E and A_F, respectively). It has been shown (34) that:

— for step index fibre, $K = 1$ always;

Table 21.3 Light sources, fibre, and detector parameters assumed for calculation of transmission distances

Sources	λ = 800–900 nm	1000–1100 nm	1300 nm	Peak* coupled power	Modulation range
LED	$\Delta\lambda$ = 45 nm	75 nm	100 nm	$P_f \simeq 200\,\mu W$ (R = 0)	0–200 Mb/s
Laser	$\Delta\lambda$ = 1.5 nm	2.5 nm	3.5 nm	$P_f \simeq 5\,mW$	0–2000 Mb/s

FIBRES		GRADED INDEX	
Attenuation (dB/km)	5	3	2
Chromatic bit rate R_c (Gb–km–nm/s)	8.5	25	∞
Multimode** dispersion Gb–km^q/s	0.5	0.5	0.5

DETECTORS

Type	PIN	APD	PIN	APD	PIN	APD
Material	Si	Si	III–V	Si (side) III–V	III–V	III–V
Quant. effic.	0.85	0.85	0.85	0.45 0.85	0.85	0.85

Notes: * Assume $P_f(\mu W) = \dfrac{200}{\sqrt{1+3(R/R_3)^2}}$, R_3 = 75 MHz

** $q = 1$, no mode mixing
$q = 1/2$, complete mode mixing

— for graded index fibre, $K = 1/2$, $(A_E \geqslant A_F)$;

$$K = 1 - \frac{A_E}{2A_F} \quad (A_E \leqslant A_F)$$

From Equation 21.5, assuming $A_E = A_F$ and $I_R = 2.5$ mW/sr, we calculate $P_f \simeq 190\,\mu W$ peak optical power coupled into graded index fibre ($NA = 0.22$). This will be the power coupled under dc conditions (i.e., bit rate $R = 0$). We have assumed $P_f = 200\,\mu W$ (i.e., 100 µW average or $dB_t = -10$ dBm) and from Equations 21.1 and 21.2 we can write down an expression for the rolloff

$$P_f(\mu W) = \frac{200}{\sqrt{1+3(f/f_3)^2}} \qquad (21.6)$$

with $f_3 = 75$ MHz expected from Table 21.2 at $I_R \simeq 2.5$ mW/sr. Although Equation 21.6 is written for analogue modulation, we will assume for our digital system that 1 MHz is equivalent to 1 Mb/s which means f can be replaced by bit rate R in Equation 21.6.

21.5.3 Power coupled into fibres from lasers

The power coupled from laser sources will be significantly higher than for LEDs because of the radiation beam directionality. In Table 21.3, we have assumed that the laser can couple 5 mW peak power (2.5 mW average or $dB_t = +4$ dBm) into the fibre. These numbers are consistent with $\simeq 50\%$ coupling efficiencies expected for fibres with bulb ends (35) or tapers (36). Thus the laser is actually operating at output $P_0 \simeq 10$ mW peak (5 mW average), a situation which is expected to be consistent with reliabilities of 10^5 h at room temperature (see Table 21.2).

21.5.4 Fibre characteristics

The system fibre is assumed to be graded index with parameters specified in Table 21.3. The total fibre loss (dB_f) is expressed by three terms as follows (37):

$$dB_f = \alpha(\lambda)L + 3\left(\frac{R\Delta\lambda}{R_c}\right)^2 L^2 + \begin{cases} 3\left(\dfrac{R}{R_m}\right)^2 L^2, & \text{no mode mixing} \\ 3\left(\dfrac{R}{R_m}\right)^2 L, & \text{full mode mixing} \end{cases} \qquad (21.7)$$

The first term is the intrinsic attenuation and includes the effects of absorption and scattering. The second term is due to chromatic dispersion effects expressed as an equivalent loss; this term depends on source properties ($\Delta\lambda$ and bit rate R) and fibre properties (L and chromatic bit rate R_c of Figure 21.2). The third term represents the effects of multimode dispersion and occurs only in multimode fibres. The equivalent attenuation is expected to vary between the two cases of no mode mixing and complete mode mixing which have L^2 and L dependences, respectively.

21.5.5 Detector/receiver characteristics

As stated previously, the three wavelength ranges identified were determined by fibre properties and availability of detectors. Table 21.3 shows those detector types which can function in the three wavelength regions. Silicon is universally used at 800–900 nm where it has peak quantum efficiencies (0.85). However, at longer wavelengths, the quantum efficiency of Si drops rapidly and is only 0.45 at 1000–1100 nm. In order to compensate, it is suggested (37) that Si detectors can still be used effectively if light is coupled into the side along the junction plane rather than the more conventional method of coupling perpendicular to junction plane. Improved performance will then occur because of the increased optical absorption path length. Alternatively, III–V detectors can be used at 1000–1100 nm. At even longer wave-lengths ($\lambda \simeq 1300$ nm), it is felt that III–V detectors will be preferred over Ge due to their expected higher quantum efficiencies and lower dark currents. The above statements apply equally well to the two detector types that are appropriate for systems (PIN or APD structures). Figure 21.9 shows the receiver

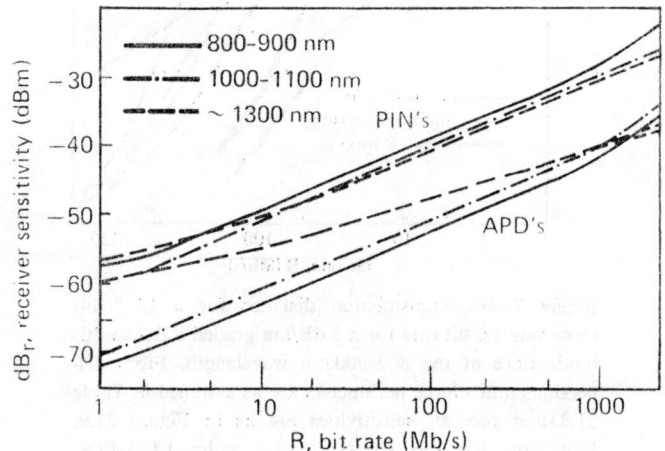

Figure 21.9 Calculated receiver sensitivities for a 10^{-9} bit error rate assuming PIN and APD detectors from Reference 37. Plotted data are for the three wavelength ranges indicated in Figure 21.2

sensitivities (dB_r) that have been used in the calculations. Several factors have been taken into account including the effects of noise currents, whether FET or bipolar transistors are used, etc. (see Reference 37). In addition, a system margin (dB_s) has been added to account for losses at optical splices, connectors, etc. This loss is assumed to increase linearly with $\log R$ from 5 dB at 1 Mb/s to 10 dB at 1 Gb/s.

21.5.6 Results of system calculations

Figures 21.10 and 21.11 illustrate some of the results which were obtained from the calculations. Figure 21.10 shows that system performance is successively improved as the source-detector combinations are varied from

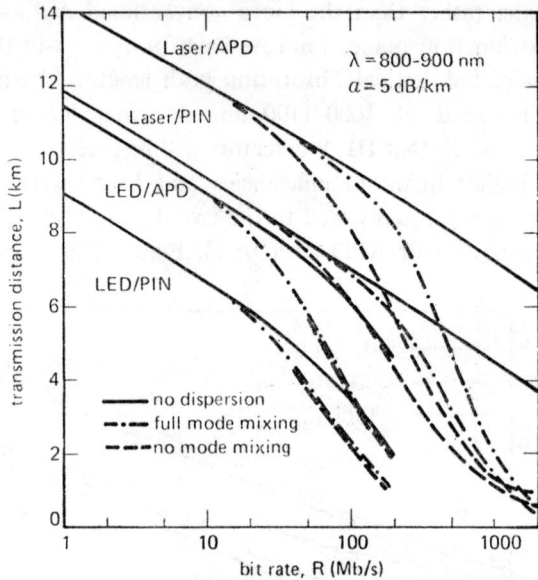

Figure 21.10 Transmission distance for a 10^{-9} bit error rate vs. bit rate for a 5 dB/km graded index multimode fibre at the 800–900 nm wavelength. Fibre dispersions and source parameters are as assumed in Table 21.3 and receiver sensitivities are as in Figure 21.9. Plots are for the source-detector pairs LED–PIN, LED–APD, Laser–PIN, Laser–APD for the following cases:
——— no fibre dispersion, wavelength attenuation only,
— — — with chromatic dispersion and no mode mixing,
—·— with chromatic dispersion and full mode mixing

LED–PIN, LED–APD, laser–PIN, and laser–APD. In the case of LED sources, the system performance is dominated by the chromatic dispersion term since $\Delta\lambda$ is so large and it is apparent that mode mixing effects play a minor role. For laser sources, the improved chromaticity (i.e., small $\Delta\lambda$) is such that the dominant contribution to the dispersion arises from multimode rather than chromatic dispersion. Furthermore, the degree of mode mixing is important in determining how rapidly the transmission distance falls off with bit rate, to the point where any substantial advantages of APDs over PIN detectors are essentially lost towards the higher bit rates.

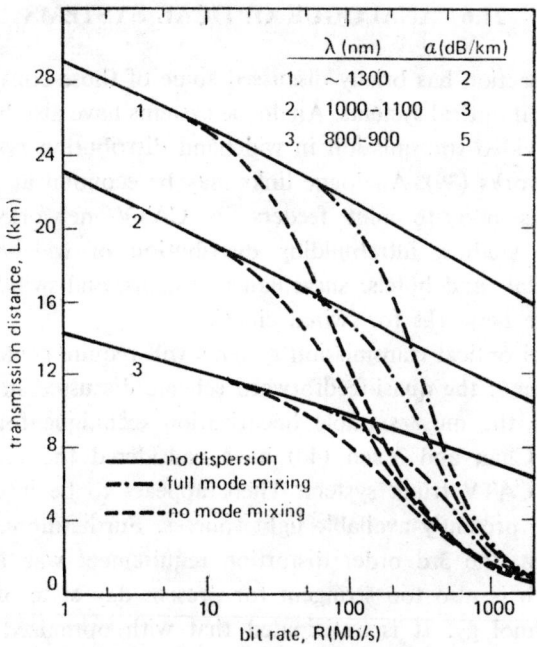

Figure 21.11 Transmission distance for a 10^{-9} bit error rate vs. bit rate for the laser/APD source-detector combination and graded index multimode fibres whose wavelength dependent attenuations at the three selected wavelength ranges are 2, 3, 5 dB/km, respectively. Fibre dispersions and source parameters are as assumed in Table 21.3 and receiver parameters are in Figure 21.9
——— no fibre dispersion, wavelength attenuation only,
— — — with chromatic dispersion and no mode mixing,
— · — with chromatic dispersion and full mode mixing

For the laser–APD combination, Figure 21.11 shows how the performance varies for the three wavelengths. At low bit rates, where dispersion effects are small, considerable increases in transmission distance can be achieved by using sources at $\lambda \simeq 1300$ nm compared to sources at $\lambda = 1000\text{--}1100$ nm or $\lambda = 800\text{--}900$ nm. Again, as in Figure 21.10, mode mixing effects reduce the transmission distances at the higher bit rates so that the large system advantages obtained at $\lambda = 1300$ nm and low bit rates have essentially vanished at higher bit rate.

21.6 ANALOGUE OPTICAL SYSTEMS

The previous section has briefly discussed some of those considerations pertinent for digital optical systems. Analogue systems have also been considered for baseband video transmission in wideband distribution systems (38) and in CATV networks (39). Analogue links may be economical in such diverse applications as point-to-point feeders in CATV networks, within and between video studios, intrabuilding distribution of wideband signals in schools, hospitals and hotels, surveillance systems, and switched broadband type interactive networks for 'wired cities'.

Multichannel optical transmission systems will require some kind of compensation scheme; the quasi-feedforward scheme discussed in Section 21.4.6 appears to be the most suitable linearization technique for multichannel transmission. Chan and Yuen (40) have considered the requirements for a 12-channel CATV trunk system. There appears to be insufficient power available from presently available light sources. Furthermore, for even a 5-channel system, the 3rd order distortion requirement was analysed to be -96 dB, which is also too stringent for present day state of the art LED and laser technology. It is anticipated that with optimized compensation schemes and improved linearity of sources, the stringent requirements for multichannel operation will be achieved.

ACKNOWLEDGEMENTS

The author has drawn extensively on the work of his colleagues at Bell-Northern Research during the preparation of this chapter.

REFERENCES

1. F. P. Kapron, D. B. Keck, R. D. Maurer, *Appl. Phys. Letts.*, **17**, 423 (1970).
2. M. B. Panish, *Proc. IEEE*, **64**, 1512 (1976).
3. F. P. Kapron, D. B. Keck, *Appl. Optics*, **10**, 1519 (1971).

4. C. J. Nuese, *J. Elect. Matl.*, **6**, 253 (1977).
5. H. Kressel, in *Fundamentals of Optical Fibre Communications*, 109, edited by M. K. Barnoski, Academic Press, New York, 1976.
6. F. D. King, A. J. SpringThorpe, O. I. Szentesi, *Int. Electron. Dev. Meeting, Wash. D.D., Tech. Dig.* 480 (1975).
7. A. W. Mabbitt, R. C. Goodfellow, *Electron. lett.*, **11**, 274 (1975).
8. T. P. Pearsall, B. I. Miller, R. J. Capik, K. J. Bachmann, *Appl. Phys. Lett.*, **28**, 499 (1976).
9. A. Y. Cho, H. C. Casey, Jr., P. W. Foy, *Appl. Phys. Lett.*, **30**, 397 (1977).
10. A. G. Dentai, T. P. Lee, C. A. Burrus, E. Buehler, *Electron. Lett.*, **13**, 484 (1977).
11. C. J. Nuese, G. H. Olsen, M. Ettenberg, *Appl. Phys. Lett.*, **29**, 54 (1976).
12. C. J. Nuese, G. H. Olsen, M. Ettenberg, J. J. Gannon, T. J. Zamerowski, *Phys. Lett.*, **29**, 807 (1976).
13. C. C. Shen, J. J. Hsieh, T. A. Lind, *Appl. Phys. Lett.*, **30**, 353 (1977).
14. R. E. Nahory, M. A. Pollack, E. D. Beebe, J. C. DeWinter, R. W. Nixon, *Appl. Phys. Lett.*, **28**, 19 (1976).
15. T. Yamamoto, K. Sakai, S. Akiba, Y. Suematsu, *Electron. Lett.*, **13**, 142 (1977).
16. J. Stone, C. A. Burrus, A. G. Dentai, B. I. Miller, *Appl. Phys. Lett.*, **29**, 37 (1976).
17. C. A. Burrus, J. Stone, A. G. Dentai, *Electron. Lett.*, **12**, 600 (1976).
18. S. Horiuchi, K. Ikeda, T. Tanaka, W. Susaki, *IEEE Trans. Electron. Devices*, **ED-24**, 986 (1977).
19. M. Abe, I. Umebu, O. Hasegawa, S. Yamakoshi, Y. Yamaoka, T. Kotani, H. Okada, H. Takanashi, *IEEE Trans. Electron Devices*, **ED-24**, 990 (1977).
20. M. Ettenberg, H. Kressel, J. P. Wittke, *IEEE J. Quant. Elect.*, **QE-12**, 360 (1976).
21. J. C. Dyment, L. A. D'Asaro, J. C. North, B. I. Miller, J. E. Ripper, *Proc. IEEE*, **60**, 726 (1972).
22. D. Kato, *J. Appl. Phys.*, **44**, 2756 (1973).
23. M. Maeda, I. Ikushima, K. Nagano, M. Tanaka, H. Nakashima, R. Itoh, *Appl. Optics*, **16**, 1966 (1977).
24. W. Harth, W. Huber, J. Heinen, *IEEE Trans. Electron Dev.*, **ED23**, 478 (1976).
25. J. E. Ripper, *J. Appl. Phys.*, **43**, 1762 (1976).
26. J. A. Rossi, S. R. Chinn, H. Heckscher, *Appl. Phys. Lett.*, **23**, 25 (1973).
27. D. R. Scifres, R. D. Burnham, W. Streifer, *IEEE Trans. Electron Devices*, **ED-22**, 609 (1975).
28. R. R. Fergusson, J. Straus, '3rd European Electro-Optics Conf.', *SPIE* **99**, 56 (1976).
29. J. Straus, O. I. Szentesi, *Proc. 1977 IEEE Int. Symp. on Circuits and Systems*, April 25–27, Pheonix, Arizona, 288 (1977).
30. H. Yonezu, Y. Matsumoto, T. Shinohara, I. Sakuma, T. Susuki, K. Kobayashi, R. Lang, Y. Nannichi, I. Hayashi, *Japan J. Appl. Phys.*, **16**, 209 (1977).
31. S. Yamakoshi, O. Hasegawa, H. Hamaguchi, M. Abe, Y. Yamaoka, *Appl. Phys. Lett.*, **31**, 627 (1977).
32. R. L. Hartman, N. E. Schumaker, R. W. Dixon, *Appl. Phys. Lett.*, **31**, 756 (1977).
33. R. L. Hartman, R. W. Dixon, *Appl. Phys. Lett.*, **26**, 239 (1975).
34. F. P. Kapron, private communication.

35. C. A. Brackett, *J. Appl. Phys.*, **45**, 2636 (1974).
36. T. Ozeki, B. S. Kawasaki, *Electronics Letts.*, **12**, 607 (1976).
37. J. Conradi, F. P. Kapron, J. C. Dyment, *IEEE Trans. Elect. Devices*, **ED-25**, 180 (1978).
38. A. Szanto, J. C. W. Taylor, *Proc. of the Int. Conf. on Communications, June 17–19, Minneapolis*, 17E1, (1974).
39. L. A. Marvin, C. K. Kao, *TV Communications*, **12**, 20 (1975).
40. D. Chan, T. M. Yuen, *IEEE Trans. on Communications*, **COMM-25**, 680 (1977).

Chapter 22

Modal Dispersion in Optical Fibres and the Influence of Mode-Coupling

Bruno Crosignani and Paolo di Porto

22.1 DISPERSION IN OPTICAL FIBRES: GENERAL REMARKS

Dispersion is one of the main characteristics of an optical fibre in that it influences its capability of carrying information, i.e. its transmissible bandwidth. In fact, this phenomenon, associated with the fact that different components of the electromagnetic wave travelling in the fibre possess different velocities, causes, for example, distortion of the shape of the pulses constituting the signal, thus giving rise to an overlapping which deteriorates it.

The different factors responsible for dispersion can be brought into evidence by examining the expression of the electromagnetic field travelling in the fibre. In general, neglecting attenuation, the field can be expressed with good approximation as a superposition of forward travelling guided modes in the form (1)

$$E(r, z, t) = \sum_{n=1}^{N} E_n(\omega_0, r) \int_{-\infty}^{+\infty} c_n(\omega, z) \exp[-i\beta_n(\omega)z + i\omega t] d\omega$$

$$\equiv \sum_{n=1}^{N} E_n(\omega_0, r) a_n(z, t) \qquad (22.1)$$

where $E_n(\omega_0, r)$ is the field configuration of the nth mode, whose propagation constant $\beta_n(\omega)$ is evaluated at the central frequency ω_0 of the exciting source, r and z represent the transverse and longitudinal coordinates of the cylindrical fibre (see Figure 22.1) and the c_n are suitable expansion coefficients. They can be z-dependent owing to possible departure of the fibre from the ideal structure to which the mode fields E_n refer. By expanding $\beta_n(\omega)$ around ω_0, Equation 22.1 yields

$$E(r, z, t) = \sum_n E_n(\omega_0, r) e^{-i\beta_n(\omega_0)z}$$
$$\times \int c_n(\omega, z) e^{-i(d\beta_n/d\omega)\omega_0(\omega-\omega_0)z} e^{-(i/2)(d^2\beta_n/d\omega^2)\omega_0(\omega-\omega_0)^2 z + i\omega t} d\omega \qquad (22.2)$$

whose validity hinges upon the small-bandwidth hypothesis $\delta\omega/\omega \ll 1$, usually verified in practice by most sources.

If just one mode is present, dispersion is connected with the appearance of the term in $(\omega-\omega_0)^2$ since its absence would imply field propagation with-

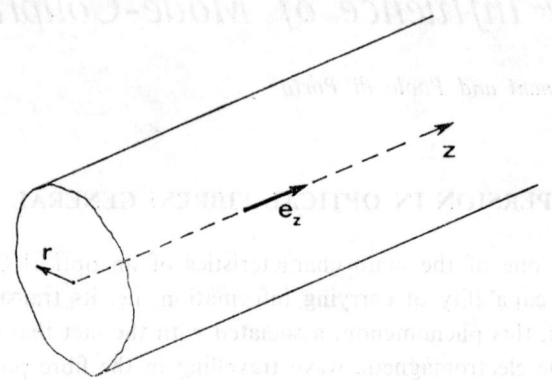

Figure 22.1 Fibre geometry

out distortion. The non-linear behaviour with frequency of $\beta_n(\omega)$ is due to the frequency dependence of the refractive index of the material constituting the guiding structure and to the intrinsic frequency dependence of the mode propagation-constant due to the fibre geometry. The two contributions may in some cases either compensate for or reinforce each other, the latter being particularly significant for modes near cutoff. This kind of dispersion (material plus waveguide) can be characterized by the time delay per unit fibre length τ_{mat}, between the largest and smallest-frequency components. Recalling that the group velocity V_n is given by

$$V_n = (d\beta_n/d\omega)^{-1}_{\omega_0} \tag{22.3}$$

one has, for material dispersion far from cutoff,

$$\tau_{mat} = (\delta\omega/\omega_0)\,(k/c)\,(d^2(kn_0)/dk^2)_{k_0} \tag{22.4}$$

where $k = \omega/c$ and $n_0(k)$ is the refractive index of the fibre material, having taken into account that, in this case,

$$d\beta_n/d\omega \simeq d(n_0(\omega)\omega/c)/d\omega \tag{22.5}$$

If the field is nearly monochromatic, then the material and waveguide dispersions are absent but still the field is distorted while propagating, if many modes are present, due to the fact that each of them possesses a different group velocity V_n. An estimate of this modal dispersion is furnished by the time delay per unit length τ_{mod}, between the lowest and fastest modes

$$\tau_{\text{mod}} = 1/V_{\text{min}} - 1/V_{\text{max}} \tag{22.6}$$

The relative importance of the two kind of dispersion can be evaluated by comparing τ_{mat} and τ_{mod}. It is worthwhile to note that τ_{mat} depends on the temporal coherence properties (namely, the bandwidth) of the source exciting the fibre, while τ_{mod} is connected with its spatial coherence properties which determine how many modes are excited. Relative bandwidths $\delta\omega/\omega_0$ of the order 5×10^{-2}, 10^{-3} and 10^{-4}, typical of a light-emitting diode (LED), of a laser diode and of a solid state laser, yield for τ_{mat} values of 5 ns/km, 0.1 ns/km and 0.01 ns/km. The τ_{mod} for a typical step-index fibre, in which all the modes have been excited, can be of the order of some tenths of ns/km, showing that this kind of dispersion is in most cases dominant over the material one. Accordingly, in the following we shall treat modal dispersion in detail, both for ideal and non-ideal fibres. In particular, in the latter situation the fibre imperfections induce a coupling mechanism between the various modes which causes modal dispersion to change qualitatively and quantitatively.

22.2 MODAL DISPERSION IN IDEAL FIBRES

Whenever fibre imperfections can be neglected, no mode-coupling is present and the c_n do not depend on z. Accordingly, Equation 22.2 furnishes

$$E(r, z, t) = \sum_n E_n(\omega_0, r) e^{-i\beta_n(\omega_0)z + i\omega_0 z/V_n} a_n(0, t - z/V_n) \tag{22.7}$$

with

$$a_n(0, t - z/V_n) = (1/2P) \int_{-\infty}^{+\infty} e_z \cdot E(r, z = 0, t - z/V_n) \times H_n^*(r, \omega_0) \, dr \tag{22.8}$$

having neglected the term in $(\omega - \omega_0)^2$ and having taken advantage of the orthogonality relation between different modes (1)

$$\int_{-\infty}^{+\infty} e_z \cdot E_n(r, \omega_0) \times H_m^*(r, \omega_0) \, dr = 2P \delta_{mn} \tag{22.9}$$

where P is a positive normalization constant, and $H_m(r, \omega_0)$ the magnetic field configuration of the mth mode.

In actual experiment, one detects the electromagnetic power carried either through the whole fibre section or a smaller area σ (see Figure 22.2). These quantities are obtained by integrating the flux of the Poynting vector (averaged over a suitable time interval)

$$S = (1/2)\overline{E \times H^*} \tag{22.10}$$

over the relevant fibre section, thus obtaining

$$P^\sigma = \text{Re} \int_\sigma c_z \cdot S \, dr \qquad (22.11)$$

In particular, in evaluating the total power P^∞ one can take advantage of Equation 22.9, thus getting

$$P^\infty(z, t) = \sum_n \bar{P}_n(z, t) \qquad (22.12)$$

where

$$P_n(z, t) = P_n(0, t - z/V_n)$$

$$= (1/4P)\left|\int_{-\infty}^{+\infty} e_z \cdot E(r, z = 0, t - z/V_n) \times H_n^*(r, \omega_0) \, dr\right|^2 \qquad (22.13)$$

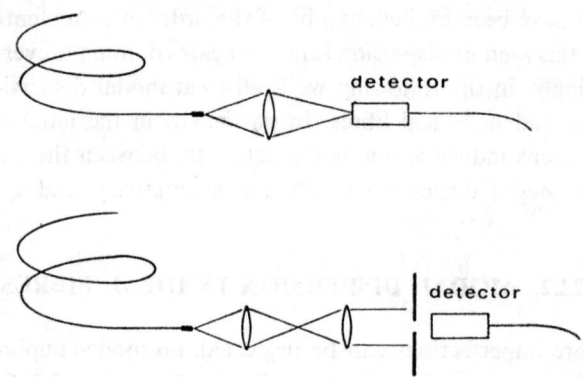

Figure 22.2 Arrangements for collecting the whole light (upper part of the figure) and the light crossing a small section σ (lower part)

represents the power carried by the nth mode. Equation 22.13 shows that the power in each mode travels with a velocity V_n, different from those of the other modes, so that a pulse possessing a given envelope at the fibre entrance undergoes a temporal broadening of the order $\tau_{\text{mod}} z$.

The situation is not so simple for a finite σ. In this case, the expression of $P^\sigma(z, t)$ contains also non-diagonal terms which do not propagate with a definite velocity, while the diagonal ones cannot be interpreted as the power carried by a given mode. More precisely, one obtains

$$P^\sigma(z, t) = \sum_n (\text{Re}\, F_{nn}/2\bar{P})\, \bar{P}_n(z, t)$$

$$+ (1/2)\text{Re} \sum_{n \neq m}\sum F_{nm} \bar{T}_{nm}(z, t)\, e^{i(\beta_m(\omega_0) - \beta_n(\omega_0))z + i\omega_0 z(1/V_n - 1/V_m)} \qquad (22.14)$$

with
$$T_{nm}(z,t) = a_n(0, t-z/V_n)a_m^*(0, t-z/V_m) \quad (22.15)$$
and
$$F_{nm} = \int_\sigma E_n(r, \omega_0) \times H_m^*(r, \omega_0) \cdot e_z \, dr \quad (22.16)$$

The form of $P^\sigma(z, t)$, and thus the pulse broadening, is similar to that of $P^\infty(z, t)$ only after a distance z over which the interference terms between the various modes disappear, that is,

$$\overline{T_{nm}(z, t)} = 0 \quad (22.17)$$

This happens, according to Equation 22.8, whenever

$$\overline{E(r, z=0, t-z/V_n) \cdot E^*(r, z=0, t-z/V_m)} = 0 \quad (22.18)$$

that is, for
$$z\tau_{\text{mod}} > \tau_{\text{coh}} \quad (22.19)$$

where $\tau_{\text{coh}} = 2\pi/\delta\omega$ is the coherence time of the source.

The analysis of the non-diagonal terms appearing in Equation 22.14 furnishes an alternative method for measuring the time delay between the various modes. In fact, if a stationary signal is sent into the fibre, $P^\sigma(z, t)$ does not depend on time and depends on z only through the contribution of the non-diagonal terms. Each of these terms goes to zero over a distance z such that

$$z|1/V_m - 1/V_n| > \tau_{\text{coh}} \quad (22.20)$$

after which the interference between the nth and the mth mode disappears. Experimental methods which have been devised (2) in order to observe this effect and thus measure the various modal delays will be presented later in this chapter.

22.3 MODAL DISPERSION AND MODE COUPLING

Whenever mode coupling is present, the expansion coefficients c_n are a function of z and their evolution is described by a set of coupled linear differential equations (1), whose coefficients are related to the departure from the ideal structure of the fibre under consideration. In most cases, this system of equations is not analytically solvable. Besides, small imperfections giving rise to weak coupling possess a random nature, so that some kind of statistical treatment appears to be the most appropriate for dealing with the problem. On the other hand, there are cases in which one can induce strong coupling in the fibre, which can be analytically approximated by a solvable model.

Let us first consider the weak-coupling case. The significant quantities are averages over an ensemble of macroscopically equivalent fibres differing among themselves for random microscopic imperfections, operations indicated by angular brackets. Accordingly, one has to consider the quantities $\langle P^\infty(z, t)\rangle$ and $\langle P^\sigma(z, t)\rangle$, that is, $\langle \overline{P}_m(z, t)\rangle$ and $\langle \overline{T}_{nm}(z, t)\rangle$. The system of equations obeyed by the $\langle \overline{P}_m \rangle$ and the $\langle \overline{T}_{nm} \rangle$ is much simpler than the one pertinent to the c_n. Under the hypothesis of imperfections homogeneously distributed along the fibre, one gets (1, 3)

$$\frac{\partial}{\partial z}\langle \overline{P}_n(z, t)\rangle + (1/V_n)\frac{\partial}{\partial t}\langle \overline{P}_n(z, t)\rangle = \sum_j h_{nj}\{\langle \overline{P}_j(z, t)\rangle - \langle \overline{P}_n(z, t)\rangle\} \quad (22.21)$$

and

$$\langle \overline{T}_{nm}(z, t)\rangle = e^{-g_{nm}z}\overline{T}_{nm}(z, t), \quad (22.22)$$

where h_{nj} and g_{nm} are constants depending on the coupling, with $\mathrm{Re}\, g_{nm} > 0$.

According to Equation 22.22 the behaviour of the non-diagonal terms is modified, with respect to the uncoupled case, only by the presence of a damping factor which does not play any role as long as $g_{nm}z \ll 1$. Conversely, the diagonal terms are deeply affected by mode coupling in that the solution of the system of Equations 22.21 furnishes for the $\langle \overline{P}_m(z, t)\rangle$ an evolution qualitatively and quantitatively different from the uncoupled case. More precisely, it can be shown (1, 4) that, for sufficiently large z, pulse distortion is no longer proportional to z, as expressed by Equation 22.6, but to $z^{1/2}$ and is reduced by the presence of coupling. This last fact can be intuitively understood if one considers how the modes, sharing progressively their energy, tend to travel at a common velocity.

It is obvious that for the statistical approach to be useful in practice when one deals with a single fibre it has to give information not only about the averages values but also about the amplitude of the fluctuations around them. Analysis of the fluctuations of $\overline{P}_m$ shows (3) that they tend to vanish over a distance proportional to the coherence time of the source. Therefore, the average value $\langle \overline{P}_m \rangle$ can be confidently assumed to represent the actual value in a single fibre, whenever the source bandwidth is large enough.

The strong coupling situation, in which the statistical approach does not work, can be analytically treated by introducing a simplified model where only two modes interact. The main results (5) are that $\overline{P}_1(z, t)$ and $\overline{P}_2(z, t)$ evolve without any substantial mutual delay for a long fibre distance and that the corresponding signal breaks up into two narrow pulses travelling with slightly different velocities, each pulse sharing contributions from both modes (see Figure 22.3). As a consequence, the interference term between the modes survives far beyond the distance over which their mutual delay exceeds τ_{coh}. In other words, mutual coherence between the modes is preserved by mode coupling, so that a measurement of coherence as a function of the travelled distance can furnish a way of evaluating the presence and the relevance of coupling.

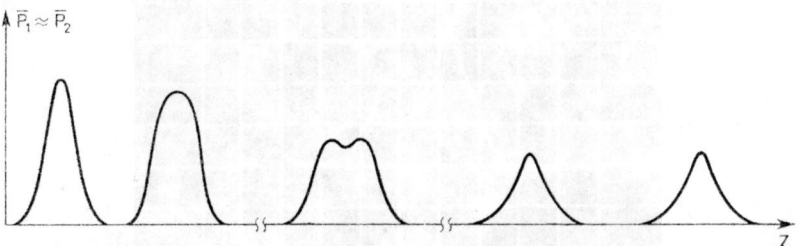

Figure 22.3 Progressive spreading of the pulse and separation in two distinct parts

22.4 MEASUREMENTS OF MODAL DISPERSION

In accordance with the above considerations the measurement of modal dispersion can be accomplished by investigating the behaviour either of the modal powers or of the interference between the propagating modes. As has

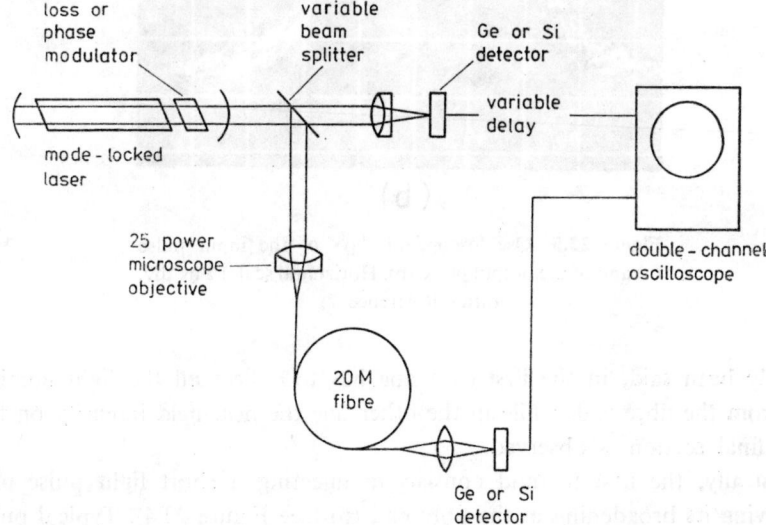

Figure 22.4 Setup to measure pulse distortion and delay in fibres (after Reference 6)

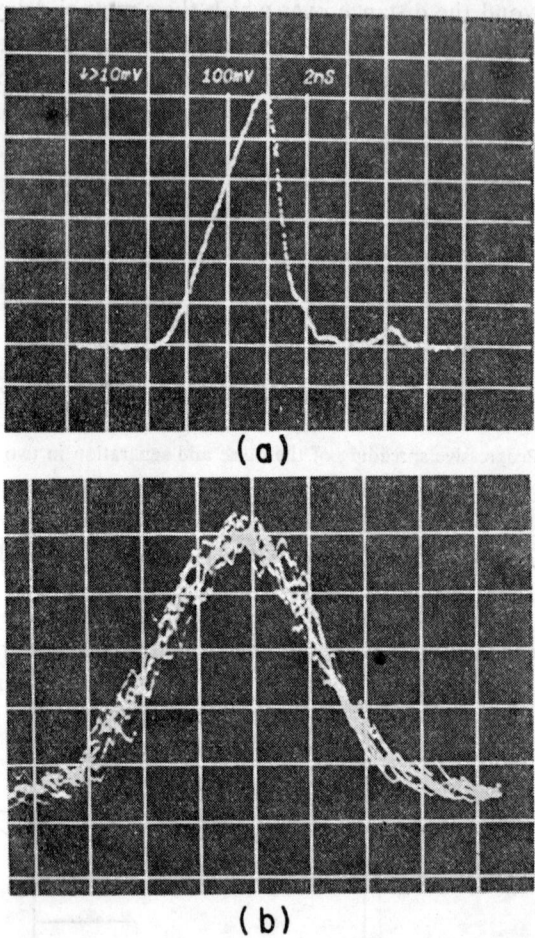

Figure 22.5 Oscilloscope displays of the input pulse (a) and of the output pulse (b). Horizontal scale: 2 ns/div.
(after Reference 7)

already been said, in the first case one has to collect all the light coming out from the fibre end, while in the other one the near-field intensity on the fibre final section is observed.

Basically, the first method consists in injecting a short light pulse and observing its broadening at the fibre end (6) (see Figure 22.4). Typical pulse broadenings for a step-index (7) and a graded-index fibre (8) are shown in Figures 22.5 and 22.6. The main limitation of the method is that rather long

Modal Dispersion in Optical Fibres

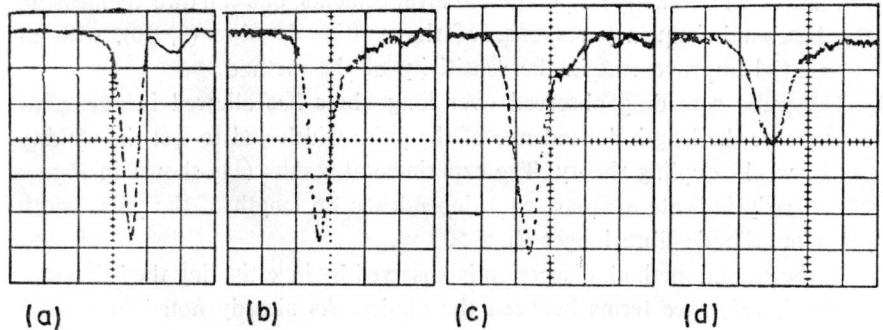

Figure 22.6 Sampling oscilloscope traces of the pulses for increasing fibre lengths. Horizontal scale: 200 ps/div. (after Reference 8)

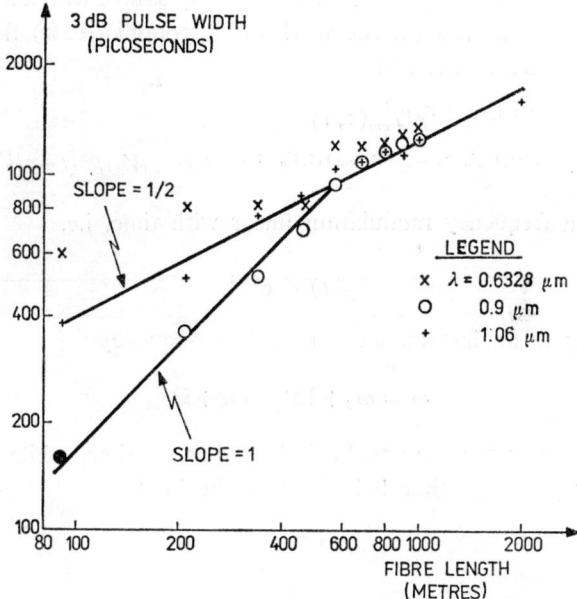

Figure 22.7 Half-power output pulsewidth versus fibre length with excitation wavelength λ as the parameter (after Reference 11)

fibre samples (hundreds of meters) are required in order to achieve a good resolution. In alternative experimental arrangements, one can directly measure the baseband frequency response of the multimode fibre (9, 10), without substantial improvement in the sensitivity of the method.

The pulse spreading observed over long fibres has allowed in particular to measure the length dependence of the pulse width and to test the validity of the weak-coupling theory. The experimental results (11) shown in Figure 22.7 clearly indicate a square-root dependence on length of the pulse width for a multimode fibre longer than 500 m.

In the second method, dispersion is observed by investigating the behaviour of the interference terms between the modes. As already noted in Section 22.2, if one uses a stationary non-modulated input, these non-diagonal terms furnish the z-dependent contribution to $P^\sigma(z, t) = P^\sigma(z, 0)$ (the diagonal terms do not depend on z), which disappear over the largest of the distances defined by Equation 22.20 (or over a larger fibre-length if strong coupling is present).

If one excites the fibre by means of a stationary source with a superimposed modulation $f(t)$, one has (in the weak or no-coupling case) that the non-diagonal terms are of the form

$$e^{i(\beta_m(\omega_0)-\beta_n(\omega_0))z+i\omega_0 z(1/V_n-1/V_m)}\overline{T}_{nm}(z, t)$$
$$\propto e^{-\vartheta_{nm}z}e^{i(\beta_m(\omega_0)-\beta_n(\omega_0))z}e^{-(z/\tau_{coh})|1/V_n-1/V_m|}f(t-z/V_n)f^*(t-z/V_m) \quad (22.23)$$

Assuming a frequency modulation linear with time, i.e.

$$f(t) = e^{i\alpha t^2} \quad (22.24)$$

corresponding to an instantaneous (angular) frequency

$$\omega = \omega_0 + 2\alpha t = \omega_0 + \Omega_t \quad (22.25)$$

the diagonal terms turn out to be independent of time, while mode interference-terms possess a time behaviour of the kind

$$e^{i\Omega_{nm}t} \quad (22.26)$$

with

$$\Omega_{nm} = 2\alpha|z/V_m - z/V_n| \quad (22.27)$$

By Fourier analysing $P^\sigma(z, t)$ one is then able to obtain all the possible mode delays

$$\tau_{nm} = z|1/V_m - 1/V_n| \quad (22.28)$$

responsible for dispersion. The intrinsic limit to the resolution of this method is furnished by the condition

$$T\Omega_{nm} \gg 2\pi \tag{22.29}$$

where T is the frequency sweeping-time, that is,

$$\tau_{nm}\Delta\Omega_t \gg 2\pi \tag{22.30}$$

$\Delta\Omega_t$ representing the spanned frequency-interval. In practice, by employing a tunable dye-laser, $\Delta\Omega_t$ can be made of the order of a hundred GHz, so that it is possible to appreciate τ_{nm} of the order of few picoseconds, corresponding to fibre lengths of few meters.

The experimental set-up pertinent to this kind of measurements (2) is shown in Figure 22.8. The beam of a commercial CW tunable dye-laser

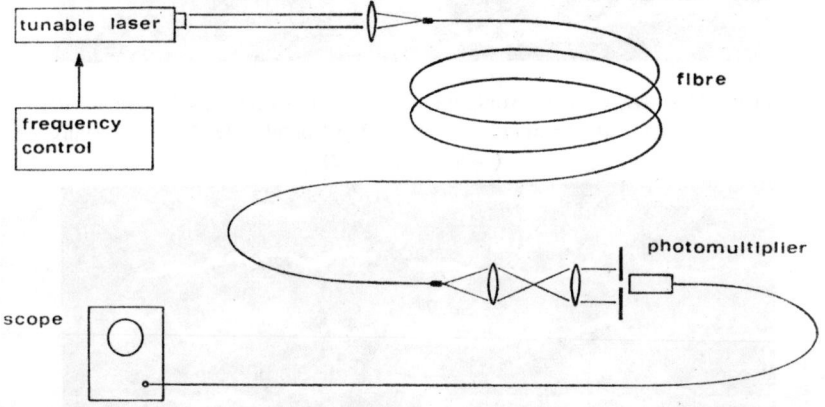

Figure 22.8 Experimental setup for measuring very short mode delays

(centred at the wavelength $\lambda_0 = 5850$ Å) travels inside the fibre for a distance of the order of a few meters. A suitable optical system furnishes a magnified view of the final section of the fibre, which can be scanned in order to obtain the intensity $P^\sigma(z, t)$ at several positions. The resulting signal is sent to the y-axis of an oscilloscope, whose x-axis is driven by the same sawtooth generator driving the laser, an arrangement which allows one to observe directly the superposition of the various terms

$$e^{i\Omega_t \tau_{nm}} \tag{22.31}$$

as a function of the instantaneous frequency Ω_t.

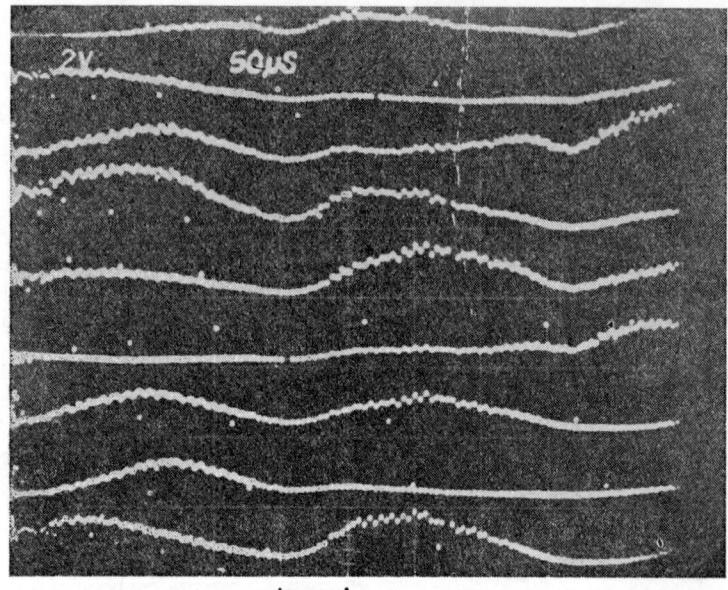

Figure 22.9 Displays of the output intensity as a function of the instantaneous frequency $\Omega_t/2\pi$ at several radial positions. Horizontal scale: 3 GHz/large div (after Reference 2)

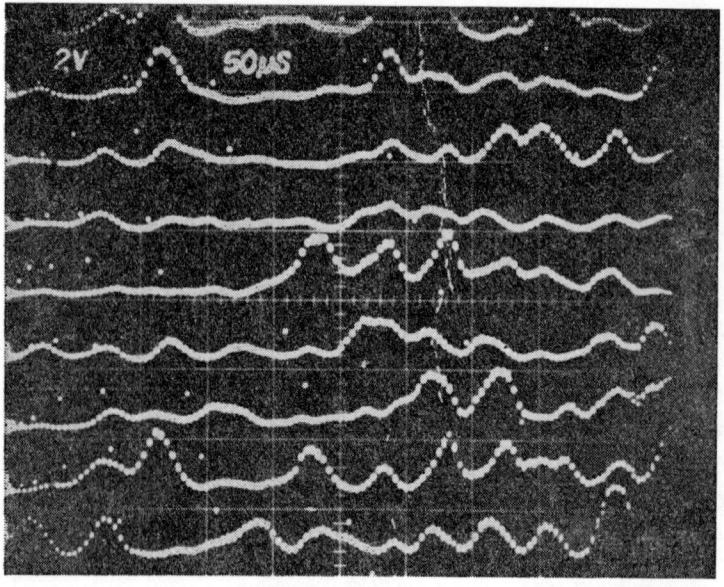

Figure 22.10 As in Figure 22.9, with horizontal scale 15 GHz/large div. (after Reference 2)

Modal Dispersion in Optical Fibres

Two typical oscilloscope displays, relative to a step-index fibre 3 meters long, are shown in Figures 22.9 and 22.10. Their analysis reveals the presence of mode-delays between 0.1 and 0.01.

22.5 THE MEASUREMENT OF SPATIAL COHERENCE ALONG THE FIBRE

We wish to conclude this review on the measurement of modal dispersion by illustrating its connection with the behaviour of spatial coherence along the fibre. As a matter of fact, the two effects are closely related since a relevant contribution to the mutual coherence function

$$\Gamma(r, r', z, t) = \overline{E(r, z, t) \cdot E(r', z, t)} \tag{22.32}$$

is given by the interference terms between the modes. If the fibre is excited by means of a stationary source (without modulation) and if no-coupling or weak coupling is present, these interference terms disappear after a distance such that all the τ_{nm} exceed the coherence time of the source τ_{coh}. Accordingly, after such a distance, the coherence function is the sum of the coherence functions of the single modes. An analysis of the behaviour of the coherence function can then furnish information about modal dispersion.

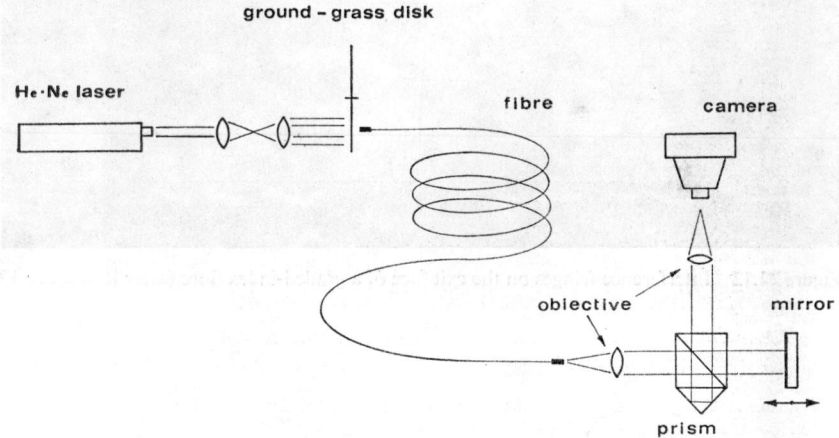

Figure 22.11 Experimental setup for measuring coherence on the exit face of the fibre

We report here some preliminary measurements on spatial coherence (12) performed using a reversing-front interferometer (13), which makes it possible to observe $\Gamma(x, y, -x, y, z, t)$ on the exit face of the fibre (see Figure 22.11).

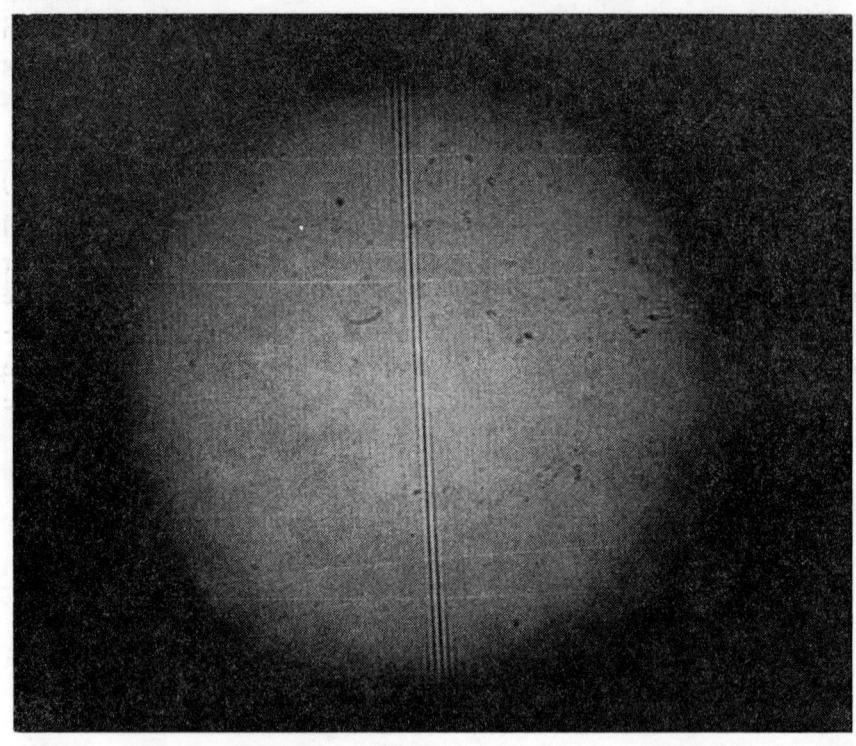

Figure 22.12 Interference fringes on the exit face of a graded-index fibre (after Reference 12)

Modal Dispersion in Optical Fibres

The exciting source, a ground-glass rotating disc illuminated by a He–Ne laser beam, is spatially incoherent, so that there are no interference terms between the modes. The interference fringes appearing in Figures 22.12 and 22.13, which refer respectively to a graded-index and to a step-index fibre

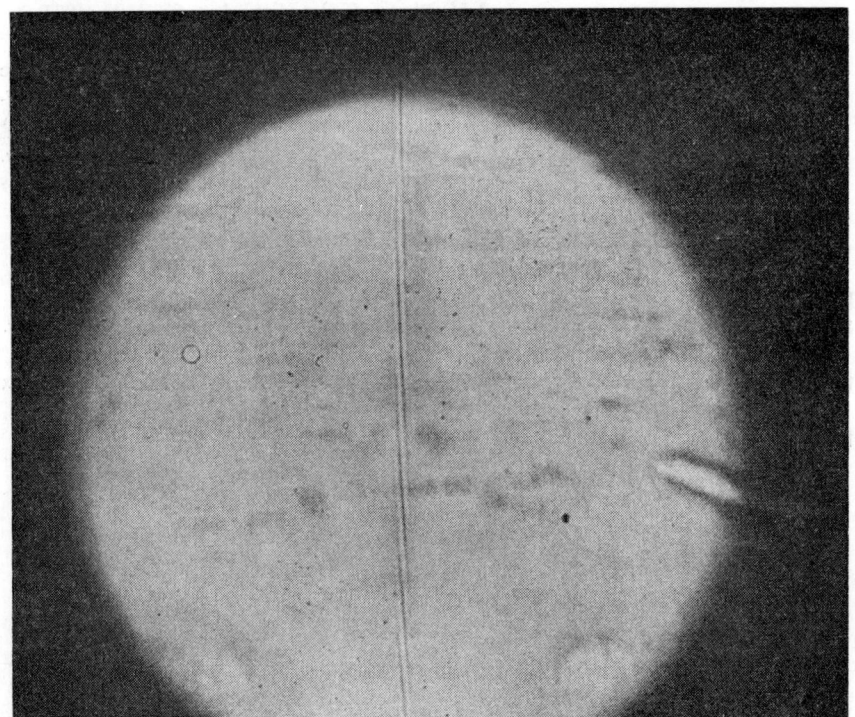

Figure 22.13 As in Figure 22.12, for a step-index fibre (after Reference 12)

a few meters long, represent the sum of the auto-correlation of the various modes. Due to the fact that the modes are uncorrelated, the mutual coherence function Γ does not depend on z and is thus equal to its value at the input face of the fibre. This situation, which is not obviously relevant to the measurement of dispersion, turns out to be useful for determining the refractive index profile, which can be proven (12) to be directly related to the modulus of the complex degree of coherence

$$\gamma(x, y) = \Gamma(x, y, -x, y, 0, 0)/\Gamma(x, y, x, y, 0, 0) \qquad (22.33)$$

REFERENCES

1. See, for example, D. Marcuse, *Theory of Dielectric Optical Waveguides*, Academic Press, New York, 1974.
2. B. Crosignani, B. Daino, and P. Di Porto, *Appl. Phys. Letters*, **27**, 237 (1975); B. Crosignani and P. Di Porto, 'Propagation of Coherence and Very High Resolution Measurements in Optical Fibres', in *Fibres & Integrated Optics, S.P.I.E.*, **77** (1976).
3. B. Crosignani, P. Di Porto, and C. H. Papas, *J. Opt. Soc. Am.*, **67**, 1300 (1977).
4. S. D. Personick, *Bell Syst. Tech. J.*, **50**, 843 (1971).
5. B. Crosignani, P. Di Porto, and C. H. Papas, 'Modal Dispersion in Lightguides in the Presence of Strong Coupling', California Institute of Technology, *Antenna Lab. Technical Report*, No. 87 (1977).
6. See, for example, D. Gloge, E. L. Chinnock, R. D. Standley, and W. S. Holden, *Electron. Letters*, **8**, 527 (1972).
7. R. W. Dawson, *Appl. Opt.*, **13**, 264 (1974).
8. D. Gloge, E. L. Chinnock, and K. Koizumi, *Electron. Letters*, **8**, 526 (1972).
9. D. Gloge, E. L. Chinnock, and D. H. Ring, *Appl. Opt.*, **11**, 1534 (1972).
10. S. D. Personick, W. M. Hubbard, and W. S. Holden, *Appl. Opt.*, **13**, 266 (1974).
11. E. L. Chinnock, L. G. Cohen, W. S. Holden, R. D. Standley, and B. B. Keck, *Proc. IEEE*, **61**, 1499 (1973).
12. B. Daino, S. Piazzolla, and A. Sagnotti, 'A New Method for Measuring the Index Profile of Optical Fibers', to be published.
13. M. Carnevale and B. Daino, *Opt. Acta*, **24**, 1099 (1977).

Chapter 23
AlGaAs Heterostructures in Integrated Optics

E. L. Portnoy

The basic concepts of integrated optics were formulated in 1969. They concern the control of light fluxes in thin-film optical waveguides and application of planar technology in the development of optical channels, integrated optical and optoelectronic circuits.

The appearance of the first publications on integrated optics coincided with creation of the first effectively injecting heterojunctions, and with the discovery of 'superinjection', and electron and optical confinement phenomena in heterostructures. It was at that time also that heterostructures were successfully applied in different optoelectronic devices, such as lasers and LEDs, photodetectors and light converters, etc. Primarily, these two trends were developing independently, although it later turned out that they were destined to be most closely connected. The possibility of controlling the main semiconductor parameters and the refraction index inside heteroepitaxial structures makes it possible to use them most widely in integrated optics. AlGaAs heterostructures are capable of fulfilling all the functions required for the operation of an integrated optical circuit. Apparently, it will be these heterostructures that will serve as the basic material for integrated optics in the same way as silicon is the basic material for integrated circuits in microelectronics.

Nowadays, the development of integrated optics is inseparably connected with semiconductor heterostructures. Among the problems solved by integrated-optics methods using heterostructures, the following three should [be distinguished:

1. Creation of thin-film waveguide analogues of different optical devices, such as modulators, deflectors, frequency upconverters, and single-mode operated thin-film lasers with highly collimated laser beams.
2. Construction of fibre-optical-coupling systems using integrated-optics elements.

3. Transmission and Fourier transformation of images in thin-film devices of analogous optical information treatment in the real-time scale.

Heterostructures with periodical optical inhomogeneities play an important role in solving these problems. Surface diffraction grating serves for input and output of radiation in waveguides, for obtaining distributed feed-back in thin-film lasers, for creation of narrow-band frequency filters, for investigation of mode characteristics and of peculiarities of refraction index distribution in waveguide heterostructures, and so forth.

The present chapter concerns only the work performed at the semiconductor-contact-phenomena laboratory of the A. F. Ioffe Physico-Technical Institute, and connected with fabrication and study of waveguide heterostructures in the AlGaAs system with periodic optical inhomogeneities.*

---

* The manuscript of this chapter presented at 'Cetniewo 78' had not been delivered in due time by the Author to the Editor of this book. A lot of ideas covered by the lecture can be found in

a. Zh. I. Alferov, E. L. Portnoy, 'AlGaAs Heterostructures in Integrated Optics. Invited paper on the 8th International Conference on Solid State Devices, Tokyo 1976', *Jap. J. Appl. Phys.*, Supplement **16-1**, 289 (1977).
b. *Pisma Zhurnal Tekhn. Phys.* (USSR), 3, 197 (1977).
c. *Pisma Zhurnal Tekhn. Phys.* (USSR), 4, 3 (1978).

Appendix

Opening Address

Nearly three years ago, in October 1975, the Institute of Physics of the Polish Academy of Sciences started organizing the Cetniewo International Schools on Semiconductor Optoelectronics. The intention was to review the current state of knowledge concerning the physical principles, technological fundamentals, and application perspectives of semiconductor optoelectronic devices and new materials used in optoelectronics by bringing together in one place a number of well-known specialists in the field, scientists and engineers engaged for many years in research on, and production of, optoelectronic materials and devices as well as newcomers to the field, and thus creating an opportunity for them all to exchange opinions on different scientific and technical topics. It is a great pleasure for me to welcome to the Second International School on Semiconductor Optoelectronics 'Cetniewo 1978' 23 lecturers from Brazil, Canada, France, Federal Republic of Germany, German Democratic Republic, Great Britain, Italy, Japan, Soviet Union, Sweden, USA and Poland as well as over 220 participants from 13 European countries. The lectures of this school cover a broad field of topics of importance for present-day semiconductor optoelectronics. The most important physical problems include the role of deep level impurities in optoelectronic devices and generation processes of dislocations in semiconductor crystals by light illumination. The stoichiometric crystallization methods applied to III–V compounds, electroepitaxy and interface morphology in liquid phase epitaxy processes represent some of the technological topics reviewed. The lectures devoted to optoelectronic devices cover MNOS structures used in optical memories, heterostructures used in a number electronic and optoelectronic devices, as well as electroluminescent image amplifiers. A large group of lectures are devoted to current problems in junction laser physics, but of special significance are the lectures on integrated optics and optical fibre communication systems. This is related to the fact that in all likelihood the above mentioned branches of optoelectronics will be the most intensively developing ones in the near future. Semiconductor optoelectronics has strong interaction with chemistry, metallurgy, electrical engineering and with many branches of physics, such as solid-state physics, statistical and quantum optics and modern information thermodynamics. The separation between basic discoveries and applications in this field of electronics is far less distinct than in some of the other fields of electronics. Semiconductor optoelectronics has a particularly effective interface with science. Let me recall the well-known sentence of Dr. Esaki at the Rome International Conference on Physics of Semiconductors in 1976 who said that 'Science is the understanding of nature, whereas engineering is the control of nature.'

Following this notion, industrial laboratories in highly developed countries appear to have played a dominant role as a junction between science and engineering in many technological developments. There may, of course, be a kind of gap between them. However, the junction between science and engineering will always be forward-biased so that electrons and holes, carrying information, can flow easily from science to engineering and vice versa. In the field of semiconductor optoelectronics one may think that the coupling between science and engineering is strong, or that the gap between them is narrow indeed. After all in semiconductor optoelectronics very often narrow gap semiconductors are used. And even when wide gap semiconductors are used, they are narrow-gap insulators.

Chairman of the Programme Committee of the International School on Semiconductor Optoelectronics 'Cetniewo 1978'

PROFESSOR JERZY KOŁODZIEJCZAK

Closing Address

Mister Chairman! Ladies and Gentlemen!

I have been greatly honoured with the opportunity of addressing you at the close of the Second International School on Semiconductor Optoelectronics. First of all, let me express our deep appreciation of the work done by the Organizing Committee and our gratitude to the host—the Institute of Physics of the Polish Academy of Sciences. We are particularly grateful to the Chairman of the Organizing Committee, Dr. Marian Herman, whose enthusiasm and energy have made the School an undoubted success.

Semiconductor optoelectronics has now become a very broad and rapidly developing area of science and technology. It is certainly the most modern branch of semiconductor physics and technology, although in a way it may also be considered as one of the oldest branches of semiconductor and solid-state physics, for it was over a hundred years ago that the first important phenomenon for semiconductor optoelectronics was discovered. In discovering photoeffect in selenium in 1876 two British scientists, Adams and Day, made the first discovery in our field—the first semiconductor material for photodetectors. It is quite possible, however, that that discovery was made too early. This seems to apply also to a number of other discoveries in our field. In the early 1920s a very talented Russian engineer and designer Losev, who worked at the Nizhnyj Novgorod Laboratory, discovered most of the devices we have been discussing here in the course of the present School. He constructed the first light emitting diodes on the basis of silicon carbide as well as the first photodetectors on the basis of the same material. But those discoveries were made too early too, and for a long time engineers relied mainly on vacuum and gas discharge electronic devices. He also worked at the Ioffe Physico-Technical Institute for a few years. Unfortunately, Losev, who was a really ingenious engineer and a very talented man, died during the Leningrad siege in 1942.

The real beginning of semiconductor optoelectronics was marked by the creation of p–n junction in germanium and silicon after World War II. During the war a big radar programme was undertaken and new semiconductor devices were developed on the basis of single crystal germanium and silicon. I think that that discovery was made in due time. But there have been a few other discoveries in our field which were also made too early. For instance, maybe only some of you know that the first suggestion to construct semiconductor injection laser was made in a private communication by von Neuman some time before the discovery of masers and lasers. The famous theoretician and mathematician suggested the use of p–n junctions in germanium to construct infrared power sources, coherent sources, by pumping current through the p–n junction. However, development of most optoelectronic devices on the basis of semiconductor materials came after the discovery of III–V semiconductor compounds, regardless of the fact that other semiconductor materials, such as II–VI or IV–VI compounds has been discovered earlier. After the technology of III–V compounds had been mastered, semiconductor optoelectronics developed very fast indeed. I am very proud to be able to say that the discovery of the first III–V compound materials was made at our Institute by Professor Goryunova. Maybe some of you knew her, even though she died so early. Later, similar investigations were carried out by Dr. Welker in Germany.

Following the successful development of III–V semiconductor materials, some very important theoretical work on how to construct p–n junction injection laser was done in the 1960s at the Lebedev Physical Institute by Basov, Popov (our chairman), and Krokhin, and also by Duraffourg and Bernard in France. I think that Dr. Nishizawa's invention of injection lasers has also been made too early. Following those investigations, stimulated emission from p–n junction in GaAs was discovered at our Institute in 1962. That was also a very important discovery. A few months later the first injection lasers were constructed at the General Electric laboratories in Syracuse and in Schenectady, independently by Professor Holonyak (on the basis of GaAsP solid solutions), and by Dr. Hall (on the basis of GaAs). Simultaneously, Dr. Marshal Nathan of the Yorktown Heights IBM Research Center constructed the first injection semiconductor laser. At the same time Professor Holonyak of General Electric, and a group of workers at our Institute led by Professor Nasledov developed the first spontaneous light sources; in the visible region—on the basis of GaAsP solid solutions (Professor Holonyak), and in the infrared—on the basis of GaAs (our Institute). And that was also a very important step in the development of our field, since these were the first LEDs of practical value. Following that, heterostructures were introduced to improve different optoelectronic devices, and especially LEDs, laser diodes, photo cells and other devices, first at our Institute, and then at the Bell Laboratories as well as at other centres. And now semiconductor optoelectronics has become a very broad area of science and technology, and I do not think that it would be illegitimate to say that in the entire field of semiconductor physics there is not one branch where the meeting ground for new and very interesting ideas of science on the one hand, and practical applications on the other, is so distinct. Optics has always been very interesting, but now in connection with semiconductors, where optics plays a very important role, it has become a meeting ground for science, for new physical ideas, for construction and for technology. In the field of semiconductor optoelectronics, physicists, chemists, and engineers work together for the construction of very complex but also very useful devices. And in this broad area we use new physical ideas, new technology, and many new materials. Semiconductor electronics in other areas usually relies on silicon and only in optoelectronics do we use a lot of materials which satisfy very difficult and widely differing requirements.

I must say that the Second Semiconductor Optoelectronics School was well organized not only because we all had a good time here—even the bad weather can be said to have contributed to our 'willingness' to study our subject—but also because the lectures presented here had been so well chosen; they covered practically all the main areas of our branch of science. First of all we heard a lot of very interesting lectures about very important physical phenomena in optoelectronic materials and devices. I am not going to review each lecture because I am sure that most of you have heard all of them, but I must admit that I myself was very much impressed by the lecture of Professor Sosnowski who presented excellent scientific results and a very short and very interesting review of technical data. The success of that presentation was due not only to the fact that Professor Sosnowski is a brilliant scientist and a very skilled man, but also to the fact that here in Poland investigations concerning IV–VI compounds have been carried out very intensively, and Polish scientists have made a major contribution to international science by providing a lot of new data and can thus be considered as leaders in the area of narrow gap semiconductors and zero gap semiconductors in general.

A very important role in our field is played by technology. It was very interesting to hear new ideas and new methods which had been developed in Japan. We have listened with great pleasure to the lecture of Professor Nishizawa as well as to some very interesting reports given by some other authors, and I must say that now use is made not only of many different semiconductor materials in semiconductor optoelectronics, but also new materials are developed using multicomponent solid solutions and heterostructures based on them. I remember that about ten years ago, when theoretical studies concerning heterostructures based on multicomponent solid solutions were conducted at our Institute, we did not anticipate as rapid a development of that branch of the heterostructures technology and of the science of solid solutions. I think that multicomponent solid solutions will play a major role in many areas of practical application. Of course, the aim of each lecture presented here, of each piece of work we have carried out and which concerned physical phenomena or technology, was to inform us about optoelectronic devices. We have heard here very interesting lectures about the properties of optoelectronic devices. Among the various semiconductor devices, to day, as well as it was many years ago, the most developed one is the injection laser. We have heard a lot of very interesting lectures about the properties of these devices. For instance, a new approach to the mode problem in semiconductor lasers has been proposed by Professor Jose Ripper. We have also heard a very interesting report on the complex properties and structure of modern stripe geometry injection lasers, and a lecture on coherence of the radiation emitted by semiconductor lasers.

Practical application of optoelectronic devices has now become a large and rapidly developing field. You will remember that about ten or fifteen years ago semiconductor lasers had practically no application; now the perspectives for their application are very wide and numerous, the most interesting of them being fibre optical systems for communication.

Another important lecture, and one containing many interesting data was that given by a friend of mine, Dr. Dyment from Bell Northern Research Laboratories. Today Dr. Dyment gave another talk, this time about the activities at Bell Northern Research Labs., where he showed us a beautiful picture and said 'you can see how happy these people are working at Bell Northern Research Labs. in Canada'. I beg to disagree with that statement; I think that those people are so very happy simply because they are fortunate enough to be working with Jack Dyment, their excellent chief of the laboratory.

Of great interest to all of us here have been the reports on the activities in different companies in different countries. I myself was greatly impressed by the talk of Professor Nishizawa today, and sometimes I did not know where the borderline between reality and fantasy was, but I know that Professor Nishizawa is absolutely sure that what may be fantasy now will become reality in the near future. Also of interest was the talk given by Dr. Mroziewicz of UNITRA-CEMI, and all of us realize now that the optoelectronics research and industry here in Poland has been developing very well, and there is no doubt that in this field the Poles can compete with other countries.

In a closing address one does not usually talk about shortcomings, but I hope that you will excuse me for saying a few words about the shortcomings of this School, although neither the Organizing Committee nor personally Dr. Herman can be blamed for them; they simply happened. This may well be only my personal opinion, but we all know that our branch of science has been developing very fast, and some recent developments in the field have not been covered by the programme of this School. In my personal opinion technology and physical ideas are very closely related in our branch of science. It seems that we missed two lectures that would have been of great value to all of us had they been delivered. One of them that had in fact been planned concerned molecular beam epitaxy. MBE plays a very important role nowadays, not so much in industry, not in the actual technology of the preparation of devices, as in theoretical research, and especially in the investigations concerning the problem of how to obtain new data about surface. With the very complex materials and very complex structures that we use we always need a knowledge of what happens at the surface and at the interfaces of our structures, and although I am sure that this method is not the best possible one for preparation of devices, it is the best method of investigating the properties of the structure during the growth and also of investigating the properties of the interface in our very complex structures in the case of most of our materials.

The other lecture that was not there had not been planned; I am thinking of the very old and at the same time very new technique which has been developed by a pupil of Professor Holonyak, Dr. Dupuis quite recently at the Rockwell International Electronics Center in the United States—the vapour phase epitaxy from metallo-organic compounds. Dr. Dupuis was the first to achieve good luminescent properties for GaAlAs system. This technology is very important because it creates the possibility of very precise control of layer thickness which makes it possible to obtain the best reproducible results for many structures. This technology competes now with more complicated techniques such as MBE, because it permits the construction of multilayer structures with layer thickness of 200 Å at the present time, and maybe less in the near future. Having missed these two very important lectures concerning technology we also missed a discussion of the physical problems connected with these two technologies. We also missed information about the very interesting physical results obtained for instance by Dr. Dingle of the

Bell Telephone Labs. during investigations of multilayer heterostructures prepared by MBE. These results concern very interesting and very special effects in these multilayer structures—quantum size effects. Investigations concerning quantum size effects in structures prepared by VPE have also been initiated quite recently, and Professor Holonyak, a very good friend of mine, has recently obtained very exciting results concerning stimulated emission and coherent light generation from the different levels at the quantum box in these double heterostructures with very thin middle layer. Quite recently, too recently for the results to have been published, he obtained continuous operation at room temperature for these structures with stimulated emission from different deep levels in the quantum box. I think that this is very interesting not only from the physical point of view, but also because it makes it possible to design tunable lasers.

These shortcomings do not diminish the overall success of the School and I am sure that at the next International School on Semiconductor Optoelectronics they will be eliminated. Let me once again express our deep gratitude to all our Polish hosts who organized such an excellent School.

Thank you for your attention.

PROFESSOR ZH. I. ALFEROV